Jan Heering Karl Meinke
Bernhard Möller Tobias Nipkow (Eds.)

Higher-Order Algebra, Logic, and Term Rewriting

First International Workshop, HOA '93
Amsterdam, The Netherlands
September 23-24, 1993
Selected Papers

Springer-Verlag
Berlin Heidelberg New York
London Paris Tokyo
Hong Kong Barcelona
Budapest

Series Editors

Gerhard Goos
Universität Karlsruhe
Postfach 69 80
Vincenz-Priessnitz-Straße 1
D-76131 Karlsruhe, Germany

Juris Hartmanis
Cornell University
Department of Computer Science
4130 Upson Hall
Ithaca, NY 14853, USA

Volume Editors

Jan Heering
Stichting Mathematisch Centrum, CWI
P. O. Box 94079, 1090 GB Amsterdam, The Netherlands

Karl Meinke
Department of Computer Science, University College of Swansea
Singleton Park, Swansea SA2 8PP, United Kingdom

Bernhard Möller
Institut für Mathematik, Universität Augsburg
Universitätsstraße 8, D-86135 Augsburg, Germany

Tobias Nipkow
Institut für Informatik, Technische Universität München
Arcisstraße 21, D-80290 München, Germany

CR Subject Classification (1991): F.3-4

ISBN 3-540-58233-9 Springer-Verlag Berlin Heidelberg New York
ISBN 0-387-58233-9 Springer-Verlag New York Berlin Heidelberg

CIP data applied for

Printed in Germany

Typesetting: Camera-ready by author
SPIN: 10131308 45/3140-543210 - Printed on acid-free paper

Lecture Notes in Computer Science 816
Edited by G. Goos and J. Hartmanis

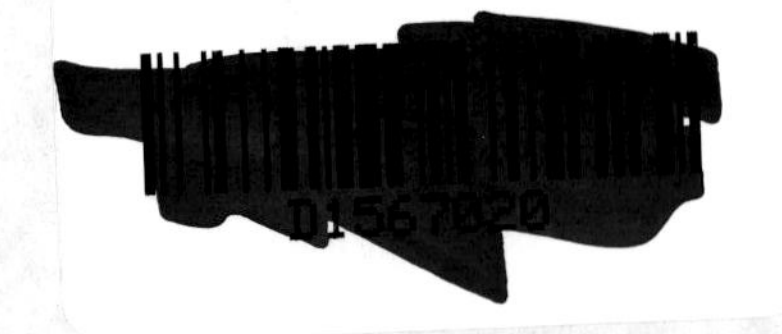

Preface

Higher-order methods are being increasingly applied in functional and logic programming languages, as well as in specification and verification of programs. The algebra and model theory of higher-order languages, computational logic techniques including resolution and term rewriting, and specification and verification case studies using higher-order techniques are particularly vigourous areas of research at present.

To provide an overview of current areas of research and to suggest new research directions, the First International Workshop on Higher-Order Algebra, Logic and Term Rewriting (HOA '93) was held at the Centrum voor Wiskunde en Informatica (CWI) in Amsterdam in September 1993 under the sponsorship of the European Association for Computer Science Logic. Its scope included higher-order aspects of:

- algebra and model theory,
- term rewriting,
- specification and verification languages,
- computational logics,
- system implementations.

The workshop was attended by 41 people. Of the 31 papers submitted, 21 were selected for presentation at the workshop and inclusion in the participant's proceedings. On the basis of 37 referee's reports, 15 of these were selected for publication in the present volume. They appear here in revised final form, ordered alphabetically according to the name of the first author. We are indebted to the referees for making this selection possible.

Finally, we would like to thank Mrs. Mieke Bruné at CWI for her help in the organization of the workshop.

April 1994

J. Heering K. Meinke B. Möller T. Nipkow

Table of Contents

Interaction Systems *

Andrea Asperti[1] and Cosimo Laneve[2]

[1] Dip. di Matematica, P.za di Porta S. Donato, 5, 40127 Bologna, Italy.
[2] INRIA Sophia Antipolis, 2004 Route des Lucioles BP 93, 06902 Valbonne, France.

Abstract. A new class of higher order rewriting systems, called Interaction Systems (IS's), was introduced by the author in [2]. Interaction Systems provide a nice integration of the functional paradigm with a rich class of data structures and basic control flow constructs such as conditionals and (primitive or general) recursion.
In this paper, we focus on the relation between Interaction Systems and Intuitionistic Logic, via the Curry-Howard analogy. Pursuing this analogy, we define an encoding of Interaction Systems into Girard's Linear Logic, passing through Danos and Regnier's *pure nets* (IS's are untyped). Using the local implementation of boxes in [9], the translation of IS's into Pure Nets provides the intermediate step towards their *optimal* implementation (in Lamping style [12]) described in [4].

1 Introduction

There are two possible ways to understand Interaction Systems. From one side they are the *intuitionistic* generalization of Lafont's Interaction Nets [11], from which we borrow the logical setting, the bipartition of operators into constructors and destructors, and the principle of binary interaction. From the other side, Interaction Systems (IS's) are the subclass of Klop's CRS [10] where the Curry-Howard (proofs as propositions) analogy still makes sense. This means that we may associate every IS with a suitable "intuitionistic" system: constructors and destructors respectively correspond to right and left introduction rules, interaction is cut, and computation is cut-elimination.

Interaction Systems have their original motivation in the problem of generalizing Lamping's optimal graph reduction technique to supersystems of λ-calculus [2, 4, 13]. From [9], it is clear that Lamping's sharing operators (fan, croissant and bracket), provide a very abstract framework for an optimal implementation of the *structural part* of Intuitionistic Systems (i.e., via the Curry-Howard analogy, the part charged of the management of resources in functional languages), passing through a local implementation of the box of Linear Logic. Therefore it seemed that Lamping-Gonthier's evaluation style could be *smoothly generalized* to a larger class of systems by just replacing the *logical part* of Intuitionistic Logic (namely, abstraction, application and their interaction through β-reduction), with a richer collection of dual operators, binarily interacting via

* Partially supported by the ESPRIT Basic Research Project 6454 - CONFER.

cuts. Note that the duality of forms comes from the logical distinction into right and left introduction rules. From the point of view of (the practice of) optimal reduction, it is important to keep close to the above logical paradigm, in order to a have a simple interface between the logical and the structural part of the system [4].

The main property of an Interaction System as a rewriting system, is its *sequential nature*; in particular, the leftmost outermost reduction only reduces needed redexes. Note that this is another main brick in the theory of optimal reductions [14]: a reduction strategy is optimal if and only if it reduces maximal families of *needed* redexes. For this reason, it is not fair to compare IS's with the full expressive power of CRS's. They can be only reasonably compared with other known sequential subclasses of the latters, such as Left Normal CRS's [10]. With respect to Left Normal CRS's, Interaction Systems have a strong logical status, a deep relation with Lafont's Interaction Nets, interesting applications to the theory of data types (due to the bipartition of operators into constructors and destructors), and nice perspectives in the typing direction. Moreover, we conjecture that every Left Normal CRS has an optimal encoding as an Interaction System.

In this paper we focus on the relation between logics and Interaction Systems. In particular, every Interaction System is associated with a suitable Intuitionistic System, which is turned into a Linear System by the usual refinement of the structural rules via the two modalities "!" (of course) and "?" (why not). Actually, since we work in an untyped setting, we shall directly define a translation of Interaction Systems into (a suitable variant of) Danos and Regnier's pure nets [6, 15]. Then we prove the correctness of the pure-nets implementation by defining a procedure reading IS-terms back from the nets. This readback and the proof of correctness are adaptations to a simple case of the approach in [3]. Indeed, the relationships between the optimal implementation in [4, 3] and the pure-net description are quite strong: the first one can be obtained from the latter by

1. implementing boxes in a "local" way, according to the principles fixed in [9];
2. keeping advantage of the sharing expressed by metavariables in the implementation of rewriting rules (see [4, 3, 13]).

The paper is structured in the following way. In Section 2 we give the formal definition of Interaction Systems, providing some examples. In Section 3 we study the *intuitionistic nature* of IS's, and their relation with Lafont's Interaction Nets. In Sections 4 and 5 we define the pure-net description of every IS. Finally, Section 6 is devoted to an analysis of the main properties of this translation.

2 Interaction Systems

An Interaction System is defined by a *signature* Σ and a set of *rewriting rules* R.

The signature Σ consists of a denumerable set of *variables* and a set of *forms*. The set of forms is partitioned into two disjoint sets Σ^+ and Σ^-, representing *constructors* (ranged over by $\mathbf{c}$) and *destructors* (ranged over by $\mathbf{d}$). Variables will be ranged over by x, y, z, $\cdots$, possibly indexed. Vectors of variables will be denoted by $\mathbf{x}_i$ where i is the length of the vector (often omitted).

Each form can work as a binder. In the arity of the form we must specify not only the number of arguments, but also, for each argument, the number of variables it is supposed to bind. Thus, the *arity* of a form $\mathbf{f}$, is a finite (possibly empty) sequence of naturals. Moreover, we have the constraint that the arity of every destructor $\mathbf{d} \in \Sigma^-$ has a leading 0 (i.e., it cannot bind over its first argument). The reason for this restriction is that, in Lafont's notation [11], at the first argument we find the *principal port* of the destructor, that is the (unique) port where we will have interaction (local sequentiality).

Expressions, ranged over by $t, t_1, \cdots$, are inductively generated as follows:

a. every variable is an expression;
b. if $\mathbf{f}$ is a form of arity $k_1 \cdots k_n$ and $t_1, \cdots, t_n$ are expressions then $\mathbf{f}(\mathbf{x}^1_{k_1}.t_1, \cdots, \mathbf{x}^n_{k_n}.t_n)$ is an expression.

Free and bound occurrences of variables are defined in the obvious way. As usual, we will identify terms up to renaming of bound variables (α-conversion).

Rewriting rules are described by using schemas or *metaexpressions*. A metaexpression is an expression with *metavariables*, ranged over by X, Y, $\cdots$, possibly indexed (see [1] for more details). Metaexpressions will be denoted by $H, H_1 \cdots$.

A *rewriting rule* is a pair of metaexpressions, written $H_1 \rightarrow H_2$, where H_1 (the *left hand side* of the rule, lhs for short) has the following format

$$\mathbf{d}(\mathbf{c}(\mathbf{x}^1_{k_1}.X_1, \cdots, \mathbf{x}^m_{k_m}.X_m), \cdots, \mathbf{x}^n_{k_n}.X_n)$$

and $i \neq j$ implies $X_i \neq X_j$ (*left linearity*). The arity of $\mathbf{d}$ is $0k_{m+1} \cdots k_n$ and that of $\mathbf{c}$ is $k_1 \cdots k_m$.

The *right hand side* H_2 (rhs, for short) is every *closed* metaexpression, whose metavariables are already in the lhs and built up by the following syntax

$$H \ ::= \ x \ \mid \ \mathbf{f}(\mathbf{x}^1_{a_1}.H_1, \cdots, \mathbf{x}^j_{a_j}.H_j) \ \mid \ X_i[^{H_1}/_{x^i_1}, \cdots, {}^{H_{k_i}}/_{x^i_{k_i}}]$$

The expression $X[^{H_1}/_{x_1}, \cdots, {}^{H_n}/_{x_n}]$ denotes a meta-operation of substitution, as in the λ-calculus.

Finally, the set of rewriting rules must be *non-ambiguous*, i.e. there exists at most one rewriting rule for every pair $\mathbf{d}$-$\mathbf{c}$.

Interaction Systems are readily a subsystem of Klop's (Orthogonal) Combinatory Reduction Systems [10, 1]. We just added a bipartition of operators into constructors and destructors, and imposed a suitable constraint on the shape of the lhs of each rule. As a subclass of non ambiguous, left-linear CRS's, Interaction Systems inherit all good properties of the formers (Church Rosser, finite development, ...).

Example 1. (**The λ-calculus**) The application @ is a destructor of arity 00, and λ is a constructor of arity 1. The only rewriting rule is β-reduction:

$$@(\lambda(x.X), Y) \rightarrow X[^Y/_x].$$

(**Primitive Recursion**) An Interaction System where no form is a binder is called *discrete*. A typical discrete IS is *Primitive Recursion*. There are only two constructors 0 and succ. Composition of two functions $\mathbf{f}(-)$ and $\mathbf{g}(-)$ is obviously expressed as $\mathbf{f}(\mathbf{g}(-))$. The primitive recursion scheme has already the correct IS-shape:

$$\begin{aligned} \mathbf{d}(0, X) &\rightarrow \mathbf{h}(X) \\ \mathbf{d}(\mathtt{succ}(X), Y) &\rightarrow \mathbf{f}(X, Y, \mathbf{g}(X, Y)) \end{aligned}$$

For instance, we may define

$$\mathtt{add}(0, X) \rightarrow X \qquad \mathtt{add}(\mathtt{succ}(X), Y) \rightarrow \mathtt{succ}(\mathtt{add}(X, Y))$$

$$\mathtt{mult}(0, X) \rightarrow 0 \qquad \mathtt{mult}(\mathtt{succ}(X), Y) \rightarrow \mathtt{add}(Y, \mathtt{mult}(X, Y))$$

In a similar way, we may define all inductive types (boolean, lists, trees, ...). Let us remark a trivial but important property of Discrete Interaction Systems, that could motivate the bipartition of forms into constructors and destructors form the point of view of the theory of data types.

Proposition 1. *If in a Discrete Interaction System we have a rewriting rule for every pair* **d**-**c**, *then all closed terms in normal form may only contain constructors.*

In other words, constructors may be used to define the "abstract data type", whose definition is then unaffected by the introduction of new destructors (provided that the definition of each destructor is complete on the data). Note also that all syntactically different terms only built up by means of constructors are eventually distinct (in the equational theory obtained by reflexive, symmetric and transitive closure of the rewriting relation), since they are normal forms (and IS-rewriting is confluent).

Example 2. (**General Recursion**) The recursion operator $\mu x.M \rightarrow M[^{\mu x.M}/_x]$ is a bit problematic, since IS's are based on a principle of *binary* interaction. The obvious idea is to introduce a suitable "dummy" operator interacting with μ. This "dummy" operator may be indifferently regarded as a constructor or a destructor, yielding the two following encodings for μ:

$$\mathbf{d}_\mu(\mu(x.X)) \rightarrow X[^{\mathbf{d}_\mu(\mu(x.X))}/_x]$$

$$\mu(\mathbf{c}_\mu, x.X) \rightarrow X[^{\mu(\mathbf{c}_\mu, x.X)}/_x]$$

3 The Intuitionistic Nature of Interaction Systems

In this section we shall investigate the logical, intuitionistic nature of Interaction Systems, and their relation with Lafont's Interaction Nets.

3.1 Statics

An Intuitionistic System, in a *sequent calculus* presentation (*à la* Gentzen) (see [8]), consists of expressions, named *sequents*, whose shape is $A_1, \cdots A_n \vdash B$ where A_i and B are formulas. Inference rules are partitioned into three groups (in order to emphasize the relationships with IS's, we write rules by assigning terms to proofs):

(Structural Rules)

$$(Exchange)\ \frac{\Gamma,\, x : A,\, y : B,\, \Delta \vdash t : C}{\Gamma,\, y : B,\, x : A,\, \Delta \vdash t : C}$$

$$(Contraction)\ \frac{\Gamma,\, x : A,\, y : A \vdash t : C}{\Gamma,\, z : A,\, \Delta \vdash t[{}^z/_x, {}^z/_y] : C} \qquad (Weakening)\ \frac{\Gamma \vdash t : C}{\Gamma,\, z : A \vdash t : C}$$

(Identity Group)

$$(Identity)\ \frac{}{x : A \vdash x : A} \qquad (Cut)\ \frac{\Gamma \vdash t : A \qquad \Delta,\, x : A \vdash t' : B}{\Gamma,\, \Delta \vdash t'[{}^t/_x] : B}$$

(Logical Rules) These are the "peculiar" operations of the systems, introducing new formulae (i.e. new types) in the proof. The unique new formula introduced by each rule is called the *principal* formula of the inference. If the principal formula is in the rhs of the final sequent, the inference rule is called *right*, and *left* otherwise. Right and left introduction rules respectively correspond with *constructors* and *destructors* in IS's. The shape of these rules is:

$$\frac{\Gamma_1,\, \mathbf{x}^1 : \mathbf{A}^1 \vdash t_1 : B_1 \quad \cdots \quad \Gamma_m,\, \mathbf{x}^m : \mathbf{A}^m \vdash t_m : B_m \quad \Delta, z : C \vdash t : D}{\Gamma_1, \cdots, \Gamma_m, \Delta, y : \mathtt{T}_{\mathbf{d}}(\mathbf{A}^1, B_1, \cdots, \mathbf{A}^m, B_m, C) \vdash t[{}^{\mathbf{d}(y, \mathbf{x}^1.t_1, \cdots, \mathbf{x}^m.t_m)}/_z] : D}$$

for left introduction rules (destructors), and

$$\frac{\Gamma_1,\, \mathbf{x}^1 : \mathbf{A}^1 \vdash t_1 : B_1 \quad \cdots \quad \Gamma_n,\, \mathbf{x}^n : \mathbf{A}^n \vdash t_n : B_n}{\Gamma_1, \cdots, \Gamma_n \vdash \mathbf{c}(\mathbf{x}^1.t_1, \cdots, \mathbf{x}^n.t_n) : \mathtt{T}_{\mathbf{c}}(\mathbf{A}^1, B_1, \cdots, \mathbf{A}^n, B_n)}$$

for right introduction rules (constructors). Above $\mathtt{T}_{\mathbf{d}}(\mathbf{A}^1, B_1, \cdots, \mathbf{A}^m, B_m, C)$ and $\mathtt{T}_{\mathbf{c}}(\mathbf{A}^1, B_1, \cdots, \mathbf{A}^n, B_n)$ are formulae (types) built up by means of the types they take as arguments and they are equal, provided they correspond to the same logical operator. The contexts Γ_i are pairwise different. We remark that the above rules have the general shape of the so-called multiplicative connectives (implication, conjunction, etc.). As in Interaction Nets, there is no means to directly represent additives (disjunction, etc.) (see Section 3.2.1 in [13]).

A standard example is implication, which gives the expressions of typed λ-calculus:

$$(\to \mathit{left})\ \frac{\Delta \vdash t : A \qquad z : B, \Gamma \vdash t' : C}{\Delta, y : A \to B, \Gamma \vdash t'[^{@(y,t)}/_z] : C} \qquad (\to \mathit{right})\ \frac{\Delta, x : A \vdash t : B}{\Delta \vdash \lambda(x.t) : A \to B}$$

An immediate consequence of the above construction is that every proof of an Intuitionistic System may be described by an IS-expression.

Following Lafont's, we may provide a graphical representation of IS-forms. This will explain some important relations between the arity of the forms and the way they link upper sequents in a proof. In the notation of Interaction Nets, a proof of a sequent $A_1, \ldots, A_n \vdash B$ is depicted as a graph with $n+1$ conclusions, n with a negative type (the inputs) and one with a positive type (the output). In particular, an axiom is just a line with one input and one output.

Every logical rule is represented by introducing a new operator in the net (a new labeled node in the graph). The operator has a *main port*, individuated by an arrow, that is associated with the *principal formula* of the logical inference. For instance, the two logical rules for implications are illustrated in Figure 1.

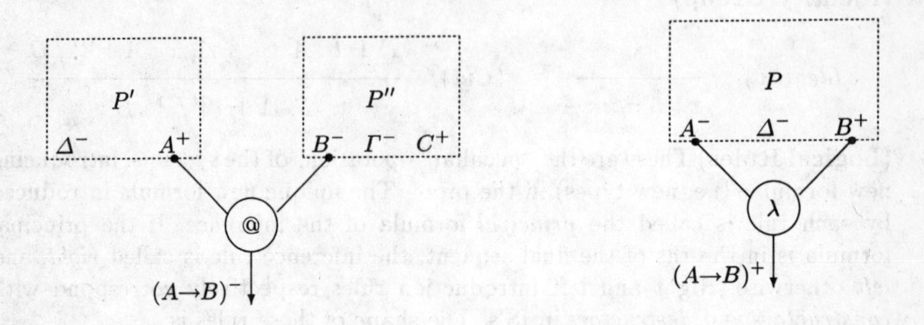

Fig. 1. The graphical representation of @ and λ.

The main port of each operator may be either an input or an output. In the first case it corresponds to a new logical assumption in the left hand side of the sequent (as for @), and the operator is a *destructor*; in the second case it corresponds to the right hand side of the sequent (as for λ), and the operator is a *constructor*. The other ports of the agents are associated with the *auxiliary* formulae of the inference rule, that is the distinguished occurrences of formulae in the upper sequents of the rule. In the two rules above, the auxiliary formulae are A and B.

The auxiliary ports of an agent may be either inputs or outputs, independently from the fact that it is a constructor or a destructor. Actually, in the general theory of Interaction Nets, which is inspired by *classical* (linear) logic, there is a complete symmetry between constructors and destructors, and no restriction at all is imposed on the type of the auxiliary ports (in other words, there is a complete symmetry between inputs and outputs). On the contrary, the fact of limiting ourselves to the intuitionistic case, imposes some obvious "functional" constraints.

Note first that auxiliary formulae may come from different upper sequents or from a single one. For instance, the auxiliary ports of @ are in different upper sequents, while those of λ are in a same one. Lafont expresses this fact by defining *partitions* of auxiliary ports. So, in the case of @, A and B are in different partitions, while in the case of λ they are in the same partition. Moreover, the polarity of an auxiliary port is the opposite of the polarity of the conclusion it has to be matched against. Then, the intuitionistic framework imposes the following constraints:

- In every partition there is at most one negative port. If a negative port exists, we shall call it an *input* partition; otherwise it is an *output* partition.
- Every agent has exactly "one output" (functionality). In particular, if the agent is a constructor, the main port is already an output, and all the partitions must be inputs. Conversely, in the case of destructors, we have exactly one output partition among the auxiliary ports, and this partition has to be a singleton.

If $k_1 \ldots k_n$ is the arity of a form, every k_i denotes the number of positive ports in the i-th *input* partition. In the case of a destructor, one input partition must be a singleton (since it corresponds to the main port), so its arity is 0. By convention, in the concrete syntax of IS's, we have supposed that this is always the first partition of the destructor (this assumption is absolutely irrelevant). Summing up, a form with arity $k_1 \ldots k_n$ is associated with an operator with $1 + \sum_{i=1}^{n}(k_i + 1)$ ports.

3.2 Dynamics

The main theorem of sequent calculi is that stating the redundancy of cut-rules, (i.e. every proof can be effectively turned into a cut-free one). From the operational point of view, this gives dynamics, since it equips the logical system with a rewriting systems on proofs. The rewritings rules must be specified *a priori*: a proof ending into a cut can be simplified into another one by means of some mechanism that is characteristic of that cut.

In order to ensure the possibility of eliminating *all* cuts, the cut-elimination process (term rewriting) must *terminate*. In general, this property does not hold in IS's. We just have a general correspondence between IS's and systems with an *intuitionistic nature*, but only *a posteriori* we may actually check if a particular system has good "logical" properties (cut-elimination, subformula property, ...).

Obviously, we could proceed the other way round, imposing some sufficient conditions on IS's (typing, first of all) to establish a tighter relation with logic. This is surely an interesting subject, but it is out of the scope of this paper.

So, we shall merely focus on the *dynamic* aspects of cut-elimination, without worrying with termination.

Foremost, there are several kinds of cuts. The most interesting case is the *logical cut*, i.e. a cut between two dual (left-right) introduction rules for the same principal formula. In all the other cases (structural cuts), Intuitionistic Systems "eliminate" the cut by lifting it in the premise(s) of one of the rules preceding the cut (that becomes the last rule of the new proof). As in the case of λ-calculus, these kind of cuts are essentially "unobservable" in IS's (they are dealt with in the metalevel definition of substitution). So, let us concentrate on logical cuts only.

Let us consider the case of implication. Here is the typical situation:

$$\dfrac{\dfrac{\Gamma, x : A \vdash t : B}{\Gamma \vdash \lambda(x.t) : A \rightarrow B} \qquad \dfrac{\Delta \vdash t' : A \qquad y : B, \Theta \vdash t'' : C}{\Delta, z : A \rightarrow B, \Theta \vdash t''[{}^{@(z,t')}/_{y}] : C}}{\Gamma, \Delta, \Theta \vdash t''[{}^{@(z,t')}/_{y}][{}^{\lambda(x.t)}/_{z}] : C}$$

The elimination of the above cut consists in introducing two cuts of lesser grade (see [8]). The rewritten proof is:

$$\dfrac{\Delta \vdash t' : A \qquad \dfrac{\Gamma, x : A \vdash t : B \qquad y : B, \Theta \vdash t'' : C}{\Gamma, x : A, \Theta \vdash t''[{}^{t}/_{y}] : C}}{\Gamma, \Delta, \Theta \vdash t''[{}^{t}/_{y}][{}^{t'}/_{x}] : C}$$

Note that this meta-operation on proofs induces a rewriting rule in the underlying IS, which is, in this case, β-reduction. Indeed, the equation

$$t''[{}^{@(z,t')}/_{y}][{}^{\lambda(x.t)}/_{z}] = t''[{}^{t}/_{y}][{}^{t'}/_{x}]$$

implies β-reduction.

Let us come to the general case. Let L and R be the left and right sequent in the cut-rule, respectively. During the process of cut-elimination, the proofs ending into the premises of L and R must be considered as "black boxes" (only their "interfaces", that is their final sequents are known). During cut-elimination, one can build new proofs out of these black boxes. The only constraint is the *prohibition of having new hypotheses* in the final sequent of rewritten proof. This has two implications:

1. the variables bound by L or R (i.e. the auxiliary formulae of the rules) must be suitably filled in (typically with cuts or introducing *new* rules binding them);
2. if a new axiom is introduced by the rewriting, then the hypothesis in the lhs must be consumed (with a cut or by binding it via a logical rule) inside the rewritten proof.

According to statics, a cut in the syntax of IS's corresponds to a term of the following kind:

$$\mathbf{d}(\mathbf{c}(\mathbf{x}^1.X_1,\cdots,\mathbf{x}^m.X_m),\cdots,\mathbf{x}^n.X_n).$$

The X_i represent the proofs ending into the upper sequents of L and R (the "black boxes" above). Following the previous discussion, in the right hand sides of a rewriting rule, we may either

- introduce new variables with axioms or with weakenings (denoted by x),

or, starting from proofs $H_1,\cdots,H_n$ that have been already built up, we may

- exploit proofs X_i, provided we fill the bound hypothesis (notation $X[^{H_1}/_{x_1},\ldots,^{H_n}/_{x_n}]$)
- or introduce a new logical rule (denoted as $\mathbf{f}(\mathbf{x}^1.H_1,\cdots,\mathbf{x}^n.H_n)$).

The other logical rules (contractions, cuts) are visible in the concrete syntax of IS's under copying or interactions. Finally, the syntactical constraint reflecting the logical absence of new hypotheses is: right hand sides of rules must be closed expressions.

Example 3. **(Naturals)** We have two constructors 0 and succ, respectively associated with the following right introduction rules:

$$(nat, right_O)\quad \vdash \mathsf{0} : nat \qquad (nat, right_S)\ \frac{\Delta, \vdash n : nat}{\Delta \vdash \mathsf{succ}(n) : nat}$$

A typical destructor is add.

$$(nat, left_{add})\ \frac{\Delta \vdash p : nat \qquad \Gamma, y : nat \vdash t : A}{\Delta, \Gamma,\ x : nat \vdash t[^{\mathsf{add}(x,p)}/_y] : A}$$

where A can be any type. The following is an example of cut:

$$\frac{\vdash \mathsf{0} : nat \qquad \dfrac{\Delta \vdash p : nat \qquad y : nat, \Theta \vdash t : A}{\Delta, x : nat, \Theta \vdash t[^{\mathsf{add}(x,p)}/_y] : A}}{\Delta, \Theta \vdash t[^{\mathsf{add}(x,p)}/_y][^{\mathsf{0}}/_x] : A}$$

that is simplified into:

$$\frac{\Delta \vdash p : nat \qquad y : nat, \Theta \vdash t : A}{\Delta, \Theta \vdash t[^p/_y] : A}$$

The above elimination induces the IS-rule $\mathsf{add}(\mathsf{0}, X) \to X$, corresponding to the equation $t[^{\mathsf{add}(x,p)}/_y][^{\mathsf{0}}/_x] = t[^p/_y]$.

Example 4. **(Lists)** Lists are defined by means of two constructors cons and nil of arity 00 and ε, respectively. The typical destructors are hd and tl of arity 0. In the case of lists of integers, we may write the following introduction rules for the type *natlist*:

$$(natlist, right_{nil}) \vdash \mathtt{nil} : natlist$$

$$(natlist, right_{cons}) \frac{\Delta \vdash n : nat \qquad \Gamma \vdash l : natlist}{\Delta, \Gamma \vdash \mathbf{cons}(n, l) : natlist}$$

$$(natlist, left_{hd}) \frac{\Gamma, y : nat \vdash t : A}{\Gamma, x : natlist \vdash t[^{\mathtt{hd}(x)}/_y] : A}$$

$$(natlist, left_{tl}) \frac{\Gamma, y : natlist \vdash t : A}{\Gamma, x : natlist \vdash t[^{\mathtt{tl}(x)}/_y] : A}$$

A typical cut is:

$$\frac{\dfrac{\Delta \vdash n : nat \qquad \Gamma \vdash l : natlist}{\Delta, \Gamma \vdash \mathbf{cons}(n, l) : natlist} \qquad \dfrac{\Theta, y : Nat \vdash t : A}{\Theta, x : natlist \vdash t[^{\mathtt{hd}(x)}/_y] : A}}{\Delta, \Gamma, \Theta \vdash t[^{\mathtt{hd}(x)}/_y][^{\mathbf{cons}(n,l)}/_x]}$$

and the obvious cut elimination rule gives:

$$\frac{\Delta \vdash n : nat \qquad \Theta, y : nat \vdash t : A}{\Delta, \Theta \vdash t[^n/_y] : A}$$

Again, by the reduction rule $\mathtt{hd}(\mathbf{cons}(X, Y)) \to X$, we have $t[^{\mathtt{hd}(l)}/_y][^{\mathbf{cons}(n,l)}/_x] = t[^n/_y]$. As an exercise the reader can provide the cut between $(natlist, right_{cons})$ and $(natlist, left_{tl})$ and verify that it induces the following rewriting:

$$\mathtt{tl}(\mathbf{cons}(X, L)) \to L$$

The following proposition sums up the previous informal discussion.

Proposition 2. *Every intuitionistic system L can be encoded by an IS S. Namely the proofs of L are expressions of S and cut-eliminations are rewriting rules of S. The encoding is correct up to structural cuts. That is let π be a proof in L which can be rewritten into π' through an elimination of a logical cut κ. Let t be the expression encoding π. Then $t \to t'$ is an instance of the rewriting encoding κ and t' encodes a proof equivalent to π' modulo structural cuts.*

In order to describe an IS as an intuitionistic system (that is the opposite of what we did so far), we must face the problem of the lack of a type discipline (as in the case of the pure λ-calculus). By adding types to IS's, we can prove their logical foundation by reverting the previous paradigm. Another possibility is to generalize intuitionistic systems by dropping any type discipline.

Instead of investigating here this (straightforward) generalization, we shall pursue an alternative approach, which is more useful in the perspective of optimal implementations. We are going to implement any IS into *pure nets*, a type-free variant of Girard's proof nets for Linear Logic, which have been already introduced in [6, 15] with the aim of encoding pure λ-calculus. The implementation, on one hand, should confirm the logical foundation of IS's; on the other hand, it provides the intermediate step towards the *optimal* implementation in [4, 13].

4 The system of pure nets

In [7] Girard proposed a refinement of Intuitionistic Logic called *Linear Logic.* In Linear Logic the structural rules of weakening and contraction are confined to a realm of exponential modalities, ! (*of course*) and ? (*why not*), as described by the following rules:

$$\text{(dereliction left)}\ \frac{A, \Delta \vdash B}{!A, \Delta \vdash B} \qquad \frac{\Delta \vdash A}{\Delta \vdash ?A}\ \text{(dereliction right)}$$

$$\text{(weakening)}\ \frac{\Delta \vdash B}{!A, \Delta \vdash B} \qquad \frac{\Delta, !A, !A \vdash B}{\Delta, !A \vdash B}\ \text{(contraction)}$$

$$\text{(? left)}\ \frac{!\Delta, A \vdash ?B}{!\Delta, ?A \vdash ?B} \qquad \frac{!\Delta \vdash A}{!\Delta \vdash !A}\ \text{(! right)}$$

Since Linear Logic essentially concerns the structural part of logical systems, we may intuitively define a linear logic version of an arbitrary intuitionistic system, by relying on the usual encoding of Intuitionistic Logic.

The correspondence between IS's and Linear Logic will be the subject of this section. We shall not consider the sequent-like formalization of Linear Logic, but its "natural deduction" version, based on *proof nets* [7]. Actually, since IS's are untyped, we shall work with Danos and Regnier's [6, 15] generalization of proof nets, named *pure nets* (pure nets are essentially obtained by proof nets by just removing types or, if you like, by using exactly one recursive type $D \equiv !D \multimap D$).

In particular, we shall define a direct encoding of Interaction Systems into pure nets. From this encoding, and by adopting the "local" implementation of boxes described in [9], one may easily derive the *optimal* implementation of Interaction Systems (in Lamping style [12]) described in [4].

Definition 3. The *formulas* are I (like *input*), O (like *output*) and $?I$ and $!O$. We say that I and O (resp. $?I$ and $!O$) are dual formulas. The *pure nets* are inductively defined over formulas by the following picture:

$$\frac{}{I \quad O}\ \text{(axiom link)} \qquad \frac{\begin{matrix}R_1 & R_2\\ \vdots & \vdots \\ X & Y\end{matrix}}{}\ \text{(cut link)} \qquad \frac{\begin{matrix}R\\ \vdots \\ I\end{matrix}}{?I}\,\delta\ \text{(der link)} \qquad \frac{\begin{matrix}R\\ \vdots \\ ?I \quad ?I\end{matrix}}{?I}\,\Delta\ \text{(co link)} \qquad \begin{matrix}R\\ \vdots\end{matrix}\ \frac{}{?I}\,\varepsilon\ \text{(weak link)} \qquad \frac{\boxed{\begin{matrix}R_1\\ \vdots \\ ?I\cdots ?I \ \ O\end{matrix}}}{?I\cdots ?I \ \ !O}\ \text{(box link)}$$

$$\frac{\begin{array}{c} R_1 \\ \vdots \\ ?I\cdots?I\;!O \end{array} \quad \cdots \quad \begin{array}{c} R_n \\ \vdots \\ ?I\cdots?I\;!O \end{array}}{O}\,(\mathbf{c}) \qquad \frac{\begin{array}{c} R_0 \\ \vdots \\ I \end{array} \quad \begin{array}{c} R_1 \\ \vdots \\ ?I\cdots?I\;!O \end{array} \quad \cdots \quad \begin{array}{c} R_n \\ \vdots \\ ?I\cdots?I\;!O \end{array}}{I}\,(\mathbf{d})$$

(constructor link) (destructor link)

where, in the case of the cut link, X and Y are dual formulas. For constructors and destructor links, the formulas in the premises are partitioned, according to the pure net R_i they come from. In every case, except the box link, the *conclusion* of the resulting pure net is the juxtaposition of the *free* conclusions of every pure net in the premise (i.e. those conclusions not involved in the premise of the rule) plus the conclusion of the rule. In the case of box link, *every* conclusion of the pure net R must be involved. Finally, if the nets R_i and R_j $(i \neq j)$ are in the premises of a rule, the two nets are supposed to be disjoint.

The names (**c**) and (**d**) in the constructor and destructor links will be replaced by the actual names of constructors and destructors.

The rewriting system for pure nets implementing IS's consists of two parts: one is common for every IS implementation (*control rules*); the other is characteristic of the particular IS under investigation (*proper rules*). We will provide a mechanism for defining the proper rules associated with a generic IS. The *control rules* are illustrated in Figure 2.

5 The implementation

In order to implement a generic IS into a pure net we must translate IS-expressions and encode IS-rewriting rules through pure net rewritings. Below we shall consider the paradigmatic case when constructors and destructors have respectively arity 1 and 01. The other cases are easily derived.

5.1 Translating the expressions

The translation $\mathcal{T}$ defined in Figure 3 gives, for any IS-expression t with n free variables, a pure net R with $n+1$ conclusions where n have the form $?I$ (the *inputs*) and one the form O (the *output*). For simplicity, in Figure 3, we have described the case when the variable bound by the destructor occurs in the body. The other case is similar to the corresponding one shown for the constructor.

Observe that in the translation of the destructor the free variables in M and those in N (except x) with the same name are contracted. For simplicity we have illustrated just one contraction.

Proper Rules The proper rules describe the interactions between destructors and constructors of the IS's. An IS-rule

$$\mathbf{d}(\mathbf{c}(\mathbf{x}^1 . X_1, \cdots, \mathbf{x}^m . X_m), \mathbf{x}^{m+1} . X_{m+1}, \cdots, \mathbf{x}^n . X_n) \rightarrow H$$

is translated in Pure nets into a rewriting rule of the following shape:

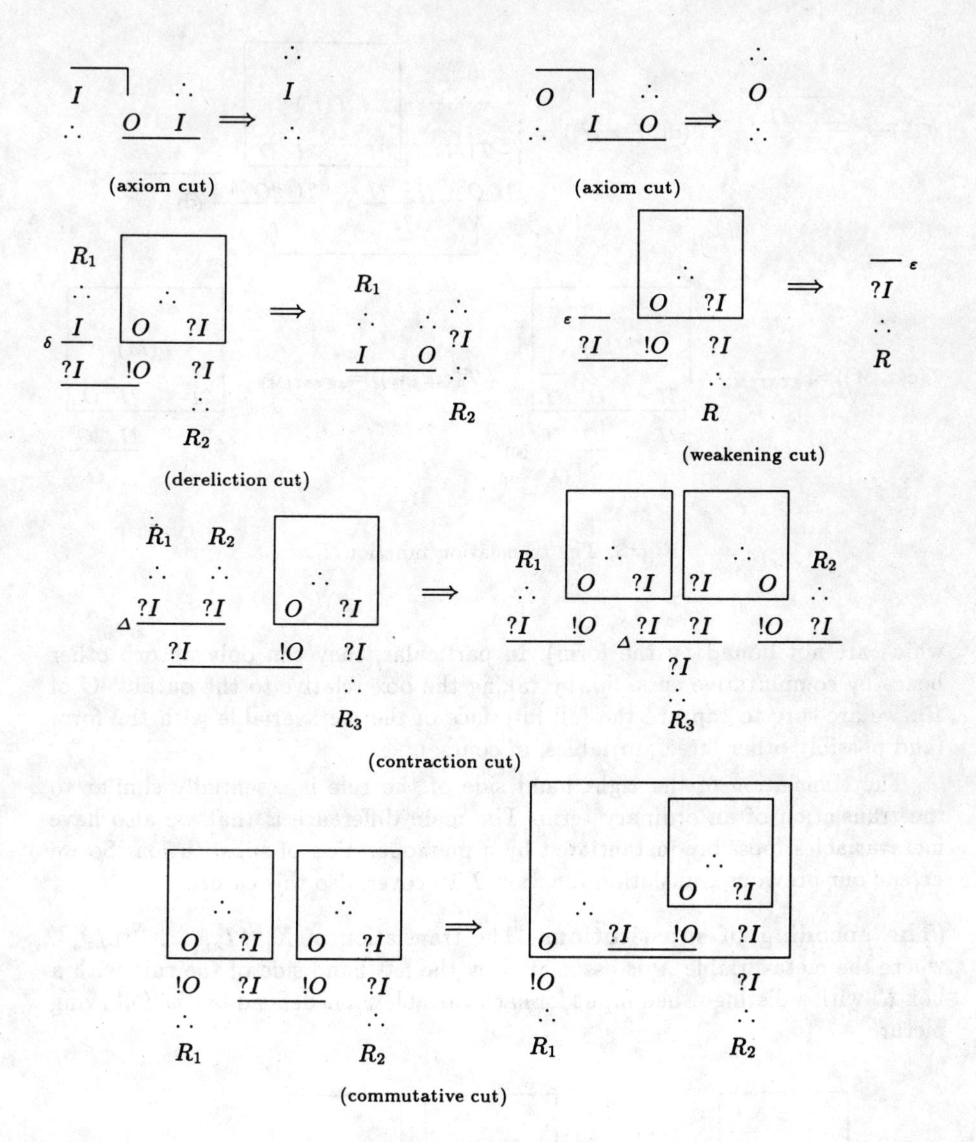

Fig. 2. Control rules

R_1 … R_m R_0 R_{m+1} … R_n

$?I \cdots ?I\ !O$ … $?I \cdots ?I\ !O$ I $?I \cdots ?I\ !O$ … $?I \cdots ?I\ !O$ $\Longrightarrow$ P_H

O I

Every box R_i in the previous picture corresponds to the metavariable X_i of the IS-rule. Note that arguments of constructors/destructors are always in boxes, according to the translation. Until the redex is fired, these boxes can be only modified along the reduction by operations at their "free" inputs (inputs

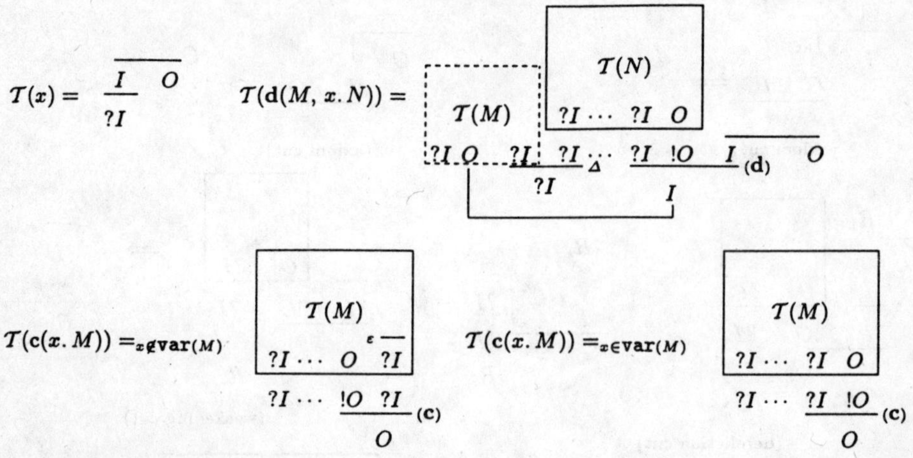

Fig. 3. The translation function $\mathcal{T}$

which are not bound by the form). In particular, they can only absorb other boxes by commutative cuts. So, by taking the box relative to the output !O of R_i, we are sure to capture the full interface of the metavariable with the form (and possibly other "free" variables, of course).

The translation of the right hand side of the rule is essentially similar to the translation of an ordinary term. The main difference is that we also have metavariables, possibly instantiated by a metaoperation of substitution. So we *extend* our previous translation function $\mathcal{T}$ to cover also this case.

(The encoding of substitutions) The translation of $X[^{N_1}/_{x_1}, \ldots, ^{N_m}/_{x_m}]$, where the metavariable X is associated by the left hand side of the rule with a box R_i with a distinguished input for each variable x_i, is defined by the following picture

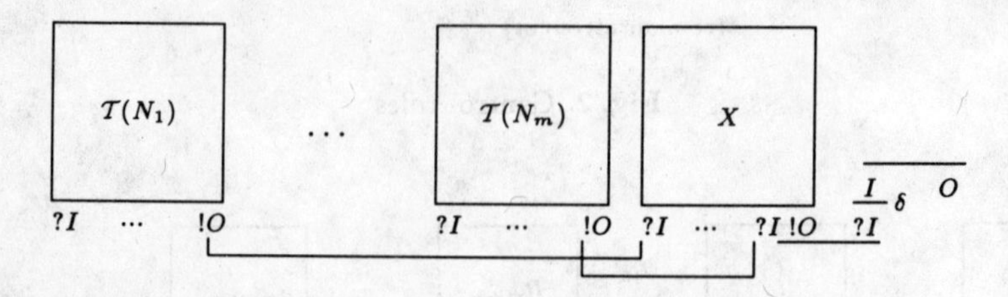

In particular, the box R_i corresponding to the metaterm X must be opened, and substitutions are obviously realized by cuts. Before doing these cuts, the arguments of the substitution must be put inside a box.

Finally, all the free variables of different instancies of a same metavariable must be shared. Note that all bound variables of metavariables are eventually "filled in" by cuts.

Example 5. The encoding of the rhs of the rewriting rule $\mathbf{d}_\mu(\mu(x.X)) \rightarrow X[{}^{\mathbf{d}_\mu(\mu(y.X[^y/_x]))}/_x]$ of Example 2 is illustrated in Figure 4.

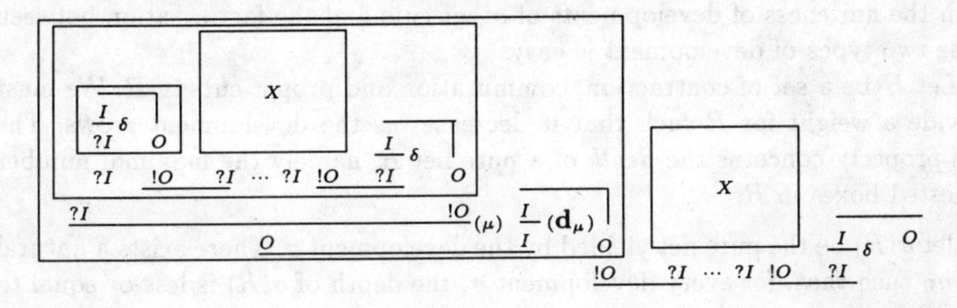

Fig. 4. The implementation of the rhs of the rule for $\mathbf{d}_\mu$-μ

Remark. For pictorial convenience, in Figure 4 we have omitted to draw the contractions over the *free* variables of the two instancies of the metavariable X. However, there is a subtle problem, here. Indeed, to be formal, we should define a different rewriting scheme for each possible set of free variables (corresponding to the contexts Δ, $\Gamma, \ldots$, in sequent calculi). This is not required in the lambda calculus, since β-reduction is *linear* in its metavariables. In [4] we adopted a different solution, applying a "linearization" procedure (essentially based on β expansion) to the rhs of each rule.

6 Properties of the implementation

Property 4. *1. The translation of every expression is a pure net;*
2. the class of pure nets is stable under reduction;
3. the rewriting system of pure nets is locally confluent.

Proof. 1. Easy check on the translation $\mathcal{T}$. 2. Examine the rhs of every rewriting rule. In every case the net replacing the redex is a pure net and is interfaced with the context by pure net links. 3. By a case analysis on the pairs of rewriting rules.

Another interesting property is the confluence of pure nets reductions. To this aim we define a reduction relation which is locally confluent, noetherian and whose transitive closure is the standard reduction of pure nets. A standard way consists of taking *developments* as basic steps of such auxiliary reduction relation. Some preliminary notions are needed.

Assume to label cuts of pure nets in order to characterize them uniquely. Then let $R \rightarrow R'$ be a pure net reduction. The *residual* of a cut labeled α in R is the set of cut labeled α in R'. A *development* of a set $H = \{\alpha, \beta, \cdots\}$ of cuts in R is a derivation $R \rightarrow R_1 \rightarrow \cdots \rightarrow R_n$ contracting at each step cuts marked by labels in H only. A development is *complete* when no label in H marks cuts in R_n.

Theorem 5. *Every development is finite.*

Proof. (*Hint*) We generalize the proof in Section 8.3 of [6]. The hard part is to prove that developments of contraction, commutation and proper cuts are finite. Both the finiteness of developments of other cuts and the factorization between these two types of development is easy.

Let H be a set of contraction, commutation and proper cuts in R. We must provide a weight for R such that it decreases as the development grows. The first property concerns the *depth* of a pure net R, namely the maximal number of nested boxes in R:

> let $\sigma(R)$ be the pure net yielded by the development σ. There exists a natural η_R such that, for every development σ, the depth of $\sigma(R)$ is less or equal to η_R.

Here, w.r.t. the corresponding proof in [6], we have to take into account proper reductions, too. However, this does not causes too much trouble.

The next step consists in defining *duplicable boxes* in $\sigma(R)$ w.r.t. cuts in the family H of R of which σ is a development. This notion is inductively defined by:

- every box in the premise of a contraction cut which is residual of a cut in H after σ is a duplication box;
- every box B connected through the principal port to a duplication box with a commutative cut residual of one in H is duplicable.

Now take the $2\eta_R$-uple defined as follows:

x_{2i} is the number of duplicable boxes at depth i in $\sigma(R)$;
x_{2i+1} is the number of boxes at depth i in $\sigma(R)$.

Then it is easy to prove that, taking the lexicographic order on tuples, the above weight decreases as the development grows. □

Theorem 6. *Pure nets rewritings are confluent.*

Proof. It is not difficult to prove that developments are locally confluent. Then the transitive closure of development-reduction gives the standard reduction. □

The main property about the implementation is the correctness, namely that $t \twoheadrightarrow t'$ implies $\mathcal{T}(t) \twoheadrightarrow \mathcal{T}(t')$. This is true up to a suitable equivalence relation over proofs. The problem is due to derelictions of bound variables that do not appear in the body of the forms. During reduction, they create "garbage" which cannot be eliminated by the rewriting system considered previously. A way out is to quotient pure nets through the following axiom:

$$\begin{array}{c} R' \\ \vdots \\ \dfrac{\overline{}^{\,\delta}\hspace{-1.5em}?I \quad ?I}{?I}\,\Delta \\ \vdots \\ R \end{array} \quad \Longrightarrow \quad \begin{array}{c} R' \\ \vdots \\ ?I \\ \vdots \\ R \end{array}$$

Up to this axiom, Danos proves the correctness in the case of λ-calculus [6]. This approach can be generalized in a straightforward way to IS's, since all the problems are related to structural rules.

Another way for proving correctness, close to the approach in [3], relies on the definition of a readback procedure for pure nets.

To this aim, we assume to have a pure net such that the free ports $?I$ keep variable names which are pairwise different. Give new variable names to ports $?I$ connected to binders such that also these variables are pairwise different. We define the readback building back the terms by traveling through the links of the pure nets. In particular, the readback $\mathcal{R}(R)$ starts from the output O of R and consists of the following steps:

1. let A be the formula under exam; if A is marked by the variable x then the algorithm gives x and stops;
2. if A is connected through an axiom to B then the algorithm continues from B towards the link whose premise is B;
3. if A is connected through a cut with B then the algorithm continues from B towards the link whose conclusion is B;
4. if A is $?I$ and
   a. the link is a weakening then $\mathcal{R}$ stops there with no output;
   b. the link is a contraction then the algorithm continues from the conclusion B of such link;
   c. the link is a box (A is internal to the box) then $\mathcal{R}$ continues from the corresponding auxiliary port;
5. if A is I and the link is a dereliction then $\mathcal{R}$ continues from $?I$;
6. if A is $!O$ and the link is a box then $\mathcal{R}$ continues from O;
7. if A is the conclusion of a constructor-link (**c**) with n formulas $!O$ in the premise, then the algorithm gives the term $\mathbf{c}(\mathbf{x}^1.t'_1, \cdots, \mathbf{x}^n.t'_n)$, where t_i are obtained by starting the algorithm from the i-th formula $!O$ in the premise and $\mathbf{x}^i$ is the vector of variables bound by the link on that formula;
8. if A is the I-premise of a destructor-link (**d**) with n formulas $!O$ in the premise, then the algorithm gives the term $\mathbf{d}(t_0, \mathbf{x}^1.t'_1, \cdots, \mathbf{x}^n.t'_n)$, where t_i are obtained by starting the algorithm from the i-th formula $!O$ in the premise and $\mathbf{x}^i$ is the vector of variables bound by the link on that formula; t_0 is the term that is readback starting from the conclusion of (**d**) toward the cut downstairs.

Let us remark some easy property of the readback.

Proposition 7. *1.* (**static correctness**) *Let t be an IS-term. Then $\mathcal{R}(\mathcal{T}(t)) = t$;*

2. (**soundness**) *let R be a pure net such that $t = \mathcal{R}(R)$ is an IS-term. If $R \to R'$ then $\mathcal{R}(R) \twoheadrightarrow \mathcal{R}(R')$ (in one or zero steps).*

Proof. 1. Trivial. 2. Boring check on the pure net rewriting rules. Note in particular that all control rules do not affect the readback. □

Note that the previous proposition does not imply the "correctness" of the implementation. In particular, every subsystem (even the empty one) of the rewriting system for pure nets would satisfy Proposition 7.(1). We should also prove that our rewriting system is powerful enough to correctly pursue β-reduction.

Theorem 8. *If R is a pure net yielded by a derivation starting at $\mathcal{T}(t)$ then:*

(A) there exists a IS-term t' such that $\mathcal{R}(R) = t'$ and $t \twoheadrightarrow t'$;
(B) if $t' \rightarrow t''$, then there exist a pure net Q such that $R \twoheadrightarrow Q$, and $\mathcal{R}(Q) = t''$
(C) if R is in normal form, than also $\mathcal{R}(R)$ is.

Proof. (*Hint*)
(A) this follows from the Proposition 7.(2);
(B) Normalizing R w.r.t. control rules and let R' be the resulting net. Note that $\mathcal{R}(R) = \mathcal{R}(R') = t'$. Now, it is not difficult to prove that every redex in t must derive, via the readback, from a proper cut in the pure net. By firing this cut, and the soundness of proper rules we get the result.
(C) If R is in normal form, it is cut-free. So we have no way to "read back" a redex. □

In [4] we proved a similar result in the case of the local implementation of "boxes". The proof, there, is much more difficult, due to the great complexity of the readback procedure caused by the presence of control operators.

7 Conclusions

Interaction Systems have been mainly introduced with the aim of generalizing Lamping's optimal graph reduction technique to a supersystem of λ-calculus [2, 4, 3, 13]. In doing this, we heavily relied on the *logical nature* of our Systems that has been discussed in this paper. In particular, the encoding of Interaction Systems into pure nets described in Section 5 provide the intermediate step to the optimal implementation in [3], by adopting the local implementation of boxes of Linear Logic defined in [9].

As a preliminary step to the optimal implementation, we have also generalized Lévy's *theory* of optimal reduction from λ-calculus to IS's (labeling, extraction process, and so on) [2, 13, 5]. These works, that revealed several, somewhat unexpected problems, have provided the first significative progress about the family relation into the realm of Higher Order Rewriting, since Lévy's PhD-thesis in 1978 [14].

Despite of their original motivations, we believe that Interaction Systems have an interest in their own: they tight together Interaction Nets and CRS's, fit well with the Proofs as Proposition analogy, have interesting applications to the theory of data types, and are conceivable of interesting extensions, especially in the typing direction.

References

1. P. Aczel. A general Church-Rosser theorem. Draft, Manchester, 1978.
2. A. Asperti and C. Laneve. Interaction Systems I: the theory of optimal reductions. Technical Report 1748, INRIA-Rocquencourt, September 1992. A revised version will appear on the journal *Mathematical Structures in Computer Science.*
3. A. Asperti and C. Laneve. Interaction Systems II: the practice of optimal reductions. Technical Report UBLCS-93-12, Laboratory for Computer Science, Universitá di Bologna, May 1993. Also Research Report n. 2001, INRIA-Sophia Antipolis.
4. A. Asperti and C. Laneve. Optimal Reductions in Interaction Systems. In *TapSoft '93*, volume 668 of *Lecture Notes in Computer Science*, pages 485 – 500. Springer-Verlag, 1993.
5. A. Asperti and C. Laneve. The family relation in Interaction Systems. to appear on Proc. of Theoretical Aspects of Computer Science 1994, Senday, Japan, 1994.
6. V. Danos. *La Logique Linéare appliquée à l'étude de divers processus de normalisation (principalement du λ-calcul).* PhD thesis, Université Paris VII, 1990.
7. J. Y. Girard. Linear Logic. *Theoretical Computer Science*, 50, 1986.
8. J. Y. Girard, P. Taylor, and Y. Lafont. *Proofs and Types.* Cambridge University Press, 1989.
9. G. Gonthier, M. Abadi, and J.J. Lévy. Linear logic without boxes. In *Proceedings 7th Annual Symposium on Logic in Computer Science*, 1992.
10. J. W. Klop. *Combinatory Reduction System.* PhD thesis, Mathematisch Centrum, Amsterdam, 1980.
11. Y. Lafont. Interaction Nets. In *Proceedings 17th ACM Symposium on Principles of Programmining Languages*, pages 95 – 108, 1990.
12. J. Lamping. An algorithm for optimal lambda calculus reductions. In *Proceedings 17th ACM Symposium on Principles of Programmining Languages*, pages 16 – 30, 1990.
13. C. Laneve. *Optimality and Concurrency in Interaction Systems.* PhD thesis, Dip. Informatica, Università di Pisa, March 1993. Technical Report TD – 8/93.
14. J.J. Lévy. *Réductions correctes et optimales dans le lambda calcul.* PhD thesis, Université Paris VII, 1978.
15. L. Regnier. *Lambda Calcul et Réseaux.* PhD thesis, Université Paris VII, 1992.

Strong Normalization of Typeable Rewrite Systems

Steffen van Bakel[1]* and Maribel Fernández[2]

[1] Afdeling Informatica,
Universiteit Nijmegen,
Toernooiveld 1, 6525 ED Nijmegen,
Nederland.
steffen@cs.kun.nl

[2] LRI Bât. 490,
CNRS / Université de Paris-Sud,
91405 Orsay Cedex,
France.
maribel@lri.fr

Abstract. This paper studies termination properties of rewrite systems that are typeable using intersection types. It introduces a notion of partial type assignment on Curryfied Term Rewrite Systems, that consists of assigning intersection types to function symbols, and specifying the way in which types can be assigned to nodes and edges between nodes in the tree representation of terms. Using a more liberal approach to recursion, a general scheme for recursive definitions is presented, that generalizes primitive recursion, but has full Turing-machine computational power. It will be proved that, for all systems that satisfy this scheme, every typeable term is strongly normalizable.

Introduction

Most functional programming languages, like Miranda [23] or ML [19] for instance, although implemented through an extended Lambda Calculus (LC) or a combinator system, allow programmers to specify an algorithm (function) as a set of equations using pattern-matching, i.e. the formal parameter of a function is allowed to have structure. Functional programs can then be seen as Term Rewriting Systems (TRS). Because of the underlying formalism, however, formal notions as type assignment (and other abstract interpretations) are studied in LC rather than on the level of the term rewriting language; the type assignment systems incorporated in most functional languages are in fact extensions of type assignment systems for LC.

It may seem straightforward to generalize formal systems defined on LC to the (significantly larger) world of TRS, but it is not evident that those ported systems have still all the properties they possessed in the world of LC. For example, type assignment in TRS in general does not satisfy the subject reduction property, i.e. types are not preserved under rewriting, as illustrated in [4]. Also, as argued in [2], not every notion of type assignment for LC can be used for TRS, and vice versa.

* Supported by the Netherlands Organisation for the advancement of pure research (N.W.O.).

To study the problem of termination a notion of types can be of significant value. For LC, there exists a well understood and well defined notion of type assignment, known as the Curry Type Assignment System [11] which expresses abstraction and application, and it can be shown that, for this notion of type assignment, all typeable terms are strongly normalizable. Curry's system forms the basis for a number of notions of type assignment used in functional programming, like for example the ones used in ML, and Miranda. In [9] the Intersection Type Discipline (the BCD-system) for LC is presented, which is a very powerful extension of Curry's system: it is closed under β-equality. Moreover, the set of terms having a head-normal form, the set of terms having a normal form, and the set of strongly normalizable terms can all be characterized by the set of assignable types. Because of this power, type assignment in this system is undecidable. In [1] a system is studied that is a variant of the BCD-system: also in this system type assignment is undecidable. That paper also contains a proof, using a Computability Predicate [22], of the statement that all typeable terms are strongly normalizable; also the converse holds.

In this paper we study the problem of termination of functional programs. Instead of studying the problem of termination using types in the world of LC, the approach taken will be to study the desired property directly on the level of a programming language with patterns, i.e. in the world of TRS. For this purpose, we introduce a notion of type assignment on Curryfied TRS (CTRS) that uses intersection types. CTRS are defined as a slight extension of the first order TRS defined in [12], in that functional types are allowed. They are restrictions of the Applicative TRS (ATRS) as defined in [4], in that the role of *Ap* in the left-hand side is restricted further.

In the past the role of types in TRS has been studied within the framework of first-order sorted rewrite systems [12], as used in the underlying model for the language OBJ (see e.g. [14]). For notions of type assignment that use the sorted approach, an enumerable collection of ***sorts*** is defined, and it is assumed that every F with arity n has a type $s_1 \times \cdots \times s_n \rightarrow s_{n+1}$, where $s_1, \ldots, s_{n+1}$ are sorts; functional types are not allowed. This implies that, by definition, there are sorts s_1, s_2 and s_3 such that (the binary operator) *Ap* has type $s_1 \times s_2 \rightarrow s_3$.

The disadvantage of this approach is, however, that the collection of typeable rewrite rules is very restricted. For example, the rewrite rules that correspond to Combinatory Logic,

$$\begin{array}{lcl} Ap(Ap(Ap(S,x),y),z) & \rightarrow & Ap(Ap(x,z),Ap(y,z)) \\ Ap(Ap(K,x),y) & \rightarrow & x \\ Ap(I,x) & \rightarrow & x. \end{array}$$

cannot be typed using the types associated to the lambda term that correspond to the combinators S, K, and I.

The notion of type assignment presented in this paper combines the approach taken in those many-sorted, first-order rewrite systems, with the one commonly used for type assignment in LC (normally defined by presenting derivation rules). First of all, by introducing *Ap* next to other function symbols, we are able to express partial applications of those symbols. Secondly, using for *Ap* the type implicitly used in the derivation rule ($\rightarrow$E), as defined in type assignment systems for LC,

$$(\rightarrow\text{E}):\ \frac{M{:}\sigma\rightarrow\tau \qquad N{:}\sigma}{MN{:}\tau}$$

i.e. $(\sigma\rightarrow\tau)\times\sigma\rightarrow\tau$ – or, in a Curryfied notation, $(\sigma\rightarrow\tau)\rightarrow\sigma\rightarrow\tau$ – invokes the possibility to assign arrow-types to all objects: we consider *higher order types*.

Our type assignment system is *partial* in the sense of [20]: we will assume that every function symbol already has a type (given in an environment), whose structure is usually motivated by a rewrite rule. In fact, the approach we take here is very much the same as the one taken by Hindley in [16], where he defines the principal Curry-type scheme of an object in Combinatory Logic. Even this notion of type assignment could be regarded as a partial one.

In [6] another partial type assignment system for higher-order rewrite systems that uses intersection types is defined. It differs from ours in that function symbols are strongly-typed with sorts only, whereas we allow for types to contain type-variables as well, and in this way we can model polymorphism (by allowing for the replacement of type-variables in types by other types).

In [4] and [2] two partial intersection type assignment systems for ATRS are presented. Apart from the difference in the syntactic definition, the system we present here is a restriction of the first one, mainly because we do not consider the type-constant ω. The second system is a *decidable* restriction of the one presented here; the restriction lies in the structure of types.

Unlike typeable terms in LC, typeable terms in CTRS need not be strongly normalizable (consider a typeable term t and a rule $t \rightarrow t$). In order to ensure strong normalization of typeable terms in CTRS we will impose some syntactical restrictions on the rewrite rules: we will present a *general scheme* for recursive definitions that generalizes primitive recursion. This kind of recursive definition was presented by Jouannaud and Okada in [17] for the incremental definition of higher order functionals based on first order definitions, so that the whole system is terminating. The general scheme of [17] was also used in [6] and [7] for defining higher order functions compatible with different lambda calculi.

We will prove (using the well-known method of Computability Predicates [15], [22]) that for all typeable CTRS satisfying this scheme, every typeable term is strongly normalizable.

In Sect. 1 we define CTRS. The type assignment system is presented in Sect. 2. In Sect. 3 we introduce the general scheme and we prove that typeable systems satisfying this scheme are strongly normalizing on typeable terms. Section 4 contains the conclusions.

1 Preliminaries

We assume the reader to be familiar with LC [8], and refer to the papers [9], [1], and [3] for an overview of intersection type assignment. For full definitions of rewrite systems we refer to [18] and [12]. Typeability of lambda terms in the system as presented in [4] is denoted in this paper by the symbol $\vdash_{\lambda\cap}$, and by $\vdash_{\lambda\cap-\omega}$ typeability in the system that is obtained from that one by removing the type-constant ω completely.

The intersection type assignment system for LC satisfies the following properties:

Theorem 1. *1.* $B \vdash_{\lambda\cap} M{:}\sigma$ *&* $M =_\beta N \Rightarrow B \vdash_{\lambda\cap} N{:}\sigma$.
2. $\exists B, \sigma$ *[*$B \vdash_{\lambda\cap} M{:}\sigma$ *&* B, σ *ω-free]* $\Leftrightarrow$ *M has a normal form.*
3. $\exists B, \sigma$ *[*$B \vdash_{\lambda\cap} M{:}\sigma$ *&* $\sigma \neq \omega$*]* $\Leftrightarrow$ *M has a head normal form.*
4. $\exists B, \sigma$ *[*$B \vdash_{\lambda\cap-\omega} M{:}\sigma$*]* $\Leftrightarrow$ *M is strongly normalizable.* ∎

1.1 Curryfied Term Rewriting Systems

In this subsection we will present Curryfied Term Rewriting Systems as an extension of first-order TRS ([18], [12]) that allow partial application of function symbols. It is easy to see that the systems presented here are equivalent to another popular way to write TRS, i.e. as the pure applicative systems, that contain only *one* function symbol (an implicit application that is normally omitted when writing terms and rules). In view of this equivalence, we have chosen to develop theory and results in the first-order setting. The language of our systems is first-order, and we *add* the binary operator *Ap* rather that restricting us to systems with only that function symbol.

CTRS are also an extension of the *function constructor* systems used in most functional programming languages. In function constructor systems the collection of function symbols is divided in two categories: constructor symbols that, given sufficient data of the right kind, create an object of a specific algebraic data-type, and function symbols that specify arbitrary operations; and in rewrite rules, function symbols are not allowed to occur in patterns, and constructor symbols can not occur at the left-most, outermost position of the left-hand side.

The systems proposed here do not discriminate in this way. The extension we made consists of allowing for not only constructor symbols in the operand space of the left-hand side of rewrite rules, but all function symbols.

Definition 2. An *alphabet* or *signature* Σ consists of:

1. A countable infinite set $\mathcal{X}$ of variables $x_1, x_2, x_3, \ldots$ (or $x, y, z, x', y', \ldots$).
2. A non-empty set $\mathcal{F}$ of *function symbols* $F, G, \ldots$, each with an 'arity'.
3. A special binary operator, called *application* (*Ap*).

Definition 3. The set $T(\mathcal{F},\mathcal{X})$ of *terms* (or *expressions*) is defined inductively:

1. $\mathcal{X} \subseteq T(\mathcal{F},\mathcal{X})$.
2. If $F \in \mathcal{F} \cup \{Ap\}$ is an n-ary symbol ($n \geq 0$), and $t_1, \ldots, t_n \in T(\mathcal{F},\mathcal{X})$, then $F(t_1,\ldots,t_n) \in T(\mathcal{F},\mathcal{X})$.

Definition 4. A *replacement* R is a map from $T(\mathcal{F},\mathcal{X})$ to $T(\mathcal{F},\mathcal{X})$ satisfying

$$\mathrm{R}(F(t_1,\ldots,t_n)) = F(\mathrm{R}(t_1), \ldots, \mathrm{R}(t_n)) .$$

So, R is determined by its restriction to the set of variables, and sometimes we will use the notation $\{x_1 \mapsto t_1, \ldots, x_n \mapsto t_n\}$ to denote a replacement. We also write t^{R} instead of $\mathrm{R}(t)$.

Definition 5. 1. A *rewrite rule* is a pair (l, r) of terms in $T(\mathcal{F}, \mathcal{X})$; we write $\mathbf{r} : l \to r$. Three conditions will be imposed:

(a) l is not a variable.
(b) The variables occurring in r are contained in l.
(c) If *Ap* occurs l, then $\mathbf{r}$ is of the shape:

$$Ap\,(F_i\,(x_1, \ldots, x_i), x_{i+1}) \;\to\; F_{i+1}\,(x_1, \ldots, x_{i+1}).$$

2. For every rewrite rule with left-hand side $F\,(t_1, \ldots, t_n)$ there are n additional rewrite rules:

$$\begin{array}{ll} Ap\,(F_{n-1}\,(x_1, \ldots, x_{n-1}), x_n) \to & F\,(x_1, \ldots, x_n), \\ \vdots & \\ Ap\,(F_0, x_1) \to & F_1\,(x_1). \end{array}$$

3. A rewrite rule $\mathbf{r} : l \to r$ determines a set of *rewrites* $l^{\mathrm{R}} \to r^{\mathrm{R}}$ for all replacements R. The left hand side l^{R} is called a *redex*; it may be replaced by its '*contractum*' r^{R} inside a context C[ ]; this gives rise to *rewrite steps*:

$$\mathrm{C}[\,l^{\mathrm{R}}\,] \to_{\mathbf{r}} \mathrm{C}[\,r^{\mathrm{R}}\,].$$

4. We call $\to_{\mathbf{r}}$ the *one-step rewrite relation* generated by $\mathbf{r}$. Concatenating rewrite steps we have (possibly infinite) *rewrite sequences* $t_0 \to t_1 \to t_2 \to \cdots$ (or *derivations* for short). If $t_0 \to \cdots \to t_n$, we write $t_0 \to^* t_n$.
5. We write $t \to_{\mathbf{R}} t'$, if there is a $\mathbf{r} \in \mathbf{R}$ such that $t \to_{\mathbf{r}} t'$.

The added rules in part 2 with $F_{n-1}, \ldots, F_1, F_0$, etc. give the 'Curried'-versions of F, and the rewrite systems are called *Curry-closed*. When presenting a rewrite system, however, we will only show the rules that are essential; we will not show the rules that define the Curried versions.

Definition 6. A *Curryfied Term Rewriting System* (CTRS) is a pair $(\Sigma, \mathbf{R})$ of an alphabet Σ and a set $\mathbf{R}$ of rewrite rules.

We take the view that in a rewrite rule a certain symbol is defined; it is this symbol to which the structure of the rewrite rule gives a type.

Definition 7. In a rewrite rule $\mathbf{r}$, the leftmost, outermost symbol in the left hand side that is not an *Ap*, is called *the defined symbol* of $\mathbf{r}$. Then $\mathbf{r}$ *defines* F, and F is *a defined symbol*. $Q \in \mathcal{F}$ is called a *constant symbol*, if there is no rewrite rule that defines Q.

We can draw the dependency-graph of the defined function-symbols, i.e. we can construct a graph whose nodes are labeled by the defined symbols of the rewrite rules, and draw an edge going from F to G if G occurs in the right hand side of one of the rules that define F. Then in that graph cycles can occur, like for the rewrite rules

$$F(x) \to G(x)$$
$$G(x) \to F(x).$$

We will call a defined symbol F a *recursive symbol* if F occurs on a cycle in the dependency-graph, and call every rewrite rule that defines F *recursive*. All function-symbols that occur on one cycle in the dependency-graph depend on each other and are, therefore, defined *simultaneously*, so we are in fact forced to give a different notion of defined symbol; the two rewrite rules above are called *mutually recursive*, and both define the symbols F and G.

It is always possible to introduce tupels into the language, and solve the problem of mutual recursion using them, so without loss of generality, we will assume that rules are *not* mutually recursive.

Definition 8. A TRS whose dependency-graph is acyclic is called a *hierarchical* TRS. The rewrite rules of a hierarchical TRS can be regrouped in such a way that they are *incremental* definitions of the defined symbols $F^1, \ldots, F^k$, so that the rules defining F^i only depend on $F^1, \ldots, F^{i-1}$.

Example 1. Our definition of recursive symbols, using the notion of defined symbols, is different from the one normally considered. Since *Ap* is never a defined symbol, the following rewrite system

$$\begin{array}{lcl} D(x) & \to & Ap(x,x) \\ Ap(D_0,x) & \to & D(x) \end{array}$$

– or, equivalently, $Ap(D,x) \to Ap(x,x)$ – is *not* considered a recursive system. Notice that, for example, the term $D(D_0)$ (or $Ap(D,D)$) has no normal form (these terms play the role of $(\lambda x.xx)(\lambda x.xx)$ in LC). This means that, in the formalism of this paper, there exist non-recursive first-order rewrite systems that are not normalizable.

Definition 9. *Applicative Combinatory Logic* (ACL) is the CTRS $(\Sigma, \mathbf{R})$, where $\mathcal{F} = \{S, S_2, S_1, S_0, K, K_1, K_0, I, I_0\}$, and $\mathbf{R}$ contains the rewrite rules

$$\begin{array}{lcl} S(x,y,z) & \to & Ap(Ap(x,z), Ap(y,z)) \\ K(x,y) & \to & x \\ I(x) & \to & x \end{array}$$

Because ACL is Curry-closed, it is in fact combinatory complete: every lambda term can be translated into a term in ACL; for details of such a translation, see [8, 13].

2 Type assignment in CTRS

The set of types that will be used in the remainder of this paper is a subset of the one used in [4]: since we are going to use types to study strong normalization, we will not consider the type constant ω. This means that the system we present here is a subsystem of that of [4], and, in particular, no properties proved there are automatically 'inherited'.

Definition 10. 1. $\mathcal{T}_{-\omega}$, the set of *strict types*, is inductively defined by:
   (a) All type-variables $\varphi_0, \varphi_1, \ldots \in \mathcal{T}_{-\omega}$.
   (b) If $\tau, \sigma_1, \ldots, \sigma_n \in \mathcal{T}_{-\omega}$ $(n \geq 1)$, then $\sigma_1 \cap \cdots \cap \sigma_n \rightarrow \tau \in \mathcal{T}_{-\omega}$.
2. $\mathcal{T}_\cap$, the set of *strict intersection types*, is defined by: If $\sigma_1, \ldots, \sigma_n \in \mathcal{T}_{-\omega}$ $(n \geq 1)$, then $\sigma_1 \cap \cdots \cap \sigma_n \in \mathcal{T}_\cap$.
3. On $\mathcal{T}_\cap$, the relation $\leq_\cap$ is defined by:
   (a) $\forall 1 \leq i \leq n$ $(n \geq 1)$ $[\sigma_1 \cap \cdots \cap \sigma_n \leq_\cap \sigma_i]$.
   (b) $\forall 1 \leq i \leq n$ $(n \geq 1)$ $[\sigma \leq_\cap \sigma_i] \Rightarrow \sigma \leq_\cap \sigma_1 \cap \cdots \cap \sigma_n$.
   (c) $\sigma \leq_\cap \tau \leq_\cap \rho \Rightarrow \sigma \leq_\cap \rho$.
4. On $\mathcal{T}_\cap$, the relation $\sim_\cap$ is defined by:
   (a) $\sigma \leq_\cap \tau \leq_\cap \sigma \Rightarrow \sigma \sim_\cap \tau$.
   (b) $\rho \sim_\cap \sigma$ & $\tau \sim_\cap \mu \Rightarrow \sigma \rightarrow \tau \sim_\cap \rho \rightarrow \mu$.

$\mathcal{T}_\cap$ may be considered modulo $\sim_\cap$. Then $\leq_\cap$ becomes a partial order, and in this paper we consider types modulo $\sim_\cap$.

Unless stated otherwise, if $\sigma_1 \cap \cdots \cap \sigma_n$ is used to denote a type, then by convention all $\sigma_1, \ldots, \sigma_n$ are assumed to be strict. Notice that $\mathcal{T}_{-\omega}$ is a proper subset of $\mathcal{T}_\cap$. The notion of type assignment as presented in [2] is a (decidable) restriction of the one presented in this paper (and of the one presented in [4]). Decidability is, in that paper, achieved by limiting the structure of types, by requiring that in $\sigma_1 \cap \cdots \cap \sigma_n \rightarrow \tau$, the types $\sigma_1, \ldots, \sigma_n$ do not contain intersections.

Definition 11. 1. A *statement* is an expression $t{:}\sigma$, with $t \in T(\mathcal{F}, \mathcal{X})$ and $\sigma \in \mathcal{T}_\cap$. t is the *subject* and σ the *predicate* of $t{:}\sigma$.
2. If B is a basis and $\sigma \in \mathcal{T}_\mathrm{S}$, then $\mathcal{T}_{\langle B, \sigma \rangle}$ is the set of all strict subtypes occurring in the pair $\langle B, \sigma \rangle$.
3. If $B_1, \ldots, B_n$ are bases, then $\Pi\{B_1, \ldots, B_n\}$ is the basis defined as follows: $x{:}\sigma_1 \cap \cdots \cap \sigma_m \in \Pi\{B_1, \ldots, B_n\}$ if and only if $\{x{:}\sigma_1, \ldots, x{:}\sigma_m\}$ is the set of all statements whose subject is x that occur in $B_1 \cup \ldots \cup B_n$.

Notice that if $n = 0$, then $\Pi\{B_1, \ldots, B_n\} = \emptyset$.

2.1 Operations on pairs

In this subsection we present two different operations on pairs of ⟨*basis*, *type*⟩, namely substitution and expansion, that are variants of similar definitions given in [4, 5, 3]. The operation of substitution deals with the replacement of type-variables by types and is the one normally used. The operation of expansion replaces types by the intersection of a number of copies of that type and coincides with the one given in [10, 21].

Definition 12. 1. The *substitution* $(\varphi \mapsto \alpha) : \mathcal{T}_\cap \rightarrow \mathcal{T}_\cap$, where φ is a type-variable and $\alpha \in \mathcal{T}_{-\omega}$, is defined by:
   (a) $(\varphi \mapsto \alpha)(\varphi) = \alpha$.
   (b) $(\varphi \mapsto \alpha)(\varphi') = \varphi'$, if $\varphi \not\equiv \varphi'$.
   (c) $(\varphi \mapsto \alpha)(\sigma \rightarrow \tau) = (\varphi \mapsto \alpha)(\sigma) \rightarrow (\varphi \mapsto \alpha)(\tau)$.
   (d) $(\varphi \mapsto \alpha)(\sigma_1 \cap \cdots \cap \sigma_n) = (\varphi \mapsto \alpha)(\sigma_1) \cap \cdots \cap (\varphi \mapsto \alpha)(\sigma_n)$.

2. If S_1, S_2 are substitutions, then so is $S_1 \circ S_2$, where $S_1 \circ S_2\,(\sigma) = S_1\,(S_2\,(\sigma))$.

The operation of expansion is an operation on types that deals with the replacement of (sub)types by an intersection of a number of copies of that type. In this process, it can be that also other types need to be copied. An expansion indicates not only the type to be expanded, but also the number of copies that has to be generated.

Definition 13. The *last type-variable* of a strict type is defined by:
1. The last type-variable of φ is φ.
2. The last type-variable of $\sigma_1 \cap \cdots \cap \sigma_n \rightarrow \tau$ is the last type-variable of τ.

Definition 14. For every $\mu \in \mathcal{T}_{-\omega}$, $n \geq 2$, basis B and $\sigma \in \mathcal{T}_{\cap}$, the quadruple $\langle \mu, n, B, \sigma \rangle$ determines an *expansion* $E_{\langle \mu,n,B,\sigma \rangle} : \mathcal{T}_{\cap} \rightarrow \mathcal{T}_{\cap}$, that is constructed as follows.
1. The set of type-variables $\mathcal{V}_\mu(\langle B, \sigma \rangle)$ is constructed by:
   (a) If φ occurs in μ, then $\varphi \in \mathcal{V}_\mu(\langle B, \sigma \rangle)$.
   (b) If the last type-variable of $\tau \in \mathcal{T}_{\langle B,\sigma \rangle}$ is in $\mathcal{V}_\mu(\langle B, \sigma \rangle)$, then all type-variables that occur in τ are in $\mathcal{V}_\mu(\langle B, \sigma \rangle)$.
2. Suppose $\mathcal{V}_\mu(\langle B, \sigma \rangle) = \{\varphi_1, \ldots, \varphi_m\}$. Choose $m \times n$ different type-variables $\varphi_1^1, \ldots, \varphi_1^n, \ldots, \varphi_m^1, \ldots, \varphi_m^n$, such that each φ_i^j $(1 \leq i \leq m, 1 \leq j \leq n)$ does not occur in $\langle B, \sigma \rangle$. Let $S_i = (\varphi_1 \mapsto \varphi_1^i) \circ \cdots \circ (\varphi_m \mapsto \varphi_m^i)$.
3. Let $\alpha \in \mathcal{T}_{\cap}$. $E_{\langle \mu,n,B,\sigma \rangle}\,(\alpha)$ is obtained by traversing α top-down and replacing, in α, a subtype β which last type-variable is an element of $\mathcal{V}_\mu(\langle B, \sigma \rangle)$, by $S_1\,(\beta) \cap \cdots \cap S_n\,(\beta)$.

Both operations are, in a natural way, extended into operations that are also defined on bases and pairs of basis and type.

For these operations, the following properties hold:

Lemma 15. *Let S be a substitution, and $E = E_{\langle \mu,n,B,\sigma \rangle}$ an expansion.*
1. *If $\tau \leq_\cap \rho$, then $S\,(\tau) \leq_\cap S\,(\rho)$, and $E\,(\tau) \leq_\cap E\,(\rho)$.*
2. *If $B \leq_\cap B'$, then $S\,(B) \leq_\cap S\,(B')$, and $E\,(B') \leq_\cap E\,(B'')$.*
3. *If $\tau \in \mathcal{T}_{\langle B,\sigma \rangle}$, then either: $E\,(\tau) = \tau_1 \cap \cdots \cap \tau_n$ where for every $1 \leq i \leq n$, τ_i is a trivial variant of τ, or $E\,(\tau) \in \mathcal{T}_{-\omega}$.*
4. *$E\,(\Pi\{B_1, \ldots, B_n\}) = \Pi\{E\,(B_1), \ldots, E\,(B_n)\}$.*

Definition 16. A *chain* is an object $<O_1, \ldots, O_n>$, with each O_i an operation of substitution or expansion, such that:

$$<O_1, \ldots, O_n>(\langle B, \sigma \rangle) = O_n\,(\cdots(O_1\,(\langle B, \sigma \rangle))\cdots).$$

2.2 Type assignment

The notion of type assignment on CTRS will in fact be defined on the tree-representation of terms and rewrite rules of these systems.

Definition 17. 1. The tree-representation of terms and rewrite rules is obtained in a straightforward way by representing a term $F(t_1, \ldots, t_n)$ by:

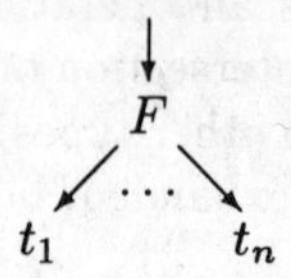

The *spine* of a term-tree is defined as usual, i.e. the root node of the term-tree is on the spine, and if a node is on the spine, then its left most descendent is on the spine. The first node on the spine of the left hand side (starting from the root node) that does not contain an *Ap* is called the *defining node* of that rule. The edge pointing to the the root of a term is called the *root edge*, and a node containing a term-variable (a function symbol $F \in \mathcal{F}$, the symbol *Ap*) will be called a *variable node* (*function node*, *application node*). Notice that if F is the defined symbol of the rule, then it occurs in the defining node.

2. Subterms can be numbered by *positions*, which are sequences of natural numbers denoting the path from the root of the term to the root of the subterms. The letters p and q stand for positions. The empty sequence (*root position*) is denoted by Λ. The subterm of t *at position* p is denoted by $t|_p$ and $t[u]_p$ is the result of replacing the subterm of t at position p by u.

Partial intersection type assignment on a CTRS (Σ, R) is defined as the labelling of nodes and edges in the tree-representation of terms and rewrite rules with types in $\mathcal{T}_\cap$. In this labelling, we use a mapping that provides a type in $\mathcal{T}_{-\omega}$ for every $F \in \mathcal{F} \cup \{Ap\}$. Such a mapping is called an environment.

Definition 18. Let $(\Sigma, \mathbf{R})$ be a CTRS.

1. An *environment* $\mathcal{E} : \mathcal{F} \cup \{Ap\} \rightarrow \mathcal{T}_{-\omega}$ is such that for every $F \in \mathcal{F}$ with arity n, $\mathcal{E}(F) = \mathcal{E}(F_{n-1}) = \cdots = \mathcal{E}(F_0)$, and $\mathcal{E}(Ap) = (\varphi_1 \rightarrow \varphi_2) \rightarrow \varphi_1 \rightarrow \varphi_2$.
2. For $F \in \mathcal{F}$ with arity $n \geq 0$, $\sigma \in \mathcal{T}_{-\omega}$, and $\mathcal{E}$ an environment, the environment $\mathcal{E}[F := \sigma]$ is defined by:
   $\mathcal{E}[F := \sigma](G) = \sigma$, if $G \in \{F, F_{n-1}, \ldots, F_0\}$,
   $\mathcal{E}[F := \sigma](G) = \mathcal{E}(G)$, otherwise.

Type assignment on CTRS is defined in two stages. In the next definition we define type assignment on terms, in Def.22 we define type assignment on term rewrite rules.

Definition 19. Let $(\Sigma, \mathbf{R})$ be a CTRS, and $\mathcal{E}$ an environment.

1. We say that $t \in T(\mathcal{F}, \mathcal{X})$ *is typeable by* $\sigma \in \mathcal{T}_\cap$ *with respect to* $\mathcal{E}$, if there exists an assignment of types to edges and nodes that satisfies the following constraints:
   (a) The root edge of t is typed with σ.
   (b) The type assigned to a function node containing $F \in \mathcal{F} \cup \{Ap\}$ (where F has arity $n \geq 0$) is $\tau_1 \cap \cdots \cap \tau_m$, if and only if, for every $1 \leq i \leq m$, there are $\sigma_1^i, \ldots, \sigma_n^i \in \mathcal{T}_\cap$, and $\sigma_i \in \mathcal{T}_{-\omega}$, such that $\tau_i = \sigma_1^i \rightarrow \cdots \rightarrow \sigma_n^i \rightarrow \sigma_i$, the

type assigned to the j-th $(1 \leq j \leq n)$ out-going edge is $\sigma_j^1 \cap \cdots \cap \sigma_j^m$, and the type assigned to the incoming edge is $\sigma_1 \cap \cdots \cap \sigma_m$.

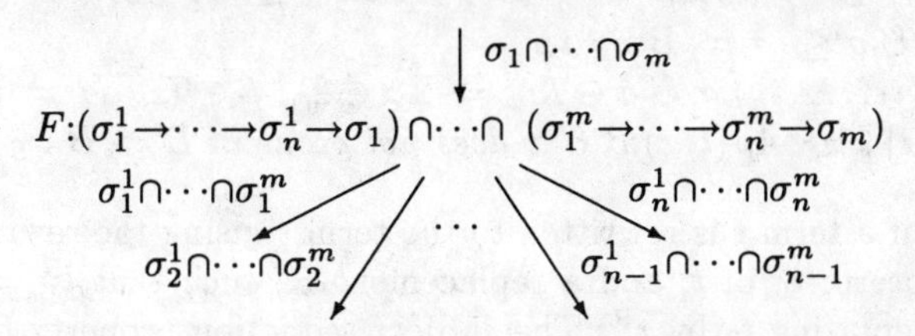

(c) If the type assigned to a function node containing $F \in \mathcal{F} \cup \{Ap\}$ is τ, then there is a chain C, such that $C(\mathcal{E}(F)) = \tau$.

2. Let $t \in T(\mathcal{F}, \mathcal{X})$ be typeable by σ with respect to $\mathcal{E}$. If B is a basis such that for every statement $x{:}\tau$ occurring in the typed term-tree there is a $x{:}\tau' \in B$ such that $\tau' \leq_\cap \tau$, we write $B \vdash_\mathcal{E} t{:}\sigma$.

Notice that if $B \vdash_\mathcal{E} t{:}\sigma$, then B can contain more statements than needed to obtain $t{:}\sigma$.

The use of an environment and part 1c of Def.19 introduce a notion of polymorphism into our type assignment system. The environment returns the 'principal type' for a function symbol; this symbol can be used with types that are 'instances' of its principal type.

A typical example for part 1b of Def.19 is the symbol Ap; for every occurrence of Ap in a term-tree, there are $\sigma_1, \ldots, \sigma_n$ and $\tau_1, \ldots, \tau_n$ such that the following is part of the term-tree.

$$\downarrow \tau_1 \cap \cdots \cap \tau_n$$
$$Ap{:}\ ((\sigma_1 \to \tau_1) \to \sigma_1 \to \tau_1) \cap \cdots \cap ((\sigma_n \to \tau_n) \to \sigma_n \to \tau_n)$$
$$\swarrow (\sigma_1 \to \tau_1) \cap \cdots \cap (\sigma_n \to \tau_n) \qquad \searrow \sigma_1 \cap \cdots \cap \sigma_n$$

Example 2. The term $S(K_0, S_0, I_0)$ can be typed with the type $\sigma \to \sigma$, under the assumption that: $\mathcal{E}(S) = (1 \to 2 \to 3) \to (4 \to 2) \to 1 \cap 4 \to 3$, $\mathcal{E}(K) = 5 \to 6 \to 5$, and $\mathcal{E}(I) = 7 \to 7$.

$$\downarrow \sigma \to \sigma$$
$$S{:}((\sigma \to \sigma) \to ((\tau \to \rho) \to (\rho \to \mu) \cap \tau \to \mu) \to \sigma \to \sigma) \to (((\rho \to \mu) \to \rho \to \mu) \to (\tau \to \rho) \to (\rho \to \mu) \cap \tau \to \mu) \to (\sigma \to \sigma) \cap ((\rho \to \mu) \to \rho \to \mu) \to \sigma \to \sigma$$
$$\swarrow K_0{:}(\sigma \to \sigma) \to ((\tau \to \rho) \to (\rho \to \mu) \cap \tau \to \mu) \to \sigma \to \sigma \qquad \searrow I_0{:}(\sigma \to \sigma) \cap ((\rho \to \mu) \to \rho \to \mu)$$
$$\downarrow S_0{:}((\rho \to \mu) \to \rho \to \mu) \to (\tau \to \rho) \to (\rho \to \mu) \cap \tau \to \mu$$

Notice that to obtain the type for S in the root-node, we have used the chain $<(1 \mapsto \sigma \to \sigma), (2 \mapsto (\tau \to \rho) \to (\rho \to \mu) \cap \tau \to \mu), (3 \mapsto \sigma \to \sigma), (4 \mapsto (\rho \to \mu) \to \rho \to \mu)>$. Notice, moreover, that to obtain the type for I_0, an expansion is needed.

If we define $D(x) \to Ap(x, x)$, then we can even check that for example $D(S(K_0, S_0, I_0))$ and $D(I_0)$ are both typeable by $\sigma \to \sigma$.

The here defined notion of type assignment satisfies the following properties:

Lemma 20. *1.* $B \vdash_{\mathcal{E}} t{:}\sigma_1 \cap \cdots \cap \sigma_n \Leftrightarrow \forall 1 \leq i \leq n\ [B \vdash_{\mathcal{E}} t{:}\sigma_i]$.
2. $B \vdash_{\mathcal{E}} t{:}\sigma\ \&\ \sigma \leq_{\cap} \tau \Rightarrow B \vdash_{\mathcal{E}} t{:}\tau$.
3. $B \vdash_{\mathcal{E}} F_n(t_1, \ldots, t_n){:}\sigma\ \&\ \sigma \in \mathcal{T}_{-\omega} \Rightarrow \exists \alpha \in \mathcal{T}_{\cap}, \beta \in \mathcal{T}_{-\omega}\ [\sigma = \alpha \rightarrow \beta]$.
4. $\Pi\{B,\{x{:}\alpha\}\} \vdash_{\mathcal{E}} Ap(t, x){:}\beta$ *& x does not occur in* $B \Rightarrow B \vdash_{\mathcal{E}} t{:}\alpha \rightarrow \beta$.

By Def.5, if a term t is rewritten to the term t' using the rewrite rule $l \rightarrow r$, there is a subterm t_0 of t, and a replacement R, such that $l^{\mathrm{R}} = t_0$, and t' is obtained by replacing t_0 by r^{R}. The subject reduction property for this notion of reduction is: If $B \vdash_{\mathcal{E}} t{:}\sigma$, and t can be rewritten to t', then $B \vdash_{\mathcal{E}} t'{:}\sigma$.

To ensure the subject reduction property, as in [4], type assignment on rewrite rules will be defined using the notion of principal pair for a typeable term.

Definition 21. Let $t \in T(\mathcal{F}, \mathcal{X})$. A pair $\langle P, \pi \rangle$ is called *a principal pair for t with respect to* $\mathcal{E}$, if $P \vdash_{\mathcal{E}} t{:}\pi$ and for every B, σ such that $B \vdash_{\mathcal{E}} t{:}\sigma$ there is a chain C such that $C(\langle P, \pi \rangle) = \langle B, \sigma \rangle$.

Notice that we do not show that every typeable term *has* a principal pair with respect to $\mathcal{E}$; at the moment we cannot give a construction of such a pair for every term. But even with this non-constructive approach we can show that the condition is sufficient with respect to the subject reduction property.

Definition 22. Let $(\Sigma, \mathbf{R})$ be a CTRS, and $\mathcal{E}$ an environment.

1. We say that $l \rightarrow r \in \mathbf{R}$ with defined symbol F *is typeable with respect to* $\mathcal{E}$, if there are basis P, type $\pi \in \mathcal{T}_{\cap}$, and an assignment of types to nodes and edges such that:
   (a) $\langle P, \pi \rangle$ is a principal pair for l with respect to $\mathcal{E}$, and $P \vdash_{\mathcal{E}} r{:}\pi$.
   (b) In $P \vdash_{\mathcal{E}} l{:}\pi$ and $P \vdash_{\mathcal{E}} r{:}\pi$, all nodes containing F are typed with $\mathcal{E}(F)$.
2. We say that $(\Sigma, \mathbf{R})$ *is typeable with respect to* $\mathcal{E}$, if every $\mathbf{r} \in \mathrm{R}$ is typeable with respect to $\mathcal{E}$.

From now on, we will only consider CTRS that are typeable with respect to a certain environment.

Condition 1b of Def.22 is in fact added to make sure that the type provided by the environment for a function symbol F is not in conflict with the rewrite rules that define F. By restricting the type that can be assigned to the defined symbol to the type provided by the environment, we are sure that the rewrite rule is typed using that type, and not using some other type. Since by part 1b of Def.22 all occurrences of the defined symbol in a rewrite rule are typed with the same type, type assignment of rewrite rules is actually defined using Milner's way of dealing with recursion [19].

Using the same technique as in [4], it is possible to show that subject reduction holds.

Theorem 23 *Subject Reduction Theorem. If* $B \vdash_{\mathcal{E}} t{:}\sigma$, *and* $t \rightarrow_{\mathbf{R}} t'$, *then* $B \vdash_{\mathcal{E}} t'{:}\sigma$. ■

It is possible to show that the two operations on pairs (substitution and expansion) are sound on typed term-trees, and that part 1c of Def.19 is sound in the following sense: if there is an operation O such that $O(\mathcal{E}(F)) = \sigma$, then for every type $\tau \in \mathcal{T}_{-\omega}$ such that $\sigma \leq_{\cap} \tau$, the rewrite rules that define F are typeable with respect to the changed environment $\mathcal{E}[F := \tau]$.

Theorem 24 *Soundness of substitution. Let S be a substitution.*

1. *If $B \vdash_{\mathcal{E}} t{:}\sigma$, then $S(B) \vdash_{\mathcal{E}} t{:}S(\sigma)$.*
2. *Let $\mathbf{r}: l \rightarrow r$ with defined symbol F be typeable with respect to $\mathcal{E}$. Then $\mathbf{r}$ is typeable with respect to $\mathcal{E}[F := S(\mathcal{E}(F))]$.* ■

Theorem 25 *Soundness of expansion. Let E be an expansion.*

1. *If $B \vdash_{\mathcal{E}} t{:}\sigma$, then $E(B) \vdash_{\mathcal{E}} t{:}E(\sigma)$.*
2. *Let $\mathbf{r}: l \rightarrow r$ with defined symbol F be typeable with respect to $\mathcal{E}$. If $E(\mathcal{E}(F)) = \tau \in \mathcal{T}_{\cap}$, then for every $\mu \in \mathcal{T}_{-\omega}$ such that $\tau \leq_{\cap} \mu$, $\mathbf{r}$ is typeable with respect to $\mathcal{E}[F := \mu]$.* ■

3 Strong Normalization and Typeability

Unlike typeable terms in $\vdash_{\lambda\cap}$, typeable terms in $\vdash_{\mathcal{E}}$ need not be normalizable. This means that the characterization of strongly normalizable terms and strongly normalizing CTRS can not be based on type conditions only, as it is possible for LC (see [1]).

In this section we analyze the relations between typeability and strong normalization in CTRS. First we show the relation between strong normalization in CTRS and normalization in $\vdash_{\lambda\cap}$; however, since the property we are characterizing is non-decidable, the condition we obtain is non-decidable as well. From a practical point of view, a decidable, sufficient (but not necessary) condition is clearly better. For this reason, in the second subsection we introduce a *general scheme*, inspired by [17], which imposes some syntactical restrictions on CTRS's rules so as to guarantee the strong normalization of all typeable terms.

From now on, we will abbreviate 't is strongly normalizable' by SN(t).

3.1 Strong Normalization of CTRS and Typeability in $\vdash_{\lambda\cap}$

Let us consider an interpreter of CTRS, i.e. a program such that, given a term t and a CTRS $(\Sigma, \mathbf{R})$, returns the set $\{t' \mid t \rightarrow_{\mathbf{R}} t'\}$ (the empty set if t is a normal form for $(\Sigma, \mathbf{R})$). In fact, we will consider a generalized interpreter P, which takes a set I of terms and a CTRS $(\Sigma, \mathbf{R})$ as input and returns the set $\bigcup_{t \in I} \{t' \mid t \rightarrow_{\mathbf{r}} t'\}$ (the empty set if I is a set of normal forms for $(\Sigma, \mathbf{R})$). We assume that the inputs and the output of P are sets of strings representing terms and rewrite rules. Besides, we assume that pure LC is used to write P, that is, P is a λ-term that will be applied to sets of strings (also coded in LC). We will use the symbol t for both the term in CTRS and its interpretation in LC. Then the λ-term $P\mathbf{R}t$ has a β-normal form which represents the set of one-step reducts of t (with respect to $(\Sigma, \mathbf{R})$).

The following property will be used to characterize strongly normalizable terms and CTRS:

Property 26. Let $(\Sigma, \mathbf{R})$ be a CTRS, and let us denote by $P^n\mathbf{R}\,t$ the term $(P\mathbf{R}\,(P\mathbf{R}\cdots(P\mathbf{R}\,t)\cdots))$, with n times P.

1. $\mathrm{SN}_{\mathbf{R}}\,(\mathrm{t}) \Leftrightarrow \exists\, n \geq 0\; [P^n\mathbf{R}\,t =_\beta \emptyset]$,
2. $\mathrm{SN}\,((\Sigma, \mathbf{R})) \Leftrightarrow \forall t \in T(\mathcal{F}, \mathcal{X})\; [\mathrm{SN}_{\mathbf{R}}\,(\mathrm{t})]$.

Given t and $(\Sigma, \mathbf{R})$, we can obtain any t' such that $t \rightarrow^*_{\mathbf{R}} t'$ using a program P^* (also in pure LC) that applies P a number of times. By definition, the result of applying P^* to $(\Sigma, \mathbf{R})$ and t is the set of strings $\{t' \mid t \rightarrow^*_{\mathbf{R}} t'\}$.

Since β-reduction is confluent, property 26 can be reformulated in this way: For any strongly normalizable term t in a CTRS $(\Sigma, \mathbf{R})$, the λ-term $P^*\mathbf{R}t$ has a β-normal form. A CTRS $(\Sigma, \mathbf{R})$ is strongly normalizing if for all t the term $P^*\mathbf{R}t$ has a β-normal form. Therefore, using Thm.1 we can characterize strongly normalizable terms and strongly normalizing systems using $\vdash_{\lambda\cap}$:

Property 27. Let $(\Sigma, \mathbf{R})$ be a CTRS.

1. $\mathrm{SN}_{\mathbf{R}}\,(t) \Leftrightarrow P^*\mathbf{R}t$ is typeable in $\vdash_{\lambda\cap}$ with a type different from ω,
2. $\mathrm{SN}\,((\Sigma, \mathbf{R})) \Leftrightarrow$ for all t, $P^*\mathbf{R}t$ is typeable in $\vdash_{\lambda\cap}$ with a type different from ω.

Note that the above results show no relation between the types assignable to a term in a CTRS, and those assignable to its interpretation in LC. In fact, Prop.27 shows only the relation between β-reduction and $\rightarrow_{\mathbf{R}}$, since we are implementing $\rightarrow_{\mathbf{R}}$ by means of $\rightarrow_\beta$.

3.2 Strong Normalization versus Typeability in CTRS

In this section we give decidable conditions for strong normalization of typeable CTRS. Since typeability alone is not sufficient, we will impose syntactic restrictions on the rules in order to get strongly normalizing systems.

We will prove that the class of typeable non-recursive CTRS is strongly normalizing. To appreciate the non-triviality of this condition, remember Ex.1: a non-recursive CTRS can be non-terminating. In fact, the main result of this section (every typeable term is strongly normalizable) shows that the term $D\,(D_0)$ (or $Ap\,(D, D)$) is not typeable.

Theorem 28. *If* $\mathbf{R}$ *contains no recursive rule and is typeable in* $\vdash_{\mathcal{E}}$ *, then any typeable term is strongly normalizable with respect to* $\mathbf{R}$.

The converse of this theorem does not hold. However, the restriction to non-recursive systems is too strong indeed. In the following we will show that there is class of recursive functions that are safe: generalized primitive recursive functions satisfying the general scheme below. This scheme for definitions is inspired by [17] where generalized primitive recursive definitions were shown strongly normalizing when combined with typed LC. The same results were shown in [6] in the context of type assignment systems for LC and in [7] for typed LC of order ω.

Definition 29. Let $\mathcal{F}_n = \mathcal{C} \cup \{Ap\} \cup \{F^1, \ldots, F^n\}$, where $F^1, \ldots, F^n$ are the defined symbols of the signature, that are not Curried-versions, and assume that $F^1, \ldots, F^n$ are defined in an incremental way. Suppose, moreover, that the rules defining $F^1, \ldots, F^n$ satisfy the *general scheme*:

$$F^i(\vec{C}[\vec{x}], \vec{y}) \to C'[F^i(\overrightarrow{C_1}[\vec{x}], \vec{y}), \ldots, F^i(\overrightarrow{C_m}[\vec{x}], \vec{y}), \vec{x}, \vec{y}],$$

where $\vec{x}$, $\vec{y}$ are sequences of variables, and $\vec{x} \subseteq \vec{y}$. Also, $\vec{C}[\,]$, $C'[\,]$, $\overrightarrow{C_1}[\,]$, and $\overrightarrow{C_m}[\,]$ are sequences of contexts in $T(\mathcal{F}_{i-1}, \mathcal{X})$, and, for $1 \leq j \leq m$, $\vec{C}[\vec{x}] \rhd_{mul} \overrightarrow{C_j}[\vec{x}]$ (where $\lhd$ is the strict subterm ordering and *mul* denotes multiset extension).

The rules defining $F^1, \ldots, F^n$ and their Curry-closure together form a *safe recursive system.*

This general scheme imposes some restrictions on the definition of functions: the terms in every $\overrightarrow{C_j}[\vec{x}]$ are subterms of terms in $\vec{C}[\vec{x}]$ (this is the 'primitive recursive' aspect of the scheme), and the variables $\vec{x}$ must also appear as arguments in the left-hand side of the rule.

It is worthwhile noting that the rewrite rules of Def. 9 are *not* recursive, so, in particular, satisfy the scheme. Therefore, although the severe restriction imposed on rewrite rules, the systems satisfying the scheme still have full Turing-machine computational power, a property that systems without *Ap* would not possess.

Example 3. The following rewrite system satisfies the general scheme:

$$\begin{array}{ll}
Add\,(Zero, y) & \to y \\
Add\,(Succ\,(x), y) & \to Succ\,(Add\,(x, y)) \\
Mul\,(Zero, y) & \to Zero \\
Mul\,(Succ\,(x), y) & \to Add\,(Mul\,(x, y), y) \\
Fac\,(Zero) & \to Succ\,(\mathrm{Zero}) \\
Fac\,(Succ\,(x)) & \to Mul\,(Succ\,(x), Fac\,(x))
\end{array}$$

Note that if we extend the definition of *Add* with the rule that expresses the associativity of *Add*,

$$Add\,(Add\,(x, y), z) \to Add\,(x, Add\,(y, z))$$

the rewrite system is no longer safe.

Theorem 28 is actually a corollary of the Strong Normalization Theorem (Thm.35). To prove this theorem, we will use the well-known method of Computability Predicates ([15], [22]). The proof will have two parts; in the first one we give the definition of a predicate *Comp* on bases, terms, and types, and prove some properties of *Comp*. The most important one states that if for a term t there are a basis B and type σ such that $Comp\,(B, t, \sigma)$ holds, then t is strongly normalizable. In the second part *Comp* is shown to hold for each typeable term.

Definition 30. 1. Let B be a basis, t a term, and σ a type, such that $B \vdash_{\mathcal{E}} t{:}\sigma$.
We define the Computability Predicate $Comp\,(B, t, \sigma)$ by induction on σ:

(a) $Comp\,(B,\,t,\,\varphi) \Leftrightarrow \mathrm{SN}(t)$.
(b) $Comp\,(B,\,t,\,\sigma\!\rightarrow\!\tau) \Leftrightarrow$
$$\forall u \in T(\mathcal{F},\mathcal{X})\ [Comp\,(B',\,u,\,\sigma) \Rightarrow Comp\,(\Pi\{B,B'\},\,Ap\,(t,u),\,\tau)].$$
(c) $Comp\,(B,\,t,\,\sigma_1\cap\cdots\cap\sigma_n) \Leftrightarrow \forall 1\leq i\leq n\ [Comp\,(B,\,t,\,\sigma_i)]$.

2. We say that t is *computable* if there exist B, σ such that $Comp\,(B,\,t,\,\sigma)$.
3. R is *computable in* $B \Leftrightarrow \exists B'\ [\forall x{:}\sigma \in B\ [Comp\,(B',\,x^{\mathrm{R}},\,\sigma)]]$.

Notice that because we use intersection types, and because of Def.11:3, in part 3 we need not consider the existence of different bases for each $x{:}\sigma \in B$.

Definition 31. A term is *neutral* if it is not of the form $F_n(t_1,\ldots,t_n)$ where F_n is a 'Curried'-version of some defined symbol F.

Property 32. *Comp* satisfies the standard properties of computability predicates:

1. *C1.* $Comp\,(B,\,t,\,\sigma) \Rightarrow \mathrm{SN}(t)$.
2. *C2.* If $Comp\,(B,\,t,\,\sigma)$, and $t \rightarrow^* t'$, then $Comp\,(B,\,t',\,\sigma)$.
3. *C3.* If t is neutral, $B \vdash_{\mathcal{E}} t{:}\sigma$ for some B, σ, and for all v such that $t \rightarrow_{\mathbf{R}} v$, $Comp\,(B,\,v,\,\sigma)$ holds, then also $Comp\,(B,\,t,\,\sigma)$ holds.

Proof: By induction on the structure of types.

1. $\sigma = \varphi$.

C1. By Def.30:1a.

C2. By Def.30:1a, using Thm. 23 and Lem. 20.

C3. By Def.30:1a, $\mathrm{SN}(v)$. Then $\mathrm{SN}(t)$ and again by 30:1a $Comp\,(B,\,t,\,\sigma)$.

2. $\sigma = \alpha\!\rightarrow\!\beta$.

C1. Let $u \equiv x$ (a new variable). x is a neutral term in normal form, and $\{x{:}\alpha\} \vdash_{\mathcal{E}} x{:}\alpha$. Then, by induction hypothesis (*C3*), $Comp\,(\{x{:}\alpha\},\,x,\,\alpha)$, and by Def.30:1b, $Comp\,(\Pi\{B,\{x{:}\alpha\}\},\,Ap\,(t,x),\,\beta)$. By induction hypothesis, $\mathrm{SN}(Ap\,(t,x))$, which implies $\mathrm{SN}(t)$.

C2. Again, let $u \equiv x$, then, as in the previous proof, $Comp\,(\{x{:}\alpha\},\,x,\,\alpha)$, and by Def.30:1b, $Comp\,(\Pi\{B,B'\},\,Ap\,(t,x),\,\beta)$. Since $Ap\,(t,x) \rightarrow^* Ap\,(t',x)$, by induction we get $Comp\,(B,\,Ap\,(t',x),\,\beta)$. Then, by Def.30:1b, $Comp\,(B,\,t',\,\sigma)$.

C3. Since $\sigma = \alpha\!\rightarrow\!\beta$, we have to prove:
$$\forall u \in T(\mathcal{F},\mathcal{X})\ [Comp\,(B',\,u,\,\alpha) \Rightarrow Comp\,(\Pi\{B,B'\},\,Ap\,(t,u),\,\beta)].$$
Since $Ap\,(t,u)$ is neutral of type β, by induction hypothesis, it is sufficient to prove that for all v' such that $Ap\,(t,u) \rightarrow_{\mathbf{R}} v'$, $Comp\,(\Pi\{B,B'\},\,v',\,\beta)$ holds. For this we apply induction on the length of the maximal derivation from u to its normal form (by Property (C1) we know that $\mathrm{SN}(u)$).

Base. If u is a normal form, because t is neutral $Ap\,(t,u)$ reduces only inside t, so $Ap\,(t,u) \rightarrow_{\mathbf{R}} Ap\,(v,u)$ and $Comp\,(B,\,v,\,\sigma)$ holds by assumption. Then, by Def.30, $Comp\,(\Pi\{B,B'\},\,Ap\,(v,u),\,\beta)$ holds.

Induction step. Consider all possible one step reductions from $Ap(t, u)$: In case $Ap(t, u) \rightarrow_{\mathbf{R}} Ap(v, u)$ we proceed as before. In case $Ap(t, u) \rightarrow_{\mathbf{R}} Ap(t, u')$, $Comp(\Pi\{B,B'\}, Ap(t, u'), \beta)$ follows by induction hypothesis. And these are all possible cases, because $Ap(t, u)$ can not be a redex itself since t is neutral and the rewrite system is safe.

3. $\sigma = \sigma_1 \cap \cdots \cap \sigma_n$.

C1. By Def.30:1c, $Comp(B, t, \sigma)$ implies $Comp(B, t, \sigma_i)$ for $1 \le i \le n$, and then by induction SN(t).

C2. As in the previous proof, $Comp(B, t, \sigma_i)$ holds for $1 \le i \le n$. By induction, for $1 \le i \le n$, $Comp(B, t', \sigma_i)$, then, by Def.30:1c, also $Comp(B, t', \sigma)$.

C3. Using Lem.20:1, we obtain $B \vdash_{\mathcal{E}} t{:}\sigma_i$, and by theorem 23, $B \vdash_{\mathcal{E}} v{:}\sigma_i$. Moreover, $Comp(B, v, \sigma)$ implies $Comp(B, v, \sigma_i)$ for $1 \le i \le n$. Then, by induction, $Comp(B, t, \sigma_i)$ for $1 \le i \le n$, and by Def.30:1c, $Comp(B, t, \sigma)$.

In order to prove the Strong Normalization Theorem we shall prove a stronger property, for which we will need the following ordering.

Definition 33. Let $(\Sigma, \mathbf{R})$ be a CTRS. Let $>_{\mathbb{N}}$ denote the standard ordering on natural numbers, $\trianglerighteq$ stand for the well-founded encompassment ordering, (i.e. $u \trianglerighteq v$ if $u \neq v$ and $u|_p = v^{\mathrm{R}}$ for some position $p \in u$ and replacement R), and *lex*, *mul* denote respectively the *lexicographic* and *multiset* extension of an ordering. Note that encompassment contains strict subterm.

We define the ordering $\gg$ on triples – consisting of a pair of natural numbers, a term, and a multiset of terms – as the object $(>_{\mathbb{N}}, \trianglerighteq, (\rightarrow_{\mathbf{R}} \cup \triangleright)_{mul})_{lex}$.

We will interpret the term u^{R} by the triple $\mathcal{I}(u^{\mathrm{R}}) = <(i, j), u, \{\mathrm{R}\}>$, where i is the maximal super-index of the function symbols belonging to u, j is the minimum of the differences $arity(F^i) - arity(F^i_j)$ such that F^i_j occurs in u, and $\{\mathrm{R}\}$ is the multiset $\{x^{\mathrm{R}} \mid x \in Var(u)\}$. These triples are compared in the ordering $\gg$.

When R is computable, then by Prop. C1 every t in the image of R is strongly normalizable, so $\rightarrow_{\mathbf{R}}$ is well-founded on the image of R. Also, because the union of the strict subterm relationship with a terminating rewrite relation is well-founded [12], the relation $(\rightarrow_{\mathbf{R}} \cup \triangleright)_{mul}$ is well-founded on $\{\mathrm{R}\}$. Hence, when restricted to computable replacements, $\gg$ is a well-founded ordering.

We use $\gg_n$ when we want to indicate that the nth element of the triple has decreased but not the n-1 first ones.

We would like to stress that we do not just interpret terms, but terms that are obtained by performing a replacement. This implies that when $t^{\mathrm{R}} = t'^{\mathrm{R}'}$, their interpretations are not necessarily equal.

We now come to the main theorem of this section, in which we show that for any typeable term and computable replacement R also the term t^{R} is computable. The strong normalization result then follows, using property C1, for any typeable term t, taking for R the identity.

In the proof, the main idea is to write a term t^{R} like $t'^{\mathrm{R}'}$ (so they are equal as terms), where $t \rhd t'$, and R' is a computable extension of R. This is accomplished by taking a computable subterm v of t, to put a new variable z in its place and to define $\mathrm{R}' = \mathrm{R} \cup \{z \mapsto v\}$.

Property 34. Let t be such that $B \vdash_{\mathcal{E}} t{:}\sigma$, and R be computable in B. Then there is a B' such that *Comp* (B', t^{R}, σ).

Proof: By noetherian induction on $\gg$. Let $B = \{x_1{:}\sigma_1, \ldots, x_n{:}\sigma_n\}$, and $\mathrm{R} = \{x_1 \mapsto u_1, \ldots, x_n \mapsto u_n\}$. Because of part 1c of Def.30, we can restrict the proof to the case $\sigma \in \mathcal{T}_{-\omega}$.

We distinguish the cases:

1. t is a neutral term. If t is a term-variable then the thesis follows trivially since R is computable. Let t be a non-variable term, so also t^{R} is neutral. If t^{R} is irreducible, then *Comp* (B', t^{R}, σ) holds by Prop.C3.
Otherwise, let $t^{\mathrm{R}} \to_{\mathbf{R}} w$ at position p. We will prove either *Comp* (B', t^{R}, σ) itself, or prove *Comp* (B', w, σ) and apply Prop.C3.
   (a) $p = qp'$, $t|_q = x_i \in \mathcal{X}$, so the rewriting takes place in a subterm of t^{R} that is introduced by the replacement. Let z be a new term-variable.
   Take $\mathrm{R}' = \mathrm{R} \cup \{z \mapsto w|_q\}$, and note that $t^{\mathrm{R}}|_q \to_{\mathbf{R}} w|_q$ at position p'. Since $t^{\mathrm{R}}|_q \in \{\mathrm{R}\}$, and R is assumed to be computable, also *Comp* (B, $t^{\mathrm{R}}|_q$, σ_i) holds. So *Comp* (B, $w|_q$, σ_i) holds by Prop.C2, hence R' is computable in $B \cup \{z_i{:}\sigma_i\}$.
   Now, if the variable x_i has exactly *one* occurrence in t, then $t = t[z]_q$ modulo renaming of term-variables, and otherwise $t \rhd t[z]_q$. In the first case (since R contains a term that is rewritten to get R') we have $\mathcal{I}(t^{\mathrm{R}}) \gg_3 \mathcal{I}(t[z]_q^{\mathrm{R}'})$, and $\mathcal{I}(t^{\mathrm{R}}) \gg_2 \mathcal{I}(t[z]_q^{\mathrm{R}'})$ in the second case. Both cases yield, by induction, *Comp* (B', $t[z]_q^{\mathrm{R}'}$, σ) and note that $t[z]_q^{\mathrm{R}'} \equiv w$.
   (b) Now assume that p is a non-variable position in t. We analyze separatedly the cases:
      i. $p \neq \Lambda$ (p is not the root position). Then $t \rhd t|_p$, hence $\mathcal{I}(t^{\mathrm{R}}) \gg_2 \mathcal{I}(t|_p^{\mathrm{R}})$, and $t|_p^{\mathrm{R}} = t^{\mathrm{R}}|_p$. Let τ be the type assigned to $t|_p$ in the derivation of $B \vdash_{\mathcal{E}} t{:}\sigma$, then *Comp* ($B$, $t^{\mathrm{R}}|_p$, τ) holds by induction.
      Let z be a new variable, and $\mathrm{R}' = \mathrm{R} \cup \{z \mapsto t^{\mathrm{R}}|_p\}$, then R' is computable in $B \cup \{z{:}\tau\}$, and $B \cup \{z{:}\tau\} \vdash_{\mathcal{E}} t[z]_p{:}\sigma$. Now $t \rhd t[z]_p$, hence $\mathcal{I}(t^{\mathrm{R}}) \gg_2 \mathcal{I}(t[z]_p^{\mathrm{R}'})$, hence *Comp* ($B$, t^{R}, σ).
      ii. $p = \Lambda$. Then the possible cases for t are:
         A. $t \equiv F(t_1, \ldots, t_n)$, where F is a defined symbol of arity n or $F \equiv Ap$ and $n = 2$, and at least one of the t_i is not a variable. Take $\mathrm{R}' = \{z_1 \mapsto t_1^{\mathrm{R}}, \ldots, z_n \mapsto t_n^{\mathrm{R}}\}$. Since $t \rhd t_i$, $\mathcal{I}(t^{\mathrm{R}}) \gg_2 \mathcal{I}(t_i^{\mathrm{R}})$. If $B \vdash_{\mathcal{E}} t_i{:}\sigma_i$, then by induction *Comp* (B, t_i^{R}, σ_i). Hence, R' is computable in $B \cup \{z_1{:}\sigma_1, \ldots, z_n{:}\sigma_n\}$. But $\mathcal{I}(t^{\mathrm{R}}) \gg_2 \mathcal{I}(F(z_1, \ldots, z_n)^{\mathrm{R}'})$, and $F(z_1, \ldots, z_n)^{\mathrm{R}'} = t^{\mathrm{R}}$ and $B \cup \{z_1{:}\sigma_1, \ldots, z_n{:}\sigma_n\} \vdash_{\mathcal{E}} F(z_1, \ldots, z_n){:}\sigma$. Hence *Comp* ($B$, t^{R}, σ).
         B. $t \equiv F^k(z_1, \ldots, z_n)$ where $z_1, \ldots, z_n$ are different term-variables. (If $z_i = z_j$ for some $i \neq j$, we can reason as in part (i.a).) Then t^{R} must be an

instance of the left hand side of a rule defining F^k: $t^{\mathrm{R}} = F^k(z_1, \ldots, z_n)^{\mathrm{R}}$ $= F^k(\vec{C}[\vec{M}], \vec{N}) \rightarrow_{\mathbf{R}} C'[F^k(\vec{C_1}[\vec{M}], \vec{N}), \ldots, F^k(\vec{C_m}[\vec{M}], \vec{N}), \vec{M}, \vec{N}] =$ w, where $\vec{C}[\vec{M}], \vec{N}$ are all terms in $\{R\}$, so are computable by hypothesis. Now, we will deduce $Comp(B, w, \sigma)$ in three steps:

Step I. Let R$'$ be the replacement that maps the left-hand side of the rewrite rule into t^{R}, so $\overrightarrow{x^{\mathrm{R}'}} = \vec{M}$. Since $\vec{M} \subseteq \vec{N}$, and all $\vec{N}$ are computable, also R$'$ is computable. For every $1 \leq j \leq m$, F^k does not occur in $\vec{C_j}$ (by definition of the general scheme), hence $\mathcal{I}(F^k(z_1, \ldots, z_n)^{\mathrm{R}}) \gg_1 \mathcal{I}(\mathrm{C}_j^{\mathrm{R}'})$, hence $\mathrm{C}_j^{\mathrm{R}'}$ is computable.

Step II. Let, for $1 \leq j \leq m$, R$_j$ be the computable replacement such that $t^{\mathrm{R}_j} = F^k(\vec{C_j}[\vec{x}], \vec{y})^{\mathrm{R}'}$. Since $\vec{C} \rhd_{mul} \vec{C_j}$, and $\rhd$ is closed under replacement, also $\vec{C}^{\mathrm{R}'} \rhd_{mul} \vec{C_j}^{\mathrm{R}'}$, hence $\mathcal{I}(F^k(z_1, \ldots, z_n)^{\mathrm{R}}) \gg_3$ $\mathcal{I}(F^k(z_1, \ldots, z_n)^{\mathrm{R}_j})$, hence $F^k(z_1, \ldots, z_n)^{\mathrm{R}_j}$ is computable.

Step III. Let v be the term obtained by replacing, in the right hand side of the rule, the terms $F^k(\vec{C_1}[\vec{M}], \vec{N}), \ldots, F^k(\vec{C_m}[\vec{M}], \vec{N}), \vec{M}$, and $\vec{N}$ by fresh variables. Let R$''$ be the replacement such that $C'[F^k(\vec{C_1}[\vec{M}], \vec{N}), \ldots, F^k(\vec{C_m}[\vec{M}], \vec{N}), \vec{M}, \vec{N}] = v^{\mathrm{R}''}$, then $t^{\mathrm{R}} \rightarrow_{\mathbf{R}} v^{\mathrm{R}''}$. Notice that above we have shown that R$''$ is computable. When F^j occurs in v, then by definition of the general scheme $j < k$, and therefore $\mathcal{I}(t^{\mathrm{R}}) \gg_1 \mathcal{I}(v^{\mathrm{R}''})$, hence $v^{\mathrm{R}''}$ is computable, and since $w = v^{\mathrm{R}''}$, we get $Comp(B, w, \sigma)$.

C. $t = Ap(z_1, z_2)$ where $z_1, z_2 \in \mathcal{X}$. By assumption, z_1^{R} and z_2^{R} are computable, and since t is well-typed, z_1 must have an arrow type. Then, by Def.30, $Ap(z_1^{\mathrm{R}}, z_2^{\mathrm{R}})$ is computable. But $Ap(z_1^{\mathrm{R}}, z_2^{\mathrm{R}})$ is the same as $Ap(z_1, z_2)^{\mathrm{R}}$.

2. Let $t \equiv F_n(t_1, \ldots, t_n)$. Again we distinguish two cases:

(a) Assume that at least one of the t_i is not a term-variable. Since $t \rhd t_i$ (for $1 \leq i \leq n$), by induction there exist B', σ_i such that $Comp(B', t_i, \sigma_i)$, and also the replacement $\mathrm{R}' = \{z_1 \mapsto t_1, \ldots, z_n \mapsto t_n\}$ is computable. Since $t \rhd F_n(z_1, \ldots, z_n)$, we have $\mathcal{I}(t^{\mathrm{R}}) \gg \mathcal{I}(t^{\mathrm{R}'})$, and $t^{\mathrm{R}'}$ is computable by induction. Note that $t^{\mathrm{R}'} = t^{\mathrm{R}}$.

(b) All t_i are term-variables. Since $B \vdash_{\mathcal{E}} t{:}\sigma$, by Lem.20:3 there exist $\alpha \in \mathcal{T}_{\cap}$, $\beta \in \mathcal{T}_{-\omega}$ such that $\sigma = \alpha \rightarrow \beta$. For all u such that $Comp(B'', u, \alpha)$, we have to prove $Comp(\Pi\{B',B''\}, Ap(t^{\mathrm{R}}, u), \beta)$. Since $Ap(t^{\mathrm{R}}, u)$ is neutral, by Prop.C3, it is sufficient to prove $Comp(\Pi\{B',B''\}, t', \beta)$ for all t' such that $Ap(t^{\mathrm{R}}, u) \rightarrow_{\mathbf{R}} t'$. This will be proved by induction on the sum of the maximal length of the derivations out of u and out of R. Note that since u and R are computable, by Prop.C1, SN(u) and SN(R).

Base. If u and R are in normal form, the only reduction step out of $Ap(t^{\mathrm{R}}, u)$ could be: $Ap(F_n(z_1, \ldots, z_n)^{\mathrm{R}}, u) \rightarrow_{\mathbf{R}} t' \equiv F_{n+1}(z_1^{\mathrm{R}}, \ldots, z_n^{\mathrm{R}}, u)$. Then, since $\mathcal{I}(t^{\mathrm{R}}) \gg_1 \mathcal{I}(F_{n+1}(z_1^{\mathrm{R}}, \ldots, z_n^{\mathrm{R}}, u))$, t' is computable.

Induction step. If the reduction step out of $Ap(t^{\mathrm{R}}, u)$ takes place inside u or inside t^{R} (in the last case it must be inside R since the rewrite system is safe) then t' is computable by induction.

If $Ap(F_n(z_1,\ldots,z_n)^{\mathrm{R}},u) \to_{\mathbf{R}} t' \equiv F_{n+1}(z_1^{\mathrm{R}},\ldots,z_n^{\mathrm{R}},u)$, then, we proceed as in the base case. ■

Theorem 35 *Strong Normalization Theorem. If $(\Sigma,\mathbf{R})$ is typeable in $\vdash_{\mathcal{E}}$ and safe, then any typeable term is strongly normalizable with respect to* **R**.

Proof: From Prop. 34 and C1, taking R such that $x^{\mathrm{R}} = x$. ■

4 Final remarks

The type assignment system defined in this paper for CTRS is undecidable: it is feasible to show that for every lambda term typeable in $\vdash_{\lambda\cap-\omega}$ there exists a term in ACL, obtained by bracket-abstraction, that is typeable as well, and vice versa. However, if we restrict the system as in [2], then typeability is decidable, and since the Strong Normalization Theorem we proved in Section 3.2 is still valid in this weaker system, any typeable CTRS satisfying the general scheme (as given in Def.29) is terminating on typeable terms.

References

1. S. van Bakel. Complete restrictions of the Intersection Type Discipline. *Theoretical Computer Science*, 102:135–163, 1992.
2. S. van Bakel. Partial Intersection Type Assignment of Rank 2 in Applicative Term Rewriting Systems. Technical Report 92-03, Department of Computer Science, University of Nijmegen, 1992.
3. S. van Bakel. Essential Intersection Type Assignment. In R.K. Shyamasunda, editor, *Proceedings of FST&TCS '93. 13th Conference on Foundations of Software Technology and Theoretical Computer Science*, *LNCS* 761, pages 13–23, Bombay, India, 1993.
4. S. van Bakel. Partial Intersection Type Assignment in Applicative Term Rewriting Systems. In M. Bezem and J.F. Groote, editors, *Proceedings of TLCA '93. International Conference on Typed Lambda Calculi and Applications*, *LNCS* 664, pages 29–44, Utrecht, the Netherlands, 1993.
5. S. van Bakel. Principal type schemes for the Strict Type Assignment System. *Logic and Computation*, 3(6):643–670, 1993.
6. F. Barbanera and M. Fernández. Combining first and higher order rewrite systems with type assignment systems. In M. Bezem and J.F. Groote, editors, *Proceedings of TLCA '93. International Conference on Typed Lambda Calculi and Applications*, *LNCS* 664, pages 60–74, Utrecht, the Netherlands, 1993.
7. F. Barbanera and M. Fernández. Modularity of Termination and Confluence in Combinations of Rewrite Systems with λ_ω. In A. Lingas, R. Karlsson, and S. Carlsson, editors, *Proceedings of ICALP '93. 20th International Colloquium on Automata, Languages and Programming*, *LNCS* 700, pages 657–668, Lund, Sweden, 1993.
8. H. Barendregt. *The Lambda Calculus: its Syntax and Semantics*. North-Holland, Amsterdam, revised edition, 1984.

9. H. Barendregt, M. Coppo, and M. Dezani-Ciancaglini. A filter lambda model and the completeness of type assignment. *Journal of Symbolic Logic*, 48(4):931–940, 1983.
10. M. Coppo, M. Dezani-Ciancaglini, and B. Venneri. Principal type schemes and λ-calculus semantics. In J.R. Hindley and J.P. Seldin, editors, *To H.B. Curry, Essays in combinatory logic, lambda-calculus and formalism*, pages 535–560. Academic press, New York, 1980.
11. H.B. Curry and R. Feys. *Combinatory Logic*, volume 1. North-Holland, Amsterdam, 1958.
12. N. Dershowitz and J.P. Jouannaud. Rewrite systems. In J. van Leeuwen, editor, *Handbook of Theoretical Computer Science*, volume B, chapter 6, pages 245–320. North-Holland, 1990.
13. M. Dezani-Ciancaglini and J.R. Hindley. Intersection types for combinatory logic. *Theoretical Computer Science*, 100:303–324, 1992.
14. K. Futatsugi, J. Goguen, J.P. Jouannaud, and J. Meseguer. Principles of OBJ2. In *Proceedings 12th ACM Symposium on Principles of Programming Languages*, pages 52–66, 1985.
15. J.Y. Girard, Y. Lafont, and P. Taylor. *Proofs and Types*. Cambridge Tracts in Theoretical Computer Science. Cambridge University Press, 1989.
16. J.R. Hindley. The principal type scheme of an object in combinatory logic. *Transactions of the American Mathematical Society*, 146:29–60, 1969.
17. J.P. Jouannaud and M. Okada. Executable higher-order algebraic specification languages. In *Proceedings of the Sixth Annual Symposium on Logic in Computer Science*, pages 350–361, 1991.
18. J.W. Klop. Term Rewriting Systems. In S. Abramsky, Dov.M. Gabbay, and T.S.E. Maibaum, editors, *Handbook of Logic in Computer Science*, volume 2, chapter 1, pages 1–116. Clarendon Press, 1992.
19. R. Milner. A theory of type polymorphism in programming. *Journal of Computer and System Sciences*, 17:348–375, 1978.
20. F. Pfenning. Partial Polymorphic Type Inference and Higher-Order Unification. In *Proceedings of the 1988 ACM conference on LISP and Functional Programming Languages*, pages 153–163, 1988.
21. S. Ronchi della Rocca and B. Venneri. Principal type schemes for an extended type theory. *Theoretical Computer Science*, 28:151–169, 1984.
22. W.W. Tait. Intensional interpretation of functional of finite types. *Journal of Symbolic Logic*, 32, 1967.
23. D.A. Turner. Miranda: A non-strict functional language with polymorphic types. In *Proceedings of the conference on Functional Programming Languages and Computer Architecture*, *LNCS* 201, pages 1–16.

A Transformation System Combining Partial Evaluation with Term Rewriting *

Françoise Bellegarde

Computer Science,
Oregon Graduate Institute of Science & Technology,
PO BOX 91000,
Portland, OREGON, 97006.
Bellegar@cse.ogi.edu

Abstract. This paper presents a new approach to optimizing functional programs based on combining partial evaluation and rewriting. Programs are composed of higher-order primitives. Partial evaluation is used to eliminate higher-order functions. First-order rewriting is used to process the transformation. Laws about the higher-order primitives that are relevant for the optimizations are automatically extracted from a library and transformed into first-order terms using partial evaluation. Such a combination of a partial evaluation system and an intrinsically first-order rewriting tool allows a form of higher-order rewriting at a first-order level. This way, it is possible to automate deforestation of higher-order programs.

Introduction

The so-called Squiggol [10] style for program construction is a high-level programming technique that consists of building a program by composing primitives or other functions while taking into account well-known laws on the primitives. Functions are usually defined according to recursion patterns that are attached to the inductive structure of the data types. These recursive patterns can be captured by higher-order functions. For example a particular kind of recursive pattern attached to the recursive data type list called catamorphism [10] can be captured by the higher-order primitive *foldr*. This higher-order primitive is provided in most functional languages. It is called *fold* in ML and *reduce* in Common Lisp.

Although constructing programs by composing higher-order primitives provides the user with a high degree of abstraction, it comes at the expense of efficiency. Indeed, compositions produce many intermediate data structures when computed in an eager (call by value) evaluation. One way to circumvent this problem is to perform *deforestation* on programs as advocated by Wadler [16]. Deforestation algorithms [16, 9], as well as algorithms based on promotion theorems [12] eliminate these useless intermediate data structures, but the optimizations they perform are limited because they do not take into account any

* The work reported here was supported in part by a contract with Air Force Material Command (F1928-R-0032) and by a grant from NSF (CCR-9101721).

particular laws about the operands of the compositions. As described in [17], laws about higher-order polymorphic primitives can be derived *for free* from their type. When appropriately defined, laws can constitute a powerful calculus to derive efficient programs from higher-level specifications. The full paper describes an automatic process that uses laws on higher-order primitives to perform an extended form of deforestation.

Related work

Wadler has proposed an algorithm for deforestation in [16]. It works intrinsically on first-order programs though it is extended to higher-order programs by treating higher-order programs as macros. His algorithm performs automatic deforestation on *treeless* terms. Chin's work on fusion [5] applies to higher-order program in general, it skips over terms to which the technique does not apply. More recently, promotion theorems have been applied to normalize programs [12]. This technique is applicable to *potentially normalizable terms* which are similar to *treeless* terms. A new automatic way to implement deforestation inside the Haskell's compiler is shown in [9].

Most of the general purpose program transformation systems are based on a folding-unfolding strategy *à la* Burstall and Darlington [4]. Deforestation is a particular instance of this strategy. In the Focus system [11], folding and unfolding are seen as term rewritings. It has been pointed out in [7] how a folding-unfolding strategy can be directed by a completion procedure. Following this idea, the transformation system Astre [1, 2] is based on *partial* completion procedures[2]. Astre takes into account of inductive laws provided by the user during the completion process. All these systems are interactive. However a currently implemented *automatic mode* of Astre performs automatically a simple [3] deforestation of a program. In this mode, it has the same functionality and the same limitations as the above deforestation algorithms. Both Focus and Astre are based on first-order term rewriting techniques therefore they are limited to first-order programs.

The paper

We describe a way to mimic an extended form of deforestation [4] of higher-order functional programs using first-order rewriting. This is achieved by using partial evaluation to transform a class of higher-order programs into first-order ones.

Partial evaluation aims at specializing a program with respect to part of its input (static parts). This process produces a specialized (residual) program [8].

[2] The completion is *partial* because it computes only part of the superpositions between rewrite rules.

[3] The deforestation is said to be *simple* if its processing does not use particular laws between the functions and primitives.

[4] The deforestation is said to be *extended* if it can be achieved only by using particular laws between the functions and primitives.

This specialized program when applied to the remaining input value parts (dynamic parts) yields the same result as the original program applied to a complete input. In this paper we are using Schism [6] a partial evaluator for pure functional programs. Our goal is to use partial evaluation to eliminate higher-order functions by specializing higher-order primitives with respect to their functional arguments.

The objective of this paper is to show how a large class of higher-order programs can be automatically improved by using powerful laws on higher-order primitives, partial evaluation, and term-rewriting. To our knowledge, no general purpose transformation system supports this kind of transformation.

1 Maxsub example

We illustrate our transformational approach with a program presented by S. Thompson [14] and first introduced by J. Bentley [3]. The problem solved by this program is stated as follows by Thompson:

> Given a finite list of numbers, find the maximum value for the sum of a (contiguous) sublist of the list.

Numbers can be positive as well as negative integers. In his book [15], S. Thompson presents formally the derivation of a functional program that achieves the computation described above. It is displayed in Figure 1. This program is quadratic in the length of the list.

In this program, [] is the empty list, :: is the constructor (often called *cons*) of the data type list and *consl* is its prefix version. The binary operation @ produces the concatenation of two lists. The operation $\circ$ is the composition of two functions.

This program also assumes that the higher-order operators *fold*, *map*, *foldr* and the function *bimax* are primitive operations. The function *bimax* is the maximum function. The operation + is prefixed. For example:

$$foldr\ +\ 0\,[2,-3,4,3] = +\,2\,(+\,(-3)\,(+\,4\,(+\,3\,0))) = 6\ .$$

The operator *fold* is similar to *foldr*, except that it only applies to non empty lists. For example:

$$fold\ bimax\,[2,-3,4,3] = bimax\,2\,(bimax\,(-3)\,(bimax\,4\,3)) = 4\ .$$

The function *sublists* returns all the contiguous sublists of a given list. The function *frontlists* returns all the sublists that are prefixes of the list. For example:

$$frontlists\,[2,-3,4,3] = [[2,-3,4,3],[2,-3,4],[2,-3],[2],[]]\ .$$

Faithful to the Squiggol method, *maxsub* takes all the contiguous sublists of the list using *sublists*, then computes all the sum of the elements of every sublists using *map sum*, finally computes their maximum using *fold bimax*.

$$
\begin{aligned}
&maxsub &&= (fold\ bimax) \circ (map\ sum) \circ sublists\\
&sum &&= foldr + 0\\
&sublists\ [] &&= [[]]\\
&sublists\ (a :: x) &&= (map\ (consl\ a)\ (frontlists\ x))\ @\ (sublists\ x)\\
&frontlists\ [] &&= [[]]\\
&frontlists\ (a :: x) &&= (map\ (consl\ a)\ (frontlists\ x))\ @\ [[]]
\end{aligned}
$$

Fig. 1. program source

$$
\begin{aligned}
&maxsub\ [] &&= 0\\
&maxsub\ (a :: x) &&= (bimax\ (+\ a\ (maxfront\ x))\ (maxsub\ x))\\
&maxfront\ [] &&= 0\\
&maxfront\ (a :: x) &&= (bimax\ (+\ a\ (maxfront\ x))\ 0)
\end{aligned}
$$

Fig. 2. final program

The final result of the transformation is the functional program shown in Figure 2 which is linear. The theorems used during the transformation process are listed in Figure 3.

The details of the manual transformations using these laws can be found in [14]. Let us now consider, a way to automate this transformation using completion and partial evaluation.

The system Astre [1, 2] based on a partial completion procedure, deals with programs presented by a set of first-order equations. A partial evaluator can automatically convert a class of higher-order programs into first-order ones.

2 Conversion to first-order programs

In order to understand the partial evaluation process, consider a functional program expressed as a *typed* λ-term M. Some of the arguments (static arguments)

$$
\begin{aligned}
&maplaw: && map\ f \circ map\ g = map\ (f \circ g)\\
&map@law: && map\ f\ (x@y) = (map\ f\ x)\ @\ (map\ f\ y)\\
&fold@law: && fold\ f\ (x@y) = f\ (fold\ f\ x)\ (fold\ f\ y)\\
& && \qquad \text{if f is associative and}\\
& && \qquad \text{x and y are non empty lists}\\
&foldmaplaw: && fold\ f \circ map\ g = g \circ fold\ f\\
& && \qquad \text{if } f\ (g\ x)\ (g\ y) = g\ (f\ x\ y)
\end{aligned}
$$

Fig. 3. laws for the transformations

of functions are substituted by values according to their respective types. Let this substitution be δ, the partial evaluation operates as if it normalizes $\delta(M)$ by β-reduction and evaluates by unfolding the fixpoint operator when possible. We denote by $M \downarrow_{beta}$ the beta-normal form of the λ-term M.

For example, consider the function *append*, that is, the prefix version of the operator @ used in the *maxsub* example. The definition is:

$$\begin{aligned} \mathit{append}\ []\ y &= y \\ \mathit{append}\ (a :: x)\ y &= a :: (\mathit{append}\ x\ y) \end{aligned}$$

A partial evaluation of the term ($\mathit{append}\ [1,2,3]\ y$), where the first argument of *append* is the static argument substituted by $[1,2,3]$, returns the specialized definition:

$$\mathit{append}1\ y = 1 :: (2 :: (3 :: y))$$

The evaluation involves unfolding the recursive definition of *append.* But the partial evaluation of the term ($\mathit{append}\ x\ [4,5,6]$), where the second argument of *append* is the static argument substituted by $[4,5,6]$, returns the specialized definition:

$$\begin{aligned} \mathit{append}2\ [] &= [4,5,6] \\ \mathit{append}2\ (a :: x) &= a :: (\mathit{append}2\ x) \end{aligned}$$

In this case unfolding the recursion is impossible, only β-reduction is involved. Similarly, consider a partial evaluation of the term ($\mathit{map}\ \mathit{sum}\ x$) given the following definition of *map*:

$$\begin{aligned} \mathit{map}\ f\ [] &= [] \\ \mathit{map}\ f\ (a :: x) &= (f\ a) :: (\mathit{map}\ f\ x) \end{aligned}$$

It returns the following specialization of *map* with respect to the static argument f substituted by the value *sum*:

$$\begin{aligned} \mathit{map1}\ [] &= [] \\ \mathit{map1}\ (a :: x) &= (\mathit{sum}\ a) :: (\mathit{map1}\ x) \end{aligned}$$

Note that the partial evaluation need not unfold the recursive definition of *map* because the recursive call applies to a variable function f. It only specializes the definition of *map f* with respect to a sustitution of f by *sum*. It is important not to forget that *map1* is a specialization of *map sum* in order to be able to know that the higher-order laws on *map* apply to *map1* during the transformation of the converted program. This information consists of a triple:

$$(\mathit{map1}, \{f \leftarrow \mathit{sum}\}, \mathit{map}\ f)$$

In the following, RHOP will denote the set of the symbols of the recursive higher-order primitives like *map*, *fold*, *foldr*..., and κ will be the substitution unfolding the definitions of the RHOP symbols. We assume that the definitions of the RHOP functions are not mutually recursive. The $\beta-$normalization of a λ-term term M is noted by $M \downarrow_{\beta}$.

Definition 1. A specialization triple is a triple (F, δ, T) where F is the symbol of a first-order specialization of a RHOP function, δ is a specialization substitution, and T is a specialized term such that a definition of F is obtained by folding every occurrence of $\delta(T)$ in the term $s = \delta(\kappa(T)) \downarrow_\beta$.

We note by $s[t \backslash F]$ the replacement of each occurrence of the subterm t in the term s by the symbol F (folding of F). Then, the definition of F according to a specialization triple (F, δ, T) is:

$$F = \delta(\kappa(T)) \downarrow_{beta} [\delta(T) \backslash F]$$

Triples (F_i, δ_i, T_i) and (F_j, δ_j, T_j) such that δ_i, δ_j are α-convertible and T_i and T_j are also α-convertible are said to be α-convertible. In this case, F_i and F_j must be the same so that double definitions are not introduced.

Let $\rightarrow^{first}$ be the relation between λ-term and set of triples defined as follows:

Definition 2. $M_{ST} \rightarrow^{first} M'_{ST'}$ if and only if there exists an occurrence in M of a subterm $F\, U_1\, U_2\, \cdots\, U_n$ where F is a symbol of a RHOP function, and U_i are terms instantiating all the functional arguments of F, then $M' = M[F\, U_1\, U_2\, \cdots\, U_n \backslash F_1]$ where F_1 is the symbol of the specialization of F with respect to the arguments $U_1, U_2, \cdots, U_n$. There can be two cases:

- either there exists in ST a specialization triple (F_1, δ, N), then $ST' = ST$,
- or, the adequate specialization triple does not exist in ST: In this case a new specialization triple is created for a new specialized function symbol F_1. The set of specialization triple ST' is then:

$$\begin{array}{l} ST \cup \{(F_1, \delta = \{(f_1 \leftarrow U_1), (f_2 \leftarrow U_2), \cdots, (f_n \leftarrow U_n)\}, \\ N = \lambda\, x_1\, x_2 \cdots x_m\, (F\, f_1\, f_2 \cdots f_n))\} \end{array}$$

  where $x_1, x_2, \cdots, x_m$ are the free variables in the λ-term terms U_i.

We will omit the subscript by the set of specialization triples ST and we will consider the updating of ST as a side effect. Then $M \downarrow_{first}$ denotes a normalization of M by the relation $\rightarrow_{first}$.

The conversion to first-order processed by the partial evaluator for the class of programs we are considering can be viewed as: $M \downarrow_{first}$. The class of higher-order programs that are tackled by our approach can be characterized as follows. These programs consist of first-order terms and *constant or variable-only* higher-order primitives. *Variable-only* higher-order primitives are functions whose higher-order arguments are solely made up of variables in each recursive call to this function. This is called the *variable-only* criterion by Chin [5]. In the context of partial evaluation this criterion ensures that specializing functions with respect to higher-order values always terminates. For example *map*

$$map\, f\, (a :: x) \;=\; (f\, a) :: (map\, f\, x)$$

is *variable only*, so are *fold* and *foldr*. A function defined by :[5]

$$mapTwo\ f\ g\ [\,] \ = \ [\,]\ ;\ mapTwo\ f\ g\ (a :: x) \ = \ (f\ a)\ ::\ mapTwo\ g\ f\ x$$

is also variable only (the specialization of *mapTwo f1 g1* into *mapTwo1* calls a specialization of *mapTwo g1 f1* into *mapTwo2* which itself calls *mapTwo1*). But a function H defined by:

$$H\ f\ (a :: x) \ = \ (f\ a) :: (H\ (f \circ f)\ x)$$

and

$$G\ f\ (a :: x) \ = \ (f\ a) :: (G\ (sqr)\ x)$$

are not. However G is *constant-only* because its functional argument in the recursive call is a constant. The variable-only criteria is different from the restrictions given to higher-order functions that can be defined using higher-order macros as advocated in [16]: the functional arguments of these higher-order primitives must be fixed or unchanged across recursion. This criteria excludes *mapTwo* and H but includes G.

Theorem 3. *Assuming that the non RHOP functions in M are first-order, that the RHOP functions are* constant or variable-only, *then M has a unique (modulo α-conversion) normal form $M \downarrow_{first}$, and all the function calls in the normal form are calls to first-order functions.*

The theorem ensures the soundness of the *conversion to first-order* process.

Proof: The only higher-order recursive functions are the RHOP functions. They are called in M by first-order functions. Let F be a RHOP function and $F\ U_1\ U_2, \cdots U_n$ be an occurrence of a call of F in M. The terms U_i instantiate all the functional arguments of F because the functions in M are first-order. This call to a higher-order function must disappear. For that, it must correspond to a potential application of the relation $\rightarrow_{first}$. Let $x_1, x_2, \cdots, x_m$ be the free variables in the U_i. Such a call must correspond to a specialization triple

$$(F_1, \delta = \{(f_1 \leftarrow U_1), (f_2 \leftarrow U_2), \cdots, (f_n \leftarrow U_n)\},$$
$$N = \lambda\, x_1\, x_2 \cdots x_m\ (F\ f_1\ f_2 \cdots f_n))$$

where $F_1 = \delta(\kappa(N)) \downarrow_{beta} [\delta(N) \backslash F_1]$ according to the definition of $\rightarrow_{first}$ We need to prove that we can define such a function F_1. The symbol F is a RHOP function. By hypothesis, a recursive call of F in the λ-term term defining F invokes variables only or constants. The functional variables are substituted by the U_i, therefore, if the variables in the recursive call appears in the same order and are not duplicated in the recursive calls, they can be *folded* into calls to F_1. If they are swapped (with eventual repetitions) like in the example of *mapTwo* above, a sequence of

[5] This example comes from an anonymous referee

specialization will eventually generate mutually recursive specializations according to the number of inversions of the permutation. If there is a contant argument, another specialization which also generates mutually recursive specializations is necessary. The *beta*-reduction itself must terminate. Therefore, the relation $\rightarrow_{first}$ applies, and the term M' such that $M \rightarrow M'$ contains one occurrence of a higher order call less than M. Therefore normalization eliminates every call to RHOP functions from M, replacing them by calls to their specializations. The result of eliminating these calls is a first-order term when the specialized functions definition (definition of F_1 for the call above) does not contain any call to a higher-order function. It is obviously the case when the definition of the higher-order function F, which is specialized, contains no calls to another RHOP function. If it does contain such a call, it instantiates all its functional arguments and this call can be the initiation of a specialization in the same way than above. The process is terminating because the RHOP functions are not mutually recursive. This proves that the specialization process is terminating and that the result is a first-order program. The normal-form $M \downarrow_{first}$ exists. Finally, the calls to RHOP functions are distinct and fully applied in M, thus the order we treat the calls does not change the result. This gives unicity of the result modulo α-conversion. □

For the *maxsub* example, the first-order conversion uses the following set of triples ST:

$$\begin{array}{lll}\{(\mathit{map1}, & \{f \leftarrow \mathit{sum}\} & ,\lambda\, x.(\mathit{map}\ f\ x)) \\ (\mathit{map2} & ,\{f \leftarrow (\mathit{consl}\ x)\} & ,\lambda\, l\ x.(\mathit{map}\ f\ l)) \\ (\mathit{fold1} & ,\{f \leftarrow \mathit{bimax}\ \} & ,\lambda\, x.(\mathit{fold}\ f\ x))\}\end{array}$$

A partial evaluator performs only the adequate specializations of the RHOP functions, there is no unfolding of recursive calls: Given M, its result is the normal form $M \downarrow_{first}$. We are currently using the partial evaluator Schism [6] for first-order conversion. Schism uses a typed dialect of Scheme as its object language. A translator from ML to Schism is available.

Modulo some renaming, the result of partial evaluation of the *translated* program for *maxsub* is shown in Figure 4. The constructor :: becomes **cons**, [] becomes '() in Scheme.

However, this program cannot be the input of a transformation system based on rewriting such as Astre. Even if we forget about the differences of syntaxes, an equation like:

$$\begin{array}{rl}(\mathit{map1}\ l) = & \mathbf{if}\,(\mathbf{null?}\ l)\ \mathbf{then}\ [] \\ & \mathbf{else}\,(\mathit{sum}\,(\mathbf{car}\ l)) :: (\mathit{map1}\,(\mathbf{cdr}\ l))\end{array}$$

gives a non terminating rewriting rule. A translator from Schism to ML is currently implemented. This translator will reintroduce the patterns that the translator from ML to Schism has eliminated. This way, the *first-order converted* program is translated into the first-order set of equations shown in Figure 5.

```
(program () () (

(define    (maxsub l)
           (fold1 (map1 (sublists l))))
(define    (sublists l)
           (if (null? l) then (cons '() '())
            else (@ (map2 (frontlists (cdr l)) (car l)) (sublists (cdr l))) ))
(define    (frontlists l)
           (if (null? l) then (cons '() '())
            else (@ (map2 (frontlists (cdr l)) (car l)) (cons '() '())) ))
(define    (map2 l x)
           (if (null? l) then '()
            else (cons (cons x (car l)) (map2 (cdr l) x)) ))
(define    (map1 l)
           (if (null? l) then '()
            else (cons (sum (car l)) (map1 (cdr l))) ))
(define    (fold1 l)
           (if (null? (cdr l)) then (car l)
            else (bimax (car l) (fold1 (cdr l))) ))
(define    (sum l)
           (if (null? l) then '0
            else (+(car l) (sum (cdr l))) ))
(definePrimitive +)
(definePrimitive bimax)
(definePrimitive @)
))
```

Fig. 4. First-Order converted program source

```
maxsub l             = fold1 (map1 (sublists l))
sublists []          = [[]]
sublists (a :: x)    = (map2 (frontlists x) a) @ (sublists x)
frontlists []        = [[]]
frontlists (a :: x)  = (map2 (frontlists x) a) @ [[]]
(map2 [] x)          = []
(map2 (l :: lx) x)   = (x :: l) :: (map2 lx x)
map1 []              = []
map1 (a :: x)        = (sum a) :: (map1 x)
fold1 [a]            = a
fold1 (a :: (b :: c)) = (bimax a (fold1 (b :: x)))
sum []               = 0
sum (a :: x)         = (+ a (sum x))
```

Fig. 5. Equations source

3 First-Order Transformation

The transformation can now begin at a first-order level using the system Astre. Let us run Astre giving the equations in Figure 5. The reader can refer to [2] to get a precise idea of the possibilities of the current version of the system. First, Astre turns the set of equations into a rewriting system, directing the equations from left to right automatically and eventually, (if it is required by the user) it verifies its ground convergence. Let us consider now the transformation process. The main function *maxsub* is defined in a Squiggol way by composition of functions and not inductively. The term

$$t = \mathit{fold1}\,(\mathit{map1}\,(\mathit{sublists}\ l))$$

is a candidate for a deforestation [16]. Deforestation algorithm of P. Wadler will not perform the deforestation of t because it requires the function definitions of *fold1*, *map1*, and *sublists* that occurs in t to be *treeless*. *fold1* and *sublists* are not *treeless* because their definitions contain applications of functions to terms that are not variables such as the application of *bimax* in $(\mathit{bimax}\ a\ (\mathit{fold1}\ (b :: x)))$ and the application of @ in $(\mathit{map2}\ (\mathit{frontlists}\ x)\ a) @ (\mathit{sublists}\, x)$. The normalization algorithm of [12] rejects also the deforestation of t which is *not potentially normalizable* for this algorithm [6].

We are currently implementing an automatic mode of Astre, we call *Automatic Astre* which discovers automatically the terms that are candidate for deforestation such as t and introduces automatically an *eureca rule* such as:

$$\mathit{fold1}\ (\mathit{map1}\,(\mathit{sublists}\ l)) \rightarrow \mathit{maxsub'}\,l \tag{1}$$

when it discovers a candidate for deforestation term. After that, the *partial* completion procedure processes the transformation. It automatically overlaps this reversed rule with the rules for *sublists* generating the following normalized equations:

$$\begin{aligned} \mathit{maxsub'}\,[] &= 0 \\ \mathit{maxsub'}\,(a :: x) &= \mathit{fold1}\,(\mathit{map1}\ (\mathit{map2}\,(\mathit{frontlists}\,x)\ a) @ (\mathit{sublists}\,x)) \end{aligned} \tag{2}$$

The transformation with pencil and paper at the higher-order level gives a similar result:

$$\begin{aligned} \mathit{maxsub}\ [] &= 0 \\ \mathit{maxsub}\ (a :: x) &= \mathit{fold}\ \mathit{bimax}\ (\mathit{map}\ \mathit{sum} \\ &\qquad (\mathit{map}\ (\mathit{consl}\ a)\ (\mathit{frontlists}\ x) @ (\mathit{sublists}\,x))) \end{aligned} \tag{3}$$

---

[6] The term t is not *potentially normalizable* because the call to *frontlists* where

$$\mathit{frontlists}\ (a :: x) \ = \ (\mathit{map2}\,(\mathit{frontlists}\ x)\ a) @ [[]]$$

applies an inner *catamorphism:map2* to the recursive call of *frontlist.*

Now it becomes interesting for the system to look for some input of theorems so that it rewrites (*folds*) the right-hand side of the above equation by the *eureca rule* in order to it introduces a recursive occurrence of *maxsub'*. In the current version of Astre, the user must provide the laws. Let us consider how they can be automatically extracted for a library of general laws on higher-order primitives and transposed at the first-order level using a *law extractor* which is based on the set of instantiation triples given by a partial evaluator. Such a combination of partial evaluation tools and intrinsically first-order rewriting tools can result in a powerful transformation system.

4 Interaction with the theorems

Let us come back to the higher-order level of the transformation. Suppose a higher-order law $L \rightarrow R$ *rewrites* the λ-term M to N. There exists a substitution σ and an occurrence in M of a subterm S $\beta\eta$-equal to $\sigma(L)$. The rewritted term N is the result of the replacement of the subterm S in M by a term $\beta\eta$-equal to $\sigma(R)$. We will denote $M[S \wr T]$ the replacement in M of a subterm S by T. Let l be $\sigma(L) \downarrow_{first}$ and r be $\sigma(R) \downarrow_{first}$. We want to show that $M \downarrow_{first}$ reduces to $N \downarrow_{first}$ by the rule $l \rightarrow r$.

As in [13], we introduce the η-expanded form of λ-term. For example the η-expanded form of $(map\ f \circ map\ g)$ of type order two is $((map\ f \circ map\ g)\ x)$ of type order one (or elementary). Then the Church-Rosser theorem for the $\beta\eta$-calculus can be expressed in the following form: For every two λ-term M and N, we have $M =_{\beta\eta} N$ if and only if their η-expanded form are β-equals. It is not a restriction to require that L and R are in β-normal form and that they are η-expanded so that their common type is elementary. For example the *maplaw*:

$$map\ f \circ map\ g\ =\ map\ (f \circ g)$$

where the type of both hand-sides is functional, can be η-expanded into:

$$((map\ f \circ map\ g)\ x\ ) = \ map\ (f \circ g)\ x$$

or into:

$$(map\ f\ (map\ g\ x)) = \ map\ (f \circ g)\ x$$

when the definition of $\circ$ is unfolded.

Theorem 4. *Assume that the non RHOP functions in M are first-order and that the RHOP functions are* constant or variable-only, *let $L \rightarrow R$ be a law that rewrites M with the substitution σ into the λ-term N, let δ be the restriction of σ to the subset of the functional variables in the domain of σ, let $m = M \downarrow_{first}$, and $n = N \downarrow_{first}$, then m reduces to n by application of the rule $l \rightarrow r$ where r is $\delta(R) \downarrow_{first}$ and l is $\delta(L) \downarrow first$.*

Proof: We require that L is in η-expanded form so that its type is elementary. Or a subterm S $\beta\eta$-equal to $\sigma(L)$ occurs in M, therefore, if L and M are in β-normal form, the subterm S is equal to $\sigma(L)$ whose type is elementary. The subterm $\sigma(R)$ in β-normal form occurs in the β-reduction of N. Consider now $m = M \downarrow_{first}$ and the corresponding set ST of triples $(F_i, \delta_i, T_i), i = 1, n$. By definition of $\rightarrow_{first}$, $m = M[\delta_i(T_i)\backslash F_i \; \forall i, i = 1, n]$, where the $\forall i = 1, n$ means that a *folding* is done for all triples in ST. In the same way, $l = \delta(l) \downarrow_{first} = \delta(L)[\delta_i(T_i)\backslash F_i \; \forall i, i = 1, n]$. Let ρ be the remaining part of the substitution δ such that $\sigma = \rho \circ \delta$ where $\circ$ is the composition of substitutions. The term $\rho(l)$ is a subterm of m. On another hand, $N = M[S \wr \sigma(R)]$, $r = \delta(R)[\delta_i(T_i)\backslash F_i \; \forall i, i = 1, n]$, and $n = N \downarrow_{first} = M[S \wr \sigma(R)][\delta_i(T_i)\backslash F_i \; \forall i, i = 1, n]$. Therefore m reduces to n by the rule $l \rightarrow r$. $\Box$

For example, the *map@law* rewrites the left hand-side of the equation 3 with the substitution σ:

$$\{f \leftarrow sum, x \leftarrow (map\ (consl\ a)\ (frontlists\ x)), y \leftarrow (sublists\ x)\}$$

The result is:

$$maxsub'(a :: x) = fold\ bimax\ ((map\ sum\ (map\ (consl\ a)\ (frontlists\ x)))\ @\ (map\ sum\ (sublists\ x))) \quad (4)$$

At the first-order level the rule:

$$map1\ ((map2\ (frontlists\ x)\ a)\ @\ (sublists\ x)) \rightarrow (map1\ (map2\ (frontlists\ x)\ a))\ @\ (map1\ (sublists\ x))$$

reduces the first-order converted equation 2. The result is:

$$maxsub'(a :: x) = fold1\ ((map1\ (map2\ (frontlists\ x)\ a))\ @\ (map1\ (sublists\ x))) \quad (5)$$

Theorem 4 shows the link between the transformation done at the higher-order level and the transformation at the first-order level.

As a consequence of Theorem 4, we can justify the following way to find a reduction of a first-order term t by a law at the first-order level:

Find a law $L \rightarrow R$, a set of specialization triples ST, a first-order substitution ρ, and a position p in t such that $t|p = \rho(L \downarrow_{first})$, then the first-order rule $L \downarrow_{first} = R \downarrow_{first}$ rewrites t at the position p with the substitution ρ.

The above can be processed by a procedure, we call a *Law-Extractor*, which returns to Astre a first-order theorem from a library of higher-order laws about the higher-order primitives.

For the *maxsub* example, the set ST contains the triple:

$$\{(map1, \{f \leftarrow sum\}, \lambda\, x.(map\ f\ x))\}$$

Considering equation 2 and the *map@law* of the library of higher-order laws in Figure 3, the *Law-Extractor* returns the first-order rule:

$$(map1\ x\ @\ y) = (map1\ x)\ @\ (map1\ y).$$

which simplifies equation 2 into equation 5.

Now, considering the *maplaw* in the library and the set *ST* containing the triples:

$$\begin{array}{l}(map1, \{f \leftarrow sum\} \quad , \lambda\, x.(map\ f\ x)) \\ (map2, \{f \leftarrow (consl\ x)\}, \lambda\, l\ x.(map\ f\ l))\end{array}$$

the *Law-Extractor* returns the first-order rule:

$$map1\ (map2\ l\ x) \rightarrow (map\ (sum\ \circ\ (consl\ x))\ l) \downarrow_{first}$$

The first-order conversion of the left hand-side processes a new specialization which adds a triple:

$$(map3, \{f \leftarrow \lambda\, l.(+\ x\ (sum\ l))\}, \lambda\, l\ x.(map\ f\ l))$$

where the term substituted for f is the η-expanded form of the β-reduction of the functional argument of *map*: $(sum\ \circ\ (consl\ x))$. This rule rewrites equation 5 into:

$$maxsub'\ (a :: x) = fold1\ (map3\ (frontlists\ x)\ a)\ @\ (map1\ (sublists\ x)) \qquad (6)$$

The interested reader can follow the remaining of the transformation in the appendix.

Notice that there is always a problem of confluence when a base of theorems is used for program transformation. The choice of a law to rewrite a term is ambiguous when two overlapping laws can be chosen. It is not a problem if the set of laws is confluent (if a notion of confluence is extended to higher-order logic). The set of laws presented in Figure 3 is not free of critical pairs. There is a unification modulo the associativity of $\circ$ between the left hand-sides *foldmaplaw* and the *maplaw*, the unified term being: $fold\ f\ \circ\ map\ f\ \circ\ map\ g$. Fortunately, by replacing the *foldmaplaw* by the *generalized foldmaplaw*

$$\begin{array}{r}foldmaplaw : fold\ f\ \circ\ map\ (g\ \circ\ h) = g\ \circ\ ((fold\ f)\ \circ\ (map\ h)) \\ \text{if } f\ (g\ x)\ (g\ y) = g\ (f\ x\ y)\end{array}$$

the library of higher-order laws becomes "confluent" so that we can use it for the transformation of the *maxsub* example.

5 Conclusion

Higher-order transformations based on well known properties of higher-order primitives are not easily automated. The paper presents a way to mimic such transformations at the first-order level by rewriting techniques. In this way, we obtain a tool that automatically implement the deforestation of higher-order Squiggol programs. Currently, the state of our technology is the following:

- a translator ML into Schism and a translator Schism to ML which restores the patterns (currently implemented).
- the partial evaluator Schism [6] which uses its own (typed) dialect of Scheme as its object language, and
- the interactive transformation system Astre [2] based on rewriting and completion procedures written in CAML. The mode "automatic Astre" for simple deforestations is currently implemented.

We have simulated the interaction of these tools to process the sketch of the transformation of the *maxsub* example. The main piece of software to add is the *Law-Extractor* defined in Section 4. We expect that all these tools will work harmoniously in the near future to fully automate deforestation of higher-order Squiggol programs. These transformations are applicable to functional programs that describe first-order functions using *constant or variable-only* higher-order primitives. For this class of programs, the paper shows that the transformation at the higher-order level can be mimicked by rewriting in first-order logic.

I would like to thank C. Consel and J. Hook for their comments and suggestions.

References

1. F. Bellegarde. Program Transformation and Rewriting. In *Proceedings of the fourth conference on Rewriting Techniques and Applications*, Springer Verlag, Lecture Notes in Computer Science 488, pages 226-239, Como, Italy, 1991.
2. F. Bellegarde. Astre, a Transformation System using Completion. Technical Report, Department of Computer Science and Engineering, Oregon Graduate Institute, 1991.
3. J. Bentley. *Programming Pearls*, Addison Wesley, 1986.
4. R. M. Burstall and J. Darlington. A Transformation System For Developing Recursive Programs. *Journal of the Association for Computing Machinery*, 24, pages 44-67, 1977.
5. W. N. Chin. *Automatic Methods for Program Transformation.* PhD thesis, University of London, Imperial College of Science, Technology and Medecine, London, UK, 1990.
6. C. Consel. A tour of Schism: A partial evaluation system for higher-order applicative languages. Research report, Pacific Software Research Center, Oregon Graduate Institute of Science and Technology, Beaverton, Oregon, USA, 1992.
7. N. Dershowitz. Completion and its Applications. *Resolution of Equations in Algebraic Structures,* Academic Press, New York, 1988.

8. A. Ershov. Mixed computation: potential applications and problem for study. *Theoritical Computer Science*, Vol. 18, pages 41-67, 1982.
9. A. Gill, J. Launchbury and S.L. Peyton Jones. A short cut to Deforestation. In *Sixth Conference on Functional Programming Languages and Computer Architecture*, Copenhagen, Denmark, pp 223-232, June 1993.
10. E. Meijer, M. Fokkinga and R. Patterson. Functional Programming with Bananas, Lenses, Envelopes and Barbed Wire. In *Conference on Functional Programming and Computer Architecture*, Lecture Notes in Computer Science 523, pages 124-144, 1991.
11. U. S. Reddy. Transformational derivation of programs using the Focus system. In *Symposium Practical Software Development Environments*, pages 163-172, ACM, December 1988.
12. T. Scheard and L. Fegaras. A fold for All Seasons. *Sixth Conference on Functional Programming Languages and Computer Architecture*, Copenhagen, Denmark, pp 233-242, June 1993.
13. W. Snyder and J. Gallier. Higher order unification revisited: complete sets of transformations. *Journal of Symbolic Computation*, Vol.8, pages 101-140, 1989. Special issue on Unification. Part two.
14. S. Thompson. Functional Programming: Executable Specifications and Program Transformation. In *Fifth International Workshop on Software Specification and Design.*, IEEE Press, 1989.
15. S. Thompson. *Type Theory and Functional Programming*, Addison Wesley, 1991.
16. P. Wadler, Deforestation: Transforming programs to eliminate trees. In *2nd European Symposium on Programming ESOP'88*, Nancy France, 1988. Lecture Notes in Computer Science 300, Springer Verlag.
17. P. Wadler. Theorem for free! In *Proc. 1989 ACM Conference on Lisp and Functional Programming*, pages 347-359, 1989.

6 Appendix

Consider following the sketch of the automatic transformation of *maxsub* starting with the equation 6 we found at the end of Section 3:

$$maxsub'\,(a :: x) = fold1\ (map3\ (frontlists\, x)\ a)\ @\ (map1\ (sublists\, x))$$

At this point, the set ST contains the triples:

$$\begin{array}{ll} \{(map1, \{f \leftarrow sum\} & , \lambda\, x.(map\ f\ x)) \\ (map2\ , \{f \leftarrow (consl\ x)\} & , \lambda\, l\, x.(map\ f\ l)) \\ (fold1\ , \{f \leftarrow bimax\ \} & , \lambda\, x.(fold\ f\ x)) \\ (map3\ , \{f \leftarrow \lambda\, l\, x.(+\ x\ (sum\ l))\} & , \lambda\, l\, x.(map\ f\ l))\} \end{array}$$

Considering the *fold@law*, the *Law-Extractor* returns the first-order rule:

$$fold1\ (x\ @\ y)\ =\ bimax\ (fold1\ x)\ (fold1\ y)$$

because *bimax* is associative. This rule rewrites equation 6 into:

$$\begin{array}{l} maxsub'\,(a :: x) = bimax\ (fold1\ (map3\ (frontlists\, x)\ a)) \\ \qquad\qquad\qquad\qquad (fold1\ (map1\ (sublists\, x))) \end{array}$$

At this point, Astre rewrites by the rule 1 which folds *maxsub'*:

$$maxsub'(a :: x) = bimax\ (fold1\ (map3\ (frontlists\ x)\ a))\ (maxsub'\ x) \qquad (7)$$

Considering the *generalized foldmaplaw* introduced in Section 4, the *Law-extractor* returns the first-order rule:

$$fold1\ (map3\ l\ x) = +\ x\ (fold1\ (map1\ l))$$

because $bimax(+\ u\ x)\ (+\ u\ y) = +\ u\ (bimax\ x\ y)$. This rule simplifies the equation 7 into:

$$maxsub'\ (a :: x) = bimax\ (+\ a\ (fold1\ (map1\ (frontlists\ x))))\ (maxsub'\ x) \quad (8)$$

Now, the composition of functions (*fold1* (*map1* (*frontlists* x))) is a candidate for a deforestation. As for *maxsub*, an *eureca rule* for a new functional symbol *maxfront* which captures this composition is introduced automatically.

$$fold1\ (map1\ (frontlists\ x)) \rightarrow maxfront\ x$$

This reversed rule simplifies equation 8 into the final result for *maxsub*:

$$maxsub'\ (a :: x) = bimax\ (+\ a\ (maxfront\ x))\ (maxsub'\ x) \qquad (9)$$

The partial completion procedure superposes this reversed rule with the rules for *frontlists*. From the above transformation of *maxsub'*, it knows the full set of first-order laws to generate the following equations for *maxfront*:

$$maxfront\ [] = 0$$
$$maxfront\ (a :: x) = bimax\ (+\ a\ (maxfront\ x))\ 0$$

Prototyping Relational Specifications Using Higher-Order Objects

Rudolf Berghammer[1], Thomas F. Gritzner[2], Gunther Schmidt[3]

[1] Institut für Informatik und Praktische Mathematik,
Christian-Albrechts-Universität Kiel, D-24105 Kiel, Germany
[2] Fakultät für Informatik, Technische Universität München,
D–80290 München, Germany
[3] Fakultät für Informatik, Universität der Bundeswehr München,
D–85577 Neubiberg, Germany

Abstract. An approach is described for the generation of certain mathematical objects (like sets, correspondences, mappings) in terms of relations using relation-algebraic descriptions of higher-order objects. From non-constructive characterizations executable relational specifications are obtained. We also show how to develop more efficient algorithms from the frequently inefficient specifications within the calculus of binary relations.

1 Introduction

During the last two decades, the axiomatic relational calculus of Tarski [28] has widely been used by computer scientists who view it as a convenient formalism for describing fundamental concepts of programming. The development starts with the work of de Bakker and de Roever in the early 70's; see [12, 13] for example. In the following decade, e.g., Hoare and He [17] related the work of Birkhoff on residuals with Dijkstra's weakest precondition approach to programming, a group in Munich (see [23, 6, 32, 31]) constructed semantic domains by relation-algebraic means and, thus, was able to treat also languages with higher-order functions, and a group in Eindhoven (see [2]) developed a theory of data types based on the calculus of relations. At this point also the approach of a group in Rio should be mentioned which was motivated mainly by the development of a relational programming calculus not bounded by lack of expressiveness; compare [29]. In program development, the relational framework has already been used. E.g., in a series of articles a group around Desharnais and Mili proposed a relational approach to the formal derivation of imperative programs from its specifications. See [19, 20, 14] for example. Recently, Möller [21] used n-ary relations between nested tuples as elements of an applicative program development language, and also the Rio group developed various case studies on formal program construction using relations (cf. [30]).

In order that investigations with relation algebra involved do not stay completely on the theoretical side, several aspects need special consideration. Firstly, relational methods are not so commonly known that competent discussion with researchers from other approaches is easy. So, there should be a tool at hand to facilitate and visualize work with relations. Secondly, some tool for rapid prototyping of program specifications expressed in a relational style should be developed. And, finally, since relational specifications have a high degree of precision and formal structure, program development methods starting from such relational specifications should be considered.

In this paper, we describe an approach to the generation of certain mathematical objects (like sets, correspondences, mappings) in terms of relations using higher-order objects. We aim at the development of executable relational specifications out of non-constructive problem descriptions, where some special functionals on relations in conjunction with a relational description of domains (including relation and function spaces, i.e., higher-order objects) play an important rôle. We present only a carefully selected couple of representative examples; for an extensive treatment of our approach, we refer to the report [8] of the same title.

2 Relation Algebraic Preliminaries

In this section, we briefly introduce the basic concepts of the algebra of relations, some special relations, and some relation-algebraic constructions (i.e., functionals on relations). For more details concerning the algebraic theory of relations, see e.g., [11, 18, 25].

2.1 Basic Operations and Relation Algebraic Laws

For two sets X and Y, a subset R of the Cartesian product $X \times Y$ is a relation between the *domain* X and the *range* Y. We call it *homogeneous* if $X = Y$, otherwise we call it *heterogeneous.* Considering the corresponding characteristic predicates instead of the set representations, a relation R between X and Y becomes a function $R : X \times Y \to \mathbb{B}$, where $\mathbb{B}$ denotes the set $\{0, 1\}$ of truth-values. Therefore, if X and Y are finite and of cardinality m and n, respectively, then we may consider R as a Boolean matrix with m rows and n columns. This matrix interpretation of relations is well-suited for a graphical representation and, e.g., used within the RELVIEW system [1, 4]. Following the notation of the specification language Z [26], we write $R : X \leftrightarrow Y$ if R is a relation between the sets X and Y. Furthermore, we use matrix notation and write R_{xy} instead of $(x, y) \in R$.

We assume the reader to be familiar with the basic operations, viz. R^{T} (transposition), $\overline{R}$ (negation), $R \cup S$ (join), $R \cap S$ (meet), RS (composition), $R \subset S$ (inclusion), and the special relations O (empty relation), L (universal relation), and I (identity relation). In this paper, we only consider relations with non-empty domain and range, and therefore $\mathsf{O} : X \leftrightarrow Y$ and $\mathsf{L} : X \leftrightarrow Y$ are distinct. The set-theoretic operations $\overline{}$, $\cup$, $\cap$, the ordering $\subset$, and the constants O, L are related as usual. Some further well-known rules concerning relations are

$$
\begin{array}{ll}
(R^{\mathrm{T}})^{\mathrm{T}} = R & R \subset S \Longrightarrow R^{\mathrm{T}} \subset S^{\mathrm{T}} \\
R^{\mathrm{T}} S^{\mathrm{T}} = (SR)^{\mathrm{T}} & \overline{R}^{\mathrm{T}} = \overline{R^{\mathrm{T}}} \\
R \subset S \Longrightarrow QR \subset QS & R \subset S \Longrightarrow RQ \subset SQ \\
Q(R \cap S) \subset QR \cap QS & Q(R \cup S) = QR \cup QS \\
(R \cap S)^{\mathrm{T}} = R^{\mathrm{T}} \cap S^{\mathrm{T}} & (R \cup S)^{\mathrm{T}} = R^{\mathrm{T}} \cup S^{\mathrm{T}} \quad ,
\end{array}
$$

where the last two lines also hold if binary meet and join are replaced by arbitrary meet and join, respectively. The theoretical framework for all these rules to hold is that of an abstract relation algebra. The axioms of this algebraic structure are the axioms of a complete atomic Boolean algebra for $\overline{}$, $\cup$, $\cap$, $\subset$, O, L, the axioms of a monoid for composition and I, the *Dedekind rule*

$$(QR \cap S) \subset (Q \cap SR^{\mathrm{T}})(R \cap Q^{\mathrm{T}} S) \ ,$$

and the *Tarski rule*

$$R \neq \mathsf{O} \iff \mathsf{L}R\mathsf{L} = \mathsf{L} \ .$$

An alternative form of the Dedekind rule are the *Schröder equivalences*, viz.

$$QR \subset S \iff Q^{\mathrm{T}}\overline{S} \subset \overline{R} \iff \overline{S}R^{\mathrm{T}} \subset \overline{Q} \ .$$

2.2 Special Relations

The basic operations and constants of the last subsection are very helpful for defining simple properties on relations in a component-free manner. In the remainder of the paper we need the following special relations:

Functions: Let $R : X \leftrightarrow Y$ be a relation. R is said to be a *partial function* or, briefly, to be *functional* if $R^{\mathrm{T}}R \subset \mathrm{I}$, and R is said to be *total* if $R\mathsf{L} = \mathsf{L}$, which is, in turn, equivalent to $\mathrm{I} \subset RR^{\mathrm{T}}$. As usual, a functional and total relation is said to be a *(total) function*. A relation R is called *injective* if R^{T} is functional, and R is called *surjective* if R^{T} is total. An injective and surjective relation is said to be *bijective*.

Partial orderings: Let $Q : X \leftrightarrow X$ be homogeneous. Q is *reflexive* if $I \subset Q$, *transitive* if $QQ \subset Q$, and *antisymmetric* if $Q \cap Q^{\mathrm{T}} \subset \mathrm{I}$. A *partial ordering* is a reflexive, antisymmetric, and transitive relation. If Q is a partial ordering, then $Q \cap \overline{\mathrm{I}}$ denotes its *irreflexive part*.

Vectors: A relation $v : X \leftrightarrow Y$ with $v = v\mathsf{L}$ is called a *(row-constant) vector* or a *predicate*. This condition means that an element x from X is either in relation v to none of the elements of Y or to all elements of Y. Vectors may be considered as subsets of X. This becomes obvious if we use the common set-theoretic notation for relations, since then v equals a Cartesian product $X' \times Y$, where X' is a subset of X. As for a vector the range is without relevance, we consider in the following only vectors $v : X \leftrightarrow \mathbb{1}$ with a singleton set $\mathbb{1} = \{u\}$ as their range and write v_x instead of v_{xu}. Then v can be considered as a Boolean matrix with exactly one column, i.e., as a Boolean column vector, and describes the subset $\{x \in X : v_x\}$ of X.

2.3 Quotients and Bounds

In this subsection, we consider some special mappings from relations to relations. To distinguish such "meta-level functions" from the relation-algebraic (or: object level) notion of functions as presented in Subsection 2.2, they are also called *functionals.* In most cases, they happen to be partial functionals, since the basic operations on relations besides the unary ones are only partially defined.

Residuals: Residuals are the greatest solutions of certain inclusions. The *left residual* of S over R (in symbols $S\,/\,R$) is the greatest relation Q such that $QR \subset S$; the *right residual* of S over R (in symbols $R \backslash S$) is the greatest relation Q such that $RQ \subset S$. Both residuals may also be represented using the basic operations: Let $R : X \leftrightarrow Z$, $S : Y \leftrightarrow Z$, $R' : Z \leftrightarrow X$, and $S' : Z \leftrightarrow Y$ be given, then, from the Schröder equivalences we obtain

$$S\,/\,R = \overline{\overline{S}R^{\mathrm{T}}} \qquad R' \backslash S' = \overline{R'^{\mathrm{T}}\overline{S'}} \ .$$

Note also that the two residuals are linked together by the relationships

$$R \backslash S = (S^{\mathrm{T}}/\,R^{\mathrm{T}})^{\mathrm{T}} \qquad \overline{R} \backslash \overline{S} = R^{\mathrm{T}}/\,S^{\mathrm{T}} \ .$$

Translating the relation-algebraic expressions into component-wise formulations, for left residual $S\,/\,R : Y \leftrightarrow X$ and right residual $R' \backslash S' : X \leftrightarrow Y$ we have the equivalences

$$(S\,/\,R)_{yx} \iff \bigwedge_z R_{xz} \rightarrow S_{yz} \qquad (R' \backslash S')_{xy} \iff \bigwedge_z R'_{zx} \rightarrow S'_{zy} \ .$$

In particular, we have the two following correspondences for single universal quantification:

$$(S\,/\,\mathsf{L})_y \iff \bigwedge_z S_{yz} \qquad (\overline{R'} \backslash \mathsf{O})_x \iff \bigwedge_z R'_{zx} \ .$$

Note that for the quantifications the domain for z may also consist of higher-order objects. The usage of this option is characteristic to our approach.

Symmetric Quotients: In the following, we will frequently need relations which are left and right residuals simultaneously, viz. symmetric quotients. The *symmetric quotient* $\mathsf{syq}(R, S)$ of two relations R and S is defined as the greatest relation Q such that $RQ \subset S$ and $QS^{\mathrm{T}} \subset R^{\mathrm{T}}$. If $R : Z \leftrightarrow X$ and $S : Z \leftrightarrow Y$, then we have $\mathsf{syq}(R, S) : X \leftrightarrow Y$ as

$$\mathsf{syq}(R, S) = \overline{R^{\mathrm{T}}\,\overline{S}} \cap \overline{\overline{R^{\mathrm{T}}}S} = (R \backslash S) \cap (R^{\mathrm{T}}/\,S^{\mathrm{T}}) \ .$$

Using a component-wise notation, the symmetric quotient $\mathsf{syq}(R, S)$ satisfies the equivalence

$$\mathsf{syq}(R, S)_{xy} \iff \bigwedge_z R_{zx} \leftrightarrow S_{zy} \ .$$

Bounds and extremal elements: Let $Q : X \leftrightarrow X$ be a partial ordering. Due to later applications, we ask for some order-theoretic concepts such as the set of lower (resp. upper) bounds of a subset wrt. Q or the set of minimal (resp. maximal) elements of a subset wrt. Q. We define four functionals dependent on Q and a further relation $R : X \leftrightarrow Y$ as follows:

$$\begin{array}{ll} \textit{Lower bounds:} & \mathsf{mi}(Q,R) = \overline{\overline{Q}R} = Q \,/\, R^{\mathrm{T}} \\ \textit{Upper bounds:} & \mathsf{ma}(Q,R) = \overline{\overline{Q^{\mathrm{T}}}R} = Q^{\mathrm{T}} \,/\, R^{\mathrm{T}} \\ \textit{Minimal elements:} & \mathsf{min}(Q,R) = R \cap \overline{(Q \cap \overline{\mathrm{I}})^{\mathrm{T}} R} \\ \textit{Maximal elements:} & \mathsf{max}(Q,R) = R \cap \overline{(Q \cap \overline{\mathrm{I}}) R}\;. \end{array}$$

Looking at the corresponding component-wise descriptions and assuming R to be a vector, it is easy to see that $\mathsf{mi}(Q,R)$ (resp. $\mathsf{ma}(Q,R)$) yields the subset of lower (resp. upper) bounds of R wrt. the partial ordering Q, while $\mathsf{min}(Q,R)$ (resp. $\mathsf{max}(Q,R)$) computes to the subset of minimal (resp. maximal) elements of R wrt. the partial ordering Q. If R is not a vector, then the functionals compute bounds and extremal elements column-wise.

3 Relational Domain Construction

To deal with composed and higher-order objects like tuples, sets, or functions, we have to explain how the corresponding domain constructions can be performed with relational algebra. Note that the constructions described in the following, after the definitions of homomorphisms (resp. isomorphisms), may or may not exist in an arbitrary model of relational algebra; however, this problem does not occur at the concrete matrix model underlying this paper.

3.1 Homomorphisms, Direct Products, and Direct Sums

As first domain constructions we consider products and sums. Furthermore, we introduce homomorphisms to show that the characterizations are monomorphic.

Homomorphisms: Let $R : X \leftrightarrow Y$ and $S : X' \leftrightarrow Y'$ be two relations and consider a pair $\mathcal{H} = (\Phi, \Psi)$ of functions $\Phi : X \leftrightarrow X'$ and $\Psi : Y \leftrightarrow Y'$. $\mathcal{H}$ is called a *homomorphism* from R to S if $R \subset \Phi S \Psi^{\mathrm{T}}$ or, equivalently, $R\Psi \subset \Phi S$ holds. If, in addition, the pair $\mathcal{H}^{\mathrm{T}} = (\Phi^{\mathrm{T}}, \Psi^{\mathrm{T}})$ is a homomorphism from S to R, then $\mathcal{H}$ is said to be an *isomorphism* between R and S. Therefore, an isomorphism $\mathcal{I} = (\Phi, \Psi)$ between R and S is a pair of bijective functions $\Phi : X \leftrightarrow X'$ and $\Psi : Y \leftrightarrow Y'$, which satisfies the condition $R\Psi = \Phi S$. If R and S are homogeneous, then Φ is briefly called a homomorphism (isomorphism) if the pair (Φ, Φ) is a homomorphism (isomorphism).

Direct products: Within the framework of abstract relation algebra it is natural to characterize direct products by means of the natural projections, see [25, 31]. Then one obtains the following specification: Let

$$\mathcal{P} = (\pi_k : PX \leftrightarrow X_k)_{1 \le k \le n}$$

be an n-tuple of $n > 0$ relations. We call $\mathcal{P}$ an *(n-ary) direct product* if

$$(P_1)\ \pi_k^{\mathrm{T}}\pi_k = \mathrm{I} \qquad (P_2)\ j \neq k \implies \pi_j^{\mathrm{T}}\pi_k = \mathsf{L} \qquad (P_3)\ \bigcap_{k=1}^{n} \pi_k\pi_k^{\mathrm{T}} = \mathrm{I}\,.$$

It is easy to verify that the natural projections from a Cartesian product $\Pi_{i=1}^{n} X_i$ to the components X_k are a model of (P_1) through (P_3) if the placeholder PX is replaced by $\Pi_{i=1}^{n} X_i$. By purely relation-algebraic reasoning, furthermore, it can be shown that the direct product is uniquely characterized up to isomorphism: If $\mathcal{Q} = (\rho_k : PY \leftrightarrow Y_k)_{1 \leq k \leq n}$ is another model of (P_1) through (P_3) and $(\Psi_k : X_k \leftrightarrow Y_k)_{1 \leq k \leq n}$ is a family of bijective functions, then we can establish an isomorphism between π_k and ρ_k by the pair $\mathcal{I}_k = (\Phi, \Psi_k)$, where the bijective function $\Phi : PX \leftrightarrow PY$ is defined as $\Phi = \bigcap_{i=1}^{n} \pi_i \Psi_i \rho_i^{\mathrm{T}}$.

Based on the binary direct products $\mathcal{P} = (\pi, \rho)$ and $\mathcal{Q} = (\sigma, \tau)$ we define the following two functionals, where the generalization to n-ary direct products $(n > 2)$ is straightforward, but not needed in the remainder of the paper:

Tupling:	$[R, S]_{\mathcal{P}} = R\pi^{\mathrm{T}} \cap S\rho^{\mathrm{T}}$
Parallel composition:	$R\,{}_{\mathcal{P}}\|_{\mathcal{Q}}\, S = \pi R \sigma^{\mathrm{T}} \cap \rho S \tau^{\mathrm{T}}$.

Direct sums: The direct sum can be defined in largely the same fashion as the direct product. Dually to the natural projections the natural injections are used, see [31]. Then one obtains the following specification: Let

$$\mathcal{S} = (\iota_k : X_k \leftrightarrow SX)_{1 \leq k \leq n}$$

be an n-tuple of $n > 0$ relations. We call $\mathcal{S}$ an *n-ary direct sum* if

$$(S_1)\ \iota_k\, \iota_k^{\mathrm{T}} = \mathrm{I} \qquad (S_2)\ j \neq k \implies \iota_j\, \iota_k^{\mathrm{T}} = \mathsf{O} \qquad (S_3)\ \bigcup_{k=1}^{n} \iota_k^{\mathrm{T}}\iota_k = \mathrm{I}\,.$$

Given sets X_k, $1 \leq k \leq n$, it is easy to verify that the injections from these sets to the direct sum $\Sigma_{k=1}^{n} X_k$ are a model of (S_1) through (S_3). Again by purely relation-algebraic reasoning it can be shown that by these laws the direct sum is uniquely characterized up to isomorphism and the relations ι_k, $1 \leq k \leq n$, are injective functions.

3.2 Powersets, Relation Spaces, and Function Spaces

Now, we present monomorphic characterizations of higher-order objects using relation algebra. The first construction formalizing the membership relation uses only a small set of set-theoretic axioms. The selection of this set of axioms such that it suffices for a monomorphic characterization has already been considered in category theory in connection with the notion of topos [9], which denotes a category such that every object has a power-object. The axiomatization presented here has been developed from the Munich group in the last decade, aiming at a relation-algebraic characterization of the kinds of function spaces used in denotational semantics. See [32, 7, 31]. Independently, an equivalent development based on the concept of topos mentioned before is provided in the book [15] by the notion of a power-allegory.

Powersets: A relation-algebraic characterization of the powerset 2^X of a set X can conveniently be done using the "is-element-of" relation. Formally, we call $\varepsilon : X \leftrightarrow PSX$ a *powerset relation* if

$$(PS_1)\quad \mathsf{syq}(\varepsilon,\varepsilon) \subset \mathrm{I} \qquad (PS_2)\quad \bigwedge_R \mathsf{L}\,\mathsf{syq}(\varepsilon, R) = \mathsf{L}\ .$$

Since every relation-algebraic equation using ε is translated into a formula with higher-order quantification (over sets), in (PS_2) the higher-order quantification (over relations) does not surprise. Again it can be shown by purely relation-algebraic reasoning that the powerset relation is uniquely characterized up to isomorphism. Indeed, if $\varepsilon' : Y \leftrightarrow PSY$ is another powerset relation, $\Phi : X \leftrightarrow Y$ is a bijective function, and one defines the bijective function $\Psi : PSX \leftrightarrow PSY$ by $\Psi = \mathsf{syq}(\varepsilon, \Phi\varepsilon')$, then $\mathcal{I} = (\Phi, \Psi)$ is an isomorphism between ε and ε'.

Now, assume the concrete case of the "is-element-of" relation $\varepsilon : X \leftrightarrow 2^X$. Then (PS_1) corresponds to the extensionality axiom, whereas (PS_2) says that every vector (set) $v : X \leftrightarrow \mathbb{1}$ has a corresponding point (i.e., a bijective vector) $\mathsf{vp}(v) := \mathsf{syq}(\varepsilon, v) : 2^X \leftrightarrow \mathbb{1}$ in the powerset. This shows that the usual "is-element-of" relation is a powerset relation. The functional vp is injective and its left-inverse on points is $\mathsf{vp}^{-1}(p) = \varepsilon p : X \leftrightarrow \mathbb{1}$. Hence, vp establishes some kind of isomorphism (resp. Galois connection) between subsets of X and elements of 2^X. For details, compare [32, 7].

Based on the relation $\varepsilon : X \leftrightarrow 2^X$, the relational specification of sets merely by equations can be established. Namely, we have:

Empty set:	$\mathbf{E} : 2^X \leftrightarrow \mathbb{1}$	$\mathbf{E} = \mathsf{syq}(\varepsilon, \mathsf{O}) = \varepsilon \setminus \mathsf{O}$
Universal set:	$\mathbf{U} : 2^X \leftrightarrow \mathbb{1}$	$\mathbf{U} = \mathsf{syq}(\varepsilon, \mathsf{L}) = \overline{\varepsilon} \setminus \mathsf{O}$
Singleton embedding:	$\mathbf{S} : X \leftrightarrow 2^X$	$\mathbf{S} = \mathsf{syq}(\mathrm{I}, \varepsilon)$
Complementation:	$\mathbf{C} : 2^X \leftrightarrow 2^X$	$\mathbf{C} = \mathsf{syq}(\varepsilon, \overline{\varepsilon})$
Meet of sets:	$\mathbf{M} : 2^X \times 2^X \leftrightarrow 2^X$	$\mathbf{M} = \mathsf{syq}([\varepsilon, \varepsilon]_{\mathcal{P}}, \varepsilon)$
Join of sets:	$\mathbf{J} : 2^X \times 2^X \leftrightarrow 2^X$	$\mathbf{J} = (\mathbf{C}\ {}_{\mathcal{P}}\Vert_{\mathcal{P}}\ \mathbf{C})\mathbf{MC}$
Inclusion of sets:	$\sqsubseteq : 2^X \leftrightarrow 2^X$	$\sqsubseteq = \varepsilon \setminus \varepsilon$

In the description of the binary operations meet and join we use a direct product $\mathcal{P} = (\pi, \rho)$, consisting of the two projections from $2^X \times 2^X$ to the first and the second component, respectively.

Relation spaces and function spaces: We are also interested in describing the set of all relations between two sets and some certain subsets by relation-algebraic means. The set of all relations between X and Y is easy to handle, since it is a powerset relation $\varepsilon_R : X \times Y \leftrightarrow 2^{X \times Y}$. Hence, $\mathcal{R} = (\pi, \rho, \varepsilon_R)$ is called a *relation space* if

(R_1) (π, ρ) is a direct product
(R_2) $\pi^{\mathrm{T}}\varepsilon_R$ and $\rho^{\mathrm{T}}\varepsilon_R$ exist
(R_3) ε_R is a powerset relation.

The transformation of a relation $R : X \leftrightarrow Y$ into a point of the relation space is described by $\mathsf{vp}(\mathsf{rv}(R))$, where $\mathsf{rv}(R) := (\pi R \cap \rho)\mathsf{L} : X \times Y \leftrightarrow \mathbb{1}$ is the vector corresponding to R; the opposite direction uses vp^{-1} and the left-inverse of rv which is $\mathsf{rv}^{-1}(v) = \pi^{\mathrm{T}}(\rho \cap v\mathsf{L}) : X \leftrightarrow Y$.

The set Y^X of all functions from X to Y may be characterized in two different ways. On one hand, it can be described by a vector $f : 2^{X \times Y} \leftrightarrow \mathbb{1}$ such that

$$\begin{aligned} f_r \iff & \bigwedge_x \bigwedge_y \bigwedge_{y'} (((x,y) \in r \wedge (x,y') \in r) \rightarrow y = y') \wedge \bigwedge_x \bigvee_y (x,y) \in r \\ \iff & \neg(\bigvee_t t \in r \wedge \bigvee_{t'} (t' \in r \wedge \pi(t') = \pi(t) \wedge \rho(t') \neq \rho(t))) \\ & \wedge \bigwedge_x \bigvee_t (t \in r \wedge \pi(t) = x) \end{aligned}$$

giving the intricate expression

$$f = \overline{[\varepsilon_R^{\mathrm{T}} \cap \varepsilon_R^{\mathrm{T}}(\pi\pi^{\mathrm{T}} \cap \overline{\rho\rho^{\mathrm{T}}})]\mathsf{L}} \cap (\varepsilon_R^{\mathrm{T}}\pi \,/\, \mathsf{L}) \ .$$

On the other hand, we can define an "is-element-of" relation $\varepsilon_F : X \times Y \leftrightarrow Y^X$ as follows: The triple $\mathcal{F} = (\pi, \rho, \varepsilon_F)$ is called a *function space* if (R_1) and (R_2) (numbered by (F_1) and (F_2), respectively) hold with ε_R replaced by ε_F and, furthermore,

$$\begin{aligned} (F_3) \quad & \mathsf{syq}(\varepsilon_F, \varepsilon_F) \subset \mathrm{I} \\ (F_4) \quad & \bigwedge_R \mathsf{L}\,\mathsf{syq}(\varepsilon_F, R) = \mathsf{L} \leftrightarrow (RR^{\mathrm{T}} \subset \overline{\pi\pi^{\mathrm{T}}} \cup \rho\rho^{\mathrm{T}} \wedge R^{\mathrm{T}}\pi = \mathsf{L}) \ . \end{aligned}$$

Now, for R being a vector $v : X \times Y \leftrightarrow \mathbb{1}$ the second condition (F_4) means that the point $\mathsf{vp}(v) = \mathsf{syq}(\varepsilon_F, v) : 2^{X \times Y} \leftrightarrow \mathbb{1}$ is in the function domain Y^X if and only if v represents a function, i.e., the relation $\mathsf{rv}^{-1}(v) : X \leftrightarrow Y$ is a function. For details, see [1].

Injections: In addition to vectors, we have injective functions as a second concept for representing subsets. Given an injective function $\imath : X' \rightarrow X$, we call X' a *subset of* X *given by* $\imath$. Clearly, if X' is a subset of X given by $\imath$, then the vector $\imath^{\mathrm{T}}\mathsf{L} : X \leftrightarrow \mathbb{1}$ describes X' in the sense of Subsection 2.2. Since we deal only with concrete relations, the transition in the other direction, i.e., the construction of an injective function $\imath : X' \rightarrow X$ from a given vector $v : X \leftrightarrow \mathbb{1}$, is also possible. Generally, we have: Let the vector $v : X \leftrightarrow \mathbb{1}$ describe the subset X' of the set X. Then, $\imath(v) : X' \leftrightarrow X$ is called an *injection of* v or an *injection of* X' *into* X if

$$(I_1) \ \ \imath(v) \text{ is an injective function} \qquad (I_2) \ \ v = \imath(v)^{\mathrm{T}}\mathsf{L} \ .$$

Clearly, it follows that X' is a subset of X given by $\imath(v)$. Again, it can be shown that injections are determined uniquely up to isomorphism by (I_1) and (I_2). Namely, if $\Psi : X \leftrightarrow Y$ is a bijective function and $v' : Y \leftrightarrow \mathbb{1}$ is a vector such that it describes a subset Y' of Y and $v = \Psi v'$ is satisfied (i.e., $v' = \Psi^{\mathrm{T}} v$), then $\mathcal{I} = (\Phi, \Psi)$ is an isomorphism between $\imath(v)$ and $\imath(v')$, where the bijective function $\Phi : X' \leftrightarrow Y'$ is given by $\Phi = \imath(v)\Psi\imath(v')^{\mathrm{T}}$.

In most cases, injections are used within our applications in the context of higher-order objects like sets of sets. Namely, if the vector $v : 2^X \leftrightarrow \mathbb{1}$ describes a subset S of sets, then it is straightforward to compute an injection $\imath(v)$ of S into 2^X. From $\imath(v)$ we obtain the elements of S (represented as vectors) as the columns of the relation $\varepsilon\imath(v)^{\mathrm{T}} : X \leftrightarrow S$, which leads to an economic representation of the set S of sets by a Boolean matrix.

4 Relational Problem Specification and Prototyping

In this section, we show how the computation of certain mathematical objects (like sets, correspondences, mappings) can be described in terms of relations. The general method is as follows. First, we specify the problem with the help of a formula φ which characterizes the objects to be computed. Using the correspondences between certain kinds of formulae and certain relation-algebraic constructions (resp. operations), we then transform φ into a component-free relational expression R such that φ is valid if and only if its free variables are related by R. Hence, this expression R can be seen as a relational problem specification (cf. [5]) which is executable as it stands, i.e., as an algorithm for computing the set of specified objects. At this place it should be mentioned that in easy cases or for people well-trained in the relational calculus the specification R can be written down immediately.

Of course, algorithms produced in the way just described may frequently be fairly inefficient compared to hand-made ones and, thus, in many cases only applicable to small or medium-sized examples. However, they are built up very quickly, which is an important factor of economy. Furthermore, they ensure correctness and their proofs of correctness are very simple. Moreover, an executable relational specification can be the starting point for the derivation of an efficient algorithm using some development method as we will show in Section 5. Hence, we have the typical situation of the rapid prototyping approach (see [10]) for a validation and analysis of specifications.

Most of the following examples of rapid prototyping using a relation-algebraic description of the given problem are borrowed from graph theory. They deal with the computation (strictly speaking: the enumeration) of higher-order objects like sets of sets (represented as vectors $v : 2^X \leftrightarrow \mathbb{1}$) or sets of functions (represented as vectors $v : Y^X \leftrightarrow \mathbb{1}$). Here our approach leads to an extensive use of the relational characterization of sets and functions as presented in Section 3. Therefore, in the following, we have to distinguish between the two meta-level symbols $\in$ and $\subseteq$ and the "is-element-of" relation and the set inclusion relation on the object-level. As in Section 3, we use in the sequel the two relations $\varepsilon : X \leftrightarrow 2^X$ and $\sqsubseteq : 2^X \leftrightarrow 2^X$ on the object level. Especially, we have $x \in s$ (resp. $s \subseteq t$) if and only if the relation ε_{xs} (resp. $\sqsubseteq_{st}$) holds.

Here, only a carefully selected couple of representative examples is presented. Further examples may be found in the report [8], so the computation of further point sets of a directed graph (like strongly connected components, point bases, or hammocks), point/edge coverings of graphs of various kinds, and so on.

4.1 Kernels

Let $\mathcal{G} = (V, B)$ be a *directed graph*, i.e., V a non-empty set of points (also: vertices, nodes) and $B : V \leftrightarrow V$ a relation between points. Furthermore, assume $\varepsilon : V \leftrightarrow 2^V$ to be the "is-element-of" relation between V and its powerset 2^V and $\sqsubseteq : 2^V \leftrightarrow 2^V$ to be the inclusion relation between point sets.

A set $a \in 2^V$ of points is said to be *absorbant* in $\mathcal{G}$ if from every point outside of a there is at least one arc leading into a, i.e., if the first-order formula

$$\bigwedge_x x \in a \vee \bigvee_y (B_{xy} \wedge y \in a)$$

holds. Furthermore, a set $s \in 2^V$ of points is called *stable* in $\mathcal{G}$ if no two points of s are related via the relation B. This situation is characterized by the first-order formula

$$\bigwedge_x x \in s \rightarrow \bigwedge_y (y \in s \rightarrow \overline{B}_{xy}) \ .$$

Finally, a *kernel* $k \in 2^V$ of $\mathcal{G}$ is a set which is at the same time absorbant and stable. The concept of a kernel plays an import rôle in combinatorial games; for an overview see e.g., [24] and [25], Sections 8.2 and 8.3.

Expressing the above two formulae in terms of the operations on relations introduced in Section 2, we get that the first one is equivalent to $(\mathsf{L} \setminus (\varepsilon \cup B\varepsilon))^{\mathrm{T}}_a$ and the second one is equivalent to $((\varepsilon \cap B\varepsilon) \setminus \mathsf{O})_s$. In the second case, for instance, we express $\bigwedge_y (y \in s \rightarrow \overline{B}_{xy})$ as $(\overline{B\varepsilon})_{xs}$ and obtain, thus, the original formula in the form $\bigwedge_x \overline{(\varepsilon \cap B\varepsilon)}_{xs}$. Now, the universal quantification can be removed using a right residual construction, cf. Section 2. Summing up, in a component-free notation, we have

$$\mathit{absorb}(B) = (\mathsf{L} \setminus (\varepsilon \cup B\varepsilon))^{\mathrm{T}}$$

as the vector $\mathit{absorb}(B) : 2^V \leftrightarrow \mathbb{1}$ describing the absorbant sets of $\mathcal{G}$ (where $\mathsf{L} : V \leftrightarrow \mathbb{1}$) and

$$\mathit{stable}(B) = (\varepsilon \cap B\varepsilon) \setminus \mathsf{O}$$

as the vector $\mathit{stable}(B) : 2^V \leftrightarrow \mathbb{1}$ of the stable sets of $\mathcal{G}$ (where $\mathsf{O} : V \leftrightarrow \mathbb{1}$). Finally, the vector $\mathit{kernel}(B) : 2^V \leftrightarrow \mathbb{1}$ describing the elements of 2^V which are kernels of $\mathcal{G}$ is given by

$$\mathit{kernel}(B) = \mathit{absorb}(B) \cap \mathit{stable}(B) \ .$$

4.2 Dedekind Cuts

Now, we deal with an order-theoretic problem; for a more visualized treatment of this example, compare [3]. Let $\mathcal{O} = (M, Q)$ be a partially ordered set, i.e., M be a non-empty set of points and $Q : M \leftrightarrow M$ be an ordering relation on points. Furthermore, assume again $\varepsilon : M \leftrightarrow 2^M$ to be the "is-element-of" relation between M and its powerset 2^M and $\sqsubseteq : 2^M \leftrightarrow 2^M$ to be the inclusion relation on point sets.

For an element $s \in 2^M$, let $Ma(s)$ denote its upper bounds wrt. Q and $Mi(s)$ denote its lower bounds wrt. Q. Then, $c \in 2^M$ is called a *Dedekind cut* of $\mathcal{O}$ if the equation $c = Mi(Ma(c))$ is valid, i.e, if the first-order formula

$$\bigwedge_x x \in c \leftrightarrow x \in Mi(Ma(c))$$

holds which in turn is equivalent to

$$\bigvee_s \bigwedge_x (x \in c \leftrightarrow x \in Mi(Ma(s))) \wedge c = s \ .$$

Obviously, for each element $x \in M$ the set $(x) = \{y \in M : Q_{yx}\}$ is a cut, called the *principal cut* generated by x. The fact that a set $p \in 2^M$ is a principal cut, hence, is described by the first-order formula

$$\bigvee_x x \in p \wedge \bigwedge_y (y \in p \leftrightarrow Q_{yx}) \ .$$

Now, let $\mathcal{C} \subseteq 2^M$ denote the set of cuts of $\mathcal{O}$ and $\mathcal{P} \subseteq 2^M$ denote the set of principal cuts of $\mathcal{O}$. Furthermore, let $\sqsubseteq_{\mathcal{C}} : \mathcal{C} \leftrightarrow \mathcal{C}$ and $\sqsubseteq_{\mathcal{P}} : \mathcal{P} \leftrightarrow \mathcal{P}$ denote the restrictions of set inclusion to the cuts and principal cuts, respectively. Then $\mathcal{O}' = (\mathcal{C}, \sqsubseteq_{\mathcal{C}})$ is a complete lattice, called the *cut completion* of $\mathcal{O}$, and the function $emb(Q) : M \to \mathcal{C}$ mapping x to the principal cut (x) is an injective order homomorphism.

Using abstract relation algebra and the above formulae (for the characterization of cuts the second version is more suited since it immediately leads to a symmetric quotient construction $\mathsf{syq}(\varepsilon, \ldots)_{cs}$) in combination with the functionals mi, ma, and syq of Section 2.3, by

$$cut(Q) = (\mathsf{syq}(\varepsilon, \mathsf{mi}(Q, \mathsf{ma}(Q, \varepsilon))) \cap \mathrm{I})\mathsf{L}$$

(where $\mathsf{L} : 2^M \leftrightarrow \mathbb{1}$ and $\mathrm{I} : 2^M \leftrightarrow 2^M$) we obtain the vector $cut(Q) : 2^M \leftrightarrow \mathbb{1}$ describing the elements of 2^M which are Dedekind cuts, and by

$$pricut(Q) = (\varepsilon^{\mathrm{T}} \cap \mathsf{syq}(\varepsilon, Q))\mathsf{L}$$

(where $\mathsf{L} : M \leftrightarrow \mathbb{1}$) we get the vector $pricut(Q) : 2^M \leftrightarrow \mathbb{1}$ describing the elements of 2^M which are principal cuts. Since the cuts are ordered by set inclusion, we consider the relation $\sqsubseteq : 2^M \leftrightarrow 2^M$ and the injection $\imath(cut(Q)) : \mathcal{C} \leftrightarrow 2^M$ in the sense of Subsection 3.2 and receive for $\sqsubseteq_{\mathcal{C}}$ the representation

$$\sqsubseteq_{\mathcal{C}} = \imath(cut(Q)) \sqsubseteq \imath(cut(Q))^{\mathrm{T}} \ .$$

Also the function $emb(Q) : M \leftrightarrow \mathcal{C}$ (in the relational sense) can be computed with the help of $\imath(cut(Q))$. From the component-wise definition of $emb(Q)_{xc}$ by the second-order formula

$$\bigwedge_y Q_{yx} \leftrightarrow \bigvee_s (y \in s \wedge \imath(cut(Q))_{cs})$$

we obtain $\bigwedge_y Q_{yx} \leftrightarrow (\varepsilon \imath(cut(Q))^{\mathrm{T}})_{yc}$ and, thus, an application of the functional syq yields the component-free relation-algebraic representation

$$emb(Q) = \mathsf{syq}(Q, \varepsilon \imath(cut(Q))^{\mathrm{T}}) \ .$$

4.3 Sets of Places of a Petri Net

A *Petri net* is a bipartite directed graph $\mathcal{N} = (X, Y, R, S)$. The point set of $\mathcal{N}$ is decomposed into the sets X of *places* and Y of *transitions* and the associated relations are $R : X \leftrightarrow Y$ and $S : Y \leftrightarrow X$. Petri nets have been widely used to design and model concurrent systems, see [22] for example. Many of the static properties of a Petri net (e.g., to be free-choice, to be conflict-free, or to contain specific sets of places/transitions) can be tested using our relational approach. In the following we show, how to compute specific sets of places. In doing so, $\varepsilon_X : X \leftrightarrow 2^X$ denotes the "is-element-of" relation between the places and the sets of places.

A set $d \in 2^X$ of places is called a *deadlock* if each of its predecessors is also a successor. A somewhat dual notion is that of a trap: $t \in 2^X$ is said to be a *trap* if its successor set is a subset of its predecessor set. Both, deadlocks and traps are of interest if one is concerned with liveness properties. E.g., a transition never can be enabled if its predecessor set contains an unmarked place of a deadlock.

Expressed by first-order formulae, we have that a set d of places is a deadlock if and only if

$$\bigwedge_y (\bigvee_x x \in d \wedge S_{yx}) \rightarrow (\bigvee_x x \in d \wedge R_{xy})$$

is valid and that a set t of places is a trap if and only if

$$\bigwedge_y (\bigvee_x x \in t \wedge R_{xy}) \rightarrow (\bigvee_x x \in t \wedge S_{yx})$$

holds. Using the operations on relations, the first formula becomes

$$\bigwedge_y (\varepsilon_X^{\mathsf{T}} S^{\mathsf{T}})_{dy} \rightarrow (\varepsilon_X^{\mathsf{T}} R)_{dy} \ , \text{ i.e. } \bigwedge_y (\overline{\varepsilon_X^{\mathsf{T}} S^{\mathsf{T}}} \cup \varepsilon_X^{\mathsf{T}} R)_{dy} \ .$$

Hence, using a left residual construction and the universal relation $\mathsf{L} : Y \leftrightarrow \mathbb{1}$ we get in a component-free notation

$$\mathit{deadlock}(R, S) = (\overline{\varepsilon_X^{\mathsf{T}} S^{\mathsf{T}}} \cup \varepsilon_X^{\mathsf{T}} R)/\mathsf{L}^{\mathsf{T}} = (S\varepsilon_X \cap \overline{R^{\mathsf{T}}\varepsilon_X}) \setminus \mathsf{O}$$

as the vector $\mathit{deadlock}(R, S) : X \leftrightarrow \mathbb{1}$ enumerating all deadlocks of the net $\mathcal{N}$. In the same way one obtains the vector $\mathit{trap}(R, S) : X \leftrightarrow \mathbb{1}$ describing all traps of $\mathcal{N}$ by exchanging the rôle of R and S^{T}, i.e., by

$$\mathit{trap}(R, S) = (\overline{\varepsilon_X^{\mathsf{T}} R} \cup \varepsilon_X^{\mathsf{T}} S^{\mathsf{T}})/\mathsf{L}^{\mathsf{T}} = (R^{\mathsf{T}}\varepsilon_X \cap \overline{S\varepsilon_X}) \setminus \mathsf{O} \ .$$

It may also be of interest to compute minimal non-empty deadlocks or traps. To this end, we assume $\sqsubseteq \ : 2^X \leftrightarrow 2^X$ to be the inclusion relation. By the min functional and the $\mathbf{E}$ vector, we obtain the minimized version of $\mathit{op} \in \{\mathit{deadlock}, \mathit{trap}\}$ from the expression $\mathsf{min}(\sqsubseteq, \mathit{op}(R, S) \cap \overline{\mathbf{E}})$.

4.4 Dicliques

In the preceding examples we have dealt with sets of points and edges, i.e., elements of the powerset of a given "base set". The example of this subsection shows how to describe the computation of pairs of elements of the powerset of a given "base set" with relation-algebraic means, i.e., the computation of a relation (correspondence) on a powerset. We consider again a directed graph $\mathcal{G} = (V, B)$ and suppose the "is-element-of" relation $\varepsilon : V \leftrightarrow 2^V$ and the inclusion relation $\sqsubseteq : 2^V \leftrightarrow 2^V$ on point sets.

A pair $(d, c) \in 2^V \times 2^V$ is called a *block* of $\mathcal{G}$ with domain d and co-domain c if the product $d \times c$ is contained in B. The (non-trivial) *dicliques* of $\mathcal{G}$ (this term is introduced in [16]) are the inclusion-maximal blocks with non-empty domains and co-domains. In other words, (d, c) is a diclique if and only if it generates an inclusion-maximal complete bipartite subgraph of $\mathcal{G}$. A decomposition of a graph into its dicliques can be very useful e.g., for storing it in a computer or for determining the essential subsystems of the system it describes. See again [16].

The description of a block $(d, c) \neq (\emptyset, \emptyset)$ by a first-order formula is

$$(\bigvee_x x \in d) \wedge (\bigvee_x x \in c) \wedge (\bigwedge_x x \in d \rightarrow \bigwedge_y (y \in c \rightarrow B_{xy})) \ .$$

Obviously, $\bigwedge_y y \in c \rightarrow B_{xy}$ is equivalent to $(\overline{\overline{B}\varepsilon})_{xc}$, i.e., to $(B/\varepsilon^{\mathsf{T}})_{xc}$, which implies the equivalence of $\bigwedge_x x \in d \rightarrow \bigwedge_y (y \in c \rightarrow B_{xy})$ and $(\varepsilon \setminus (B/\varepsilon^{\mathsf{T}}))_{dc}$. This immediately leads to

$$block(B) = \varepsilon^{\mathsf{T}}\mathsf{L} \cap \mathsf{L}^{\mathsf{T}}\varepsilon \cap (\varepsilon \setminus (B/\varepsilon^{\mathsf{T}})) = \overline{\mathbf{E}} \cap \overline{\mathbf{E}}^{\mathsf{T}} \cap \overline{\varepsilon^{\mathsf{T}}\overline{\overline{B}\varepsilon}}$$

(where $\mathsf{L} : V \leftrightarrow 2^V$) as the relation $block(B) : 2^V \leftrightarrow 2^V$ of the blocks with non-empty domain and co-domain. I.e., $block(B)_{dc}$ holds if and only if $(d, c) \neq (\emptyset, \emptyset)$ is a block. To describe dicliques, we use that for a relation $R : 2^V \leftrightarrow 2^V$ and a pair $(s, t) \in 2^V \times 2^V$ we have $\mathsf{max}(\sqsubseteq, R)_{st}$ if and only if s is inclusion-maximal in the set $\{s' \in 2^V : R_{s't}\}$. Hence, a two-fold maximalization via the functional max yields

$$diclique(B) = \mathsf{max}(\sqsubseteq, \mathsf{max}(\sqsubseteq, block(B))^{\mathsf{T}})^{\mathsf{T}}$$

as the relation $diclique(B) : 2^V \leftrightarrow 2^V$ of the dicliques of the directed graph $\mathcal{G}$.

Now, let $D \subseteq 2^V$ (resp. $C \subseteq 2^V$) be the set of domains (resp. co-domains) of the dicliques of $\mathcal{G}$. Then we have the two injections

$$\imath(diclique(B)\mathsf{L}) : D \leftrightarrow 2^V \qquad \imath(diclique(B)^{\mathsf{T}}\mathsf{L}) : C \leftrightarrow 2^V$$

for embedding D and C, respectively, into 2^V. Based on these injections, finally, we are able to define a relation $Diclique(B) : D \leftrightarrow C$ describing the correspondence between the domain and the co-domain of a diclique by

$$Diclique(B) = \imath(diclique(B)\mathsf{L})diclique(B)\imath(diclique(B)^{\mathsf{T}}\mathsf{L})^{\mathsf{T}} \ .$$

This means: A pair $(d, c) \in D \times C$ is a diclique of $\mathcal{G}$ if and only if $Diclique(B)_{cd}$ holds.

4.5 Homomorphisms

This example will show how to describe the computation of sets of functions from a set V to a set W with relation-algebraic means. As already mentioned, we consider the set W^V of functions from V to W as the set of the functional and total relations between V and W, i.e., as a subset of the powerset $2^{V \times W}$. Therefore, suppose $\varepsilon_F : V \times W \leftrightarrow W^V$ to be the "is-element-of" relation between $V \times W$ and the set W^V.

Let $\mathcal{G} = (V, B)$ and $\mathcal{H} = (W, C)$ be directed graphs and $\Phi \in 2^{V \times W}$ be a function in the relational sense (see Subsection 2.2). Furthermore, assume (π, ρ) to be the projections of the direct product $V \times W$ to the first and the second component, respectively. If we use the common function notation for π and ρ, the fact that Φ is a *homomorphism* means exactly that the first-order formula

$$\bigwedge_u u \in \Phi \rightarrow \bigwedge_v (v \in \Phi \rightarrow (B_{\pi(u)\pi(v)} \rightarrow C_{\rho(u)\rho(v)}))$$

is valid; see Subsection 3.1. Now, we use the relational notation also for the two projections π and ρ, i.e., write $(\pi B \pi^{\mathrm{T}})_{uv}$ (resp. $(\rho C \rho^{\mathrm{T}})_{uv}$) instead of $B_{\pi(u)\pi(v)}$ (resp. $C_{\rho(u)\rho(v)}$). This leads to the equivalent version

$$\bigwedge_u u \in \Phi \rightarrow \bigwedge_v (v \in \Phi \rightarrow (\overline{\pi B \pi^{\mathrm{T}}} \cup \rho C \rho^{\mathrm{T}})_{uv}) .$$

Hence, we have again the pattern of the formula defining a set to be stable and, therefore, we get the vector $hom(B, C) : W^V \leftrightarrow \mathbb{1}$ describing the homomorphisms from B to C as

$$hom(B, C) = (\varepsilon_F \cap \overline{\overline{\overline{\pi B \pi^{\mathrm{T}}} \cup \rho C \rho^{\mathrm{T}}}}\varepsilon_F) \setminus \mathsf{O} = (\varepsilon_F \cap (\pi B \pi^{\mathrm{T}} \cap \overline{\rho C \rho^{\mathrm{T}}})\varepsilon_F) \setminus \mathsf{O} \ ,$$

where the typing of the empty relation is $\mathsf{O} : V \times W \leftrightarrow \mathbb{1}$.

5 Development of Efficient Algorithms

The execution of a specification produced in the way described in Section 4, frequently may be fairly inefficient. In this subsection we demonstrate by means of an example how to develop more efficient algorithms from the original inefficient specifications using the relational calculus.

We consider again the problem of computing the kernels of a directed graph $\mathcal{G} = (V, B)$. In contrast with Subsection 4.1, however, in the following we do not consider sets as points $p : 2^V \leftrightarrow \mathbb{1}$ but as vectors $v : V \leftrightarrow \mathbb{1}$. In doing so, $x \in p$ will be replaced by v_x. So, the first-order formulae of Subsection 4.1 defining absorbant and stable sets become

$$\bigwedge_x \overline{a}_x \rightarrow \bigvee_y (a_y \wedge B_{xy}) \qquad\qquad \bigwedge_x s_x \rightarrow \bigwedge_y (s_y \rightarrow \overline{B}_{xy}) .$$

Translating these formulae into a notation without components, we get the inclusions $\overline{a} \subset Ba$ and $s \subset \overline{Bs}$. As a consequence, a vector $k : V \leftrightarrow \mathbb{1}$ is a kernel of $\mathcal{G}$ if and only if $\overline{k} = Bk$, i.e., if and only if it is a fixpoint of the functional $\tau(v) = \overline{Bv}$.

This example shows also that the change of set-representation does not eliminate the use of higher-order structures. Instead of the relation ε, in the new specification now a functional τ mapping vectors to vectors is used. However, for specific classes of graphs the fixpoint specification enables the development of efficient algorithms as will be shown now.

5.1 Progressively Finite Graphs

The functional τ just defined is antitonic, so the fixpoint theorem for monotone functions on complete lattices cannot be applied. We therefore study the fixpoints of $\tau^2(v) := \tau(\tau(v)) = \overline{B\overline{Bv}}$ which is monotonic. Suppose m_{τ^2} and M_{τ^2} to denote the least resp. greatest fixpoint of τ^2. Then we have for each kernel k that

$$\mathsf{O} \subset \tau^2(\mathsf{O}) \subset \tau^4(\mathsf{O}) \subset \ldots \subset m_{\tau^2} \subset k \subset M_{\tau^2} \subset \ldots \subset \tau^4(\mathsf{L}) \subset \tau^2(\mathsf{L}) \subset \mathsf{L} \ .$$

Also the two equations

$$\text{(i)} \ \ \tau(m_{\tau^2}) = M_{\tau^2} \qquad \text{(ii)} \ \ \tau(M_{\tau^2}) = m_{\tau^2}$$

easily can be shown. Hence, if the gap between the lower bound m_{τ^2} and the upper bound M_{τ^2} of the set of all kernel closes, then the uniquely determined fixpoint is a kernel of $\mathcal{G}$:

Theorem. *If the functional τ^2 has exactly one fixpoint (which is equivalent to $M_{\tau^2} \subset \tau(M_{\tau^2})$ or to $\tau(m_{\tau^2}) \subset m_{\tau^2}$), then $\mathcal{G}$ has precisely one kernel.* □

Using this theorem, for instance, it can be shown that a progressively finite graph $\mathcal{G} = (V, B)$ (i.e., a graph in which all paths have finite lengths) has exactly one kernel. When specifying progressive finiteness relationally, we obtain

$$(*) \quad \bigwedge_v (v = v\mathsf{L} \wedge v \subset Bv \rightarrow v = \mathsf{O}) \ .$$

Now, if we use the Schröder equivalences, we obtain $B^{\mathrm{T}} M_{\tau^2} \subset BM_{\tau^2}$ from $M_{\tau^2} \subset \tau^2(M_{\tau^2})$. Next, the Dedekind rule yields

$$\begin{aligned} BM_{\tau^2} \cap M_{\tau^2} &\subset (B \cap M_{\tau^2} M_{\tau^2}^{\mathrm{T}})(M_{\tau^2} \cap B^{\mathrm{T}} M_{\tau^2}) \\ &\subset B(M_{\tau^2} \cap B^{\mathrm{T}} M_{\tau^2}) \\ &\subset B(M_{\tau^2} \cap BM_{\tau^2}) \end{aligned}$$

and, finally, in combination with $(*)$ we obtain $BM_{\tau^2} \cap M_{\tau^2} = \mathsf{O}$, which implies $M_{\tau^2} \subset \tau(M_{\tau^2})$. Hence, the functional τ has precisely one fixpoint.

For the point set being finite, we have that a directed graph is progressively finite if and only if it is circuit-free. Therefore, we can compute the only kernel of a finite circuit-free graph $\mathcal{G} = (V, B)$ by the iteration $\mathsf{O} \subset \tau^2(\mathsf{O}) \subset \tau^4(\mathsf{O}) \subset \ldots$ which takes at most $|V|$ steps.

At this place it should be mentioned that in the case of a transitive relation B the iteration stops after one step. This is due to the fact that in the case $B = B^+$ and $B = B^*\overline{B\mathsf{L}}$ (this latter condition follows from $(*)$ and means that from each point there is a path to a terminal one) the only kernel of $\mathcal{G}$ is the least absorbant set which equals the set $\overline{B\mathsf{L}} : V \leftrightarrow \mathbb{1}$ of all terminal points:

$$\tau^2(\mathsf{O}) = \overline{B\overline{B\mathsf{O}}} = \overline{B\mathsf{L}} = \overline{B\overline{B\mathsf{L}}} = \tau^2(\mathsf{L})$$

follows from $B\overline{B\mathsf{L}} \subset B\mathsf{L}$ and $\mathsf{L} = B^*\overline{B\mathsf{L}} = \overline{B\mathsf{L}} \cup B^+\overline{B\mathsf{L}} = \overline{B\mathsf{L}} \cup B\overline{B\mathsf{L}}$ which in turn is equivalent to $B\mathsf{L} \subset B\overline{B\mathsf{L}}$.

5.2 Bipartite Graphs

Now, assume $\mathcal{G} = (X, Y, R, S)$ to be a bipartite graph in the sense of Subsection 4.3, i.e., we have $R : X \leftrightarrow Y$ and $S : Y \leftrightarrow X$. Furthermore let $\iota : X \leftrightarrow X + Y$ and $\kappa : Y \leftrightarrow X + Y$ be the natural injections into the binary direct sum $X + Y$. Then

$$\mathcal{H} = (V, B), \text{ where } V := X + Y \text{ and } B := \iota^{\mathrm{T}} R \kappa \cup \kappa^{\mathrm{T}} S \iota \ ,$$

is the "ordinary" directed graph corresponding to $\mathcal{G}$. Generalizing the above technique of the composition of the antitone functional τ with itself to pairs of antitone functionals and using the laws of the direct sum, in the following we will show that $\mathcal{H}$ has at least the kernels according to the following two characterizations. To this end, we consider the functionals $\alpha(v) = \overline{Rv}$ and $\beta(w) = \overline{Sw}$ and obtain for the least and greatest fixpoints of the compositions

$$\text{(i) } \alpha(M_{\beta o \alpha}) = m_{\alpha o \beta} \qquad \text{(ii) } \alpha(m_{\beta o \alpha}) = M_{\alpha o \beta}$$
$$\text{(iii) } \beta(M_{\alpha o \beta}) = m_{\beta o \alpha} \qquad \text{(iv) } \beta(m_{\alpha o \beta}) = M_{\beta o \alpha} \ .$$

If we use (only for explanatory purposes) 2×2-matrices and 2-vectors with relations and vectors, respectively, as coefficients, then we have B as matrix

$$B = \begin{pmatrix} \mathsf{O} & R \\ S & \mathsf{O} \end{pmatrix}$$

and obtain two kernels of $\mathcal{H}$ by the vectors

$$k_1 = \begin{pmatrix} M_{\alpha o \beta} \\ m_{\beta o \alpha} \end{pmatrix} \qquad k_2 = \begin{pmatrix} m_{\alpha o \beta} \\ M_{\beta o \alpha} \end{pmatrix} .$$

A component-free proof of this fact based on (i) through (iv) and on the relational characterization of the direct sum is given in the following. Define two vectors

$$k_1 := \iota^{\mathrm{T}} M_{\alpha o \beta} \cup \kappa^{\mathrm{T}} m_{\beta o \alpha} : X + Y \leftrightarrow \mathbb{1}$$
$$k_2 := \iota^{\mathrm{T}} m_{\alpha o \beta} \cup \kappa^{\mathrm{T}} M_{\beta o \alpha} : X + Y \leftrightarrow \mathbb{1} \ .$$

Then, we obtain the equations $Bk_1 = \overline{k_1}$ and $Bk_2 = \overline{k_2}$. E.g., a proof of $Bk_1 = \overline{k_1}$ proceeds as follows: We use an immediate consequence of (S_1) through (S_3), viz. $\iota^{\mathrm{T}}\mathsf{L} = \overline{\kappa^{\mathrm{T}}\mathsf{L}}$, and obtain

$$\kappa^{\mathrm{T}}\mathsf{L} = \overline{\iota^{\mathrm{T}}\mathsf{L}} \subset \overline{\iota^{\mathrm{T}} M_{\alpha o \beta}} \qquad \iota^{\mathrm{T}}\mathsf{L} = \overline{\kappa^{\mathrm{T}}\mathsf{L}} \subset \overline{\kappa^{\mathrm{T}} m_{\beta o \alpha}} \ .$$

From the axioms of the direct sum we get also that both ι^{T} and κ^{T} are partial functions and, due to Proposition 4.2.2.v of [25] (saying that $Q\overline{R} = \overline{QR} \cap Q\mathsf{L}$ for Q being a partial function), we have

$$\iota^{\mathrm{T}}\overline{M_{\alpha\circ\beta}} = \overline{\iota^{\mathrm{T}} M_{\alpha\circ\beta}} \cap \iota^{\mathrm{T}}\mathsf{L} \qquad \kappa^{\mathrm{T}}\overline{m_{\beta\circ\alpha}} = \overline{\kappa^{\mathrm{T}} m_{\beta\circ\alpha}} \cap \kappa^{\mathrm{T}}\mathsf{L}\ .$$

Now, we combine these properties and arrive at

$$\begin{aligned}
\iota^{\mathrm{T}}\overline{M_{\alpha\circ\beta}} \cup \kappa^{\mathrm{T}}\overline{m_{\beta\circ\alpha}} &= (\overline{\iota^{\mathrm{T}} M_{\alpha\circ\beta}} \cap \iota^{\mathrm{T}}\mathsf{L}) \cup (\overline{\kappa^{\mathrm{T}} m_{\beta\circ\alpha}} \cap \kappa^{\mathrm{T}}\mathsf{L}) \\
&= (\overline{\iota^{\mathrm{T}} M_{\alpha\circ\beta}} \cup \overline{\kappa^{\mathrm{T}} m_{\beta\circ\alpha}}) \cap (\overline{\iota^{\mathrm{T}} M_{\alpha\circ\beta}} \cup \kappa^{\mathrm{T}}\mathsf{L}) \\
&\quad \cap (\iota^{\mathrm{T}}\mathsf{L} \cup \overline{\kappa^{\mathrm{T}} m_{\beta\circ\alpha}}) \cap (\iota^{\mathrm{T}}\mathsf{L} \cup \kappa^{\mathrm{T}}\mathsf{L}) \\
&= (\overline{\iota^{\mathrm{T}} M_{\alpha\circ\beta}} \cup \overline{\kappa^{\mathrm{T}} m_{\beta\circ\alpha}}) \cap \overline{\iota^{\mathrm{T}} M_{\alpha\circ\beta}} \cap \overline{\kappa^{\mathrm{T}} m_{\beta\circ\alpha}} \\
&= \overline{\iota^{\mathrm{T}} M_{\alpha\circ\beta}} \cap \overline{\kappa^{\mathrm{T}} m_{\beta\circ\alpha}}\ ,
\end{aligned}$$

which in turn implies the desired result as follows:

$$\begin{aligned}
Bk_1 &= (\iota^{\mathrm{T}} R\kappa \cup \kappa^{\mathrm{T}} S\iota)(\iota^{\mathrm{T}} M_{\alpha\circ\beta} \cup \kappa^{\mathrm{T}} m_{\beta\circ\alpha}) \\
&= \iota^{\mathrm{T}} R m_{\beta\circ\alpha} \cup \kappa^{\mathrm{T}} S M_{\alpha\circ\beta} && \text{axioms of the direct sum} \\
&= \iota^{\mathrm{T}}\overline{\alpha(m_{\beta\circ\alpha})} \cup \kappa^{\mathrm{T}}\overline{\beta(M_{\alpha\circ\beta})} \\
&= \iota^{\mathrm{T}}\overline{M_{\alpha\circ\beta}} \cup \kappa^{\mathrm{T}}\overline{m_{\beta\circ\alpha}} && \text{due to (ii) and (iii)} \\
&= \overline{\iota^{\mathrm{T}} M_{\alpha\circ\beta}} \cap \overline{\kappa^{\mathrm{T}} m_{\beta\circ\alpha}} \\
&= \overline{\iota^{\mathrm{T}} M_{\alpha\circ\beta} \cup \kappa^{\mathrm{T}} m_{\beta\circ\alpha}} \\
&= \overline{k_1}\ .
\end{aligned}$$

Altogether, we have shown the following generalization of Richardson's theorem an early version of which goes back to [27].

Theorem. *If $\mathcal{G}$ is a bipartite graph, then the corresponding directed graph $\mathcal{H}$ has (not necessarily distinct) kernels k_1 and k_2.* □

Namely, these two kernels can be computed with the help of the iterations

$$\mathsf{O} \subset \alpha\circ\beta(\mathsf{O}) \subset \alpha\circ\beta(\alpha\circ\beta(\mathsf{O})) \subset \ldots \qquad \mathsf{O} \subset \beta\circ\alpha(\mathsf{O}) \subset \beta\circ\alpha(\beta\circ\alpha(\mathsf{O})) \subset \ldots$$

to obtain $m_{\alpha\circ\beta}$ and $m_{\beta\circ\alpha}$, since by applying (iv) and (ii), respectively, from above we can obtain the missing $M_{\beta\circ\alpha}$ and $M_{\alpha\circ\beta}$.

Continuing the preceding treatment, the report [8] also deals with the evaluation of graph games, where moves are one-step walks along the edges, due to loss, draw (by infinite repetition), and win: loss is obtained as the greatest stable set and win is described as the complement of the smallest absorbant set.

6 Concluding Remarks

In this paper we have described a rapid prototyping approach for the enumeration of certain mathematical objects in terms of relations. This has lead to an extensive use of a relational characterization of higher-order objects like sets, sets of sets, or functions. We have also shown how to develop more efficient algorithms from inefficient specifications within the abstract relational framework.

Let us close with a few remarks about the execution of relational specifications. All examples given in this paper have been performed using RELVIEW, a totally interactive and completely video-oriented computer system for the manipulation of concrete relations which are considered as Boolean matrices (see [1, 4]). The system does not only provide commands implementing the basic operations on relations, but also commands for residuals, quotients, and closures, commands for certain tests on relations, and commands which implement the operations important in relation-algebraic domain description (direct product, direct sum, injections from vectors, powersets, and function spaces). And, finally, RELVIEW allows the user to define and apply his own functionals on relations, where in the case of a unary functional with identical domain and range even repeated application is possible by an iteration command. A useful fact in applications is that the latter command can be used to compute fixpoints of monotone functionals, as for the efficient computations of Section 5. Of course, computation with RELVIEW is limited in space and time. The limit, however, depends heavily on the type of problem handled. In general, it is difficult to treat the powerset of a set or function spaces since this means the computation of vectors $v : 2^V \leftrightarrow \mathbb{1}$ or $v : W^V \leftrightarrow \mathbb{1}$ and consumes a lot of space and time. However, the handling of relations $R : V \leftrightarrow V$, vectors $v : V \times V \leftrightarrow 1$, or vectors $v : V \leftrightarrow \mathbb{1}$ is of such a small complexity that admits a wide range of RELVIEW applications. E.g., on our installation (SUN SPARCstation 10) we have treated relations with domain/range up to 5000 elements. As the computation of kernels shows, it seems promising to apply the relational calculus to obtain efficient algorithms from inefficient specifications.

Acknowledgement: We wish to thank the referees for their helpful comments and hints.

References

1. Abold-Thalmann H., Berghammer R., Schmidt G.: Manipulation of concrete relations: The RELVIEW-system. Report Nr. 8905, Fakultät für Informatik, Universität der Bundeswehr München (1989)
2. Backhouse R.C., Hoogendijk P., Voermans E., van der Woude J.C.S.P: A relational theory of datatypes. Eindhoven University of Technology, Dept. of Mathematics and Computer Science (1992)
3. Berghammer R.: Computing the cut completion of a partially ordered set – An example for the use of the RELVIEW-system. Report Nr. 9205, Fakultät für Informatik, Universität der Bundeswehr München (1992)
4. Berghammer R., Schmidt G.: The RELVIEW-system. In: Choffrut C., Jantzen M. (eds.): Proc. STACS '91, Lect. Notes Comput. Sci. 480, Springer, 535–536 (1991)
5. Berghammer R., Schmidt G.: Relational specifications. In: Rauszer C. (ed.): Algebraic Methods in Logic and Computer Science, Banach Center Publications, Volume 28, Institute of Mathematics, Polish Academy of Sciences, 167–190 (1993)
6. Berghammer R., Zierer H.: Relational algebraic semantics of deterministic and nondeterministic programs. Theoret. Comput. Sci. 43, 123–147 (1986)

7. Berghammer R., Schmidt G., Zierer H.: Symmetric quotients and domain constructions. Inform. Proc. Letters 33, 3, 163–168 (1989/90)
8. Berghammer R., Gritzner T.F., Schmidt G.: Prototyping relational specifications using higher-order objects. Report Nr. 9304, Fakultät für Informatik, Universität der Bundeswehr München (1993)
9. Brook T.: Order and recursion in topoi. Notes on Pure Mathematics, Vol. 9, Australian National University of Canberra (1977)
10. Budde R., Kuhlenkamp K., Matthiassen H., Züllinghoven H. (eds.): Approaches to prototyping. Springer (1984)
11. Chin L.H., Tarski A.: Distributive and modular laws in the arithmetic of relation algebras. University of California Publications in Mathematics (new series) 1, 341–384 (1951)
12. De Bakker J.W., de Roever W.P.: A calculus for recursive program schemes. In: Nivat M. (ed.): Proc. ICALP 73, North-Holland, 167–196 (1973)
13. De Roever W.P.: Recursion and parameter mechanisms: An axiomatic approach. In: Loeckx J. (ed.): Proc. ICALP 74, Lect. Notes Comput. Sci. 14, Springer, 34-65 (1974)
14. Desharnais J., Jaoua A., Mili F., Boudriga N., Mili A.: Conjugate kernels: An operator for program construction. Theoret. Comput. Sci., to appear
15. Freyd P.J., Ščedrov A.: Categories, allegories. Mathematical Library, Vol. 39, North-Holland (1990)
16. Haralick R.M.: The diclique representation and decomposition of binary relations. J. ACM 21, 3, 356–366 (1974)
17. Hoare C.A.R., He J.: The weakest prespecification, Parts I&II, Fundamenta Informaticae IX, 51–84 & 217–252 (1986)
18. Jónsson B., Tarski A.: Boolean algebras with operators, Part II. Amer. J. Math. 74, 127–167 (1952)
19. Mili A.: A relational approach to the design of deterministic programs. Acta Informatica 20, 315–328 (1983)
20. Mili A., Desharnais J., Mili F.: Relational heuristics for the design of deterministic programs. Acta Informatica 24, 239–276 (1987)
21. Möller B.: Relations as a program development language. In: Möller B. (ed.): Proc. IFIP TC2/WG2.1 Working Conference on Constructing Programs from Specifications, North-Holland, 373–397 (1991)
22. Reisig W.: Petri nets – An introduction. EATCS Monographs on Theoret. Comput. Sci., Springer (1985)
23. Schmidt G.: Programs as partial graphs I: Flow equivalence and correctness. Theoret. Comput. Sci. 15, 1–25 (1981)
24. Schmidt G., Ströhlein T.: On kernels of graphs and solutions of games: A synopsis based on relations and fixed points. SIAM J. Alg. Disc. Meth. 6,1, 54–65 (1985)
25. Schmidt G., Ströhlein T.: Relationen und Graphen. Springer (1989); English version: Relations and graphs. Discrete Mathematics for Computer Scientists, EATCS Monographs on Theoret. Comput. Sci., Springer (1993)
26. Spivey J.M.: The Z notation: A reference manual. Prentice Hall (1989)
27. Ströhlein T.: Untersuchungen über kombinatorische Spiele. Doctoral Thesis, Technische Universität München (1970)
28. Tarski A.: On the calculus of relations. Journal of Symbolic Logic 6, 73–89 (1941)
29. Veloso P., Haeberer A.: A finitary relational algebra for classical first-order logic. Bull. of the Section on Logic of the Polish Academy of Sciences 20, 52–62 (1991)

30. Veloso P., Haeberer A., Baum G.: Formal program construction within an extended calculus of binary relations. J. Symbolic Comp., to appear
31. Zierer H.: Relation algebraic domain constructions. Theoret. Comput. Sci. 87, 163–188 (1991)
32. Zierer H., Schmidt G., Berghammer R.: An interactive graphical manipulation system for higher objects based on relation algebra. In: Tinhofer G., Schmidt G. (eds.): Proc. 12th International Workshop on Graph-Theoretic Concepts in Computer Science (WG 86), Bernried/Starnberger See, 17.6.-19.6. 1986, Lect. Notes Comput. Sci. 246, Springer: Berlin-Heidelberg-New York, 68-81 (1987)

Origin Tracking for Higher-Order Term Rewriting Systems

Arie van Deursen * and T.B. Dinesh **

CWI - P.O. Box 94079, 1090 GB Amsterdam, The Netherlands
{arie,dinesh}@cwi.nl

Abstract. *Origin Tracking* is a technique which, in the framework of first-order term rewriting systems, establishes relations between each subterm t of a normal form and a set of subterms, the *origins of* t, in the initial term. Origin tracking is based on the notion of residuals. It has been used successfully for the generation of error handlers and debuggers from algebraic specifications of programming languages. Recent experiments with the use of higher-order algebraic specifications for the definition of programming languages revealed a need to extend origin tracking for higher-order term rewriting systems.
In this paper, we discuss how origin information can be maintained for $\beta\eta$ reductions and expansions, during higher-order rewriting. We give a definition of higher-order origin tracking. The suitability of this definition is illustrated with a small, existing specification.

1 Origin Tracking

When algebraic specifications are executed as term rewriting systems (TRSs), computations are performed by reducing an initial term to its result value — its normal form. Often, it is enough just to compute this result value, but in many cases it is useful to have some additional information. For instance, one may wish to know how the initial term influenced the normal form; are there perhaps parts of the initial term that were copied without a change to the result term? Or, if a subterm of the normal form does not *literally* recur in the initial term, is it possible to identify a set of subterms in the initial term which in some sense were *responsible* for its creation?

Trying to capture how intermediate and final terms originate from the initial term is formalized in a notion called *origin tracking* [4, 5, 10]. Origin tracking is based on so-called *residuals*. Residuals have been used successfully in more theoretical work [15, 21, 23], for reasoning about optimal reduction strategies in TRSs.

* Supported by the European Communities, ESPRIT project 2177: Generation of Interactive Programming Environments II — GIPE II; and by the Netherlands Organization for Scientific Research - NWO, project *Incremental Program Generators.*

** Supported by the European Communities, ESPRIT project 5399: Compiler Generation for Parallel Machines - COMPARE

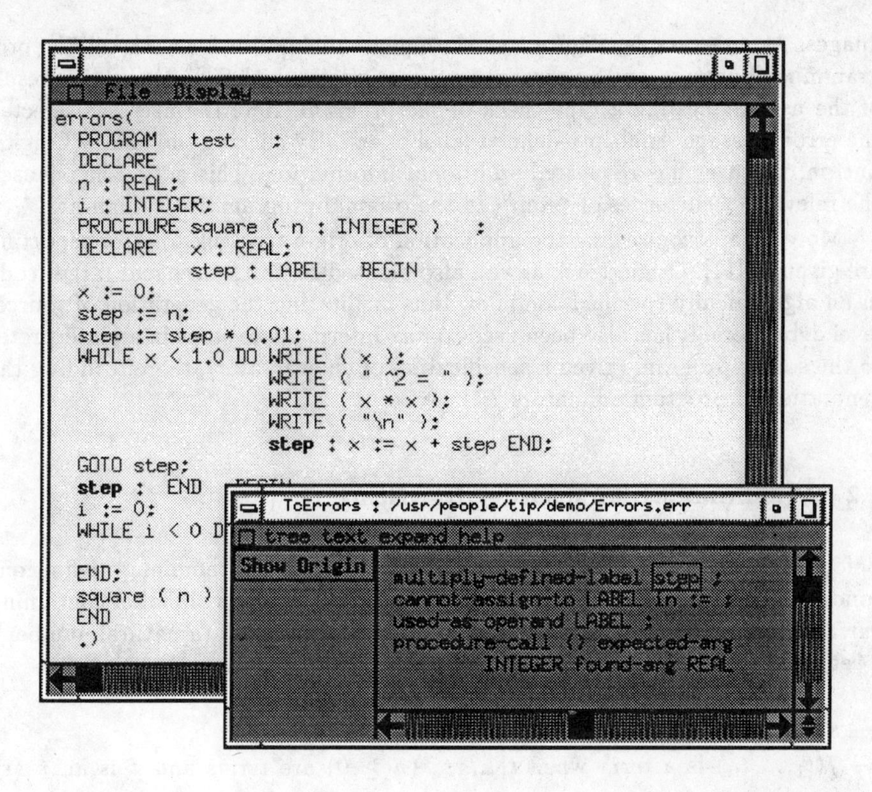

Fig. 1. Example of a generated environment using origin tracking.

1.1 Applications

Our motivation for working on origin tracking is its applicability to the automatic generation of tools from algebraic specifications of programming languages. As an example, let us take an algebraic specification of a type checker for some programming language. Assume that this specification can be executed using rewriting, and that the type check function is called *tc*. In order to type check a program P, a term p is constructed representing P. The term $tc(p)$ is reduced to its normal form, which is a list $[E_1, ..., E_n]$ of error messages. Just carrying out the reduction will only give such a list, whereas when reduced in combination with origin tracking we get additional information. Namely, for each error E_i, a relation to the part in the initial term $tc(p)$, e.g., a statement, an expression or an identifier, responsible for causing E_i is established.

In the ASF+SDF programming environment generator[3] [3, 17] origin tracking has been implemented. This implementation has been used to derive error reporters from algebraic specifications of static semantics of programming lan-

[3] ASF+SDF is the name of the formalism used to specify programming languages; it originated from combining the Algebraic Specification Formalism ASF and the Syntax Definition Formalism SDF.

guages. As an example, Figure 1 shows a generated editor for a Pascal-like programming language, and an error reporting window which is displayed as a result of the user requesting a type-check of the program. Here the user has selected the error message "multiply-defined-label `step`". By clicking the "Show Origin" button, the user has requested additional information. This action has caused the relevant occurrences of "step", in the original program, to be high-lit.

More details concerning the application of origin tracking to error reporting are given in [11]. Origin tracking can also be used to link source and target code in an algebraically specified compiler, thus facilitating the generation of source-level debuggers. It has also been used to link intermediate steps in an interpreter to the source program (given a specification of an evaluator), thereby aiding the generation of program animators [25].

1.2 Preliminaries: First-Order Rewriting

Before defining origins more rigorously, we borrow some preliminary definitions concerning first-order term rewriting from [19, 15]. Given an alphabet containing *variables* and *function symbols* each equipped with an *arity* (a natural number), a set of *terms* is constructed by considering

- all variables as terms.
- $f(t_1, ..., t_n)$ is a term when $t_1, ..., t_n$ $(n \geq 0)$ are terms and f is an n-ary function symbol.

A term t can be *reduced* to a term t' according to a *rewrite rule* $r : p \rightarrow q$ by identifying a context $C[]$ and a subterm s in t such that $t \equiv C[s]$, and by finding a substitution σ such that $s \equiv p^\sigma$. Then $t \equiv C[p^\sigma]$ rewrites to $C[q^\sigma] \equiv t'$ by one *elementary reduction*, written $t \rightarrow t'$. We call p^σ the *redex*, and q^σ the *contractum*. The concatenation of reduction steps $t_0 \rightarrow t_1 \rightarrow ... \rightarrow t_n$ is also written $t_0 \rightarrow^* t_n$ $(n \geq 0)$.

Subterms are characterized by *occurrences* (paths), which are either equal to $[\,]$ for the entire term or to a sequence of integers (the *branches*) $[n_1, ..., n_m]$ $(m \geq 0)$ representing the access path to the subterm. The occurrence $[1, 2]$ denotes the second son of the first son of the root, i.e., for term $f(g(a, b), c)$ it denotes subterm b. The subterm in t at occurrence u is written t/u. Paths are concatenated by the (associative) operator "$\cdot$". If u, v, w are occurrences and $u = v \cdot w$, then v is *above* u, written $v \preceq u$, or $v \prec u$ if $w \neq []$. If neither $u \preceq v$ nor $v \preceq u$ then u and v are *disjoint*, written $u \mid v$. The set of all occurrences in a particular term t is identified by $\mathcal{O}(t)$, which we furthermore partition into a set $\mathcal{O}_{var}(t)$ denoting variable occurrences and a set $\mathcal{O}_{fun}(t)$ denoting function symbol occurrences.

When we wish to identify the redex, rule, and substitution explicitly, we will write $t \xrightarrow{u,\sigma}_r t'$ for the one-step rewrite relation, indicating that rule r is applied at occurrence u in term t under substitution σ.

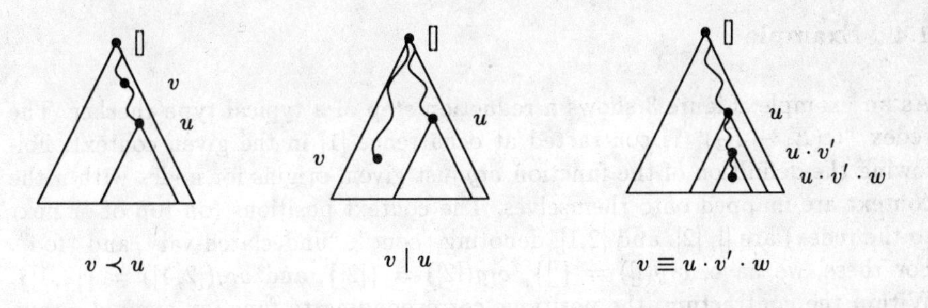

Fig. 2. Relative positions of v with respect to contractum position u

1.3 Definition of the Origin Function

We give the definition of origins as described in [10], following the presentation of [4]. Let $t \xrightarrow{u,\sigma}_r t'$, where r is a rule $p \rightarrow q$, be an elementary reduction step. With each step we associate a function $org : \mathcal{O}(t') \rightarrow \mathcal{P}(\mathcal{O}(t))$ mapping occurrences in t' to sets of occurrences in t. Let $v \in \mathcal{O}(t')$. We define org by distinguishing the following cases (see Figure 2):

- (Context)
  If $v \prec u$ or $v \mid u$ then $org(v) = \{v\}$;
- (Common Variables)
  If $v \equiv u \cdot v' \cdot w$ with $v' \in \mathcal{O}_{var}(q)$ denoting some variable X in the right-hand side, and $w \in \mathcal{O}(X^\sigma)$ an occurrence in the instantiation of that variable, then
  $$org(v) = \{u \cdot v'' \cdot w \mid p/v'' \equiv X\}$$
  Hence, $v'' \in \mathcal{O}_{var}(p)$ denotes an occurrence of X in the left-hand side p.

For the time being, we will assume that $org(v) = \emptyset$ for the remaining case, i.e., where v denotes a function symbol in the right-hand side (see also Section 1.5).

Function org covers one-step reductions. It is generalized to a function org^* for a multistep reduction $t_1 \rightarrow^* t_n$ $(n \geq 0)$ by considering the origin functions for the individual steps in the complete reduction $t_0 \rightarrow t_1 \rightarrow \cdots \rightarrow t_n$. Let $o_i : \mathcal{O}(t_i) \rightarrow \mathcal{P}(\mathcal{O}(t_{i-1}))$ be the origin function associated with rewrite step $t_{i-1} \rightarrow t_i$ $(0 < i \leq n)$. Recursively define $org^j : \mathcal{O}(t_j) \rightarrow \mathcal{P}(\mathcal{O}(t_0))$ for $0 \leq j \leq n$, and $v \in \mathcal{O}(t_j)$:

- $j = 0$: $org^j(v) = \{v\}$.
- $0 < j \leq n$: $org^j(v) = \{w \mid w \in org^{j-1}(w'),\ w' \in o_j(v)\}$

Then org^* is equal to org^n for multistep reduction $t_0 \rightarrow^* t_n$ $(n \geq 0)$.

For orthogonal (left-linear and non-overlapping) TRSs the origin function is the reversal of the well-known notion of *descendant* or *residual* [15]; origins "point backward", whereas residuals indicate what remains of a term during rewriting. In the orthogonal case, the org^* function always yields a set consisting of at most one element.

1.4 Example

As an example, Figure 3 shows a reduction step of a typical type checker. The redex "tc($E_1 + E_2$)" is contracted at occurrence [1] in the given context. Following the definition of the function *org* just given, origins for nodes within the context are mapped onto themselves. The context positions (on top of or next to the redex) are [], [2], and [2,1], denoting "conc", "undeclared-var", and "foo". For these, we have, $org([]) = \{[]\}$, $org([2]) = \{[2]\}$, and $org([2,1]) = \{[2,1]\}$. Within the contractum, the positions corresponding to function symbol occurrences in the right-hand side obtain the empty set as origin. These positions are [1], [1,1] and [1,2], denoting "conc", "tc", and "tc" respectively, for which we have $org([1]) = org([1,1]) = org([1,2]) = \emptyset$. Finally, the origins within the contractum corresponding to variable occurrences receive an origin to the recurrences of these variables. From variable E_1 we have $org([1,1,1]) = \{[1,1,1]\}$ and $org([1,1,1,1]) = \{[1,1,1,1]\}$, and from E_2 we have $org([1,2,1]) = \{[1,1,2]\}$.

In this example, the origins are sets of at most one element. Sets with more elements can be caused by non-linearity. E.g., rule "and$(X, X) \rightarrow X$" will cause X to have origins to both occurrences of X in the left-hand side.

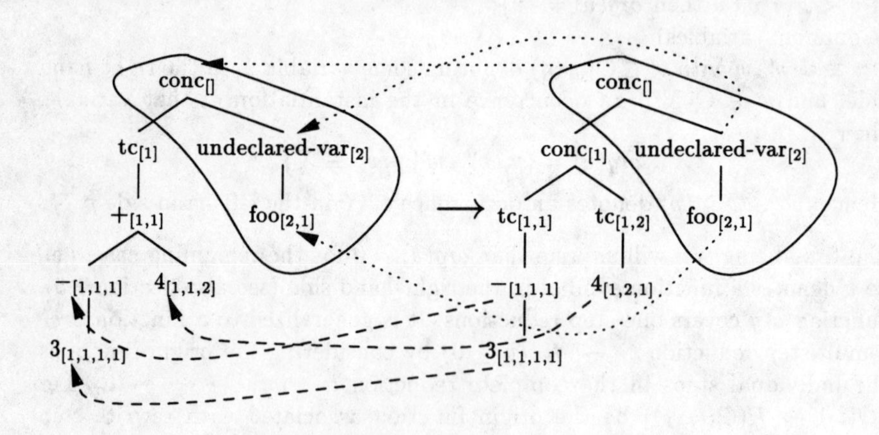

Rewrite rule: $\text{tc}(E_1 + E_2) \rightarrow \text{conc}(\text{tc}(E_1), \text{tc}(E_2))$
Substitution: $\{E_1 \mapsto -3,\ E_2 \mapsto 4\}$
Context: $\text{conc}(\Box, \text{undeclared-var}(\text{foo}))$

Dashed Lines: Origins for Common Variables
Dotted Lines: Context Origins.

Fig. 3. Origins established for one rewrite step.

1.5 Discussion

Are the origins in the previous example the ones we were looking for? The origin of "4" to "4" was good, but it is doubtful that the empty set is the best origin for the two occurrences of "tc". Here we summarize some issues we should be aware of when dealing with (extensions of) origins.

Typically, having origins based only on the Common Variables case is insufficient. These will only establish origins for literal recurrences of terms and not for any function symbols introduced. Therefore, in addition to relations based on common variables, relations following from function symbol occurrences in left- and right-hand sides of rewrite rules are needed.

Blindly relating any symbol in the right-hand side to all symbols in the left-hand side will not do either, since this would result in origin sets that are too big to give accurate information. On the other hand, it should not be too restrictive. An error message indicating a discrepancy between declaration and use of an identifier should have an origin containing at least two paths: one to the use and one to the declaration. In general, however, we will try to keep the origin sets small.

We will refer to the origins based only on Contexts / Common Variables as *primary origins.* These are clearly necessary and are useful in all applications. Moreover, we will deal with *secondary origins*, where the emphasis is on relations established due to function symbols occurring in left- and right-hand sides of rewrite rules. Proposals for secondary origins may be biased towards particular applications, with emphasis on, e.g., error handling or debugger generation.

1.6 Goal of this Paper

Recent experiments by Heering demonstrated that the use of higher-order algebraic specifications can be advantageous for the definition of programming languages [14]. These experiments, however, also revealed that rapid prototyping of these specifications using higher-order term rewriting would only be of limited use unless some form of origin tracking were available [14, Section 2.2]. Moreover, they suggest that a simple origin scheme based only on the primary origins rule would be inadequate.

This paper addresses these problems. First, we briefly summarize the definitions of higher-order rewriting in Section 2, along with a small example. Next, we present primary origins for the higher-order case in Section 3, and extensions to secondary origins in Section 4. In Sections 5 and 6 we mention related work and draw some conclusions.

2 Higher-Order Term Rewriting

For the definition of Higher-Order Term Rewriting Systems (HRSs), we follow [26, 22, 24]. The main difference from the first-order case is that terms in HRSs are constructed according to the simply-typed λ-calculus [7].

2.1 The Simply-Typed λ-Calculus

The set of *type symbols* T consists of *elementary* type symbols from T_0 and of *functional* type symbols $(\alpha \to \beta)$, where $\alpha, \beta \in T$. We may abbreviate a type $(\alpha_1 \to (\alpha_2 \to (\cdots \to (\alpha_n \to \beta)\cdots)))$ to $(\alpha_1, \ldots, \alpha_n \to \beta)$. *Terms* are built using *constants* and *variables*, each of which has an associated type symbol. The type of t is written $\tau(t)$. If x is a variable with $\tau(x) = \alpha$, and t a term with $\tau(t) = \beta$, then the *abstraction* $(\lambda x.t)$ is a term of type $(\alpha \to \beta)$. If t, t' are terms with $\tau(t) = (\alpha \to \beta)$ and $\tau(t') = \alpha$, then the *application*[4] $(t\ t')$ is a term of type β. When omitting brackets, application is left-associative.

Occurrences in λ-terms are defined as for the first-order terms, by representing abstraction as a node with 1 son and application as a node with 2 sons. As an example, Figure 4 shows all occurrences in the term (add $((\lambda N.N)$ zero) zero).

All occurrences of x in $(\lambda x.t)$ are said to be *bound*. Non-bound occurrences are *free*. A term is *closed* if it does not contain free variables, *open* otherwise. Bound variables can be renamed according to the rule of α-conversion. A *replacement* of a term t at occurrence u by subterm s is denoted by $t[u \leftarrow s]$. A *substitution* σ is a mapping from variables to terms. Application of a substitution σ to a term t, written t^σ, has the effect that all free occurrences of variables in the domain of σ are replaced by their associated term. Following the *variable convention* [2], bound variables are renamed if necessary.

Let x be a variable, t_1, t_2 terms, and let substitution $\sigma = \{x \mapsto t_2\}$. Then the term $((\lambda x.t_1)\ t_2)$ is a *β-redex* and can be transformed to t_1^σ by β-reduction. A term without β-redex occurrences is said to be in *β-normal form*. All typed λ-terms have a β-normal form, which is unique up to α-conversion. A β-normal form always has the form

$$(\lambda x_1.(\lambda x_2.\cdots(\lambda x_n.\{(\cdots((H\ t_1)\ t_2)\cdots\ t_m)\})\cdots))$$

where $x_1, \ldots, x_n$ are variables, $t_1, \ldots, t_m$ terms in β-normal form, H a constant or a variable, $m, n \geq 0$. We will sometimes write this as $\lambda x_1 \cdots x_n.H(t_1, \ldots, t_m)$. In such a term, H is called the *head*, $H(t_1, \ldots, t_m)$ is called the *matrix*, and $\lambda x_1 \cdots x_n$ is called the *binder*.

The rule of η-reduction states that terms of the form $\lambda x.(t\ x)$ can be transformed to just t, provided that x does not occur freely in t. Its counterpart is $\overline{\eta}$-expansion: if a head H of a β-normal form $\lambda x_1 \cdots x_n.H(t_1, \ldots, t_m)$ is of type $(\alpha_1, \ldots, \alpha_{m+k} \to \beta)$ $(k > 0)$, then clearly as H expects more arguments, we can add these as extra abstractions. The term above can be $\overline{\eta}$-expanded to $\lambda x_1 \cdots x_n y.H(t_1, \ldots, t_m, y)$, where y is a fresh variable of type α_{m+1}. Every term has a $\overline{\eta}$-normal form.

Let χ be any of $\{\alpha, \beta, \eta, \overline{\eta}\}$. If t can be transformed to t' by performing a χ-reduction at occurrence u, we write this as $t \triangleright_{\chi,u} t'$, or alternatively as $t' \triangleleft_{\chi,u} t$, where we may omit occurrence u. Repeated χ-reduction is written $t \triangleright_\chi^* t'$. Since

[4] We use $@(t, t')$ alternatively, when there is a need to make the application operator explicit, as in Figure 4. We also use $t(t')$ in the context of algebraic specification, as in Figure 5.

$\triangleright_\alpha^*$ is a symmetric relation, we will sometimes write it as $=_\alpha$. The $\beta\overline{\eta}$-normal form of t is indicated by $t\downarrow_{\beta\overline{\eta}}$. The relation $t =_{\beta\overline{\eta}} t'$ holds if and only if $t\downarrow_{\beta\overline{\eta}} =_\alpha t'\downarrow_{\beta\overline{\eta}}$.

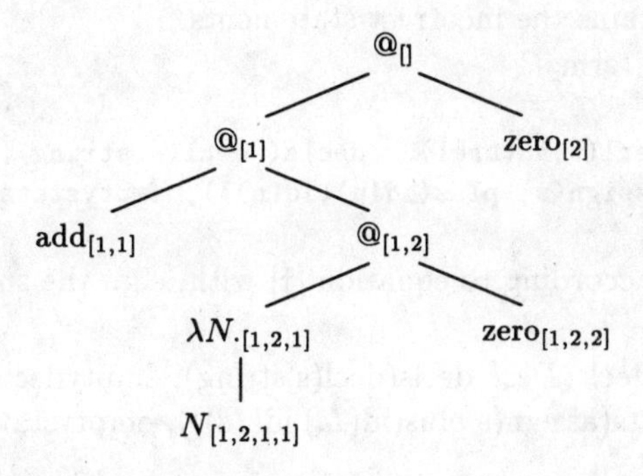

Fig. 4. Occurrences in the term "(add (($\lambda N.N$) zero)) zero".

2.2 Higher-Order Rewrite Steps

If p, q are open simply-typed λ-terms of the same type and in $\beta\overline{\eta}$-normal form, and if every free variable in q also occurs in p, then $p \rightarrow q$ is a (higher-order) rewrite rule. A reduction $t \xrightarrow{u,\sigma}_r t'$, where t, t' are closed λ-terms in $\beta\overline{\eta}$-normal form, σ is a substitution, and u is an occurrence in $\mathcal{O}(t)$ denoting the redex position, is possible if:

- The types of the redex and the left-hand side of the rule are the same: $\tau(t/u) = \tau(p)$
- The instantiated left-hand side is $\beta\overline{\eta}$-equal to the redex: $\{p^\sigma\}\downarrow_{\beta\overline{\eta}} =_\alpha \{t/u\}\downarrow_{\overline{\eta}}$
- Replacement of the redex by the instantiated right-hand side followed by $\beta\overline{\eta}$-normalization yields the result t': $\{t[u \leftarrow q^\sigma]\}\downarrow_{\beta\overline{\eta}} =_\alpha t'$

Notice the variety of $\{\alpha, \beta, \overline{\eta}\}$-conversions involved in the application of one rule. This turns out to have consequences for the definition of origins. Also note that matching the redex against a left-hand side may yield more than one substitution. For origin tracking purposes, however, we are not concerned with finding matches; we assume that in some way it has been decided to apply a rewrite rule under a given substitution (see also Section 4.3).

2.3 Example

Consider the second-order algebraic specification of a simple type checker shown in Figure 5, which was taken from [14]. The objective of this specification is to replace all simple expressions (identifiers, string or natural constants) by a term "tp(τ)", where τ is the type of that simple expression (see equations [1], [2], and [3]). Next, type correct expressions are reduced to their type (equation [4]). Finally, type correct statements are eliminated (equation [5]). The resulting normal form only contains the incorrect statements.

Take the initial term P_1:

```
program( decls( decl(n,natural),  decls( decl(s,string), emptydecls) ),
        stats( assign(s, plus(id(n),id(n))), emptystats )    )
```

It can be reduced according to equation [1] with, e.g., the substitution[5] σ_1:

$$\{\ \mathcal{D} \mapsto \lambda Decl.\ \text{decls}(Decl, \text{decls}(\text{decl}(\text{s},\text{string}), \text{emptydecls})),$$
$$\mathcal{S} \mapsto \lambda Id.\ \text{stats}(\text{assign}(\text{s},\text{plus}(\text{id}(Id),\text{id}(Id))), \text{emptystats}),$$
$$X \mapsto \text{n},$$
$$\tau \mapsto \text{natural}\quad \}$$

Applying this rule replaces occurrences of "n" by "tp(natural)", which results in a term P_2:

```
program( decls( decl(n,natural),  decls( decl(s,string), emptydecls) ),
        stats( assign(s, plus(id( tp(natural) ),
                              id( tp(natural) ))), emptystats )  )
```

Next, equation [1] can be applied again, this time replacing "s" by "tp(string)", yielding a P_3. Finally, equation [4] can be used to replace the "plus" expression by a representation of its type (natural) resulting in P_4, which is the normal form of P_1.

Initially, we are allowed to apply equation [1] on P_1, since under substitution σ_1, the left-hand side of equation [1] produces a new term P_1'', which after two β-reductions (one for $\mathcal{D}$ and one for $\mathcal{S}$) is exactly equal to term P_1.

To construct the result P_2 of this one-step reduction, we first apply σ_1 to the right-hand side of equation [1], producing some term P_2''. Then two more β-reductions transform P_2'' to its β-normal form, which results in the desired P_2. We can summarize this first single-step rewrite as follows:

$$P_1 \lhd_\beta P_1' \lhd_\beta P_1'' \equiv l_1^{\sigma_1} \rightsquigarrow r_1^{\sigma_1} \equiv P_2'' \rhd_\beta P_2' \rhd_\beta P_2$$

where $\rightsquigarrow$ denotes the replacement of the instantiated left-hand side by the instantiated right-hand side, and l_1 and r_1 are the left and right-hand side of equation [1]. Our definition of origins also follows this "flow" where origins between P_2 and P_1 are defined using elementary origin definitions between the pairs $P_2 - P_2'$, $P_2' - P_2''$, etc.

[5] It is necessary to avoid vacuous abstraction of *Decl* in the assignments of $\mathcal{D}$ [14].

sorts: PROG DECLS DECL STAT STATS ID TYPE EXP ...
functions:

program	: DECLS, STATS	→ PROG
decls	: DECL, DECLS	→ DECLS
emptydecls	:	→ DECLS
decl	: ID, TYPE	→ DECL
natural	:	→ TYPE
string	:	→ TYPE
stats	: STAT, STATS	→ STATS
emptystats	:	→ STATS
assign	: ID, EXP	→ STAT
plus	: EXP, EXP	→ EXP
id	: ID	→ EXP
nat	: NAT	→ EXP
str	: STRING	→ EXP
...		
tp	: TYPE	→ ID

variables:

$\mathcal{D}$	: DECL → DECLS	X	: ID
τ	: TYPE	$\mathcal{S}$	: ID → STATS
S	: STATS	N	: NAT
R	: STRING		

equations:

[1] $\text{program}(\mathcal{D}(\text{decl}(X,\tau)), \mathcal{S}(X)) = \text{program}(\mathcal{D}(\text{decl}(X,\tau)), \mathcal{S}(\text{tp}(\tau)))$
[2] $\text{nat}(N) = \text{id}(\text{tp}(\text{natural}))$
[3] $\text{str}(R) = \text{id}(\text{tp}(\text{string}))$
[4] $\text{plus}(\text{id}(\text{tp}(\text{natural})), \text{id}(\text{tp}(\text{natural}))) = \text{id}(\text{tp}(\text{natural}))$
[5] $\text{stats}(\text{assign}(\text{tp}(\tau), \text{id}(\text{tp}(\tau))), S) = S$

Fig. 5. Part of the static semantics specification

3 Higher-Order Origins

We define origins for higher-order rewriting by (i) indicating how origins are to be established for $\triangleright_\alpha$, $\triangleright_\beta$, $\triangleright_\eta$, and $\triangleright_{\overline{\eta}}$ conversion; then (ii) describing how the inverses $\triangleleft_\beta$ and $\triangleleft_{\overline{\eta}}$ can be dealt with; and (iii) explaining how origin relations can be set up between the left- and right-hand side of a rewrite rule. In this section we give a very basic definition, which we refer to as *primary origins*. In the next section we discuss various proposals and heuristics to extend these origins.

We use the following notational conventions. For a term t and variable x, we write $\mathcal{O}_{fvars}(t)$ for all free variable occurrences in t, $\mathcal{O}_{fvars(x)}(t)$ for the occurrences of x in t that are free, and $\mathcal{O}_{bfun}(t)$ for the application, abstraction, or constants as well as the bound variable occurrences in t. Moreover, we abbreviate occurrences of a series of n b-branches as $[b^n]$. For example, for a β-normal form $\lambda x_1 \cdots x_n.H(t_1, \ldots, t_m)$, the path to λx_j is $[1^{j-1}]$ $(1 \leq j \leq n)$ and the path to

t_i is $[1^n] \cdot [1^{m-i}] \cdot [2]$. The left side of Figure 6 shows a term in β normal form, and some path abbreviations.

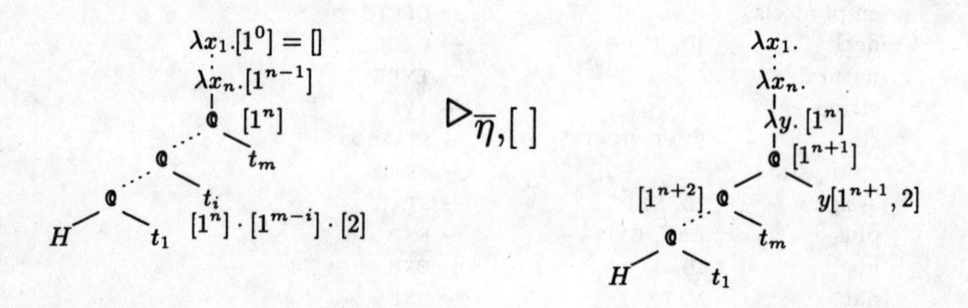

Fig. 6. $\overline{\eta}$-Expansion.

3.1 Conversions

Let t, t' be terms, $u \in \mathcal{O}(t)$, and let χ be any of $\{\alpha, \beta, \eta, \overline{\eta}\}$. Given $t \rhd_{\chi,u} t'$, we define $org(v)$ for $v \in \mathcal{O}(t')$. First, if $v \mid u$ or $v \prec u$ then $org(v) = \{v\}$. Otherwise,

- $\chi = \alpha$:
  α-Conversion does not change the term structure, so we simply have $org(v) = \{v\}$.
- $\chi = \beta$:
  Since t/u is a β-redex, we have $t/u \equiv ((\lambda x.t_1)\ t_2)$. Note that the path to t_1 is $[1, 1]$, and to t_2 is $[2]$. Now let $w_1 \in \mathcal{O}(t_1), w_2 \in \mathcal{O}(t_2)$. We distinguish two cases:
  1. $v \equiv u \cdot w_1$: Then $org(v) = \{u \cdot [1, 1] \cdot w_1\}$.
  2. $v \equiv u \cdot w_1 \cdot w_2$, and $w_2 \succ [\,]$: then $org(v) = \{u \cdot [2] \cdot w_2\}$.

     The condition $w_2 \succ [\,]$ avoids overlap with the former case.

  Thus, origins in the body t_1 "remain the same"; origins for the top node of an instantiated variable have an origin to their corresponding variable position in the body t_1, which is indicated by the dashed lines in Figure 7; and origins to non-top nodes of an instantiated variable have an origin to their position in the actual parameter t_2, which is indicated by the dotted lines.
- $\chi = \eta$:
  In η-reduction one λ is eliminated. Since t/u is an η-redex, we can assume $t/u = \lambda x.(t_1\ x)$. Realizing that the path to t_1 is $[1, 1]$, we simply have: $org(u \cdot v') = \{u \cdot [1, 1] \cdot v'\}$.
- $\chi = \overline{\eta}$:
  In η-expansion, an extra λ is added. The origins of the old parts point to the same old parts, while the origin of the new λ is the empty set:

Since t/u is an $\overline{\eta}$-redex, we have $t/u = \lambda x_1 \cdots x_n.H(t_1, \ldots, t_m)$. We distinguish three cases for $v = u \cdot v'$:

1. For $v' \preceq [1^{n-1}]$, $org(u \cdot v') = \{u \cdot v'\}$.
2. For $v' \in \{[1^n], [1^{n+1}], [1^{n+1}, 2]\}$, $org(u \cdot v') = \emptyset$.
   Figure 6 shows, using tree representations, the occurrences $[1^n]$, $[1^{n+1}]$ and $[1^{n+1}, 2]$ introduced by $\overline{\eta}$-expansion.
3. For $v' \succeq [1^{n+2}]$, $org(u \cdot [1^{n+2}] \cdot v'') = \{u \cdot [1^{n+1}] \cdot v''\}$ where $v' \equiv [1^{n+2}] \cdot v''$.

Assume that we have an origin function O mapping occurrences of t' to sets of occurrences in t. Then O is said to be *unitary* if its result values are always sets containing exactly one element, and *unique* if they contain at most one element. If an occurrence can have the empty set as origin, we say O is *forgetful*. If several occurrences in t' have an origin to the same node in t, we may refer to O as *many-to-one*, while its counterpart, where an origin set can contain more than one path, is called *one-to-many*. Finally, if for every $v \in \mathcal{O}(t')$ we have $O(v) = \{v\}$, then we say O is *identical*.

Thus, the origin function is identical for α, is unitary for η, is forgetful for $\overline{\eta}$ and finally, is unitary and many-to-one for β. None of these is one-to-many, which is fortunate, since in Section 1.5 we concluded that it was advisable to keep the origin sets small.

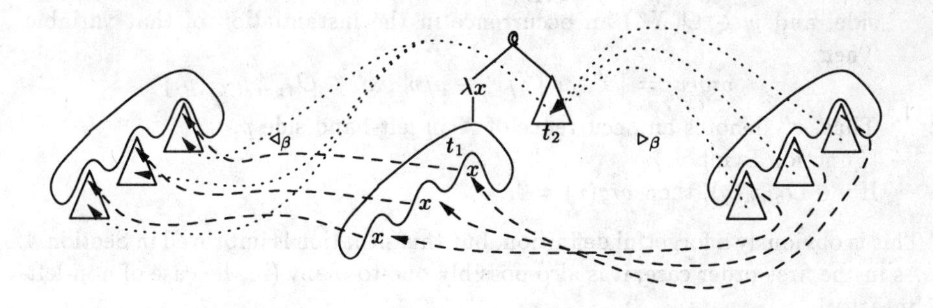

Fig. 7. β-reduction in both directions.

3.2 Equality modulo $\beta\overline{\eta}$-conversions

In Section 2.3 we discussed, reversed β and $\overline{\eta}$-reductions that need to take place. The origin functions for $\triangleright_{\{\alpha,\beta,\eta,\overline{\eta}\}}$ defined in the previous section can easily be inverted, thus yielding origin functions for $\triangleleft_{\{\alpha,\beta,\eta,\overline{\eta}\}}$. Note that, from an origin tracking point of view, the inverse of η-reduction is $\overline{\eta}$-expansion.

Since the origin function for α-conversion is identical, performing several α-conversions in one direction or another does not affect the origins. This is not the case for $\overline{\eta}$ or β reduction. Since β-reduction is many-to-one, its inverse must be one-to-many. As can be seen from Figure 7, this may lead to a growth of the origin sets. Consider a reduction $t \lhd_\beta t' \rhd_\beta t''$, where $t' = ((\lambda x.t_1)\ t_2)$, and $t, t'' = t_1^{\{x \mapsto t_2\}}$, then the origins from t'' to t' will cause all instantiated occurrences of x to be related to the same t_2 in t'; the origins of t' to t in turn will link this t_2 to all instantiated occurrences of x in t. Thus, transitively, one occurrence of t_2 in t'' has origins to *all* occurrences of t_2 in t. This is illustrated by the dotted lines of Figure 7. Note that the definition of the origin function for the β reduction (case 1), relates the *top node* of t_2, via the xs occurring in t_1 to its position in t (dashed lines of Figure 7).

Since the origins for $\overline{\eta}$ conversions are unique this problem does not arise for $\overline{\eta}$ conversions. However, the $\rhd_{\overline{\eta}}$ are forgetful, so checking for $\overline{\eta}$-equality may result in loss of some origin information (in particular in the binders).

3.3 Left- and Right-Hand Sides

We define the relations between the instantiated left and right-hand side of a rewrite rule, where we assume that these are instantiated but not yet $\beta\overline{\eta}$-normalized. We closely follow the first-order case defined in Section 1.3.

Let $p \to q$ be a rewrite rule, and σ a substitution. The function $org : \mathcal{O}(q^\sigma) \to \mathcal{P}(\mathcal{O}(p^\sigma))$, for a path $v \in \mathcal{O}(q^\sigma)$, is defined as follows:

- (Common Free Variables)
  If $v \equiv v' \cdot w$ with $v' \in \mathcal{O}_{fvars}(q)$ denoting some variable X in the right-hand side, and $w \in \mathcal{O}(X^\sigma)$ an occurrence in the instantiation of that variable. Then:
  $$org(v) = \{v'' \cdot w \mid q/v' \equiv p/v'',\ v'' \in \mathcal{O}_{fvars(X)}(p)\}$$
  Thus, v'' denotes an occurrence of X in left-hand side p.
- (Function Symbols)
  If $v \in \mathcal{O}_{bfun}(q)$, then $org(v) = \emptyset$.

This is obviously a forgetful definition, but this situation is improved in Section 4. As in the first-order case, it is also possibly one-to-many (in the case of non-left-linearity).

Note that the common free variables case results in the same origins as in the common variables case of Section 1.3, when the specification does not use the higher-order features. The Context case will be dealt with in the next section.

3.4 Rewrite Steps

Knowing how to both establish origins for α-, β-, and $\overline{\eta}$-conversions in either direction and to set up origins between the instantiated left- and right-hand side, we can obtain the origins for one complete reduction step $t_1 \to t_2$. Figure 8 summarizes the work to be done for one reduction, following as described in Section 2.2.

Note that in general the situation is slightly more complicated than in the example of Section 2.3

$$P_1 \lhd_\beta P_1' \lhd_\beta P_1'' \equiv l_1^{\sigma_1} \rightsquigarrow r_1^{\sigma_1} \equiv P_2'' \rhd_\beta P_2' \rhd_\beta P_2$$

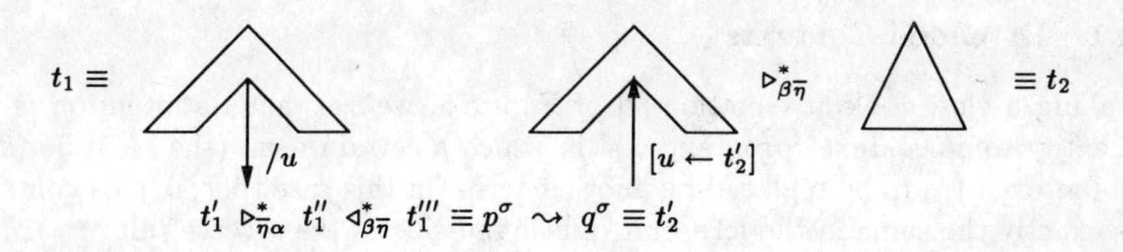

Fig. 8. All conversions for one reduction step $t_1 \rightarrow t_2$, applying rule $p \rightarrow q$ at occurrence u in t_1 under substitution σ.

where the rewrite rule is applied at the root of P_1 which has the effect that Figure 8 can be reduced to just "one level": The context is empty ($u = []$), and consequently the term t/u is already a $\overline{\eta}$-normal form, hence the result need not be put back into the context (in the figure, $[\,[] \leftarrow t'_2]$ is just equal to t'_2).

3.5 Example

Consider reduction $P_1 \rightarrow P_2$ as presented in Section 2.3. Most occurrences in P_2 have their intuitive origin; mainly because they also occur in bodies of the instantiations of $\mathcal{D}$ and $\mathcal{S}$ in substitution σ_1. However, some origins are lost; in particular for nodes occurring in the right-hand side of rule [1]. Thus, symbols "program", "decl" (for the declaration of n), and "tp" do not have an origin. Moreover, rule [1] is non-linear in X, and therefore the X-occurrence in the declaration at the right-hand side has an origin to the occurrence in the statement as well as in the declaration. Thus, the single n in P_2 has origins to all n occurrences in P_1 (this does not seem intuitive). All occurrences of "natural" in P_2 have their origin to the declaration it came from (seems reasonable).

Now consider the entire reduction $P_1 \rightarrow^* P_4$, where normal form P_4 is:

```
program( decls( decl(n,natural),  decls( decl(s,string), emptydecls) ),
         stats( assign( tp(string),
                   plus(id( tp(natural) ),id( tp(natural) ))),
              emptystats )  )
```

In this case, more origins are lost. In particular, the two "decl" nodes have an empty origin, and the reduction according to rule [4] did not establish any origins, so "tp(natural)" does not have any origins.

4 Extensions

The origins in the previous example were nice, but still not sufficient for using them in practice. In this section we present some extensions of the origin function. Some of these extensions are of a heuristic nature, based on frequently occurring forms of (higher-order) rewrite rules.

4.1 Extended Contexts

Taking a close look at equation [1] of Figure 5, we see that its intention is to identify some context "program(...)" in which a certain term (the identifier denoted by X) is to be replaced by another term (in this case tp(τ)). This context is exactly the same in the left- and right-hand side of the rewrite rule.

It seems reasonable to extend the notion of a context to cover such similarities within rewrite rules as well. Considering a rewrite rule $p \to q$, we can look for a (possibly empty) *common context* C and *holes* (terms) $h_1, \ldots, h_m$ and $h'_1 \ldots h'_m$ ($m \geq 0$) such that $p =_\alpha C[h_1, \ldots, h_m]$ and $q =_\alpha C[h'_1, \ldots, h'_m]$, where $h_j \neq_\alpha h'_j$ for all $1 \leq j \leq m$. We are actually looking for the biggest of such contexts which contain the smallest possible number of holes where none of the holes h_j, h'_j ($1 \leq j \leq m$) start with a non-empty context $\overline{C}$ such that $h_j =_\alpha \overline{C}[\overline{h}_1, \ldots \overline{h}_n]$ and $h'_j =_\alpha \overline{C}[\overline{h'}_1, \ldots \overline{h'}_n]$. As an example, equation [1] of Figure 5 has a common context $C =$ "program($\mathcal{D}$(decl(X,τ)), $\mathcal{S}$($\Box$))", where the hole h_1 at the left is equal to "X", and h'_1 at the right to "tp(τ)".

For every node in this extended context, the origin should point only to its corresponding occurrence in that same context at the left-hand side. Note that, as a consequence, the common variables case should *not* apply to variables occurring in the common context. For example, in equation [1], the origin of X at the right will only point to its counterpart under the "decl" at the left and not to the X in the statements. Moreover, when trying to find origins for a node in a hole h'_j, it seems reasonable to focus on origins that can be found within the corresponding hole h_j. Only if it is impossible to find origins there, an origin can be looked for in the rest of the left-hand side.

There is, however, a minor catch in this. If two consecutive holes h_j and h_{j+1} are only separated by an application in the context C, i.e. they actually occur as $@(h_j, h_{j+1})$ at the left and as $@(h'_j, h'_{j+1})$ at the right, then it is more natural to regard these two as one hole ($H = @(h_j, h_{j+1})$ instead of h_j and h_{j+1}). As an example, equation [2] in applicative form reads as @(nat,N) = @(id, @(tp, natural)). It would be counter-intuitive to regard the top-application as a common context $@(\Box, \Box)$ with two holes: $h_1 =$ nat, $h'_1 =$ id, and $h_2 = N$, $h'_2 =$ @(tp, natural).

Note that this new extended context case would be useful in the first-order case as well.

4.2 Origins for Constants

Let $p \equiv C[h_1, \ldots, h_m] \to C[h'_1, \ldots, h'_m] \equiv q$ be a rewrite rule with the common context C and m ($m \geq 0$) holes. We define origins for constants occurring in the h'_j ($1 \leq j \leq m$) according to the following three cases:

1. Head-to-Head
   The origin for the occurrence of the head symbol of a hole h'_j at the right is the occurrence of the head symbol of that same hole h_j at the left. For example, the "tp" symbol in equation [1] is linked to the occurrence of X in

the statements at the left. This head-to-head rule corresponds to the "redex-contractum" rule of the first-order origins as described in [10]. Note that if the head symbol at the right is a free variable, the common variables case is applicable as well. This can, in general, have the effect that the origin set for the head symbols consist of more than one path.

2. Common Subterms.
If a term s is a subterm of both h'_j and h_j, then these occurrences of s are related. For example, the subterm "tp(natural)" at the right of equation [4] (Figure 5) is related to both occurrences of "tp(natural)" at the left. Note that these common subterms are identified in the *un-instantiated* left- and right-hand side. This rule can in some cases lead to seemingly wild connections, but has already proven its usefulness for the first-order case [10, 11].
The common subterms behave slightly different in the higher-order case, due to the applicative form of the λ-terms. In the first-order case, function symbols were only related if all arguments were identical at the left and right. In the higher-order case, function symbols are constants. Each constant F in h'_j is related to all occurrences of F in h_j. This effect is similar to the *tokenization* discussed in [11].
If for a subterm s of h'_j no occurrences of s can be found in h_j, then the entire left-hand side p can be used to find a common subterm occurrence of s.
3. Any to All.
If after application of the head-to-head and common subterms case there are still constants in h'_j with an empty origin, the set of all constant occurrences at the left-hole h_j is defined as its origin set. For example, in equation [2], the subterms "tp(natural)" and "natural" relate to both "nat(N)" and "N".

4.3 Abstraction and Concretization Degree

Let us end our discussion with an interesting observation. Recall from Section 3.2 that $\triangleleft_\beta$ conversions are one-to-many. Assume that $t' \triangleleft_\beta t$ with $t \equiv ((\lambda x.t_1)\ t_2)$. It would be useful to call the number of free occurrences of x in t_1 the *abstraction degree* of $\lambda x.t_1$, and the number of occurrences of term t_2 in t_1 the *concretization degree*. When trying to find a matching substitution σ in order to apply a rewrite rule, freedom exists concerning the abstraction and concretization degree. For example, if σ assigns F a value T with abstraction degree $N > 0$ and concretization degree $M \geq 0$, then an alternative match σ' can also be possible which assigns F a term T' with abstraction degree $N - 1$ and concretization degree $M + 1$. The problems with $\triangleleft_\beta$ are minimized if matches with abstraction degree 1 are preferred over those with a higher abstraction degree.

In practice, however, such a preference may be somewhat problematic. Firstly, a substitution with a lower degree of abstraction may not even exist. Secondly, the repeated application of a substitution with abstraction degree 1 need not yield the same result as a single application with a higher abstraction degree. Finally, repeated applications may be more expensive in terms of run time behavior, than a single application with a high abstraction degree.

4.4 Example

With these extensions, suitable origins for the example in Section 2.3 are obtained. We assume that equation [1] is applied with substitutions of abstraction degree 1 only. The extended contexts assure that "program" and "decl" are linked. Moreover, the effect of linking variables in contexts only to the same occurrence in the context, guarantees that the n and s in the declaration have the proper unitary origin. Furthermore, relating heads of holes guarantees that the "tp" nodes get the right origin to the variable they were substituted for. Likewise, the application of equation [4] results in "plus" as the origin of "tp". Finally, common subterms results in "tp(natural)" to be linked to both occurrences of "tp(natural)" in the "plus" expression (equation [4]).

The example given here is only part of the specification discussed in [14]. The origins with extensions create the proper relations for the full specification as well.

5 Related Work

The current document is part of a series of papers studying origins and their applications to the automatic generation of parts of compilers or programming environments – in particular error handlers, symbolic debuggers, and animators. The extensions to primary origins studied in [10] establishes relations between *common subterms* in left and right-hand side of rewrite rules, as well as a link between the top-node of the *redex* and the *contractum*. Moreover, origins are defined for *conditional* rewrite rules. Several issues related to the efficient implementation of origin tracking in the ASF+SDF Meta-Environment [17] are discussed in [10]. The applicability of origins in practice, using a specification of the semantics of a subset of Pascal, is studied by Dinesh and Tip where the static semantics and generated error handler is covered in [11] and the dynamic semantics and generated animator is described in [25]. In order to improve origin tracking for *syntax-directed* specifications (typically translators or type checkers), an extension for *primitive recursive schemes* is proposed in [9]. An origin-like relation, called *dynamic-dependence* relation is studied by Field and Tip [12]. They show that the dependence tracking technique is useful in the context of program slicing.

The study of origins was pioneered by Bertot [4, 5], who was concerned with origins in natural semantics, (orthogonal) term rewriting, and the (untyped) λ-calculus. He describes a language for the definition and representation of origins. In his setting, origins are unitary (consisting of at most one path). Secondary origins are represented by *marking functions*. This work was done in the framework of the CENTAUR system [6]. In particular, the specification language TYPOL [16] has been extended with *subject tracking* [8].

Closely related to origins are *residual maps*, *descendants*, or *labelings* [20, 15, 21, 13], which are used to study reduction strategies. Residuals indicate which redexes survive if a particular redex is contracted. One can think of this as giving interesting parts in the initial term a particular color, and then looking how this color survives during reduction. An interesting combination of origins and labeling systems is presented by Bertot [5] where he investigates how origins for TRSs can be used to simulate labeling systems for the λ-calculus. The labels of [20] suggest that alternative representations for origins containing more structure than the (simple) sets of paths could be fruitful.

Nipkow's definition of higher-order TRSs requires the rewrite rules to satisfy several syntactic constraints [22]. We have discussed origins using the more liberal setting of Wolfram [26]. Obviously, the same origins can be established for Nipkow's HRSs. The nicer matching behavior of Nipkow's HRSs will probably have a favorable effect on the origins. The mapping between Nipkow's HRSs and Klop's combinatory reduction systems (CRSs) [18] as described in [24] can be the basis for a definition of origins for CRSs.

Another issue is the study of origins as transformations on HRSs. Tip has conducted such experiments for the first-order case. For the higher-order case, it may be useful to use specifications of the λ-calculus with explicit substitutions as in [1].

6 Conclusions

Origin tracking for higher-order specifications is considerably more difficult than establishing origin relations for the first-order case. Various conversions to be performed, both as reductions and as expansions, have to be taken into account. Nevertheless, we have found a satisfactory origin scheme, which is applicable to arbitrary higher-order term rewriting systems

There is, however, still some future work to do. The most important thing is to gain experience with these origins. More specifications of realistic problems and their applicability for origin tracking should be studied.

Finally, after having seen many variants of origin tracking, it may be worthwhile to investigate the possibility of generalizing to some kind of origin *scheme*. This may clarify and ease future discussions of further extensions of origin tracking.

Acknowledgments This paper would not have been possible without Jan Heering's support and advice. Comments of Jan Heering, Femke van Raamsdonk, Susan Üsküdarlı, Machteld Vonk and the reviewers have helped improve the presentation.

References

1. M. Abadi, L. Cardelli, P.-L. Currien, and J.-J. Lévy. Explicit substitutions. In *Proceedings of the 17th conference on Principles of Programming Languages*, pages 31–46, 1990.
2. H.P. Barendregt. *The Lambda Calculus; its Syntax and Semantics*, volume 103 of *Studies in Logic and the Foundations of Mathatematics.* North-Holland, 1984.

3. J.A. Bergstra, J. Heering, and P. Klint, editors. *Algebraic Specification.* ACM Press Frontier Series. The ACM Press in co-operation with Addison-Wesley, 1989.
4. Y. Bertot. *Une Automatisation du Calcul des Résidus en Sémantique Naturelle.* PhD thesis, INRIA, Sophia-Antipolis, 1991. In French.
5. Y. Bertot. Origin functions in lambda-calculus and term rewriting systems. In J.-C. Raoult, editor, *Proceedings of the 17th Colloquium on Trees in Algebra and Programming (CAAP '92)*, volume 581 of *LNCS*. Springer-Verlag, 1992.
6. P. Borras, D. Clément, Th. Despeyroux, J. Incerpi, B. Lang, and V. Pascual. CENTAUR: the system. In *Proceedings of the ACM SIGSOFT/SIGPLAN Software Engineering Symposium on Practical Software Development Environments*, pages 14–24, 1989. Appeared as *SIGPLAN Notices* 24(2).
7. A. Church. A formulation of a Simple Theory of Types. *Journal of Symbolic Logic*, 5:56–68, 1940.
8. Th. Despeyroux. Typol: a formalism to implement natural semantics. Technical Report 94, INRIA, 1988.
9. A. van Deursen. Origin tracking in primitive recursive schemes. In H.A. Wijshoff, editor, *Conference Proceedings Computing Science in the Netherlands CSN'93*, pages 132–143, 1993.
10. A. van Deursen, P. Klint, and F. Tip. Origin tracking. *Journal of Symbolic Computation*, 15:523–545, 1993. Special Issue on Automatic Programming.
11. T.B. Dinesh. Type checking revisited: Modular error handling. In *Proceedings of the Workshop on Semantics of Specification Languages*, Utrecht, 1993. Springer-Verlag, LNCS. To Appear.
12. J. H. Field and F. Tip. Dynamic dependence in term rewriting systems and its application to program slicing. Technical report, Centrum voor Wiskunde en Informatica (CWI), 1994. To appear.
13. J.H. Field. *Incremental Reduction in the Lambda Calculus and Related Reduction Systems.* PhD thesis, Cornell University, 1991.
14. J. Heering. Second-order algebraic specification of static semantics. Technical Report CS-R9254, Centrum voor Wiskunde en Informatica (CWI), 1992. Extented version to appear, 1994.
15. G. Huet and J.-J. Lévy. Computations in orthogonal rewriting systems part I and II. In J.-L. Lassez and G. Plotkin, editors, *Computational Logic; essays in honour of Alan Robinson*, pages 395–443. MIT Press, 1991.
16. G. Kahn. Natural semantics. In F.J. Brandenburg, G. Vidal-Naquet, and M. Wirsing, editors, *Fourth Annual Symposium on Theoretical Aspects of Computer Science*, volume 247 of *LNCS*, pages 22–39. Springer-Verlag, 1987.
17. P. Klint. A meta-environment for generating programming environments. *ACM Transactions on Software Engineering and Methodology*, 2(2):176–201, 1993.
18. J.W. Klop. *Combinatory Reduction Systems.* Number 127 in Mathematical Center Tracts. Mathematisch Centrum, Amsterdam, 1980.
19. J.W. Klop. Term rewriting systems. In S. Abramsky, D. Gabbay, and T. Maibaum, editors, *Handbook of Logic in Computer Science, Volume 2. Background: Computational Structures*, pages 1–116. Oxford University Press, 1992.
20. J.-J. Lévy. An algebraic interpretation of the $\lambda\beta$K-calculus and a labelled λ-calculus. In C. Böhm, editor, *λ-Calculus and Computer Science Theory*, number 37 in LNCS. Springer-Verlag, 1975.
21. L. Maranget. Optimal derivations in weak lambda-calculi and in orthogonal term rewriting systems. In *Proceedings of the Eighteenth conference on Principles of Programming Languages POPL '91*, pages 225–269, 1991.

22. T. Nipkow. Higher-order critical pairs. In *Proceedings of the Sixth Annual IEEE Symposium on Logic in Computer Science*, pages 342–349. IEEE Computer Society Press, 1991.
23. V. van Oostrom. *Confluence for Abstract and Higher-Order Rewriting.* PhD thesis, Vrije Universiteit, Amsterdam, March 1994.
24. V. van Oostrom and F. van Raamsdonk. Comparing combinatory reduction systems and higher-order rewrite systems, 1994. This proceedings.
25. F. Tip. Animators for generated programming environments. In P. Fritzson, editor, *Proceedings of the First International Workshop on Automated and Algorithmic Debugging AADEBUG'93*, LNCS. Springer-Verlag, 1993.
26. D.A. Wolfram. *The Clausal Theory of Types*, volume 21 of *Cambridge Tracts in Theoretical Computer Science.* Cambridge University Press, 1993.

Theory Interpretation in Simple Type Theory*

William M. Farmer

The MITRE Corporation
202 Burlington Road
Bedford, MA 01730-1420, USA

farmer@mitre.org

Abstract. Theory interpretation is a logical technique for relating one axiomatic theory to another with important applications in mathematics and computer science as well as in logic itself. This paper presents a method for theory interpretation in a version of simple type theory, called LUTINS, which admits partial functions and subtypes. The method is patterned on the standard approach to theory interpretation in first-order logic. Although the method is based on a nonclassical version of simple type theory, it is intended as a guide for theory interpretation in classical simple type theories as well as in predicate logics with partial functions.

1 Introduction

Theory interpretation—in which one theory is interpreted in another via a syntactic mapping—is a fundamental logical technique which has important applications in mathematics and computer science as well as in logic itself. An *interpretation*[1] of a theory[2] T_1 in a theory T_2 is a mapping from the expressions of T_1 to the expressions of T_2 which preserves the validity of sentences. (T_1 and T_2 are called the *source theory* and the *target theory* of the interpretation, respectively.) In logic, interpretations are used to prove metamathematical properties about theories and to compare theories in terms of their "strength". In mathematics, theorems and problems are transported from one context to another via interpretations. In computer science, interpretations are a rigorous tool for documenting and verifying that one system specification is a refinement of another.

Until recently, interpretations have been almost exclusively employed by theoreticians. However, implementors are now discovering that interpretations are useful for organizing and supporting mathematical reasoning in automated reasoning systems such as mechanical theorem provers and computer system specification and verification environments. Interpretations are used extensively with

* Supported by the MITRE-Sponsored Research program.

[1] Theory interpretations of this kind are also called *translations*, *theory morphisms*, *immersions*, and *realizations*.

[2] We take a *theory* to be a set of sentences in a formal language (that is not necessarily closed under logical consequence). The sentences are called the *axioms* of the theory.

success in the IMPS Interactive Mathematical Proof System [10, 11, 12]. They are also a fundamental component in the following programming and verification environments: EHDM [27], m-EVES [5] and EVES [6], IOTA [24], and OBJ3 [14].

Theory interpretation has primarily been studied and applied in the context of first-order predicate logic. Logic textbooks like Enderton [7], Monk [22], and Shoenfield [28] present a fairly standard approach to theory interpretation in first-order logic. The approach revolves around a special class of interpretations that are well behaved both syntactically and semantically. Suppose Φ is an interpretation of T_1 in T_2 which is in this class. Then Φ will be a kind of homomorphism which preserves the structure of terms and formulas and which is completely determined by how it associates the sorts (if there are any) and constants of T_1 with objects of T_2. Moreover, Φ will define a way of extracting a model for T_1 from any model for T_2.

Although there is a wealth of writing on theory interpretation in first-order logic, the subject is only beginning to be seriously explored in higher-order logic (and type theory) [9, 17, 36]. There are, however, at least two good reasons to study theory interpretation in higher-order logic. First, higher-order logic is becoming increasingly important in computer science and mechanized mathematics. Second, since higher-order logic is much more expressive than first-order logic, the space of interpretations is much richer in higher-order logic than in first-order logic. This means that some techniques based on theory interpretation are more powerful in a higher-order logic than in a first-order logic (e.g., the technique of verifying that a theory T' is a model conservative extension of T by exhibiting an interpretation of T' in T which fixes T).

The most well-known and widely used form of higher-order logic is simple type theory [4, 1]. Since it has built-in support for functions—a hierarchy of function types, full quantification over functions, and (usually) λ-notation for specifying functions—it is a convenient logic for formalizing mathematics. For this reason, it is the logical basis for several automated reasoning systems, including EHDM, HOL [15], IMPS, Isabelle [25], PVS [26], and TPS [2]. In spite of its popularity and utility, there is not a well-developed approach to theory interpretation in simple type theory.

The goal of this paper is to develop a method for theory interpretation in simple type theory patterned on the standard first-order approach. We want the method to handle interpretations in which a base type (i.e., a type of individuals) of the source theory can be associated with either a (possibly higher-order) *type* or *subtype* of the target theory.

In first-order logic, an interpretation which associates base types with types is merely an interpretation which associates the universe (i.e., the implicit type of individuals) of the source theory with the universe of the target theory. An interpretation of this kind does not alter the quantifiers in expressions of the source theory. An interpretation which associates base types with subtypes is one which associates the universe of the source theory with a unary predicate of the target theory. An interpretation of this kind "relativizes" the quantifiers in expressions of the source theory. For example, if Φ is an interpretation which associates the universe of the source theory with the predicate φ, then

$$\Phi((\forall x)\psi) = (\forall x)(\varphi(x) \rightarrow \Phi(\psi)).$$

Many natural theory interpretations associate a base type with a subtype (i.e., part of a type). For example, suppose G is a theory of an abstract group in which α is a base type denoting the set of group elements; F is a theory of an abstract field in which β is a base type denoting the set of field elements; and Φ is the interpretation of G in F in which the group structure of G is "interpreted" as the structure of the multiplicative group of F. Then Φ would associate α with the subtype of β consisting of the nonzero field elements. Moreover, the most natural translation of the group operation of G via Φ would be an expression denoting the multiplication operation of F restricted to the nonzero field elements. Thus we see that associating base types with subtypes leads to functions with restricted domains. (This example is worked out in detail in Section 7.)

If only interpretations which associate base types with full types are considered, it is easy to lift the first-order notion of a theory interpretation to simple type theory. On the other hand, associating base types with subtypes is messy in simple type theory since one must deal with functions with restricted domains, as we have seen above. Restricting the domain of a function is unproblematic in informal mathematics, but there is no completely satisfactory way that it can be done in classical predicate logic since expressions cannot directly denote partial functions. (See [8] for a discussion on the various ways of dealing with partial functions in predicate logic.) In first-order logic, partial functions are avoided by relativizing quantifiers. This approach would be more complicated in simple type theory because more than just quantifiers would have to be relativized; in particular, all predicates on functions (such as those corresponding to universal and existential quantification) would have to be relativized.

Our method for theory interpretation is formulated in a version of simple type theory, called LUTINS[3] [8, 9, 16], which supports both partial functions and subtypes. We have chosen LUTINS over a classical simple type for three *pragmatic* reasons. First, as we have pointed out, partial functions naturally arise from interpretations that associate base types with subtypes. Consequently, interpretations of this kind can be formalized more directly in a logic which admits partial functions like LUTINS. Second, since LUTINS contains subtypes, an interpretation in LUTINS does not have to relativize quantifiers and other variable binders, provided appropriate subtypes are defined. Finally, as the logic of the IMPS interactive theorem proving system, LUTINS has been implemented and rigorously and extensively tested [12]. It is clearly an effective logic for formalizing a wide range of mathematics. Although the method is based on a nonclassical form of simple type theory, we expect it to be useful as a guide for theory interpretation in classical simple type theories as well as in predicate logics which admit partial functions.

The paper is organized as follows. An overview and discussion via examples of the standard approach to theory interpretation in first-order logic is given in

[3] Pronounced as the word in French.

Sections 2 and 3. Section 4 gives a quick introduction to **PF***, an austere version of simple type theory with partial functions and subtypes on which LUTINS is based. The syntax and semantics of LUTINS are then presented in Section 5. The notion of a theory interpretation in LUTINS is defined in Section 6. Section 7 contains some examples in LUTINS of interpretations of groups in fields. The interpretation and relative satisfiability theorems for LUTINS are proved in Section 8. And a brief conclusion is found in Section 9.

Comparisons between our method of theory interpretation and the standard approach in first-order logic are made at several places in the paper.

2 Theory Interpretation in First-Order Logic

This section presents an outline of the standard approach to theory interpretation in first-order logic [7, 22, 28]. For the most part, we shall adopt in this section the definitions and notation of first-order logic (with equality) presented in [3]. An *expression* of a first-order language $\mathcal{L}$ or theory T is a term or a formula of $\mathcal{L}$ or T. An *n-ary expression function* is a λ-expression of the form $\lambda\{x_1,\ldots,x_n\,.\,E\}$ where E is an expression. Let $\theta = \lambda\{x_1,\ldots,x_n\,.\,E\}$ be an expression function. θ is a *term* [respectively, *formula*] *function* if E is a term [respectively, formula]. Given terms $t_1,\ldots,t_n$, $\theta(t_1,\ldots,t_n)$ denotes the result of simultaneously substituting t_i for all free occurrences of x_i in E, for all i with $1 \le i \le n$.

Let T_i be a first-order theory for $i = 1, 2$. A *standard translation from T_1 to T_2* is a pair (U, ν) where U is a closed formula function of the form $\lambda\{x\,.\,\varphi\}$ which represents a unary predicate and ν is a function from the nonlogical constants of T_1 to the nonlogical constants, expressions, and expression functions of T_2 such that:

1. If c is an individual constant symbol of T_1, then $\nu(c)$ is either an individual constant symbol or a closed term.
2. If F is an n-ary function symbol of T_1, then $\nu(F)$ is either an n-ary function symbol or a closed n-ary term function.
3. If P is an n-ary relation symbol of T_1, then $\nu(P)$ is either an n-ary relation symbol or a closed n-ary formula function.
4. $\nu(\equiv) = \equiv$.[4]

Let $\Phi = (U, \nu)$ be a standard translation from T_1 to T_2 throughout the rest of this section. For an expression E of T_1, the *translation of E via Φ*, written $\Phi(E)$, is the expression of T_2 defined inductively by:

1. $\Phi(x) = x$, if x is a variable.
2. $\Phi(c) = \nu(c)$, if c is an individual constant symbol.
3. $\Phi(S(t_1,\ldots,t_n)) = \nu(S)(\Phi(t_1),\ldots,\Phi(t_n))$, if S is an n-ary function or relation symbol.
4. $\Phi(\neg\varphi) = \neg\Phi(\varphi)$.

[4] In [3], the binary relation symbol $\equiv$ denotes the equality relation.

5. $\Phi(\varphi \mathbin{\Box} \psi) = \Phi(\varphi) \mathbin{\Box} \Phi(\psi)$, if $\Box \in \{\wedge, \vee, \rightarrow, \leftrightarrow\}$.
6. $\Phi((\Box x)\varphi) = (\Box x)\Phi(\varphi)$, if $\Box \in \{\forall, \exists\}$ and $U = \lambda\{x \,.\, x \equiv x\}$.
7. $\Phi((\Box x)\varphi) = \begin{cases} (\forall x)(U(x) \rightarrow \Phi(\varphi)) \text{ if } \Box = \forall \\ (\exists x)(U(x) \wedge \Phi(\varphi)) \text{ if } \Box = \exists \end{cases}$
   if $U \neq \lambda\{x \,.\, x \equiv x\}$.

A standard translation thus associates the universe of its source theory with a closed unary predicate of its target theory; the nonlogical constants of its source theory with closed expressions (of appropriate "type") of its target theory; and the variables and logical connectives with themselves. The quantifiers are relativized to the unary predicate if it is not $\lambda\{x \,.\, x \equiv x\}$. Except for the relativization of quantifiers, a standard translation preserves the structure of first-order syntax. Hence, a standard translation can be viewed as a homomorphism from the expressions of its source theory to the expressions of its target theory.

Φ is a *standard interpretation of T_1 in T_2* if $\Phi(\varphi)$ is valid in T_2 for each sentence φ which is valid in T_1. That is, Φ is an interpretation if it maps valid sentences to valid sentences. The theorem below gives a sufficient condition for a standard translation to be a standard interpretation.

An *obligation* of Φ is any one of the following sentences of T_2:

1. $\Phi(\varphi)$ for each axiom φ of T_1.
2. $(\exists x)U(x)$.
3. $\Phi((\exists y)c \equiv y)$ for each individual constant symbol c of T_1.
4. $\Phi((\forall x_1 \cdots x_n)(\exists y)F(x_1, \ldots, x_n) \equiv y)$ for each function symbol F of T_1.

The four kinds of obligations are called, in order, *axiom*, *universe nonemptiness*, *individual constant symbol*, and *function symbol* obligations. The meaning of an individual constant symbol obligation is that the interpretation of the universe contains the interpretation of the individual constant symbol, and the meaning of a function symbol obligation is that the interpretation of the universe is closed under the interpretation of the function symbol. Note: The last three kinds of obligations are trivially valid in T_2 if $U = \lambda\{x \,.\, x \equiv x\}$.

Theorem 2.1 (Standard Interpretation Theorem) *A standard translation from T_1 to T_2 is a standard interpretation if each of its obligations is valid in T_2.*

T_1 is *interpretable in T_2 (in the standard sense)* if there is a standard interpretation of T_1 in T_2. The next theorem is the most important consequence of interpretability.

Theorem 2.2 (Standard Relative Satisfiability) *If T_1 is interpretable in T_2 and T_2 is satisfiable, then T_1 is satisfiable.*

The key idea in the proof of this theorem is to use the standard interpretation of T_1 in T_2 to extract a model of T_1 from a model of T_2. The Standard Interpretation Theorem and the Standard Relative Satisfiability theorem are the chief theorems of the standard approach to theory interpretation in first-order logic.

By virtue of being a validity preserving homomorphism, a standard interpretation syntactically and semantically embeds its source theory in its target theory. Standard interpretations are used to compare the strength of theories: T_2 is at least as strong as T_1, if T_1 is interpretable in T_2. Also, standard interpretations have long been used in logic to prove metamathematical properties about first-order theories, mainly relative consistency, decidability, and undecidability. For example, the classic work of Tarski, Mostowski, and Robinson [32] illustrates how the undecidability of T_1 can be reduced to the undecidability of T_2 by constructing an appropriate standard interpretation of T_2 in T_1. For other references on the theory and use of standard interpretations, see [13, 23, 30, 31, 34].

3 Some Simple Examples

This section contains three examples of standard first-order interpretations. Although the examples are very simple, they illustrate some of the power and versatility of theory interpretation.

The first example is an interpretation of a theory of an abstract nonstrict partial order in a theory of an abstract strict total order.

Example 3.1 Let PO be the theory consisting of the following three sentences in the first-order language of a binary relation symbol $\leq$:

1. *Reflexivity.* $(\forall x)(x \leq x)$.
2. *Transitivity.* $(\forall xyz)((x \leq y \land y \leq z) \rightarrow x \leq z)$.
3. *Antisymmetry.* $(\forall xy)((x \leq y \land y \leq x) \rightarrow x \equiv y)$.

PO clearly specifies $\leq$ to be a nonstrict partial order.

Similarly, let TO be the theory consisting of the following three sentences in the first-order language of a binary relation symbol $<$:

1. *Irreflexivity.* $(\forall x)\neg(x < x)$.
2. *Transitivity.* $(\forall xyz)((x < y \land y < z) \rightarrow x < z)$.
3. *Trichotomy.* $(\forall xy)(x < y \lor y < x \lor x \equiv y)$.

TO clearly specifies $<$ to be a strict total order.

Let Φ_1 be the standard translation (U_1, ν_1) from PO to TO where $U_1 = \lambda\{x \,.\, x \equiv x\}$ and $\nu_1(\leq) = \lambda\{x, y \,.\, (x < y \lor x \equiv y)\}$. Φ_1 is clearly an interpretation of PO in TO by the Standard Interpretation Theorem.□

Φ_1 links two *different* axiomatizations—different axioms in different formal languages—of the same mathematical domain (i.e., the notion of an order relation). Via Φ_1, theorems proved in PO about the nonstrict partial order $\leq$ can be transformed into corresponding theorems in TO about the strict total order $<$. In other words, Φ_1 serves as a conduit through which theorems can be freely "transported" from one mathematical formalization to another. Interpretations are used in this manner, at least informally, in many areas of mathematics to transport theorems from abstract to more concrete contexts or to move a problem to a more convenient setting (for examples, see [18]). Moreover, interpretations of this sort are fundamental to the "little theories" version of the axiom method in which mathematical reasoning is performed over a network of theories linked by interpretations—instead of entirely within one single "big theory" such as Zermelo-Fraenkel set theory. (In [11], we describe the little theories approach and argue in favor of its implementation in mechanical theorem provers.)

Φ_1 also establishes that TO is a "refinement" of PO. Actually, TO refines PO in two ways: (1) the models of TO are a special kind of partial order, namely a total order, and (2) the atomic relation $\leq$ is decomposed into a compound relation built from $<$ and $\equiv$ via Φ_1. (This latter sort of refinement is much like a definition done in reverse.) Refinement is a basic technique in computer science for specifying and building computer systems. Theory interpretation is an excellent tool for rigorizing computer system development based on refinement (for examples, see [20, 21, 29, 33, 35]).

The interpretation in the next example formalizes the symmetry between left and right multiplication in a monoid.

Example 3.2 Let M be the theory consisting of the following three sentences in the first-order language of a binary function constant $*$ and an individual constant e:

1. *Associativity.* $(\forall xyz)((x * y) * z \equiv x * (y * z))$.
2. *Left Identity.* $(\forall x)(e * x \equiv x)$.
3. *Right Identity.* $(\forall x)(x * e \equiv x)$.

M is a very standard formulation of a theory of an abstract monoid.

Now let Φ_2 be the standard translation (U_2, ν_2) from M to itself where $U_2 = \lambda\{x \,.\, x \equiv x\}$, $\nu_2(*) = \lambda\{x, y \,.\, y * x\}$, and $\nu_2(e) = e$. Φ_2 is a homomorphism from the expressions (terms and formulas) of M to the expressions of M, but (unlike Φ_1) Φ_2 alters terms. Each expression A of M is mapped by Φ_2 to an expression $\Phi_2(A)$ which is symmetric to A with respect to reversing the argument order of $*$. For instance, Φ_2 maps the Left Identity axiom to the Right Identity axiom and vice versa. Consequently, Φ_1 is an interpretation of M in itself by the Standard Interpretation Theorem.□

Interpretations like Φ_2 allow one to reason "by symmetry" in mechanized mathematics systems. For an illustration, suppose A and B are two sentences we want to prove which are symmetric to each other in some way. In informal mathematics, we would first construct a proof of A, and then we would prove B by simply noting that there is a proof of B that is symmetric to the proof of A. We would not have bothered to construct this proof of B, for nothing new would be learned. In a mechanized mathematics system with interpretations, after proving A, we would find an interpretation which formalizes the symmetry between A and B and which maps A to B. Just as in informal mathematics, we would not bother to produce a proof of B, but we would know that B is valid because it is the image of a theorem by an interpretation. Moreover, the verification that the mapping is really an interpretation would effectively be a verification that this particular kind of reasoning by symmetry is valid.

The interpretation in the last example associates the universe of an abstract monoid with the singleton set of the identity element.

Example 3.3 Let Φ_3 be the standard translation (U_3, ν_3) from M to itself where $U_3 = \lambda\{x \,.\, x \equiv e\}$, $\nu(*) = *$, and $\nu_3(e) = e$. Since $(\{e\}, *, e)$ has the structure of a monoid, it is easy to see that Φ_3 is an interpretation of M in itself.□

Since U_3 denotes a proper subset of the universe of M, Φ_3 relativizes quantifiers. For example,

$$\Phi_3((\forall x)(e * x \equiv x)) = (\forall x)(x \equiv e \rightarrow e * x \equiv x).$$

Also, notice that Φ_3 fixes the primitive constants of M (i.e., $*$ and e). The act of verifying that Φ_3 is an interpretation is equivalent to proving that $\{e\}$ is a submonoid of M. A standard interpretation that fixes the primitive constants of the source theory but restricts the universe of the source theory is called a *relativization*. Relativizations are commonly used in set theory to establish relative consistency [19].

4 PF*

PF* is a version of simple type theory with partial functions and subtypes. This section gives a quick introduction to the syntax and semantics of **PF***. The next section introduces a "user-friendly" version of **PF*** called LUTINS. See [9] for the full, definitive presentation of **PF***.

Preliminary Definitions

We begin by defining the machinery we will use to specify the various symbols of a **PF*** language. In particular, the set of type or subtype symbols built from a set S of base type or subtype symbols will be the set $\Omega(S)$ defined below.

Let S be a set of symbols containing the symbol $*$. $\Omega(S)$ is the set defined inductively by:

1. $S \subseteq \Omega(S)$.
2. If $\alpha_1, \ldots, \alpha_n, \alpha_{n+1} \in \Omega(S)$ $(n \geq 1)$, then $[\alpha_1, \ldots, \alpha_n, \alpha_{n+1}] \in \Omega(S)$.

$\Omega^*(S)$ is the subset of $\Omega(S)$ defined inductively by:

1. $* \in \Omega^*(S)$.
2. If $\alpha_1, \ldots, \alpha_n \in \Omega(S)$ and $\alpha_{n+1} \in \Omega^*(S)$ $(n \geq 1)$, then $[\alpha_1, \ldots, \alpha_n, \alpha_{n+1}] \in \Omega^*(S)$.

Given a total function $f : S \to \Omega(S)$, $\preceq_f$ is the smallest binary relation on $\Omega(S)$ such that:

1. If $\alpha \in S$, then $\alpha \preceq_f f(\alpha)$.
2. $\preceq_f$ is reflexive, i.e., for all $\alpha \in \Omega(S)$, $\alpha \preceq_f \alpha$.
3. $\preceq_f$ is transitive, i.e., for all $\alpha, \beta, \gamma \in \Omega(S)$, if $\alpha \preceq_f \beta$ and $\beta \preceq_f \gamma$, then $\alpha \preceq_f \gamma$.
4. If $\alpha_1 \preceq_f \beta_1$, ..., $\alpha_n \preceq_f \beta_n$, $\alpha_{n+1} \preceq_f \beta_{n+1}$, then $[\alpha_1, \ldots, \alpha_n, \alpha_{n+1}] \preceq_f [\beta_1, \ldots, \beta_n, \beta_{n+1}]$.

$\preceq_f$ is *noetherian* if every ascending sequence of members of $\Omega(S)$,

$$\alpha_1 \preceq_f \alpha_2 \preceq_f \alpha_3 \preceq_f \cdots,$$

is eventually stationary, i.e., there is some m such that $\alpha_i = \alpha_m$ for all $i \geq m$. If $\preceq_f$ is noetherian, then $\preceq_f$ is obviously antisymmetric (i.e., for all $\alpha, \beta \in \Omega(S)$, if $\alpha \preceq_f \beta$ and $\beta \preceq_f \alpha$, then $\alpha = \beta$). Hence, $\preceq_f$ is a partial order if it is noetherian.

An *S-tagged symbol* is a symbol tagged with a member of $\Omega(S)$. A tagged symbol whose symbol is a and whose tag is α is written as a_α. A tagged symbol a_α, where $a = b_i$, will be written as b^i_α (instead of as $(b_i)_\alpha$). Two tagged symbols a_α and b_β are *distinct* if $a \neq b$.

Sort Systems

The types and subtypes of a **PF*** language are called "sorts". Syntactically, they form a partial order. The semantics of the partial order of sorts is a mapping that takes each sort to a nonempty set and takes the order relation to set inclusion.

A *sort system* of **PF*** is a pair $(\mathcal{A}, \xi)$ where $\mathcal{A}$ is a finite set of symbols such that, for some $m \geq 1$, $\mathcal{B}_m = \{*, \iota_1, \ldots, \iota_m\} \subseteq \mathcal{A}$ and ξ is a total function from $\mathcal{A}$ to $\Omega(\mathcal{A})$ such that:

1. For all $\alpha \in \mathcal{A}$, $\alpha \in \mathcal{B}_m$ iff $\xi(\alpha) = \alpha$.
2. For all $\alpha \in \mathcal{A}$, $\xi(\alpha) \in \Omega^*(\mathcal{A})$ iff $\alpha = *$.
3. $\preceq_\xi$ is noetherian.

A *sort* of the system is any member of $\Omega(\mathcal{A})$, and a *type* of the system is any member of $\Omega(\mathcal{B}_m)$. The sorts in $\mathcal{A}$, $\Omega(\mathcal{A}) \setminus \mathcal{A}$, $\mathcal{B}_m$, and $\Omega(\mathcal{B}_m) \setminus \mathcal{B}_m$ are called the *atomic sorts*, *compound sorts*, *base types*, and *function types* of the system, respectively. ($*$ is the type of propositions, and each ι_k is a type of individuals.) The *enclosing sort* of $\alpha \in \mathcal{A}$ is the sort $\xi(\alpha)$.

Let $\mathcal{S}$ be a sort system $(\mathcal{A}, \xi)$. By $\alpha \in \mathcal{S}$ we shall mean $\alpha \in \Omega(\mathcal{A})$. Also, let $\mathcal{T}$ be the set of types of $\mathcal{S}$. The definition of a sort system implies that, for each $\alpha \in \mathcal{S}$, there is a unique $\beta \in \mathcal{T}$, called the *type* of α, such that $\alpha \preceq_\xi \beta$. The type of α is denoted by $\tau(\alpha)$. A sort α is of *kind* $*$ [respectively, ι] if $\alpha \in \Omega^*(\mathcal{A})$ [respectively, $\alpha \in \Omega(\mathcal{A}) \setminus \Omega^*(\mathcal{A})$]. Let $\mathcal{S}^*$ [$\mathcal{S}^\iota$] denote the set of all $\alpha \in \mathcal{S}$ of kind $*$ [ι]. The *least upper bound* of α and β, written $\alpha \sqcup_\xi \beta$, is the least upper bound of α and β in the partial order $\preceq_\xi$. Since each atomic sort has just a single enclosing sort, the least upper bound of two sorts with the same type is always defined.

A *standard sort frame* for $\mathcal{S}$ is a set $\{\mathcal{D}_\alpha : \alpha \in \mathcal{S}\}$ of nonempty domains (sets) such that:

1. $\mathcal{D}_* = \{\text{T}, \text{F}\}$. ($\text{T} \neq \text{F}$.)
2. If $\alpha \preceq_\xi \beta$, then $\mathcal{D}_\alpha \subseteq \mathcal{D}_\beta$.
3. If $\alpha = [\alpha_1, \ldots, \alpha_n, \alpha_{n+1}]$ is of kind ι, then $\mathcal{D}_\alpha$ is the set of all *partial* functions $f : \mathcal{D}_{\alpha_1} \times \cdots \times \mathcal{D}_{\alpha_n} \to \mathcal{D}_{\alpha_{n+1}}$.
4. If $\alpha = [\alpha_1, \ldots, \alpha_n, \alpha_{n+1}]$ is of kind $*$, then $\mathcal{D}_\alpha$ is the set of all *total* functions $f : \mathcal{D}_{\tau(\alpha_1)} \times \cdots \times \mathcal{D}_{\tau(\alpha_n)} \to \mathcal{D}_{\alpha_{n+1}}$ such that, for all $\langle b_1, \ldots, b_n \rangle \in \mathcal{D}_{\tau(\alpha_1)} \times \cdots \times \mathcal{D}_{\tau(\alpha_n)}$, $f(b_1, \ldots, b_n) = \text{F}_{\tau(\alpha_{n+1})}$ (see definition below) whenever $b_i \notin \mathcal{D}_{\alpha_i}$ for at least one i with $1 \leq i \leq n$.

For a type $\alpha \in \mathcal{S}^*$, F_α is defined inductively by:

1. $\text{F}_* = \text{F}$.
2. If $\alpha = [\alpha_1, \ldots, \alpha_n, \alpha_{n+1}]$, then F_α is the function which maps every n-tuple $\langle a_1, \ldots, a_n \rangle \in \mathcal{D}_{\alpha_1} \times \cdots \times \mathcal{D}_{\alpha_n}$ to $\text{F}_{\alpha_{n+1}}$.

It follows from the definition of a standard sort frame that $\text{F}_{\tau(\alpha)} \in \mathcal{D}_\alpha$ for all $\alpha \in \mathcal{S}^*$.

Languages

A *language* of $\mathbf{PF}^*$ is a tuple $(\mathcal{A}, \xi, \mathcal{V}, \mathcal{C})$ such that:

1. $(\mathcal{A}, \xi)$ is a sort system of $\mathbf{PF}^*$.
2. $\mathcal{V}$ and $\mathcal{C}$ are disjoint countable sets of pairwise distinct $\mathcal{A}$-tagged symbols, whose members are called *variables* and *constants*, respectively.
3. For each $\alpha \in \Omega(\mathcal{A})$, there is an infinite subset $\mathcal{V}_\alpha$ of $\mathcal{V}$ such that $x_\beta \in \mathcal{V}_\alpha$ iff $\beta = \alpha$.
4. For each $\alpha \in \Omega(\mathcal{A})$, $=_{[\alpha,\alpha,*]}$ and $\iota_{[[\alpha,*],\alpha]}$ are members of $\mathcal{C}$ called *logical constants*.

Variables are denoted by $f_\alpha, g_\alpha, h_\alpha, x_\alpha, y_\alpha, z_\alpha$, etc. For the rest of this section, let $\mathcal{L} = (\mathcal{A}, \xi, \mathcal{V}, \mathcal{C})$ be a language of $\mathbf{PF}^*$, $\mathcal{S}$ be the sort system $(\mathcal{A}, \xi)$, and $\mathcal{T}$ be the set of types of $\mathcal{S}$.

An *expression of* $\mathcal{L}$ *of sort* α is a finite sequence of members of $\mathcal{V} \cup \mathcal{C} \cup \{@, \lambda, \text{“\{”}, \text{“\}”}, \text{“,”}, \text{“.”}\}$ defined inductively by:

1. Each $x_\alpha \in \mathcal{V}$ and $c_\alpha \in \mathcal{C}$ is an expression of sort α.
2. If F is an expression of sort $[\alpha_1, \ldots, \alpha_n, \beta]$; $A_1, \ldots, A_n$ are expressions of sort $\alpha'_1, \ldots, \alpha'_n$; and $\tau(\alpha_1) = \tau(\alpha'_1), \ \ldots, \ \tau(\alpha_n) = \tau(\alpha'_n)$, then $@\{F, A_1, \ldots, A_n\}$ is an expression of sort β.
3. If $x^1_{\alpha_1}, \ldots, x^n_{\alpha_n}$ are distinct variables and B is an expression of sort β, then $\lambda\{x^1_{\alpha_1}, \ldots, x^n_{\alpha_n} \cdot B\}$ is an expression of sort $[\alpha_1, \ldots, \alpha_n, \beta]$.

Expressions are denoted by A, B, C, etc., and expressions of sort α are denoted by $A_\alpha, B_\alpha, C_\alpha$, etc. "Free variable", "closed expression", and similar notions are defined in the obvious way. A *formula* is an expression of sort $*$. A *sentence* is a closed formula.

Intuitively, when an expression $@\{F, A_1, \ldots, A_n\}$ has a denotation, it denotes the value of the function denoted by F applied to the arguments denoted by $A_1, \ldots, A_n$. Also, an expression $\lambda\{x_1, \ldots, x_n \,.\, B\}$ denotes the function whose value, when given arguments $x_1, \ldots, x_n$, is the denotation of B (which generally depends on $x_1, \ldots, x_n$) if B has a denotation and is undefined otherwise.

The following abbreviations provide notation for the major logical operations of **PF***:

$F(A_1, \ldots, A_n)$ for $@\{F, A_1, \ldots, A_n\}$

$(A_\alpha =_\gamma B_\beta)$ for $=_{[\gamma,\gamma,*]} (A_\alpha, B_\beta)$

$(A_* \equiv B_*)$ for $(A_* =_* B_*)$

T_* for $(=_{[*,*,*]} \; =_{[*,*,*]} \; =_{[*,*,*]})$

F_* for $(\lambda\{x_* \,.\, \mathsf{T}_*\} =_{[*,*]} \lambda\{x_* \,.\, x_*\})$

$\neg_{[*,*]}$ for $\lambda\{x_* \,.\, (x_* \equiv \mathsf{F}_*)\}$

$(\neg A_*)$ for $\neg_{[*,*]}(A_*)$

$\wedge_{[*,*,*]}$ for $\lambda\{x_*, y_* \,.\, (\lambda\{g_{[*,*,*]} \,.\, g_{[*,*,*]}(\mathsf{T}_*, \mathsf{T}_*)\} =_{[[*,*,*],*]} \lambda\{g_{[*,*,*]} \,.\, g_{[*,*,*]}(x_*, y_*)\})\}$

$(A_* \wedge B_*)$ for $\wedge_{[*,*,*]}(A_*, B_*)$

$\vee_{[*,*,*]}$ for $\lambda\{x_*, y_* \,.\, (\neg((\neg x_*) \wedge (\neg y_*)))\}$

$(A_* \vee B_*)$ for $\vee_{[*,*,*]}(A_*, B_*)$

$\supset_{[*,*,*]}$ for $\lambda\{x_*, y_* \,.\, (x_* \equiv (x_* \wedge y_*))\}$

$(A_* \supset B_*)$ for $\supset_{[*,*,*]} (A_*, B_*)$

$\Pi_{[[\alpha_1,\ldots,\alpha_n,*],*]}$ for $\lambda\{p_{[\alpha_1,\ldots,\alpha_n,*]} \,.\, (p_{[\alpha_1,\ldots,\alpha_n,*]} =_{[\alpha_1,\ldots,\alpha_n,*]} \lambda\{x^1_{\alpha_1}, \ldots, x^n_{\alpha_n} \,.\, \mathsf{T}_*\})\}$

$\forall\{x^1_{\alpha_1}, \ldots, x^n_{\alpha_n} \,.\, A_*\}$ for $\Pi_{[[\alpha_1,\ldots,\alpha_n,*],*]}(\lambda\{x^1_{\alpha_1}, \ldots, x^n_{\alpha_n} \,.\, A_*\})$

$\exists\{x^1_{\alpha_1}, \ldots, x^n_{\alpha_n} \,.\, A_*\}$ for $(\neg\forall\{x^1_{\alpha_1}, \ldots, x^n_{\alpha_n} \,.\, (\neg A_*)\})$

$\mathrm{I}\{x_\alpha \,.\, A_*\}$ for $\iota_{[[\alpha,*],\alpha]}(\lambda\{x_\alpha \,.\, A_*\})$

$(A_\alpha \downarrow \beta)$ for $\lambda\{x_\beta \,.\, \mathsf{T}_*\}(A_\alpha)$

$(A_\alpha \uparrow \beta)$ for $(\neg(A_\alpha \downarrow \beta))$

$(A_\alpha\downarrow)$ for $(A_\alpha \downarrow \alpha)$

$(A_\alpha\uparrow)$ for $(\neg(A_\alpha\downarrow))$

$(A_\alpha \simeq_\gamma B_\beta)$ for $((A_\alpha\downarrow) \vee (B_\beta\downarrow)) \supset (A_\alpha =_\gamma B_\beta)$

$(A_\alpha \neq_\gamma B_\beta)$ for $(\neg(A_\alpha =_\gamma B_\beta))$

$(A_\alpha \not\simeq_\gamma B_\beta)$ for $(\neg(A_\alpha \simeq_\gamma B_\beta))$

$\bot_\alpha$ for $\mathrm{I}\{x_\alpha \,.\, \mathsf{F}_*\}$ where $\alpha \in \mathcal{S}^\iota$

$\mathsf{F}_{[\alpha_1,\ldots,\alpha_n,\beta]}$ for $\lambda\{x^1_{\alpha_1}, \ldots, x^n_{\alpha_n} \,.\, \mathsf{F}_\beta\}$ where $\beta \in \mathcal{S}^*$

$\alpha \subseteq \beta$ for $\forall\{x_\alpha \,.\, (x_\alpha \downarrow \beta)\}$

$\mathrm{if}\{A_*, B_\beta, C_\gamma\}$ for $\mathrm{I}\{x_\alpha \,.\, ((A_* \supset (x_\alpha =_\alpha B_\beta)) \wedge ((\neg A_*) \supset (x_\alpha =_\alpha C_\gamma)))\}$ where $\alpha = \beta \sqcup_\xi \gamma$ and x_α does not occur in A_*, B_β, or C_γ.

Parentheses in expressions may be suppressed when meaning is not lost.

Standard Models

In [9] we define two semantics for **PF***, one based on "general models" and the other based on "standard models". In this paper we are interested only in the standard models semantics. We begin by describing the objects that the logical constants are intended to denote.

Let $\{\mathcal{D}_\alpha : \alpha \in \mathcal{S}\}$ be a standard sort frame for $\mathcal{S}$. Fix $\beta \in \mathcal{S}$ with $\gamma = \tau(\beta)$. The *identity relation for* $\mathcal{D}_\beta$ is the total function $p : \mathcal{D}_\gamma \times \mathcal{D}_\gamma \to \mathcal{D}_*$ such that, for all $a, b \in \mathcal{D}_\gamma$, $p(a,b) = \mathrm{T}$ iff $a, b \in \mathcal{D}_\beta$ and $a = b$. Given $b \in \mathcal{D}_\gamma$, the *b-descriptor for* $\mathcal{D}_\gamma$ is the total function $p : \mathcal{D}_\gamma \to \mathcal{D}_*$ such that, for all $a \in \mathcal{D}_\gamma$, $p(a) = \mathrm{T}$ iff $a = b$. When $\beta \in \mathcal{S}^\iota$, the *definite description function for* $\mathcal{D}_\beta$ is the partial function $f : \mathcal{D}_{[\gamma,*]} \to \mathcal{D}_\gamma$ such that $f(p) = b$ if p is the b-descriptor for $\mathcal{D}_\gamma$ for some $b \in \mathcal{D}_\beta$, and $f(p)$ is undefined otherwise. When $\beta \in \mathcal{S}^*$, the *definite description function for* $\mathcal{D}_\beta$ is the total function $f : \mathcal{D}_{[\gamma,*]} \to \mathcal{D}_\gamma$ such that $f(p) = b$ if p is the b-descriptor for $\mathcal{D}_\gamma$ for some $b \in \mathcal{D}_\beta$, and $f(p) = \mathrm{F}_\gamma$ otherwise.

A *standard model for* $\mathcal{L}$ is a pair $\mathcal{M} = (\{\mathcal{D}_\alpha : \alpha \in \mathcal{S}\}, I)$ where $\{\mathcal{D}_\alpha : \alpha \in \mathcal{S}\}$ is a standard sort frame for $\mathcal{S}$ such that $\bot \notin \mathcal{D}_\alpha$ for all $\alpha \in \mathcal{S}$ and I is a function which maps each $c_\alpha \in \mathcal{C}$ to an element of $\mathcal{D}_\alpha$ such that, for each $\alpha \in \mathcal{S}$, (1) $I(=_{[\alpha,\alpha,*]})$ is the identity relation for $\mathcal{D}_\alpha$ and (2) $I(\iota_{[[\alpha,*],\alpha]})$ is the definite description function for $\mathcal{D}_\alpha$. A *$\mathcal{V}$-assignment* into $\mathcal{M}$ is a function which maps each $x_\alpha \in \mathcal{V}$ to an element of $\mathcal{D}_\alpha$. Given a $\mathcal{V}$-assignment φ into $\mathcal{M}$; distinct variables $x^1_{\alpha_1}, \ldots, x^n_{\alpha_n} \in \mathcal{V}$ ($n \geq 1$); and $a_1 \in \mathcal{D}_{\alpha_1}, \ldots a_n \in \mathcal{D}_{\alpha_n}$, let $\varphi[x^1_{\alpha_1} \mapsto a_1, \ldots, x^n_{\alpha_n} \mapsto a_n]$ be the $\mathcal{V}$-assignment ψ such that $\psi(x^i_{\alpha_i}) = a_i$ for all i with $1 \leq i \leq n$ and $\psi(y_\beta) = \varphi(y_\beta)$ for all $y_\beta \notin \{x^1_{\alpha_1}, \ldots, x^n_{\alpha_n}\}$.

Let $\mathcal{M} = (\{\mathcal{D}_\alpha : \alpha \in \mathcal{S}\}, I)$ be a standard model for $\mathcal{L}$. Then there exists a binary function $V = V^{\mathcal{M}}$ such that: (1) for each $\mathcal{V}$-assignment φ into $\mathcal{M}$ and each expression A_α, $V_\varphi(A_\alpha) \in \mathcal{D}_\alpha \cup \{\bot\}$ if $\alpha \in \mathcal{S}^\iota$ and $V_\varphi(A_\alpha) \in \mathcal{D}_\alpha$ if $\alpha \in \mathcal{S}^*$, and (2) the following conditions are satisfied for all $\mathcal{V}$-assignments φ into $\mathcal{M}$ and all expressions:

1. $V_\varphi(x_\alpha) = \varphi(x_\alpha)$ if $x_\alpha \in \mathcal{V}$.
2. $V_\varphi(c_\alpha) = I(c_\alpha)$ if $c_\alpha \in \mathcal{C}$.
3. Let $A_\alpha = @\{F, A_1, \ldots, A_n\}$ with $\alpha \in \mathcal{S}^\iota$. If each of $V_\varphi(F)$, $V_\varphi(A_1), \ldots,$ $V_\varphi(A_n)$ does not equal $\bot$ and $V_\varphi(F)$ is defined at $\langle V_\varphi(A_1), \ldots, V_\varphi(A_n)\rangle$, then
$$V_\varphi(A_\alpha) = (V_\varphi(F))(V_\varphi(A_1), \ldots, V_\varphi(A_n));$$
otherwise $V_\varphi(A_\alpha) = \bot$.
4. Let $A_\alpha = @\{F, A_1, \ldots, A_n\}$ with $\alpha = [\alpha_1, \ldots, \alpha_n, \beta] \in \mathcal{S}^*$. If each of $V_\varphi(A_1), \ldots, V_\varphi(A_n)$ does not equal $\bot$, then
$$V_\varphi(A_\alpha) = (V_\varphi(F))(V_\varphi(A_1), \ldots, V_\varphi(A_n));$$
otherwise $V_\varphi(A_\alpha) = \mathrm{F}_{\tau(\beta)}$.

5. Let $A_\alpha = \lambda\{x^1_{\alpha_1}, \ldots, x^n_{\alpha_n} \,.\, B_\beta\}$ with $\beta \in \mathcal{S}^\iota$. $V_\varphi(A_\alpha)$ is the partial function $f : \mathcal{D}_{\alpha_1} \times \cdots \times \mathcal{D}_{\alpha_n} \to \mathcal{D}_\beta$ such that

$$f(a_1, \ldots, a_n) = V_{\varphi[x^1_{\alpha_1} \mapsto a_1, \ldots, x^n_{\alpha_n} \mapsto a_n]}(B_\beta)$$

if $V_{\varphi[x^1_{\alpha_1} \mapsto a_1, \ldots, x^n_{\alpha_n} \mapsto a_n]}(B_\beta)$ is defined; otherwise $f(a_1, \ldots, a_n)$ is undefined.
6. Let $A_\alpha = \lambda\{x^1_{\alpha_1}, \ldots, x^n_{\alpha_n} \,.\, B_\beta\}$ with $\beta \in \mathcal{S}^*$. $V_\varphi(A_\alpha)$ is the total function $f : \mathcal{D}_{\tau(\alpha_1)} \times \cdots \times \mathcal{D}_{\tau(\alpha_n)} \to \mathcal{D}_\beta$ such that

$$f(a_1, \ldots, a_n) = V_{\varphi[x^1_{\alpha_1} \mapsto a_1, \ldots, x^n_{\alpha_n} \mapsto a_n]}(B_\beta)$$

if $\langle a_1, \ldots, a_n \rangle \in \mathcal{D}_{\alpha_1} \times \cdots \times \mathcal{D}_{\alpha_n}$; otherwise $f(a_1, \ldots, a_n) = \mathrm{F}_{\tau(\beta)}$.

When $V_\varphi(A_\alpha) \neq \bot$, $V_\varphi(A_\alpha)$ is called the *value* of A_α in $\mathcal{M}$ with respect to φ. Intuitively, $V_\varphi(A_\alpha) = \bot$ means that A_α has no denotation or value in $\mathcal{M}$ with respect to φ. That is, V is intuitively a *partial* valuation function. For a closed expression A_α, it is clear that $V_\varphi(A_\alpha)$ does not depend on φ and thus $V(A_\alpha)$ is meaningful.

A formula A_* of $\mathcal{L}$ is *valid in* $\mathcal{M}$ if $V_\varphi(A_*) = \mathrm{T}$ for every $\mathcal{V}$-assignment φ into $\mathcal{M}$.

Theories

A *theory* of **PF*** is a pair $T = (\mathcal{L}, \Gamma)$ where $\mathcal{L}$ is a language of **PF*** and Γ is a set of sentences of $\mathcal{L}$. The members of Γ are called the *axioms* of T. T, T', etc. denote theories.

Let $T = (\mathcal{L}, \Gamma)$. A *standard model for* T is a standard model for $\mathcal{L}$ in which each member of Γ is valid. T is *satisfiable (in the standard sense)* if there is some standard model for T. A sentence A_* is a *(semantic) theorem of* T *(in the standard sense)*, written $T \models A_*$, if A_* is valid in every standard model for T.

5 LUTINS

LUTINS (Logic of Undefined Terms for Inference in a Natural Style) is the logic of IMPS. It is essentially the same as **PF*** plus several additional expression constructors, which are defined in terms of the primitive notions of **PF***: function application, λ-abstraction, equality, and definite description. This section gives a brief presentation of the syntax and semantics of LUTINS. A more detailed presentation of a slightly different version of LUTINS is found in [16].[5] We assume that each concept and notation defined in the previous section for **PF*** is defined for LUTINS in the obvious way if it is not explicitly defined in this section.

[5] The reader should be aware that in [16] LUTINS is called **PF**.

Syntax

A *language* of LUTINS is a tuple $(\mathcal{A}, \xi, \mathcal{V}, \mathcal{C})$ such that $(\mathcal{A}, \xi, \mathcal{V}, \widetilde{\mathcal{C}})$ is a language of **PF***, where

$$\widetilde{\mathcal{C}} = \mathcal{C} \cup \{=_{[\alpha,\alpha,*]}, \iota_{[[\alpha,*],\alpha]} : \alpha \in \Omega(\mathcal{A})\}.$$

For any language $\mathcal{L} = (\mathcal{A}, \xi, \mathcal{V}, \mathcal{C})$ of LUTINS, let $\widetilde{\mathcal{L}}$ be the language $(\mathcal{A}, \xi, \mathcal{V}, \widetilde{\mathcal{C}})$ of **PF***.

For the remainder of this section, let $\mathcal{L} = (\mathcal{A}, \xi, \mathcal{V}, \mathcal{C})$ be a language of LUTINS, $\mathcal{S}$ be the sort system $(\mathcal{A}, \xi)$, and $\mathcal{T}$ be the set of types of $\mathcal{S}$. When there is no possibility of confusion, the components of a language $\mathcal{L}_i$ will be denoted by $\mathcal{A}_i, \xi_i, \mathcal{V}_i, \mathcal{C}_i$, and the sort system and set of types of $\mathcal{L}_i$ will be denoted by $\mathcal{S}_i$ and $\mathcal{T}_i$, respectively.

Our version of LUTINS has 21 expression constructors: **apply**, **lambda**, **equals**, **quasi-equals**, **iota**, **iota-p**, **the-true**, **the-false**, **not**, **and**, **or**, **implies**, **iff**, **if-form**, **forall**, **forsome**, **defined-in**, **is-defined**, **if**, **undefined**, and **falselike**. An *expression of* $\mathcal{L}$ *of sort* $\alpha \in \mathcal{S}$ is a finite sequence of variables, constants, expression constructors, sorts, and punctuation symbols defined inductively by:

1. Each $x_\alpha \in \mathcal{V}$ and $c_\alpha \in \mathcal{C}$ is an expression of sort α.
2. If F is an expression of sort $[\alpha_1, \ldots, \alpha_n, \beta]$; $A_1, \ldots, A_n$ are expressions of sort $\alpha'_1, \ldots, \alpha'_n$; and $\tau(\alpha_1) = \tau(\alpha'_1), \ldots, \tau(\alpha_n) = \tau(\alpha'_n)$, then $\mathtt{apply}\{F, A_1, \ldots, A_n\}$ is an expression of sort β.
3. If $x^1_{\alpha_1}, \ldots, x^n_{\alpha_n}$ are distinct variables and B is an expression of sort β, then $\mathtt{lambda}\{x^1_{\alpha_1}, \ldots, x^n_{\alpha_n} \,.\, B\}$ is an expression of sort $[\alpha_1, \ldots, \alpha_n, \beta]$.
4. If A and B are expressions of sort α and β, respectively, with $\tau(\alpha) = \tau(\beta)$, then $\mathtt{equals}\{A, B\}$ and $\mathtt{quasi\text{-}equals}\{A, B\}$ are expressions of sort $*$.
5. If x_α, y_β are variables; α, β are of kind $\iota, *$, respectively; and C is an expression of sort $*$, then $\mathtt{iota}\{x_\alpha \,.\, C\}$ and $\mathtt{iota\text{-}p}\{y_\beta \,.\, C\}$ are expressions of sort α and β, respectively.
6. If $A, B, C, A_1, \ldots, A_n$ are expressions of sort $*$ with $n \geq 0$, then $\mathtt{the\text{-}true}\{\}$, $\mathtt{the\text{-}false}\{\}$, $\mathtt{not}\{A\}$ $\mathtt{and}\{A_1, \ldots, A_n\}$, $\mathtt{or}\{A_1, \ldots, A_n\}$, $\mathtt{implies}\{A, B\}$, $\mathtt{iff}\{A, B\}$, and $\mathtt{if\text{-}form}\{A, B, C\}$ are expressions of sort $*$.
7. If $x^1_{\alpha_1}, \ldots, x^n_{\alpha_n}$ are distinct variables and B is an expression of sort $*$, then $\mathtt{forall}\{x^1_{\alpha_1}, \ldots, x^n_{\alpha_n} \,.\, B\}$ and $\mathtt{forsome}\{x^1_{\alpha_1}, \ldots, x^n_{\alpha_n} \,.\, B\}$ are expressions of sort $*$.
8. If A is an expression of sort α and $\tau(\alpha) = \tau(\beta)$, then $\mathtt{defined\text{-}in}\{A, \beta\}$ and $\mathtt{is\text{-}defined}\{A\}$ are expressions of sort $*$.
9. If A, B, C are expressions of sort $*, \beta, \gamma$, respectively, with $\tau(\beta) = \tau(\gamma)$, then $\mathtt{if}\{A, B, C\}$ is an expression of sort $\beta \sqcup_\xi \gamma$.
10. If α is of kind ι, then $\mathtt{undefined}\{\alpha\}$ is an expression of sort α.
11. If α is of kind $*$, then $\mathtt{falselike}\{\alpha\}$ is an expression of sort α.

Expressions that are not variables or constants are called *compound expressions*. The *length* of an expression A, written $|A|$, is the number of occurrences of expression constructors in A. The set of expressions of $\mathcal{L}$ [respectively, $\mathcal{L}_i$] is denoted by $\mathcal{E}$ [respectively, $\mathcal{E}_i$], and the set of expressions of $\widetilde{\mathcal{L}}$ [$\widetilde{\mathcal{L}_i}$] is denoted by $\widetilde{\mathcal{E}}$ [$\widetilde{\mathcal{E}}_i$].

Semantics

The semantics of LUTINS is defined in terms of the standard models semantics of **PF***.

Let $\zeta : \mathcal{E} \rightarrow \widetilde{\mathcal{E}}$ be the mapping defined by the following statements:

1. $\zeta(a_\alpha) = a_\alpha$ where $a_\alpha \in \mathcal{V} \cup \mathcal{C}$.
2. $\zeta(\mathtt{apply}\{F, A_1, \ldots, A_n\}) = @\{\zeta(F), \zeta(A_1), \ldots, \zeta(A_n)\}$.
3. $\zeta(\mathtt{lambda}\{x^1_{\alpha_1}, \ldots, x^n_{\alpha_n} \,.\, B\}) = \lambda\{x^1_{\alpha_1}, \ldots, x^n_{\alpha_n} \,.\, \zeta(B)\}$.
4. $\zeta(\mathtt{equals}\{A_\alpha, B_\beta\}) = (\zeta(A_\alpha) =_\gamma \zeta(B_\beta))$ where $\gamma = \tau(\alpha) = \tau(\beta)$.
5. $\zeta(\mathtt{quasi\text{-}equals}\{A_\alpha, B_\beta\}) = (\zeta(A_\alpha) \simeq_\gamma \zeta(B_\beta))$ where $\gamma = \tau(\alpha) = \tau(\beta)$.
6. $\zeta(\Box\{x_\alpha \,.\, A_*\}) = \mathrm{I}\{x_\alpha \,.\, \zeta(A_*)\}$ where $\Box$ is **iota** or **iota-p**.
7. $\zeta(\Box\{\}) = \mathsf{T}_*$ where $\Box$ is **the-true** or **and**.
8. $\zeta(\Box\{\}) = \mathsf{F}_*$ where $\Box$ is **the-false** or **or**.
9. $\zeta(\mathtt{not}\{A\}) = \neg\zeta(A)$.
10. $\zeta(\mathtt{and}\{A_1, \ldots, A_n\}) = (\zeta(A_1) \wedge (\zeta(A_2) \wedge \cdots))$ where $n \geq 1$.
11. $\zeta(\mathtt{or}\{A_1, \ldots, A_n\}) = (\zeta(A_1) \vee (\zeta(A_2) \vee \cdots))$ where $n \geq 1$.
12. $\zeta(\mathtt{implies}\{A, B\}) = (\zeta(A) \supset \zeta(B))$.
13. $\zeta(\mathtt{iff}\{A, B\}) = (\zeta(A) \equiv \zeta(B))$.
14. $\zeta(\Box\{A, B, C\}) = \mathrm{if}\{\zeta(A), \zeta(B), \zeta(C)\}$ where $\Box$ is **if-form** or **if**.
15. $\zeta(\mathtt{forall}\{x^1_{\alpha_1}, \ldots, x^n_{\alpha_n} \,.\, B\}) = \forall\{x^1_{\alpha_1}, \ldots, x^n_{\alpha_n} \,.\, \zeta(B)\}$.
16. $\zeta(\mathtt{forsome}\{x^1_{\alpha_1}, \ldots, x^n_{\alpha_n} \,.\, B\}) = \exists\{x^1_{\alpha_1}, \ldots, x^n_{\alpha_n} \,.\, \zeta(B)\}$.
17. $\zeta(\mathtt{defined\text{-}in}\{A, \beta\}) = (\zeta(A) \downarrow \beta)$.
18. $\zeta(\mathtt{is\text{-}defined}\{A\}) = \zeta(A)\downarrow$.
19. $\zeta(\mathtt{undefined}\{\alpha\}) = \perp_\alpha$.
20. $\zeta(\mathtt{falselike}\{\alpha\}) = \mathsf{F}_\alpha$.

Notice that A_α and $\zeta(A_\alpha)$ have the same sort for every expression A_α of $\mathcal{L}$.

A *model for* $\mathcal{L}$ is a standard model for $\widetilde{\mathcal{L}}$. Let $\mathcal{M} = (\{\mathcal{D}_\alpha : \alpha \in \mathcal{S}\}, I)$ be a model for $\mathcal{L}$, and let $V = V^{\mathcal{M}}$ be the valuation function for $\widetilde{\mathcal{L}}$ with respect to $\mathcal{M}$. $U = U^{\mathcal{M}}$ is the binary function, on $\mathcal{V}$-assignments φ into $\mathcal{M}$ and expressions A_α of $\mathcal{L}$, defined by

$$U_\varphi(A_\alpha) = V_\varphi(\zeta(A_\alpha)).$$

Clearly, for each $\mathcal{V}$-assignment φ into $\mathcal{M}$ and each expression A_α of $\mathcal{L}$, $U_\varphi(A_\alpha) \in \mathcal{D}_\alpha \cup \{\perp\}$ if $\alpha \in \mathcal{S}^\iota$ and $U_\varphi(A_\alpha) \in \mathcal{D}_\alpha$ if $\alpha \in \mathcal{S}^*$. Thus U is a (partial) valuation function for $\mathcal{L}$ with respect to $\mathcal{M}$.

A formula A_* of $\mathcal{L}$ is *valid in* $\mathcal{M}$ if $U_\varphi(A_*) = \mathrm{T}$ for every $\mathcal{V}$-assignment φ into $\mathcal{M}$.

Let $\mathcal{M}_i = (\{\mathcal{D}^i_\alpha : \alpha \in \mathcal{S}_i\}, I_i)$ be a model for $T_i = (\mathcal{L}_i, \Gamma_i)$ for $i = 1, 2$. $\mathcal{L}_2$ is an *expansion* of $\mathcal{L}_1$, written $\mathcal{L}_1 \leq \mathcal{L}_2$, if $\mathcal{A}_1 \subseteq \mathcal{A}_2$, ξ_1 is a subfunction of ξ_2, $\mathcal{V}_1 \subseteq \mathcal{V}_2$, and $\mathcal{C}_1 \subseteq \mathcal{C}_2$. T_2 is an *extension* of T_1, written $T_1 \leq T_2$, if $\mathcal{L}_1 \leq \mathcal{L}_2$ and $\Gamma_1 \subseteq \Gamma_2$. $\mathcal{M}_2$ is an *expansion* of $\mathcal{M}_1$ to $\mathcal{L}_2$ (and $\mathcal{M}_2$ is the *reduct* of $\mathcal{M}_1$ to $\mathcal{L}_1$) if $\mathcal{D}^1_\alpha = \mathcal{D}^2_\alpha$ for all $\alpha \in \mathcal{S}_1$ and I_1 is a subfunction of I_2.

6 Theory Interpretation in LUTINS

We have seen that, in first-order predicate logic, an interpretation of one theory in another is a certain kind of homomorphism on expressions that preserves the validity of sentences. In this section we define an interpretation of one LUTINS theory in another which is very similar to the notion of a standard first-order interpretation. We will first define a notion of a *translation* from a source theory to a target theory. A translation is determined by how the primitive symbols of the source theory (i.e., the atomic sorts, variables, and constants of the source theory) are associated with objects of the target theory. In the definition given here, atomic sorts are associated with sorts and closed unary predicates; variables are associated with variables; and constants are associated with closed expressions (of appropriate sort). Then an *interpretation* is defined to be a translation which maps every theorem of the source theory to a theorem of the target theory.

Our notion of a translation in LUTINS extends the notion of a translation in **PF*** defined in [9] in two ways:

- A LUTINS translation directly handles all 21 expression constructors of LUTINS, while a **PF*** translation handles just the constructors corresponding to `apply` and `lambda`.
- A LUTINS translation can associate an atomic sort with either a sort or a closed unary predicate, but a **PF*** translation can associate an atomic sort with only a sort.

This section is the heart of the paper. Most of the complexity in the definitions is a result of allowing sorts to be associated with unary predicates.

Quasi-Sorts

Let us define a *quasi-sort* to be any closed unary predicate, i.e., any expression whose sort is of the form $[\alpha, *]$. Then let $\mathcal{Q}$ $[\mathcal{Q}_i]$ denote the set of quasi-sorts of the language $\mathcal{L}$ $[\mathcal{L}_i]$.

For $q \in \mathcal{S} \cup \mathcal{Q}$ and $A_\alpha \in \mathcal{E}$, the *domain* of q, written $\delta(q)$, is defined by

$$\delta(q) = \begin{cases} q \text{ if } q \in \mathcal{S} \\ \alpha \text{ if } q \in \mathcal{Q} \text{ and } [\alpha, *] \text{ is the sort of } q \end{cases}$$

and, provided $\delta(q) = \alpha$, the *condition of q on A_α*, written $\kappa(q, A_\alpha)$, is defined by

$$\kappa(q, A_\alpha) = \begin{cases} \texttt{the-true\{\}} & \text{if } q \in \mathcal{S} \\ \texttt{apply}\{q, A_\alpha\} & \text{if } q \in \mathcal{Q} \end{cases}$$

We will need a quasi-sort constructor (denoted by $[\![\cdots]\!]$) for building "compound" quasi-sorts that is analogous to the $[\cdots]$ sort constructor for building compound sorts. Let $q_1, \ldots, q_{n+1} \in \mathcal{S} \cup \mathcal{Q}$ with $n \geq 1$. When $\delta(q_{n+1}) \in \mathcal{S}^\iota$, $[\![q_1, \ldots, q_{n+1}]\!]$ will be a quasi-sort whose extension (in a model) is the set of partial functions $f : \mathcal{D}_1 \times \cdots \times \mathcal{D}_n \to \mathcal{D}_{n+1}$, where $\mathcal{D}_i$ is the extension of q_i for i with $1 \leq i \leq n+1$.

$[\![q_1, \ldots, q_{n+1}]\!]$ is defined as follows. Let $\gamma = [\delta(q_1), \ldots, \delta(q_{n+1})]$. Choose distinct variables $f_\gamma, x^1_{\delta(q_1)}, \ldots, x^n_{\delta(q_n)}$ which do not occur in $q_1, \ldots, q_{n+1}$. Make the following definitions:

$A = \texttt{and}\{\kappa(q_1, x^1_{\delta(q_1)}), \ldots, \kappa(q_n, x^n_{\delta(q_n)})\}$.
$B = \texttt{apply}\{f_\gamma, x^1_{\delta(q_1)}, \ldots, x^n_{\delta(q_n)}\}$.

$$C = \begin{cases} \texttt{if-form}\{A, & \text{if } \gamma \text{ is of kind } \iota \\ \quad \texttt{implies}\{\texttt{is-defined}\{B\}, \kappa(q_{n+1}, B)\}, & \\ \quad \texttt{not}\{\texttt{is-defined}\{B\}\}\} & \\ \texttt{if-form}\{A, & \text{if } \gamma \text{ is of kind } * \\ \quad \kappa(q_{n+1}, B), & \\ \quad \texttt{equals}\{B, \texttt{falselike}\{\delta(q_{n+1})\}\}\} & \end{cases}$$

Then define $[\![q_1, \ldots, q_{n+1}]\!]$ to be

$$\texttt{lambda}\{f_\gamma \,.\, \texttt{forall}\{x^1_{\delta(q_1)}, \ldots, x^n_{\delta(q_n)} \cdot C\}\}.$$

Clearly, for all $q_1, \ldots, q_{n+1} \in \mathcal{S} \cup \mathcal{Q}$, $[\![q_1, \ldots, q_{n+1}]\!]$ is a well-formed member of $\mathcal{E}$.

Proposition 6.1

1. *If* $\alpha_1, \ldots, \alpha_{n+1} \in \mathcal{S}$, *then* $\delta([\alpha_1, \ldots, \alpha_{n+1}]) = [\delta(\alpha_1), \ldots, \delta(\alpha_{n+1})]$.
2. *If* $q_1, \ldots, q_{n+1} \in \mathcal{S} \cup \mathcal{Q}$, *then* $\delta([\![q_1, \ldots, q_{n+1}]\!]) = [\delta(q_1), \ldots, \delta(q_{n+1})]$.

Translations

Let $T_i = (\mathcal{L}_i, \Gamma_i)$ be a theory of LUTINS for $i = 1, 2$. Given a function $\mu : \mathcal{A}_1 \to \mathcal{S}_2 \cup \mathcal{Q}_2$, $\bar{\mu} : \mathcal{S}_1 \to \mathcal{S}_2 \cup \mathcal{Q}_2$ is the canonical extension of μ defined inductively by:

1. If $\alpha \in \mathcal{A}_1$, then $\bar{\mu}(\alpha) = \mu(\alpha)$.
2. If $\alpha = [\alpha_1, \ldots, \alpha_{n+1}] \in \mathcal{S}_1$ and $\{\bar{\mu}(\alpha_1), \ldots, \bar{\mu}(\alpha_{n+1})\} \subseteq \mathcal{S}_2$, then $\bar{\mu}(\alpha) = [\bar{\mu}(\alpha_1), \ldots, \bar{\mu}(\alpha_{n+1})]$.
3. If $\alpha = [\alpha_1, \ldots, \alpha_{n+1}] \in \mathcal{S}_1$ and $\{\bar{\mu}(\alpha_1), \ldots, \bar{\mu}(\alpha_{n+1})\} \not\subseteq \mathcal{S}_2$, then $\bar{\mu}(\alpha) = [\![\bar{\mu}(\alpha_1), \ldots, \bar{\mu}(\alpha_{n+1})]\!]$.

Given $\alpha \in \mathcal{A}_1$, let $\mu[\alpha] = \delta(\mu(\alpha))$, and given $\alpha \in \mathcal{S}_1$, let $\bar{\mu}[\alpha] = \delta(\bar{\mu}(\alpha))$.

A *translation from* T_1 *to* T_2 is a pair (μ, ν), where $\mu : \mathcal{A}_1 \to \mathcal{S}_2 \cup \mathcal{Q}_2$ and $\nu : \mathcal{V}_1 \cup \mathcal{C}_1 \to \mathcal{E}_2$, such that:

1. $\mu(*) = *$.
2. For all $\alpha \in \mathcal{A}_1$ with $\alpha \neq *$, $\mu[\alpha]$ is a sort of type $\tau(\bar{\mu}[\tau(\alpha)])$ of kind ι.
3. For all $x_\alpha \in \mathcal{V}_1$, $\nu(x_\alpha)$ is a variable of sort $\bar{\mu}[\alpha]$.
4. ν is injective on $\mathcal{V}_1$.
5. For all $c_\alpha \in \mathcal{C}_1$, $\nu(c_\alpha)$ is a closed expression of type $\tau(\bar{\mu}[\alpha])$.

(μ, ν) is *normal* if $\mu(\alpha) \in \mathcal{S}_2$ for all $\alpha \in \mathcal{A}_1$. Obviously, $\bar{\mu}[\alpha] = \bar{\mu}(\alpha)$ for all $\alpha \in \mathcal{S}_1$ if (μ, ν) is normal.

Let $\Phi = (\mu, \nu)$ be a translation from T_1 to T_2 throughout the rest of this section. For $A \in \mathcal{E}_1$, the *translation of* A *via* Φ, written $\Phi(A)$, is the member of $\mathcal{E}_2$ defined inductively by:

1. $\Phi(a) = \nu(a)$ if $a \in \mathcal{V}_1 \cup \mathcal{C}_1$.
2. $\Phi(\Box\{A_1, \ldots, A_n\}) = \Box\{\Phi(A_1), \ldots, \Phi(A_n)\}$ if $\Box$ is apply, equals, quasi-equals, the-true, the-false, not, and, or, implies, iff, if-form, is-defined, or if.
3. $\Phi(\Box\{x^1_{\alpha_1}, \ldots, x^n_{\alpha_n} \,.\, B\}) = \Box\{\Phi(x^1_{\alpha_1}), \ldots, \Phi(x^n_{\alpha_n}) \,.\, \Phi(B)\}$ if $\Box$ is lambda, iota, iota-p, forall, or forsome and $\{\bar{\mu}(\alpha_1), \ldots, \bar{\mu}(\alpha_n)\} \subseteq \mathcal{S}_2$.
4. $\Phi(\Box\{x^1_{\alpha_1}, \ldots, x^n_{\alpha_n} \,.\, B\}) =$

$$\begin{cases} \Box\{\Phi(x^1_{\alpha_1}), \ldots, \Phi(x^n_{\alpha_n}) \,.\, \mathtt{if}\{C, \Phi(B), \Box'\{\gamma\}\}\} & \text{if } \Box \text{ is lambda} \\ \Box\{\Phi(x^1_{\alpha_1}), \ldots, \Phi(x^n_{\alpha_n}) \,.\, \mathtt{implies}\{C, \Phi(B)\}\} & \text{if } \Box \text{ is forall} \\ \Box\{\Phi(x^1_{\alpha_1}), \ldots, \Phi(x^n_{\alpha_n}) \,.\, \mathtt{and}\{C, \Phi(B)\}\} & \text{if } \Box \text{ is forsome,} \\ & \text{iota, or iota-p} \end{cases}$$

if $\{\bar{\mu}(\alpha_1), \ldots, \bar{\mu}(\alpha_n)\} \not\subseteq \mathcal{S}_2$, where

$$C = \mathtt{and}\{\kappa(\bar{\mu}(\alpha_1), \Phi(x^1_{\alpha_1})), \ldots, \kappa(\bar{\mu}(\alpha_n), \Phi(x^n_{\alpha_n}))\},$$

γ is the sort of $\Phi(B)$, and $\Box'$ is undefined [falselike] if γ is of kind ι [$*$].
5. $\Phi(\Box\{\alpha\}) = \Box\{\bar{\mu}[\alpha]\}$ if $\Box$ is undefined or falselike.
6. $\Phi(\mathtt{defined\text{-}in}\{A, \beta\}) = \begin{cases} \mathtt{defined\text{-}in}\{\Phi(A), \bar{\mu}(\beta)\} & \text{if } \bar{\mu}(\beta) \in \mathcal{S}_2 \\ \mathtt{apply}\{\bar{\mu}(\beta), \Phi(A)\} & \text{if } \bar{\mu}(\beta) \in \mathcal{Q}_2 \end{cases}$

Proposition 6.2 *Let $\alpha, \beta \in \mathcal{S}_1$.*

1. *α and $\bar{\mu}[\alpha]$ are of the same kind.*
2. *$\tau(\bar{\mu}[\alpha]) = \tau(\bar{\mu}[\tau(\alpha)])$.*
3. *$\tau(\alpha) = \tau(\beta)$ implies $\tau(\bar{\mu}[\alpha]) = \tau(\bar{\mu}[\beta])$.*

Proof The same as the proof of Proposition 11.1 in [9].□

Proposition 6.3 *Let $A_\alpha \in \mathcal{E}_1$.*

1. *$\Phi(A_\alpha)$ is a (well-formed) member of $\mathcal{E}_2$ of type $\tau(\bar{\mu}[\alpha])$.*
2. *If the sort of $\nu(c_\gamma)$ is $\bar{\mu}[\gamma]$ for each constant c_γ occurring in A_α, then the sort of $\Phi(A_\alpha)$ is $\bar{\mu}[\alpha]$.*

Proof By induction on $|A_\alpha|$.□

Before proceeding to the notion of an interpretation, we shall make a few comments about this definition of a translation:

- The mapping μ plays the role of U in a standard translation.
- The variables of the source theory are injectively mapped to the variables of the target theory by both a standard and LUTINS translation. In (single-sorted) first-order logic, variables all have effectively the same sort. Consequently, they can be uniformly mapped to themselves and need not be in the domain of ν in a standard translation. Although variables have different sorts in a sorted logic such as LUTINS, one might think that a variable of the form x_α could be uniformly mapped to $x_{\bar{\mu}[\alpha]}$. However, this mapping would

clearly not be injective if there exists variables $x_\alpha, x_\beta \in \mathcal{V}_1$ such that $\alpha \neq \beta$ and $\bar{\mu}[\alpha] = \bar{\mu}[\beta]$. Therefore, in our definition of Φ, the mapping of variables is specified explicitly by ν.

- Since simple type theory has a more uniform syntax than first-order logic, the definition of ν on constants is less complicated in a LUTINS translation than in a standard translation.
- A normal LUTINS translation preserves the structure of LUTINS syntax, as can be seen clearly by the definition given above.
- In some cases, a nonnormal LUTINS translation relativizes the constructor `defined-in` and the variable-binding constructors: `lambda`, `iota`, `iota-p`, `forall`, `forsome`. For each of these expression constructors, the relativization is defined in the obvious way. In particular, the relativization of `lambda` in an expression $A = \texttt{lambda}\{x^1_{\alpha_1}, \ldots, x^n_{\alpha_n} \,.\, B\}$ produces an expression beginning with `lambda` which is the appropriate restriction of the unrelativized translation of A.

Interpretations

Φ is an *interpretation of* T_1 *in* T_2 if $\Phi(A_*)$ is a theorem of T_2 for each theorem A_* of T_1. T_1 is *interpretable in* T_2 if there is an interpretation of T_1 in T_2.

A *pre-obligation* of Φ is any one of the following theorems of T_1:

1. An axiom of T_1.
2. $\texttt{forsome}\{x_\alpha \,.\, \texttt{the-true}\{\}\}$ where $\alpha \in \mathcal{A}_1$.
3. $\texttt{defined-in}\{c_\alpha, \alpha\}$ where $c_\alpha \in \mathcal{C}_1$.
4. $\texttt{forall}\{x_\alpha \,.\, \texttt{defined-in}\{x_\alpha, \xi_1(\alpha)\}\}$ where $\alpha \in \mathcal{A}_1$.

An *obligation* of Φ is any sentence $\Phi(A_*)$ where A_* is a pre-obligation of Φ. The four kinds of pre-obligations [obligations] are called, in order, *axiom*, *sort nonemptiness*, *constant sort*, and *sort inclusion* pre-obligations [obligations]. Note: The sort nonemptiness obligations of Φ are trivially theorems of T_2 if Φ is normal.

The first three kinds of obligations correspond very closely to the four kinds of obligations for a standard first-order translation (constant sort obligations correspond to both individual constant symbol and function symbol obligations). There are no standard translation obligations corresponding to sort inclusion obligations because there are no subtypes in first-order logic. Obligations serve the same purpose for LUTINS translations as they do for standard translations: if all of the obligations of a translation are theorems of the target theory, then the translation is an interpretation. This result is proved for LUTINS in Section 8.

7 Some Examples Involving Groups and Fields

In this section we construct some interpretations of a theory G of an abstract group in a theory F of an abstract field. We will employ the following abbreviations:

- $F(A_1, \ldots, A_n)$ for $\mathtt{apply}\{F, A_1, \ldots, A_n\}$.
- $A = B$ for $\mathtt{equals}\{A, B\}$.
- $A \neq B$ for $\neg\{\mathtt{equals}\{A, B\}\}$.

Define $G = ((\mathcal{A}_G, \xi_G, \mathcal{V}_G, \mathcal{C}_G), \Gamma_G)$ to be a theory of an abstract group where:

1. $\mathcal{A}_G = \{*, \alpha\}$.
2. ξ_G is the identity function.
3. $\mathcal{C}_G = \{e_\alpha, \mathrm{mul}_{[\alpha,\alpha,\alpha]}, \mathrm{inv}_{[\alpha,\alpha]}\}$.
4. Γ_G is the usual set of axioms for a group.

Define $F = ((\mathcal{A}_F, \xi_F, \mathcal{V}_F, \mathcal{C}_F), \Gamma_F)$ to be a theory of an abstract field where:

1. $\mathcal{A}_F = \{*, \beta\}$.
2. ξ_F is the identity function.
3. $\mathcal{C}_F = \{0_\beta, 1_\beta, +_{[\beta,\beta,\beta]}, \times_{[\beta,\beta,\beta]}\}$.
4. Γ_F is the usual set of axioms for a field.

Except for $\mathcal{V}_G$ and $\mathcal{V}_F$, the languages of G and F are determined by the conditions given above.

Example 7.1 Let $\Phi_1 = (\mu_1, \nu_1)$ be an interpretation of G in F such that:

1. $\mu_1(\alpha) = \beta$.
2. $\nu_1(e_\alpha) = 0_\beta$.
3. $\nu_1(\mathrm{mul}_{[\alpha,\alpha,\alpha]}) = +_{[\beta,\beta,\beta]}$.
4. $\nu_1(\mathrm{inv}_{[\alpha,\alpha]}) = \mathtt{lambda}\{x_\beta \,.\, \mathtt{iota}\{y_\beta \,.\, +_{[\beta,\beta,\beta]}(x_\beta, y_\beta) = 0_\beta\}\}$.

Φ_1 shows that the elements of a field form a group under addition.□

Example 7.2 Let $\Phi_2 = (\mu_2, \nu_2)$ be an interpretation of G in F such that:

1. $\mu_2(\alpha) = \mathtt{lambda}\{x_\beta \,.\, x_\beta \neq 0_\beta\}$.
2. $\nu_2(e_\alpha) = 1_\beta$.
3. $\nu_2(\mathrm{mul}_{[\alpha,\alpha,\alpha]}) = \mathtt{lambda}\{x_\beta, y_\beta \,.\, \mathtt{if}\{\mathtt{and}\{x_\beta \neq 0_\beta, y_\beta \neq 0_\beta\},$
   $\times_{[\beta,\beta,\beta]}(x_\beta, y_\beta),$
   $\mathtt{undefined}\{\beta\}\}\}$.
4. $\nu_2(\mathrm{inv}_{[\alpha,\alpha]}) = \mathtt{lambda}\{x_\beta \,.\, \mathtt{iota}\{y_\beta \,.\, \times_{[\beta,\beta,\beta]}(x_\beta, y_\beta) = 1_\beta\}\}$.

Φ_2 shows that the nonzero elements of a field form a group under multiplication. Since α is associated with a quasi-sort, Φ_2 is nonnormal and thus relativizes the quantifiers in expressions of G. Notice that $\mathrm{mul}_{[\alpha,\alpha,\alpha]}$ is associated with the restriction of $\times_{[\beta,\beta,\beta]}$ to the nonzero elements of β, and $\nu_2(\mathrm{inv}_{[\alpha,\alpha]})$ is undefined at 0_β.□

Now define $F' = ((\mathcal{A}_F \cup \{\gamma\}, \xi_{F'}, \mathcal{V}_{F'}, \mathcal{C}_F), \Gamma_F \cup \{A_*\})$ to be an extension of F such that:

1. $\xi_{F'}(\gamma) = \beta$.
2. $A_* = \texttt{forall}\{x_\beta\, .\, \texttt{iff}\{\texttt{defined-in}\{x_\beta, \gamma\}, x_\beta \neq 0_\beta\}\}$.

Thus F' is obtained by defining in F an atomic sort γ with the same extension as the quasi-sort $\texttt{lambda}\{x_\beta\, .\, x_\beta \neq 0_\beta\}$.

Example 7.3 Let $\Phi_3 = (\mu_3, \nu_3)$ be an interpretation of G in F' such that:

1. $\mu_3(\alpha) = \gamma$.
2. $\nu_3(e_\alpha) = 1_\beta$.
3. $\nu_3(\mathrm{mul}_{[\alpha,\alpha,\alpha]}) = \texttt{lambda}\{x_\gamma, y_\gamma\, .\, \times_{[\beta,\beta,\beta]}(x_\gamma, y_\gamma)\}$.
4. $\nu_3(\mathrm{inv}_{[\alpha,\alpha]}) = \texttt{lambda}\{x_\gamma\, .\, \texttt{iota}\{y_\gamma\, .\, \times_{[\beta,\beta,\beta]}(x_\gamma, y_\gamma) = 1_\beta\}\}$.

Like Φ_2, Φ_3 shows that the nonzero elements of a field form a group under multiplication. In a certain sense, Φ_2 and Φ_3 are two renditions of the same interpretation. However, since Φ_3 associates α with a sort, Φ_3 is syntactically more economical than Φ_2. In particular, Φ_3 is normal while Φ_2 is nonnormal.□

Examples 7.2 and 7.3 together illustrate how a nonnormal translation can be transformed into a normal translation. Since LUTINS admits subtypes, any nonnormal translation can be "normalized" in this way as long as one is willing to define new atomic sorts. The general idea is as follows. Suppose the translation Φ is nonnormal. For each $\alpha \in \mathcal{A}_1$ such that $\mu(\alpha) \notin \mathcal{S}_2$, (1) add to T_2 a new atomic sort α' and a new axiom which says that $\mu(\alpha)$ and α' are coextensional and (2) redefine μ so that $\mu(\alpha) = \alpha'$. The resulting translation is obviously normal. (This construction is described in detail in the next section.) Hence, nonnormal translations are unnecessary in LUTINS if there is no restriction on defining atomic sorts. However, in practice the strict avoidance of nonnormal translations can easily lead to theories with a large number of atomic sorts which are almost never used.

8 Interpretation and Satisfiability Theorems

We prove in this section the interpretation and relative satisfiability theorems for LUTINS. The former establishes a sufficient condition for a translation to be an interpretation, and the latter says that a theory which is interpretable in a satisfiable theory is itself satisfiable. They correspond, respectively, to the Standard Interpretation Theorem (Theorem 2.1) and Standard Relative Satisfiability (Theorem 2.2) for first-order logic. The proofs are based on the interpretation and relative satisfiability theorems for **PF*** proved in [9].

Let $T_i = (\mathcal{L}_i, \Gamma_i)$ be a LUTINS theory for $i = 1, 2$.

PF* Theories and Translations

Given a LUTINS theory $T = (\mathcal{L}, \Gamma)$, let $\widetilde{T} = (\widetilde{\mathcal{L}}, \{\zeta(A_*) : A_* \in \Gamma\})$, which is a **PF*** theory.

Proposition 8.1 *Let $T = (\mathcal{L}, \Gamma)$ be a* LUTINS *theory and $\mathcal{M}$ be a model for $\mathcal{L}$. Then:*

1. *$\mathcal{M}$ is a model for T iff $\mathcal{M}$ is a standard model for $\widetilde{T}$.*
2. *$T \models A_*$ iff $\widetilde{T} \models \zeta(A_*)$, for all sentences $A_* \in \mathcal{E}$.*

Proof Let φ be a $\mathcal{V}$-assignment into $\mathcal{M}$ and A_* be a formula of $\mathcal{L}$. Then, by definition, $U_\varphi^{\mathcal{M}}(A_*) = V_\varphi^{\mathcal{M}}(\zeta(A_*))$. This identity implies (1), and (1) and the identity imply (2).□

Suppose $\Phi_{\mathrm{n}} = (\mu, \nu)$ is a normal LUTINS translation from T_1 to T_2. Let $\widetilde{\Phi_{\mathrm{n}}} = (\mu, \widetilde{\nu})$ where $\widetilde{\nu} : \mathcal{V}_1 \cup \widetilde{\mathcal{C}}_1 \to \widetilde{\mathcal{E}}_2$ is defined by:

1. For all $x_\alpha \in \mathcal{V}_1$, $\widetilde{\nu}(x_\alpha) = \nu(x_\alpha)$.
2. For all $c_\alpha \in \mathcal{C}_1$, $\widetilde{\nu}(c_\alpha) = \zeta(\nu(c_\alpha))$.
3. For all $\alpha \in \mathcal{S}_1$, $\widetilde{\nu}(=_{[\alpha,\alpha,*]}) = =_{[\bar{\mu}(\alpha),\bar{\mu}(\alpha),*]}$ and $\widetilde{\nu}(\iota_{[[\alpha,*],\alpha]}) = \iota_{[[\bar{\mu}(\alpha),*],\bar{\mu}(\alpha)]}$.

$\widetilde{\Phi_{\mathrm{n}}}$ is easily verified to be a **PF*** translation from $\widetilde{T_1}$ to $\widetilde{T_2}$. For $E \in \widetilde{\mathcal{E}_1}$, let $\widetilde{\Phi_{\mathrm{n}}}(E) = \widetilde{\nu}(E)$, the translation of E via $\widetilde{\Phi_{\mathrm{n}}}$.

The key lemma of this section is

Lemma 8.2 *Let Φ be a normal* LUTINS *translation from T_1 to T_2 such that each of its constant sort and sort inclusion obligations is a theorem of T_2. Also, let $\mathcal{M}_2$ be a model for T_2 and φ be a $\mathcal{V}_2$-assignment into $\mathcal{M}_2$. Then:*

1. *$V_\varphi^{\mathcal{M}_2}(\zeta(\Phi(E))) = V_\varphi^{\mathcal{M}_2}(\widetilde{\Phi}(\zeta(E)))$, for all expressions $E \in \mathcal{E}_1$.*
2. *$\widetilde{T_2} \models \zeta(\Phi(A_*))$ iff $\widetilde{T_2} \models \widetilde{\Phi}(\zeta(A_*))$, for all sentences $A_* \in \mathcal{E}_1$.*

The proof of this lemma requires the following technical lemma:

Lemma 8.3 *Let $\Phi = (\mu, \nu)$ be a normal* LUTINS *translation from T_1 to T_2 such that each of its constant sort and sort inclusion obligations is a theorem of T_2. Also, let $\mathcal{M}_2$ be a model for T_2, φ be a $\mathcal{V}_2$-assignment into $\mathcal{M}_2$, and $A_\alpha, B_\beta, C_\gamma, X_* \in \mathcal{E}_1$. Then:*

1. *If $\Box \in \{=, \simeq\}$, δ is the type of $\Phi(A_\alpha)$, and $\epsilon = \bar{\mu}(\tau(\alpha))$, then $V_\varphi^{\mathcal{M}_2}(E^\delta) = V_\varphi^{\mathcal{M}_2}(E^\epsilon)$ where*
$$E^\theta = \zeta(\Phi(A_\alpha)) \,\Box_\theta\, \zeta(\Phi(B_\beta)).$$
2. *If δ is the least upper bound in $\preceq_{\xi_2}$ of the sorts of $\Phi(B_\beta)$ and $\Phi(C_\gamma)$, and $\epsilon = \bar{\mu}(\beta \sqcup_{\xi_2} \gamma)$, then $V_\varphi^{\mathcal{M}_2}(E^\delta) = V_\varphi^{\mathcal{M}_2}(E^\epsilon)$ where*
$$E^\theta = \mathrm{I}\{x_\theta \,.\, (\zeta(\Phi(X_*)) \supset (x_\theta =_\theta \zeta(\Phi(B_\beta)))) \wedge (\neg\zeta(\Phi(X_*)) \supset (x_\theta =_\theta \zeta(\Phi(C_\gamma))))\}.$$

3. If δ is the sort of $\Phi(A_\alpha)$ and $\epsilon = \bar{\mu}(\alpha)$, then $V_\varphi^{\mathcal{M}_2}(E^\delta) = V_\varphi^{\mathcal{M}_2}(E^\epsilon)$ where

$$E^\theta = (\zeta(\Phi(A_\alpha)) \downarrow \theta).$$

Proof Follows from the lemmas in Section 11 of [9].□

Proof of Lemma 8.2 Part (2) follows from part (1) by Proposition 8.1.

Let $\Phi = (\mu, \nu)$ and $E \in \mathcal{E}_1$. The proof of (1) is by induction on $|E|$.

Basis. Assume $|E| = 0$; then $E \in \mathcal{V}_1 \cup \mathcal{C}_1$.

Let $E \in \mathcal{V}_1$. Then

$$\begin{aligned} V_\varphi^{\mathcal{M}_2}(\zeta(\Phi(E))) &= V_\varphi^{\mathcal{M}_2}(\Phi(E)) & (1)\\ &= V_\varphi^{\mathcal{M}_2}(\widetilde{\Phi}(E)) & (2)\\ &= V_\varphi^{\mathcal{M}_2}(\widetilde{\Phi}(\zeta(E))) & (3) \end{aligned}$$

(1) and (3) hold since ζ is the identity on $\mathcal{V}_1$, and (2) holds since $\Phi = \widetilde{\Phi}$ on $\mathcal{V}_1$.

Let $E \in \mathcal{C}_1$. Then

$$\begin{aligned} V_\varphi^{\mathcal{M}_2}(\zeta(\Phi(E))) &= V_\varphi^{\mathcal{M}_2}(\widetilde{\Phi}(E)) & (4)\\ &= V_\varphi^{\mathcal{M}_2}(\widetilde{\Phi}(\zeta(E))) & (5) \end{aligned}$$

(4) holds since $\zeta \circ \Phi = \widetilde{\Phi}$ on $\mathcal{C}_1$, and (5) holds since ζ is the identity on $\mathcal{C}_1$.

Induction step. Assume $|E| > 0$; then E is a compound expression beginning with the expression constructor □.

Let $E = \texttt{defined-in}\{A_\alpha, \beta\}$. Then

$$\begin{aligned} &V_\varphi^{\mathcal{M}_2}(\zeta(\Phi(\texttt{defined-in}\{A_\alpha, \beta\}))) & \\ &= V_\varphi^{\mathcal{M}_2}(\zeta(\texttt{defined-in}\{\Phi(A_\alpha), \bar{\mu}(\beta)\})) & (6)\\ &= V_\varphi^{\mathcal{M}_2}(\zeta(\Phi(A_\alpha)) \downarrow \bar{\mu}(\beta)) & (7)\\ &= V_\varphi^{\mathcal{M}_2}(\widetilde{\Phi}(\zeta(A_\alpha)) \downarrow \bar{\mu}(\beta)) & (8)\\ &= V_\varphi^{\mathcal{M}_2}(\widetilde{\Phi}(\zeta(A_\alpha) \downarrow \beta)) & (9)\\ &= V_\varphi^{\mathcal{M}_2}(\widetilde{\Phi}(\zeta(\texttt{defined-in}\{A_\alpha, \beta\}))) & (10) \end{aligned}$$

(6) is by the definition of the application of a normal LUTINS translation to an expression; (7) and (10) are by the definition of ζ; (8) is by the induction hypothesis; and (9) is by Lemma 11.5 in [9].

The derivation is very similar when □ is any one of the other expression constructors except for `equals`, `quasi-equals`, `if`, or `is-defined`. The derivations for these last four expression constructors require Lemma 8.3.

This completes the proof.□

A Special Case of the Interpretation Theorem

Lemma 8.2 and the Semantic Interpretation Theorem for **PF*** imply the Interpretation Theorem for LUTINS restricted to normal translations:

Theorem 8.4 *A normal* LUTINS *translation from T_1 to T_2 is an interpretation if each of its obligations is a theorem of T_2.*

Proof Let Φ be a normal LUTINS translation from T_1 to T_2 such that each of its obligations is a theorem of T_2.

We begin by showing that $\widetilde{\Phi}$ is an interpretation. Let O be an obligation of $\widetilde{\Phi}$. Then O has the form $\widetilde{\Phi}(\zeta(A_*))$ where A_* is an axiom, constant sort, or sort inclusion pre-obligation of Φ. By hypothesis, $T_2 \models \Phi(A_*)$. Then

$$T_2 \models \Phi(A_*) \text{ iff } \widetilde{T_2} \models \zeta(\Phi(A_*)) \quad (11)$$

$$\text{iff } \widetilde{T_2} \models \widetilde{\Phi}(\zeta(A_*)) \quad (12)$$

(11) holds by Proposition 8.1, and (12) holds by Lemma 8.2. Hence, each obligation of $\widetilde{\Phi}$ is a theorem of $\widetilde{T_2}$, and so $\widetilde{\Phi}$ is a **PF*** interpretation by the Semantic Interpretation Theorem for **PF*** [9, Theorem 12.4].

Let A_* be a theorem of T_1. Then

$$T_1 \models A_* \quad \text{iff} \quad \widetilde{T_1} \models \zeta(A_*) \quad (13)$$

$$\text{implies } \widetilde{T_2} \models \widetilde{\Phi}(\zeta(A_*)) \quad (14)$$

$$\text{iff} \quad \widetilde{T_2} \models \zeta(\Phi(A_*)) \quad (15)$$

$$\text{iff} \quad T_2 \models \Phi(A_*) \quad (16)$$

(13) and (16) hold by Proposition 8.1; (14) holds because $\widetilde{\Phi}$ is a **PF*** interpretation; and (15) holds by Lemma 8.2. Therefore, the sequence of implications shows that Φ maps each theorem of T_1 to a theorem of T_2, that is, that Φ is an interpretation.□

The Translation Φ' and Its Properties

Before we can strengthen Theorem 8.4 to the full interpretation theorem for LUTINS, we must show how an arbitrary LUTINS translation can be transformed into a normal translation.

Let $\Phi = (\mu, \nu)$ be a LUTINS translation from T_1 to T_2. The key idea is to add new atomic sorts to T_2. Let $T_2' = (\mathcal{L}_2', \Gamma_2')$ be the theory

$$((\mathcal{A} \cup \mathcal{A}_2, \xi \cup \xi_2, \mathcal{V} \cup \mathcal{V}_2, \mathcal{C}_2), \Gamma \cup \Gamma_2)$$

such that:

1. $\mathcal{A} = \{\gamma^{\mu(\alpha)} : \alpha \in \mathcal{A}_1 \text{ and } \mu(\alpha) \in \mathcal{Q}_2\}$.
2. $\xi : \mathcal{A} \to \mathcal{A}_2$ and $\xi(\gamma^{Q[\alpha,*]}) = \alpha$.
3. $\mathcal{V}$ is an appropriate set of new variables.

4. $\Gamma = \{A^{Q_{[\alpha,*]}} : \gamma^{Q_{[\alpha,*]}} \in \mathcal{A}\}$ where $A^{Q_{[\alpha,*]}}$ is

$$\texttt{forall}\{x_{\tau(\alpha)}\cdot \texttt{iff}\{\texttt{defined-in}\{x_{\tau(\alpha)}, \gamma^{Q_{[\alpha,*]}}\}, \texttt{apply}\{Q_{[\alpha,*]}, x_{\tau(\alpha)}\}\}\}.$$

T_2' is a "definitional" extension of T_2; its new atomic sorts are axiomatically defined to be coextensional with the quasi-sorts in the range of μ. Consequently, the following proposition is easy to prove.

Proposition 8.5 *Suppose each sort nonemptiness obligation of Φ is a theorem of T_2. Then:*

1. *Each model for T_2 expands to a model for T_2'.*
2. *Each model for T_2' reduces to a model for T_2.*

Let $\Phi' = (\mu', \nu')$ be a translation from T_1 to T_2' such that:

1. If $\mu(\alpha) \in \mathcal{S}_2$, then $\mu'(\alpha) = \mu(\alpha)$.
2. If $\mu(\alpha) \in \mathcal{Q}_2$, then $\mu'(\alpha) = \gamma^{\mu(\alpha)}$.
3. If $c_\alpha \in \mathcal{C}_1$, then $\nu'(c_\alpha) = \nu(c_\alpha)$.

Φ' is clearly normal. Notice that $\Phi \neq \Phi'$ on $\mathcal{V}_1$ unless Φ is normal.

Lemma 8.6 *Let $\mathcal{M}_2$ and $\mathcal{M}_2'$ be models for T_2 and T_2', respectively, such that $\mathcal{M}_2'$ is an expansion of $\mathcal{M}_2$. Then*

$$U_\varphi^{\mathcal{M}_2}(\Phi(E)) = U_{\varphi'}^{\mathcal{M}_2'}(\Phi'(E))$$

for all $E \in \mathcal{E}_1$, $\mathcal{V}_2$-assignments φ into $\mathcal{M}_2$, and $\mathcal{V}_2'$-assignments φ' into $\mathcal{M}_2'$ such that $\varphi(\Phi(x_\alpha)) = \varphi'(\Phi'(x_\alpha))$ for all $x_\alpha \in \mathcal{V}_1$.

Proof By induction on $|E|$.□

Lemma 8.7 *Suppose each sort nonemptiness obligation of Φ is a theorem of T_2. Then*

$$T_2 \models \Phi(A_*) \text{ iff } T_2' \models \Phi'(A_*)$$

for all sentences $A_* \in \mathcal{E}_1$.

Proof By Proposition 8.5 and Lemma 8.6.□

The Theorems

Theorem 8.8 (Interpretation Theorem) *A translation from T_1 to T_2 is an interpretation if each of its obligations is a theorem of T_2.*

Proof Let Φ be a LUTINS translation from T_1 to T_2 such that each of its obligations is a theorem of T_2. As above, construct the theory T_2' from T_2 and the normal translation Φ' from Φ. Since Φ and Φ' have the same source theory, they also have the same pre-obligations. Hence, each obligation of Φ' is a theorem of T_2' by Lemma 8.7. Φ' is normal, so Φ' is an interpretation by Theorem 8.4. This implies that Φ is an interpretation by Lemma 8.7.□

Theorem 8.9 (Relative Satisfiability) *If T_1 is interpretable in T_2 and T_2 is satisfiable, then T_1 is satisfiable.*

Proof Let Φ be a LUTINS interpretation of T_1 in T_2 and $\mathcal{M}_2$ be a model for T_2. As above, construct the theory T_2' from T_2 and the normal translation Φ' from Φ. By Proposition 8.5, there is a model $\mathcal{M}_2'$ for T_2' which is an expansion of $\mathcal{M}_2$ since Φ is an interpretation. $\mathcal{M}_2'$ is also a standard model for $\widetilde{T_2'}$ by Proposition 8.1. By Lemma 8.7, Φ' is an interpretation, and so by the proof of Theorem 8.4, $\widetilde{\Phi}'$ is a also an interpretation. Therefore, there is a standard model $\mathcal{M}_1$ for $\widetilde{T_1}$ by Relative Satisfiability for **PF*** [9, Theorem 12.3]. $\mathcal{M}_1$ is also a model for T_1 by Proposition 8.1, so T_1 is satisfiable.□

9 Conclusion

In this paper we have developed a method for theory interpretation in LUTINS, a version of simple type theory which supports partial functions and subtypes. The method embodies the principal characteristics of the standard approach to theory interpretation in first-order logic:

- A translation is a kind of homomorphism that is determined by how the primitive symbols of the source theory are associated with objects of the target theory.
- When sort symbols are associated with unary predicates, the translation of an expression may involve the relativization of variable-binding constructors such as the universal and existential quantifiers.
- A translation is an interpretation if the "obligations" of the translation are theorems of the target theory (interpretation theorem).
- A theory which is interpretable in a satisfiable theory is itself satisfiable (relative satisfiability).

Since the method is based on a logic with partial functions and subtypes, it has two advantages over the standard first-order approach:

- The functions with restricted domains that naturally arise from interpretations which associate base types with subtypes are handled directly as partial functions.
- Relativization of variable-binding constructors can be avoided completely, provided that appropriate sort symbols are defined.

The method is intended as a guide for theory interpretation in classical simple type theories as well as in predicate logics which admit partial functions.

Acknowledgments

I am grateful for the suggestions received from the referees. Many of the ideas in this paper grew out of conversations I have had with Dr. Joshua Guttman and Dr. Javier Thayer in the course of developing IMPS.

References

1. P. B. Andrews. *An Introduction to Mathematical Logic and Type Theory: To Truth through Proof.* Academic Press, 1986.
2. P. B. Andrews, S. Issar, D. Nesmith, and F. Pfenning. The TPS theorem proving system (system abstract). In M. E. Stickel, editor, *10th International Conference on Automated Deduction*, volume 449 of *Lecture Notes in Computer Science*, pages 641–642. Springer-Verlag, 1990.
3. C. C. Chang and H. J. Keisler. *Model Theory.* North-Holland, 1990.
4. A. Church. A formulation of the simple theory of types. *Journal of Symbolic Logic*, 5:56–68, 1940.
5. D. Craigen, S. Kromodimoeljo, I. Meisels, A. Neilson, B. Pase, and M. Saaltink. m-EVES: A tool for verifying software. In *11th International Conference on Software Engineering (ICSE'11)*, Singapore, 1988.
6. D. Craigen, S. Kromodimoeljo, I. Meisels, B. Pase, and M. Saaltink. EVES: An overview. Technical Report CP-91-5402-43, ORA Corporation, 1991.
7. H. B. Enderton. *A Mathematical Introduction to Logic.* Academic Press, 1972.
8. W. M. Farmer. A partial functions version of Church's simple theory of types. *Journal of Symbolic Logic*, 55:1269–1291, 1990.
9. W. M. Farmer. A simple type theory with partial functions and subtypes. *Annals of Pure and Applied Logic*, 64:211–240, 1993.
10. W. M. Farmer, J. D. Guttman, and F. J. Thayer. IMPS: System description. In D. Kapur, editor, *Automated Deduction—CADE-11*, volume 607 of *Lecture Notes in Computer Science*, pages 701–705. Springer-Verlag, 1992.
11. W. M. Farmer, J. D. Guttman, and F. J. Thayer. Little theories. In D. Kapur, editor, *Automated Deduction—CADE-11*, volume 607 of *Lecture Notes in Computer Science*, pages 567–581. Springer-Verlag, 1992.
12. W. M. Farmer, J. D. Guttman, and F. J. Thayer. IMPS: an Interactive Mathematical Proof System. *Journal of Automated Reasoning*, 11:213–248, 1993.
13. F. Giunchiglia and T. Walsh. A theory of abstraction. *Artificial Intelligence*, 57:323–389, 1992.
14. J. A. Goguen and T. Winkler. Introducing OBJ3. Technical Report SRI-CSL-99-9, SRI International, August 1988.
15. M. J. C. Gordon. HOL: A proof generating system for higher-order logic. In G. Birtwistle and P. A. Surahmanyam, editors, *VLSI Specification, Verification, and Synthesis*, pages 73–128. Kluwer, 1987.
16. J. D. Guttman. A proposed interface logic for verification environments. Technical Report M91-19, The MITRE Corporation, 1991.
17. R. W. Harper and F. Pfenning. A module system for a programming language based on the LF logical framework. *Journal of Functional Programming.* Forthcoming.
18. D. Hilbert. *The Foundations of Geometry.* Open Court, Chicago, 1902.
19. K. Kunen. *Set Theory: An Introduction to Independence Proofs.* North-Holland, 1980.
20. B. H. Levy. *An Approach to Compiler Correctness Using Interpretation Between Theories.* PhD thesis, University of California, Los Angeles, 1986. Also Technical Report ATR-86(8454)-4, The Aerospace Corporation, El Segundo, California.
21. T. S. E. Maibaum, P. A. S. Veloso, and M. R. Sadler. A theory of abstract data types for program development: Bridging the gap? In H. Ehrig, C. Floyd, M. Nivat, and J. Thatcher, editors, *Formal Methods and Software Development*,

Volume 2, volume 186 of *Lecture Notes in Computer Science*, pages 214–230. Springer-Verlag, 1985.

22. J. D. Monk. *Mathematical Logic*. Springer-Verlag, 1976.
23. J. Mycielski. A lattice of interpretability types of theories. *Journal of Symbolic Logic*, 42:297–305, 1977.
24. R. Nakajima and T. Yuasa, editors. *The* IOTA *Programming System*, volume 160 of *Lecture Notes in Computer Science*. Springer-Verlag, 1982.
25. T. Nipkow and L. C. Paulson. Isabelle-91. In D. Kapur, editor, *Automated Deduction—CADE-11*, volume 607 of *Lecture Notes in Computer Science*, pages 673–676. Springer-Verlag, 1992.
26. S. Owre, J. M. Rushby, and N. Shankar. PVS: A prototype verification system. In D. Kapur, editor, *Automated Deduction—CADE-11*, volume 607 of *Lecture Notes in Computer Science*, pages 748–752. Springer-Verlag, 1992.
27. J. Rushby, F. von Henke, and S. Owre. An introduction to formal specification and verification using EHDM. Technical Report SRI-CSL-91-02, SRI International, 1991.
28. J. R. Shoenfield. *Mathematical Logic*. Addison-Wesley, 1967.
29. D. R. Smith and M. R. Lowry. Algorithmic theories and design tactics. *Science of Computer Programming*, 14:305–321, 1990.
30. L. W. Szczerba. Interpretability of elementary theories. In R. E. Butts and J. Hintikka, editors, *Logic, Foundations of Mathematics, and Computability Theory*, pages 129–145. Reidel, 1977.
31. L. W. Szczerba. Interpretability and axiomatizability. *Bulletin de L'Académie Polonaise des Sciences*, 27:425–429, 1979.
32. A. Tarski, A. Mostowski, and R. M. Robinson. *Undecidable theories*. North-Holland, 1953.
33. W. M. Turski and T. S. E. Maibaum. *The Specification of Computer Programs*. Addison-Wesley, 1987.
34. J. van Bentham and D. Pearce. A mathematical characterization of interpretation between theories. *Logica Studia*, 43:295–303, 1984.
35. P. J. Windley. Formal modeling and verification of microprocessors. *IEEE Transactions on Computers*. Forthcoming.
36. P. J. Windley. Abstract theories in HOL. In L. Claesen and M. J. C. Gordon, editors, *Proceedings of the 1992 International Workshop on the* HOL *Theorem Prover and its Applications*. North-Holland, November 1992.

The Semantics of SPECTRUM*

Radu Grosu and Franz Regensburger

Fakultät für Informatik, Technische Universität München
80290 München, Germany

E–Mail: spectrum@informatik.tu-muenchen.de

Abstract. The SPECTRUM project concentrates on the process of developing well-structured, precise system specifications. SPECTRUM is a specification language, with a deduction calculus and a development methodology. An informal presentation of the SPECTRUM language with many examples illustrating its properties is given in [2, 3]. The purpose of this article is to describe its formal semantics.

1 Introduction

The SPECTRUM specification language is axiomatic and borrows concepts both from algebraic languages (e.g. LARCH [12]) as well as from type theoretic languages (e.g. LCF [4]). An informal presentation with many examples illustrating its properties is given in [2, 3]. We briefly summarize its principal characteristics.

Influences from algebra. In SPECTRUM specifications the influence of algebraic techniques is evident. Every specification consists of a signature and an axioms part. However, in contrast to most algebraic specification languages, the semantics of a specification in SPECTRUM is loose, i.e. it is not restricted to initial models or even term generated ones. Moreover, SPECTRUM is not restricted to equational or conditional-equational axioms, since it does not primarily aim at executable specifications. One can use full first order logic to write very abstract and non-executable specifications or only use its constructive part to write specifications which can be understood and executed as programs.

Loose semantics leaves a large degree of freedom for later implementations. It also allows the simple definition of refinement as the reduction of the class of models. This reduction is achieved by imposing new axioms which result from design decisions occurring in the stepwise development of the data structures and algorithms.

Since writing well-structured specifications is one of our main goals, a flexible language for structuring specifications has been designed for SPECTRUM. This

* This work is sponsored by the German Ministry of Research and Technology (BMFT) as part of the compound project "KORSO - Korrekte Software" and by the German Research Community (DFG) project SPECTRUM.

structuring is achieved by using so-called specification building operators which map a list of argument specifications into a result specification. The language for these operators was originally inspired by ASL [19]. The current version borrows concepts also from Haskell [13], LARCH and PLUSS [7].

Influences from type theory. The influence from type theory is twofold. On the type level SPECTRUM uses shallow predicative polymorphism with type classes in the style of Isabelle [16]. The theory of type classes was introduced by Wadler and Blott [22] and originally realized in the functional programming language Haskell. Type classes may be used both to model overloading [6, 21] as well as many instances of parameterized specifications. Like in object oriented languages type classes can be organized in hierarchies such that every class inherits properties from its parent classes. This gives our language a weak object oriented flavor.

The other influence of type theory can be seen in the language of terms and their underlying semantics. SPECTRUM incorporates the entire notation for typed λ-terms. The definition of the semantics and the proof system was heavily influenced by LCF. Therefore SPECTRUM supports a notion for partial and non–strict functions as well as higher–order functions in the sense of domain theory. The models of SPECTRUM specifications are assumed to be certain continuous algebras. All the statements about the expressiveness of LCF due to its foundation in domain theory carry over to SPECTRUM.

Beside type classes there are also two features in the SPECTRUM logic which distinguish SPECTRUM from LCF. SPECTRUM uses three–valued logic and also allows in a restricted form the use of non–continuous functions for specification purposes. These non–continuous functions are an extension of predicates. They allow the specifier to express facts in a functional style which otherwise he would have to encode as a relation. The practical usefulness of these features has to be proved in case studies.

In conclusion, all the above features make SPECTRUM a very powerful general purpose specification language. It can be used successfully in data base applications, computationally intensive applications or even distributed applications since it can easily incorporate a theory for streams and stream processing functions [1].

The purpose of this article is to describe the formal semantics of the language SPECTRUM. This semantics incorporates in a uniform and coherent way the properties already mentioned. In comparison with other logics for higher order functions (e.g. LCF family) our main contributions are:

- a denotational semantics based on order sorted algebras for type classes (we are only aware of an operational semantics for Haskell),
- the use of non–continuous functions for specification purposes,
- the identification of predicates with (strong) boolean functions in the context of a three valued logic.

The paper is organized as follows. In section 2 we present some examples that show the use of the specification language SPECTRUM. In sections 3 and 4 we introduce the polymorphic signatures and the well–formed terms. In section 5 we describe the polymorphic algebras. Section 6 is devoted to the interpretation of terms in these algebras and to the notions of satisfaction and model. Finally we draw some conclusions in section 7.

Note that space limitations caused us to leave out the treatment of generation constraints and of constructs related to specifying in the large (e.g. signature morphisms, reducts and logical relations between algebras). A full treatment is given in [10].

2 Some Motivating Examples

Before discussing the more involved technicalities of SPECTRUM we present some motivating examples. They will help to better appreciate the design decisions made in SPECTRUM and to get more intuition about its syntax and semantics. We start by giving a polymorphic specification of lists.

```
LIST = {
    sort List α;                                  --Sort constructor List
    nil: List α;                                  --Constructors
    cons: α × List α → List α;
    first: List α → α;                            --Selectors
    rest: List α → List α;
    cons, first, rest strict;                     --Strictness & Totality
    cons total;
    List α freely generated by nil, cons;         --Generation
    axioms ∀ a : α, l : List α in
        first(nil) = ⊥;
        first(cons(a,l)) = a;
        rest(nil) = ⊥;
        rest(cons(a,l)) = l;
    endaxioms; }
```

The signature of this specification consists of the sort constructor List, the value constructors nil and cons and the selectors first and rest. The sort variable α is used both to indicate the unarity of the sort constructor and to abstract from a concrete element sort. As a consequence, all functions in the signature are defined polymorphically i.e. they can be used on all lists List τ where τ is an instance sort of α. Similarly, the use of the sort variable in the axioms part indicates their validity for all instantiations τ of this variable.

Although not explicitly in the **axioms** part, the **strict** and **total** declarations as well as the **freely generated by** declaration are actually first and respectively second order axioms. They have a significant influence on the structure of lists and their associated limits. Making cons strict we have obtained the *finite* ML lists. Had we not declared cons strict, we had obtained the Haskell lazy lists which can be infinite or finite.

Beside the explicitly declared signature, each specification also has an implicit, predefined one. This contains for example the non-continuous polymorphic strong equality function (or mapping) $=\ :\ \alpha \times \alpha$ **to** Bool. Note the use of **to** instead of $\rightarrow$ to mark this distinction.

A class is used in SPECTRUM to group together sorts which own a given set of functions which in turn satisfy a given set of axioms. For example the class EQ of all sorts owning a weak equality "predicate" == can be defined as follows:

```
Equality = {
    class EQ;
    .==. : α :: EQ ⇒ α × α → Bool;
    .==. strict total;
    axioms α :: EQ ⇒ ∀ a, b : α in
        (a == b) = (a = b);
    endaxioms; }
```

For each sort constructor (and in particular a nullary one) one can declare the domain and range classes. For example the following specification:

```
EqInstances = {
    enriches Equality + LIST + NAT;
    Nat :: EQ;
    List :: (EQ)EQ; }
```

declares that the sort Nat belongs to the class EQ. It also states that each sort in the class EQ is also mapped by List into a sort from EQ. The keywords **enriches** and + are "specifying-in-the-large" constructs. The **enriches** construct includes the signature and the axioms of the argument specification into the current definition. The + construct takes the union (on signatures and axioms) of the argument specifications.

Classes are used to restrict the range of the sort variables. For example suppose that we want to extend the specification LIST with a "predicate" .∈., testing whether an element is contained in a list. This can be done by using the above class EQ as follows:

```
LISTI = {
    enriches LIST + Equality;
    .∈. : α :: EQ ⇒ α × List α → Bool;
    .∈. strict total;
    axioms α :: EQ ⇒ ∀a, x : α, l : List α in
        ¬(a ∈ nil);
        a ∈ cons(x,l) ⇔ (a == x) ∨ a ∈ l;
    endaxioms; }
```

The use of the class EQ is vital here. On the one hand a total .∈. function is not monotonic and as a consequence not a continuous function on non-flat sorts. On the other hand it is implicitly declared as a continuous function in the signature. Hence, allowing α to range over all possible sorts would make the above specification inconsistent. A similar problem occurs in ML where equality sorts are syntactically distinguished from the ones ranging over all possible sorts.

Classes can be built hierarchically. For example we can reuse the definition of the class EQ in defining a more restrictive class TOrder of total orders as follows:

TOrder = {
  enriches Equality;
  class TO **subclass of** EQ;
  .≤.: α :: TO ⇒ $\alpha \times \alpha \rightarrow$ Bool;
  .≤. **strict total**;
  axioms α :: TO ⇒ ∀ a, b, c: α **in**
    a ≤ a; —— *reflexivity*
    a ≤ b ∧ b ≤ c ⇒ a ≤ c; —— *transitivity*
    a ≤ b ∧ b ≤ a ⇒ a == b; —— *antisymmetry*
    a ≤ b ∨ b ≤ a; —— *totality*
  endaxioms; }

Each sort in TOrder is required by the **subclass of** declaration to be also contained in the class EQ.

By allowing arbitrary class definitions we can achieve a certain degree of function overloading and of specification parameterization similar to OBJ. Moreover, since we can tune the extent of polymorphism for each function separately, we achieve a considerable degree of specification reuse. For example, we can easily extend the specification LISTI with a function min which takes the minimum of a list provided the elements are totally ordered as follows:

LISTM = {
  enriches LISTI + TOrder;
  min : α :: TO ⇒ List $\alpha \rightarrow \alpha$;
  min **strict**;
  axioms α :: TO ⇒ ∀ e : α, s : List α **in**
    s ≠ nil ⇒ min(s) ∈ s ∧ (e ∈ s ⇒ min(s) ≤ e);
  endaxioms; }

This ends the section with examples about SPECTRUM. Now we present the technical details of the core language of SPECTRUM.

3 Signatures

As an abstraction from the concrete syntax a specification $S = (\Sigma, E)$ is a pair where $\Sigma = (\Omega, F, O)$ is a polymorphic signature and E is a set of Σ-formulae.

The definitions for Σ and for its components Ω, F and O are sketched in the text below. For a detailed presentation we refer to [10].

Definition 1 Sort Signature.
A sort–signature $\Omega = (K, \leq, SC)$ is an order sorted signature[2], where

- $(K, \leq)$ is a partial order on *kinds*,
- $SC = \{SC_{w,k}\}_{w \in (K \setminus \{map\})^*, k \in K}$ is an indexed set of *sort constructors* with monotonic functionalities i.e.:

$$(sc \in SC_{w,k} \cap SC_{w',k'}) \wedge (w \leq w') \Rightarrow (k \leq k')$$

A sort–signature must satisfy the following additional constraints:

- It is *regular, coregular* and *downward complete*. These properties[3] guarantee the existence of principal kinds.
- It includes the standard sort–signature (see below).
- All kinds except *map* and *cpo*, which are in the standard signature, are below *cpo* with respect to $\leq$. In other words, *cpo* is the top kind for all kinds a user may introduce.

Definition 2 The standard (predefined) sort–signature.
The standard sort–signature

$$\begin{aligned} \Omega_{standard} = (\ & \{cpo, map\}, \emptyset, \\ & \{\{\mathsf{Bool}\}_{cpo}, \\ & \ \{\rightarrow\}_{cpo\ cpo,\ cpo}, \\ & \ \{\mathbf{to}\}_{cpo\ cpo,\ map}, \\ & \ \{\times_n\}_{\underbrace{cpo \ldots cpo}_{n\ times},\ cpo} \\ & \} \\ ) & \end{aligned}$$

contains two kinds and four sort constructors (actually, we have for each n a sort constructor $\times_n$):

- *cpo* represents the kind of *all complete partial orders*, *map* represents the kind of *all full function spaces*[4],

---

[2] Order kinded would be more precise; see [9, 8] for order sorted algebras.

[3] Regularity guarantees least kinds for every sort term. Coregularity and downward completeness guarantee unitary unification of sort terms. See [16, 20] for details.

[4] Complete partial orders are used to model continuous functions and full function spaces are used to model non-continuous functions. The latter ones are never implemented but are extremely useful for specification purposes. An alternative approach is to use only full function spaces in the semantics and to encode continuity of functions in the logic. In [18] HOLCF a higher order version of LCF is embedded into the logic HOL using the generic framework of Isabelle. In this thesis it is shown that the full function space and its subspace of continuous functions over cpo's can live together in one type frame without problems.

- Bool is the type of booleans, $\longrightarrow$ is the constructor for lifted continuous function spaces, **to** is the constructor for full function spaces and $\times_n$ for $n \geq 2$ is the constructor for Cartesian product spaces.

The sort–signatures together with a disjoint family χ of sort variables indexed by kinds (a *sort context*) allows us to define the set of sort terms.

Definition 3 Sort Terms.
$T_\Omega(\chi)$ is the freely generated order kinded term algebra over χ.

Example 1 Some sort terms. Let Set $\in SC_{cpo,\ cpo}$. Then the following terms are valid sort terms:

$$\mathsf{Set}\ \alpha \longrightarrow \mathsf{Bool},\ \ \mathsf{Bool} \times \mathsf{Bool} \in T_\Omega(\{\alpha\}_{cpo})$$

The idea behind polymorphic elements is to describe families of non-polymorphic elements. In the semantics this is represented with the concept of the generalized cartesian product. For the syntax however there are several techniques to indicate this fact. E.g. in HOL ([5]) the type of a polymorphic constant in the signature is treated as a template that may be arbitrarily instantiated to build terms. This technique is also used for the concrete syntax of SPECTRUM. In the technical paper [10] we decided to make this mechanism explicit in the syntax, too, and introduced a binding operator Π for sorts and an application mechanism on the syntactic level. For a system with simple predicative polymorphism this is just a matter of taste. For a language with local polymorphic elements (ML–polymorphism) or even deep polymorphism such binding mechanisms are essential.

Definition 4 Π – Sort Terms.
$\Pi\alpha_1 : k_1, \ldots, \alpha_n : k_n.e \in T_\Omega^\Pi$ if:

- $e \in T_\Omega(\chi)$
- $\mathsf{Free}(e) \subseteq \{\alpha_1, \ldots, \alpha_n\}$
- $k_i \leq cpo \quad$ for $k_i \in K$, $1 \leq i \leq n$

Note that the third condition rules out bound sort variables of kind *map*.

Example 2 A Π – Sort Term.

$$\Pi\alpha : cpo.\mathbf{Set}\ \alpha \longrightarrow \mathsf{Bool} \in T_\Omega^\Pi$$

The idea of the template and its instantiation is made precise by Π–abstraction and application of such Π–sorts to non–polymorphic sorts $s \in T_\Omega(\chi)$.

In a signature every constant or mapping will have a sort without free sort variables. This motivates the following definition.

Definition 5 Closed Sort Terms.

$$T_\Omega = T_\Omega(\emptyset)$$
$$T_\Omega^{closed} = T_\Omega \cup T_\Omega^\Pi$$

Note that $T^{closed}_{\Omega,cpo}$ will contain valid sorts for constants while $T^{closed}_{\Omega,map}$ will contain valid sorts for mappings.
Now we are able to define polymorphic signatures.

Definition 6 Polymorphic Signature.
A polymorphic signature $\Sigma = (\Omega, F, O)$ is a triple where:

- $\Omega = (K, \leq, SC)$ is a sort–signature.
- $F = \{F_\mu\}_{\mu \in T^{closed}_{\Omega,cpo}}$ is an indexed set of constant symbols.
- $O = \{O_\nu\}_{\nu \in T^{closed}_{\Omega,map}}$ is an indexed set of mapping symbols.

It must include the standard signature
$\Sigma_{standard} = (\Omega_{standard}, F_{standard}, O_{standard})$ which is defined as follows:

- Predefined Constants ($F_{standard}$):
  - $\{\mathsf{true}, \mathsf{false}\} \subseteq F_{\mathsf{Bool}}$, $\{\neg\} \subseteq F_{\mathsf{Bool} \rightarrow \mathsf{Bool}}$, $\{\wedge, \vee, \Rightarrow\} \subseteq F_{\mathsf{Bool} \times \mathsf{Bool} \rightarrow \mathsf{Bool}}$ are the boolean constants and connectives.
  - $\{\perp\} \subseteq F_{\Pi\alpha : cpo.\ \alpha}$ is the polymorphic bottom symbol.
  - $\{\mathsf{fix}\} \subseteq F_{\Pi\alpha : cpo.\ (\alpha \rightarrow \alpha) \rightarrow \alpha}$ is the polymorphic fixed point operator.
- Predefined mappings ($O_{standard}$):
  - $\{=, \sqsubseteq\} \subseteq O_{\Pi\alpha : cpo.\ \alpha \times \alpha\ \mathbf{to}\ \mathsf{Bool}}$ are the polymorphic equality and approximation predicates.
  - $\{\delta\} \subseteq O_{\Pi\alpha : cpo.\ \alpha\ \mathbf{to}\ \mathsf{Bool}}$ is the polymorphic definedness predicate.

4 The language of Terms

In the previous section we introduced the polymorphic signatures which serve to construct terms in the object language. The construction itself is the purpose of this section.
Like in [15] the core language used to define the semantics of SPECTRUM is explicitly typed i.e. the application of polymorphic constants to sort terms is explicit and the λ–bounded variables are written together with their sorts. This assures that every well formed term has a unique sort in a given context and that the semantics of this term, although given with respect to one of its derivations, is independent from the particular derivation if the sorts of the free variables are the same in all derivation contexts.

For convenience, the *concrete* language of SPECTRUM is like ML, HOL, LCF and Isabelle implicitly typed i.e. the type information is erased from terms. However, like all the above languages, SPECTRUM has principles types i.e. every implicitly typed term t has a corresponding explicitly typed term t' such that erasing all type information from t' yields again t and for every other explicitly typed term t'' having the above property, the type of t'' is an instance of the type of t' for some special notion of instance. Having principles types guaranteed, the semantics of an implicitly typed term t is simply defined to be the semantics of t'[5]. The set of well formed terms is defined in two steps. First we define the context free syntax of pre terms via a BNF like grammar. In the second step we introduce a calculus for well formed terms that uses formation rules to express the context sensitive part of the syntax.

4.1 Context Free Language (Pre Terms)

<term> ::= ψ	(*Variables*)
| <id>	(*Constants*)
| <Πid> [{<sortexp> //,}$^+$]	(*Polyconstant-Inst*)
| <map> <term>	(*Mapping application*)
| <Πmap> [{<sortexp> //,}$^+$]<term>	(*Polymapping-Inst*)
| ⟨{<term> //,}$^{2+}$⟩	(*Tuple* $n \geq 2$)
| λ <pattern> . <term>	(λ-*abstraction*)
| <term> <term>	(*Application*)
| **Q** <tid> . <term>	($Q \in \{\forall^{\perp}, \exists^{\perp}\}$)
| (<term>)	(*Priority*)

<tid> ::= ψ : <sortexp>   <sortexp> ::= $T_{\Omega}\ (\chi)$

<pattern> ::= <tid> | ⟨{<tid> //,}$^{2+}$⟩

<id> ::= $F_{T_{\Omega,cpo}}$ <map> ::= $O_{T_{\Omega,map}}$

<Πid> ::= $F_{T^{\Pi}_{\Omega,cpo}}$ <Πmap> ::= $O_{T^{\Pi}_{\Omega,map}}$

In addition all object variables $x \in \psi$ are different from sort variables $\alpha \in \chi$ and all variables are different from identifiers in F and O.

4.2 Context Sensitive Language

With the pre terms at hand we can now define the well formed terms. We use a technique similar to [15] and give a calculus of formation rules. Since for sort variables there is only a binding mechanism in the language of sort terms but not in the language of object terms, we need no dynamic context for sort variables. The disjoint family χ of sort variables (the sort context) carries

[5] The advantage of this technique is that the problem of defining and finding (rsp. deciding) the principal type property is separated from the definition of the semantics. The drawback is the introduction of two languages namely the one with implicit typing and the one with explicit types. An alternative would be to define the semantics directly on well formed derivations for implicitly typed terms avoiding the introduction of an explicitly typed language. However, since the type system of SPECTRUM is an instance of the type system of Isabelle, we preferred to use an explicit type system and refer to [16, 17] for results about principal typings.

enough information. For the object variables, however, there are several binders and therefore we need an explicit variable context.

Definition 7 Sort Assertions.
The set of sort assertions $\rhd$ consists of tuples (χ, Γ, e, τ) where:

- χ is a sort context.
- $\Gamma = \{x_1 : \tau_1, \dots, x_n : \tau_n\}$ is a set of sort assumptions (a variable context), such that $\tau_i \in T_{\Omega,cpo}(\chi)$ and no x_i occurs twice in the sort assumptions contained in Γ (valid context condition). This prohibits overloading of variables in one scope.
- e is the pre term to be sorted.
- $\tau \in T_{\Omega,cpo}(\chi)$ is the derived sort for e.

We define:
$(\chi, \Gamma, e, \tau) \in \rhd$ if and only if there is a finite proof tree D for this fact according to the natural deduction system below.

When we write $\Gamma \rhd_\chi e :: \tau$ in the text we actually mean that there is a proof tree (sort derivation) for $(\chi, \Gamma, e, \tau) \in \rhd$. If we want to refer to a special derivation D we write $D : \Gamma \rhd_\chi e :: \tau$. The intuitive meaning of the sort assertion (χ, Γ, e, τ) with $\Gamma = \{x_1 : \tau_1, \dots, x_n : \tau_n\}$ is that if the variables $x_1, \dots, x_n$ have sorts $\tau_1, \dots, \tau_n$ then the pre term e is well formed and has sort τ.

Formation rules for well formed terms

Axioms:

$$(\text{var})\frac{}{x : \tau \rhd_\chi x :: \tau} \qquad (\text{const})\frac{}{\emptyset \rhd_\chi c :: \tau}\left\{ c \in F_\tau \right.$$

$$(\Pi\text{-inst})\frac{}{\emptyset \rhd_\chi f[\tau_1, \dots, \tau_n] :: \tau[\tau_1/\alpha_1, \dots, \tau_n/\alpha_n]}\left\{ \begin{array}{l} f \in F_{\Pi\alpha_1:k_1,\dots,\alpha_n:k_n.\tau} \\ \alpha_i : k_i \Rightarrow \tau_i : k_i \end{array} \right.$$

Note that in the above axiom $f[\tau_1, \dots, \tau_n]$ is part of the syntax whereas $\tau[\tau_1/\alpha_1, \dots, \tau_n/\alpha_n]$ is a meta notation for this presentation of the calculus. The axiom states that given a polymorphic constant $f \in F_{\Pi\alpha_1:k_1,\dots,\alpha_n:k_n.\tau}$ every instance of f via the sort expressions $\tau_i : k_i$ yields an explicitly typed term $f[\tau_1, \dots, \tau_n]$ of sort $\tau[\tau_1/\alpha_1, \dots, \tau_n/\alpha_n]$ which is τ after simultaneous replacement of all sort variables α_i by sort expressions τ_i of appropriate kind.

Inference Rules:

$$(\text{weak})\frac{\Gamma \rhd_\chi e :: \tau}{\Gamma \cup \{x_1 : \tau_1, \dots, x_n : \tau_n\} \rhd_\chi e :: \tau}$$

The 'valid context condition' in the rule (weak) prevents us from building contexts Γ with $x : \tau, x : \sigma \in \Gamma$ and $\tau \neq \sigma$

$$\text{(map-appl)}\ \frac{\Gamma \rhd_\chi e :: \tau_1}{\Gamma \rhd_\chi oe :: \tau_2} \Big\{\, o \in \mathrm{O}_{\tau_1 \mathbf{to} \tau_2}$$

$$(\Pi\text{map-appl}) \frac{\Gamma \rhd_\chi e :: \sigma_1[\tau_1/\alpha_1, \ldots, \tau_n/\alpha_n]}{\Gamma \rhd_\chi o[\tau_1, \ldots, \tau_n]e :: \sigma_2[\tau_1/\alpha_1, \ldots, \tau_n/\alpha_n]} \{\star$$

where

$$\star = \begin{cases} o \in \mathrm{O}_{\Pi\alpha_1:k_1,\ldots,\alpha_n:k_n.\sigma_1 \mathbf{to} \sigma_2} \\ \alpha_i : k_i \Rightarrow \tau_i : k_i \end{cases}$$

The rules (map-appl) and (Πmap-appl) are the formation rules for application of (polymorphic) mappings to terms. They ensure that a symbol for a mapping alone is not a well formed term which means that mappings may only occur in application context.

$$\text{(tuple)}\ \frac{\Gamma \rhd_\chi e_1 :: \tau_1 \ldots \Gamma \rhd_\chi e_n :: \tau_n}{\Gamma \rhd_\chi \langle e_1, \ldots, e_n \rangle :: \tau_1 \times \ldots \times \tau_n} \Big\{\, n \geq 2$$

$$\text{(abstr)} \frac{\Gamma, x{:}\tau_1 \rhd_\chi e :: \tau_2}{\Gamma \rhd_\chi \lambda x{:}\tau_1.e :: \tau_1 \rightarrow \tau_2} \{ e \dagger x$$

$$\text{(patt-abstr)} \frac{\Gamma, x_1{:}\tau_1, \ldots, x_n{:}\tau_n \rhd_\chi e :: \tau}{\Gamma \rhd_\chi \lambda\langle x_1{:}\tau_1, \ldots, x_n{:}\tau_n\rangle.e :: \tau_1 \times \ldots \times \tau_n \rightarrow \tau} \begin{cases} e \dagger x_i \\ 1 \leq i \leq n \end{cases}$$

where $e \dagger x$ is a property of pre terms. A calculus for $e \dagger x$ is presented below.

$$\text{(appl)} \frac{\Gamma \rhd_\chi e_1 :: \tau_1 \rightarrow \tau_2 \qquad \Gamma \rhd_\chi e_2 :: \tau_1}{\Gamma \rhd_\chi e_1 e_2 :: \tau_2}$$

Note that formations for (map-appl) (Πmap-appl) and (appl) use implicit but different application mechanisms. There is no problem in determining the last step in a derivation for a term $e_1 e_2$. If e_1 is not a constant then rule (appl) must be used since there are no variables or composed terms for mappings. If on the other hand e_1 is a constant then the choice is also clear since F and O are disjoint. Of course there remains the problem of guessing the right type τ_1 for the term e_1 in rule (appl) if e_1 is a composed term. But this is another problem of type inference not concerning the distinction between mappings and functions in application context.

$$\text{(quantifier)}\ \frac{\Gamma, x{:}\tau \rhd_\chi e :: \mathsf{Bool}}{\Gamma \rhd_\chi Qx{:}\tau.e :: \mathsf{Bool}} \Big\{\, Q \in \{\forall^\perp, \exists^\perp\}$$

$$\text{(priority)} \frac{\Gamma \rhd_\chi e :: \tau}{\Gamma \rhd_\chi (e) :: \tau}$$

This concludes the definition of sort derivations. We now present the calculus for $e \dagger x$. The purpose of this side condition is to prohibit the building of λ-terms that do not have a continuous interpretation. Consider the term:

$$\lambda x : \mathsf{Bool}.{=}[\mathsf{Bool}]\langle x, x\rangle$$

In our semantics the interpretation of the symbol = is the polymorphic identity which is by definition not monotonic. If we allowed the above expression as a well formed term its interpretation would have to be a non-monotonic function.

The property $e \dagger x$ is recursively defined on the structure of the pre term e. It's reading is 'e dagger x' and means 'e is continuous in x'. In the calculus below the set $\Phi(e)$ represents the set of free variables with respect to the binders $\forall^{\perp}$, $\exists^{\perp}$ and λ with the obvious definition.

$$(\dagger - \text{var})\frac{}{x \dagger x}$$

$$(\dagger - \text{notfree})\frac{x \notin \Phi(e)}{e \dagger x}$$

$$(\dagger - \text{tuple})\frac{e_1 \dagger x \qquad \dots \qquad e_n \dagger x}{\langle e_1, \dots, e_n\rangle \dagger x}$$

$$(\dagger - \text{abstr})\frac{e \dagger x \qquad e \dagger y}{\lambda y : \tau . e \dagger x}$$

$$(\dagger - \text{patt-abstr})\frac{e \dagger x \qquad e \dagger x_1 \qquad \dots \qquad e \dagger x_n}{\lambda\langle x_1 : \tau_1, \dots, x_n : \tau_n\rangle . e \dagger x}$$

$$(\dagger - \text{appl})\frac{e_1 \dagger x \qquad e_2 \dagger x}{e_1 e_2 \dagger x}$$

$$(\dagger - \text{quant})\ \frac{e \dagger x \qquad e \dagger y}{Qy : \tau . e \dagger x}\left\{ Q \in \{\forall^{\perp}, \exists^{\perp}\}\right.$$

$$(\dagger - \text{prio})\frac{e \dagger x}{(e) \dagger x}$$

As we will see later in section 6 the quantifiers get a three valued Kleene interpretation. If e is continuous in x and y also $\forall^{\perp} y : \tau . e$ and $\exists^{\perp} y : \tau . e$ are continuous in x. Therefore we can allow terms like $\lambda x : \sigma . \forall^{\perp} y : \tau . e$ provided the dagger test $\forall^{\perp} y : \tau . e \dagger x$ succeeds. For example the test $\exists^{\perp} y : \tau . \delta y \wedge e \dagger x$ will fail since $\delta y \wedge e \dagger y$ fails.

In the report [2, 3] we used the phrase 'where x is not free on a mappings argument position' as a context condition for the formation rules (abstr) and (patt-abstr). Looking at the example $\lambda x : \sigma . \exists^{\perp} y : \tau . \delta y \wedge e$ we see that this is too weak for terms with quantifiers inside.

4.3 Well formed Terms and Sentences

With the context sensitive syntax of the previous paragraph we are now able to define the notion of well formed terms over a polymorphic signature. Since we use an explicitly typed system, a well formed term is a pre term e together with a sort context χ, a variable context Γ and a sort τ.

Definition 8 Well formed terms.
Let Σ be a polymorphic signature. The set of well formed terms over Σ in sort context χ and variable context Γ with sort τ is defined as follows:

$$T_{\Sigma,\tau}(\chi,\Gamma) = \{(\chi,\Gamma,e,\tau) \mid \Gamma \rhd_\chi e :: \tau\}$$

The set of all well formed terms in context (χ,Γ) is defined to be the family

$$T_\Sigma(\chi,\Gamma) = \{T_{\Sigma,\tau}(\chi,\Gamma)\}_{\tau\in T_\Omega(\chi)}$$

In addition we define the following abbreviations:

$$T_\Sigma(\chi) = T_\Sigma(\chi,\emptyset) \quad \text{(closed object terms)}$$
$$T_\Sigma = T_\Sigma(\emptyset) \quad \text{(non-polymorphic closed object terms)}$$

Considering a well formed term $(\chi,\Gamma,e,\tau) \in T_{\Sigma,\tau}(\chi,\Gamma)$ we see that all the sort derivations $D : \Gamma \rhd_\chi e :: \tau$ for this term can only differ in the applications of the formation rule (weak). Due to the vast type information contained in our pre terms e there are no other possibilities for different sort derivations.

In section 6 we will define the interpretation of a well formed term (χ,Γ,e,τ) in set $T_{\Sigma,\tau}(\chi,\Gamma)$ with respect to the inductive structure of some sort derivation for this term. To guarantee the uniqueness of our definition we now distinguish the unique and always existing normal form of a sort derivation.

Definition 9 Normal Sort Derivation.
Let $(\chi,\Gamma,e,\tau) \in T_{\Sigma,\tau}(\chi,\Gamma)$ be a well formed term. The Normal Sort Derivation $ND : \Gamma \rhd_\chi e :: \tau$ is that derivation where introductions of sort assumptions via the formation rule (weak) occur as late as possible.

A formal definition of the normal form together with a proof for the existence and uniqueness result is pretty obvious. A thorough discussion of a slightly different technique containing all the definitions and proofs can be found in [14]. Next we define formulae $\mathbf{Form}(\Sigma,\chi,\Gamma)$ and sentences $\mathbf{Sen}(\Sigma,\chi)$ over a polymorphic signature Σ and sort context χ. In SPECTRUM the set of formulae $\mathbf{Form}(\Sigma,\chi,\Gamma)$ is the set of well formed terms in context (χ,Γ) of sort Bool. This leads to a three valued logic. The sentences are as usual the closed formulae.

Definition 10 Formulae and Sentences.

$$\mathbf{Form}(\Sigma,\chi,\Gamma) = T_{\Sigma,\mathsf{Bool}}(\chi,\Gamma)$$
$$\mathbf{Sen}(\Sigma,\chi) = \mathbf{Form}(\Sigma,\chi,\emptyset) \quad \text{(closed formulae are sentences)}$$
$$\mathbf{Sen}(\Sigma) = \mathbf{Sen}(\Sigma,\emptyset) \quad \text{(non-polymorphic sentences)}$$

Example 3 Some formulae.

$$\forall^{\perp} x:\alpha.{=}[\alpha]\langle x, x\rangle \in \mathbf{Sen}(\Sigma, \{\alpha\})$$

$$\forall^{\perp} x:\mathbf{Nat}.{=}[\mathbf{Nat}]\langle x, x\rangle \in \mathbf{Sen}(\Sigma)$$

Definition 11 Specifications.
A polymorphic specification $S = (\Sigma, E)$ is a pair where $\Sigma = (\Omega, F, O)$ is a polymorphic signature and $E \subseteq \mathbf{Sen}(\Sigma, \chi)$ is a set of sentences for some sort context χ.

5 Algebras

The following definitions are standard definitions of domain theory (see [11]). We include them here to get a self-contained presentation.

Definition 12 Partial Order.
A partial order $\mathcal{A}$ is a pair $(A, \leq)$ where A is a set and $(\leq) \subseteq A \times A$ is a reflexive, transitive and antisymmetric relation.

Definition 13 Chain Complete Partial Order.
A partial order $\mathcal{A}$ is ω-chain complete iff every chain $a_1 \leq \ldots \leq a_n \leq \ldots, n \in \mathbb{N}$ has a least upper bound in $\mathcal{A}$. We denote it by $\sqcup_{i \in \mathbb{N}} x_i$.

Definition 14 Pointed Chain Complete Partial Order (PCPO).
A chain complete partial order $\mathcal{A}$ is pointed iff it has a least element. In the sequel we denote this least element by uu_A.

Definition 15 Monotonic Functions.
Let $\mathcal{A} = (A, \leq_A)$ and $\mathcal{B} = (B, \leq_B)$ be two PCPOs. A function[6] $f \in B^A$ is *monotonic* iff

$$d \leq_A d' \ \Rightarrow\ f(d) \leq_B f(d')$$

Definition 16 Continuous Functions.
A monotonic function f between PCPOs $\mathcal{A}$ and $\mathcal{B}$ is continuous iff for every ω-chain $a_1 \leq \ldots \leq a_n \leq \ldots$ in $\mathcal{A}$:

$$f(\bigsqcup_{i \in \mathbb{N}} a_i) = \bigsqcup_{i \in \mathbb{N}} f(a_i)$$

Since f is monotonic and $\mathcal{A}$ and $\mathcal{B}$ are PCPOs the least upper bound on the right hand side exists.

Definition 17 Product PCPO.
If $\mathcal{A} = (A, \leq_A)$ and $\mathcal{B} = (B, \leq_B)$ are two PCPOs then the product PCPO $\mathcal{A} \times \mathcal{B} = (A \times B, \leq_{A \times B})$ is defined as follows:

[6] We write B^A for all functions from A to B.

- $A \times B$ is the usual cartesian product of sets,
- $(d,e) \leq_{A\times B} (d',e') \;\; iff \;\; (d \leq_A d) \wedge (e \leq_B e')$,
- $uu_{A\times B} = (uu_A, uu_B)$

This definitions may be generalized to n-ary products in a straight forward way.

Definition 18 Function PCPO.
If $\mathcal{A} = (A, \leq_A)$ and $\mathcal{B} = (B, \leq_B)$ are two PCPOs then the function PCPO $\mathcal{A}\xrightarrow{c}\mathcal{B} = (A\xrightarrow{c}B, \leq_{A\xrightarrow{c}B})$ is defined as follows:

- $A\xrightarrow{c}B$ is the set of all continuous functions from A to B,
- $f \leq_{A\xrightarrow{c}B} g \;\; iff \;\; \forall a \in A.f(a) \leq_B g(a)$,
- $uu_{A\xrightarrow{c}B} = \lambda x{:}A.uu_B$

Definition 19 Lift PCPO.
If $\mathcal{A} = (A, \leq_A)$ is a PCPO then the lifted PCPO $\mathcal{A}\,\mathbf{lift} = (A\,\mathbf{lift}, \leq_{A\,\mathbf{lift}})$ is defined as follows:

- $A\,\mathbf{lift} = (A \times \{0\}) \cup \{uu_{A\,\mathbf{lift}}\}$ where $uu_{A\,\mathbf{lift}}$ is a new element which is not a pair.
- $(x,0) \leq_{A\,\mathbf{lift}} (y,0) \;\; iff \;\; x \leq_A y$

  $\forall z \in A\,\mathbf{lift}.uu_{A\,\mathbf{lift}} \leq_{A\,\mathbf{lift}} z$
- We also define an extraction function $\downarrow$ from $A\,\mathbf{lift}$ to A such that

$$\downarrow uu_{A\,\mathbf{lift}} = uu_A \quad ; \quad \downarrow (x,0) = x$$

We will call the PCPOs also domains (note that in the literature domains are usually algebraic directed complete po's [11]).

5.1 The Sort Algebras

Definition 20 Sort–Algebras.
Let $\Omega = (K, \leq, SC)$ be a sort–signature. An Ω–algebra $\mathcal{SA}$ is an order sorted algebra[7] of domains i.e.:

- For the kind $cpo \in K$ we have a set of domains $cpo^{\mathcal{SA}}$. For the kind $map \in K$ we have a set of full functions spaces $map^{\mathcal{SA}}$.
- For all kinds $k \in K$ with $k \leq cpo$ we have a nonempty subset $k^{\mathcal{SA}} \subseteq cpo^{\mathcal{SA}}$.
- For all kinds $k_1, k_2 \in K$ with $k_1 \leq k_2$ we have $k_1^{\mathcal{SA}} \subseteq k_2^{\mathcal{SA}}$.
- For each sort constructor $sc \in SC_{k_1 \ldots k_n, k}$ there is a domain constructor $sc^{\mathcal{SA}} : k_1^{\mathcal{SA}} \times \ldots \times k_n^{\mathcal{SA}} \to k^{\mathcal{SA}}$ such that if $sc \in SC_{w,s} \cap SC_{w',s'}$ and $w \leq w'$ then

$$sc^{\mathcal{SA}}_{w',s'} \mid_{w^{\mathcal{SA}}} = sc^{\mathcal{SA}}_{w,s}$$

  In other words overloaded domain constructors must be equal on the smaller domain $w^{\mathcal{SA}} = k_1^{\mathcal{SA}} \times \ldots \times k_n^{\mathcal{SA}}$ where $w = k_1 \ldots k_n$.

---
[7] See [9, 8].

We further require the following interpretation for the sort constructors occurring in the standard sort–signature:

- $\mathsf{Bool}^{\mathcal{SA}} = (\{\omega_{\mathrm{Bool}}, \mathit{ff}, \mathit{tt}\}, \leq_{\mathrm{Bool}})$ is the flat three–valued boolean domain.
- For $\times_n^{\mathcal{SA}} \in cpo^{\mathcal{SA}} \times \ldots \times cpo^{\mathcal{SA}} \to cpo^{\mathcal{SA}}$:
$$\times_n^{\mathcal{SA}}(d_1, \ldots, d_n) = d_1 \times \ldots \times d_n,\ n \geq 2$$
is the n-ary cartesian product of domains.
- For $\to^{\mathcal{SA}} \in cpo^{\mathcal{SA}} \times cpo^{\mathcal{SA}} \to cpo^{\mathcal{SA}}$:
$$\to^{\mathcal{SA}}(d_1, d_2) = (d_1 \overset{c}{\to} d_2)\mathbf{lift}$$
is the lifted domain of continuous functions. We lift this domain because we want to distinguish between $\perp$ and $\lambda x.\ \perp$.
- For $\mathbf{to}^{\mathcal{SA}} \in cpo^{\mathcal{SA}} \times cpo^{\mathcal{SA}} \to map^{\mathcal{SA}}$
$$\mathbf{to}^{\mathcal{SA}}(d_1, d_2) = d_2^{d_1}$$
is the full space of functions between d_1 and d_2.

Definition 21 Interpretation of sort terms.
Let $\nu : \chi \to \mathcal{SA}$ be a sort environment and $\nu^* : T_\Omega(\chi) \to \mathcal{SA}$ its homomorphic extension. Then $\mathcal{SA}[.]\nu$ is defined as follows:

- $\mathcal{SA}[e]\nu = \nu^*(e)$ **if** $e \in T_\Omega(\chi)$
- $\mathcal{SA}[\Pi\alpha_1 : k_1, \ldots, \alpha_n : k_n.e] =$
$$\{f \mid f(\nu(\alpha_1), \ldots, \nu(\alpha_n)) \in \mathcal{SA}[e]\nu \quad \textbf{for all} \quad \nu\}$$

For closed terms we write for $\mathcal{SA}[e]$ also $e^{\mathcal{SA}}$.

Sort terms in T_Ω^Π are interpreted as generalized cartesian products (dependent products). By using *n-ary* dependent products we can interpret Π–terms in one step. This leads to simpler models as the ones for the polymorphic λ–calculus.

5.2 Polymorphic Algebras

Definition 22 Polymorphic Algebra.
Let $\Sigma = (\Omega, F, O)$ be a polymorphic signature with $\Omega = (K, \leq, SC)$ the sort–signature. A polymorphic Σ–algebra $\mathcal{A} = (\mathcal{SA}, \mathcal{F}, \mathcal{O})$ is a triple where:

- $\mathcal{SA}$ is an Ω sort algebra,
- $\mathcal{F} = \{\mathcal{F}_\mu\}_{\mu \in T_{\Omega,cpo}^{closed}}$ is an indexed set of constants (or functions), with:
$$\mathcal{F}_\mu = \{f^{\mathcal{A}} \in \mu^{\mathcal{SA}} \mid f \in F_\mu\}$$
such that if $f \in F_\mu$ is not the constant $\perp \in \Pi\alpha : cpo.\alpha$ then its interpretation $f^{\mathcal{A}}$ is different from ω in $\mu^{\mathcal{SA}}$. If f is polymorphic then all its instances must be different from the corresponding least element.

– $\mathcal{O} = \{\mathcal{O}_\nu\}_{\nu \in T^{closed}_{\Omega,map}}$ is an indexed set of mappings, with:

$$\mathcal{O}_\nu = \{o^{\mathcal{A}} \in \nu^{\mathcal{SA}} \mid o \in O_\nu\}$$

We further require a fixed interpretation for the symbols in the standard signature. In order to simplify notation we will write $f^{\mathcal{A}}_{d_1,\ldots,d_n}$ for the instance $f^{\mathcal{A}}(d_1,\ldots,d_n)$ of a polymorphic function and $o^{\mathcal{A}}_{d_1,\ldots,d_n}$ for the instance of a polymorphic mapping $o^{\mathcal{A}}(d_1,\ldots,d_n)$.

- Predefined Mappings ($O_{standard}$):
  - $\{=, \sqsubseteq\} \subseteq O_{\Pi\alpha:cpo.\ \alpha\times\alpha \text{ to Bool}}$ are interpreted as *identity* and *partial order*. More formally, for every domain $d \in cpo^{\mathcal{A}}$ and $x, y \in d$:

$$x =^{\mathcal{A}}_d y := \begin{cases} tt \text{ if } x \text{ is identical to } y \\ ff \text{ otherwise} \end{cases}$$

$$x \sqsubseteq^{\mathcal{A}}_d y := \begin{cases} tt \text{ if } x \leq_d y \\ ff \text{ otherwise} \end{cases}$$

  - $\{\delta\} \subseteq O_{\Pi\alpha:cpo.\ \alpha \text{ to Bool}}$ is the polymorphic definedness predicate. For every $d \in cpo^{\mathcal{A}}$ and $x \in d$:

$$\delta^{\mathcal{A}}_d(x) := \begin{cases} tt \text{ if } x \text{ is different from } uu_d \\ ff \text{ otherwise} \end{cases}$$

- Predefined Constants ($F_{standard}$):
  - $\{\mathsf{true}, \mathsf{false}\} \subseteq F_{\mathsf{Bool}}$ are interpreted in the $\mathsf{Bool}^{\mathcal{SA}}$ domain as follows:

$$\mathsf{true}^{\mathcal{A}} = tt \quad ; \quad \mathsf{false}^{\mathcal{A}} = ff$$

  - The interpretations of $\{\neg\} \subseteq F_{\mathsf{Bool}\to\mathsf{Bool}}$, $\{\wedge, \vee, \Rightarrow\} \subseteq F_{\mathsf{Bool}\times\mathsf{Bool}\to\mathsf{Bool}}$ are pairs in the lifted function spaces such that the function components behave like three–valued Kleene connectives on $\mathsf{Bool}^{\mathcal{SA}}$ as follows:

x	y	$(\downarrow \neg^{\mathcal{A}})(x)$	$x(\downarrow \wedge^{\mathcal{A}})y$	$x(\downarrow \vee^{\mathcal{A}})y$	$x(\downarrow \Rightarrow^{\mathcal{A}})y$
tt	tt	ff	tt	tt	tt
tt	ff	ff	ff	tt	ff
ff	tt	tt	ff	tt	tt
ff	ff	tt	ff	ff	tt
uu	tt	uu	uu	tt	tt
uu	ff	uu	ff	uu	uu
uu	uu	uu	uu	uu	uu
tt	uu	ff	uu	tt	uu
ff	uu	tt	ff	uu	tt

  - $\{\bot\} \subseteq F_{\Pi\alpha:cpo.\ \alpha}$ is interpreted in each domain as the least element of this domain. For every $d \in cpo^{\mathcal{SA}}$:

$$\bot^{\mathcal{A}}_d := uu_d$$

- $\{\mathsf{fix}\} \subseteq F_{\Pi\alpha:cpo.\,(\alpha\to\alpha)\to\alpha}$ is interpreted for each domain d as a pair $\mathsf{fix}^{\mathcal{A}}_d \in (d \to^{\mathcal{SA}} d) \to^{\mathcal{SA}} d$ such that the function component behaves as follows:

$$(\downarrow \mathsf{fix}^{\mathcal{A}}_d)(f) := \bigsqcup_{i\in\mathbb{N}} f^n(uu_d)$$

where:

$$f^0(uu_d) := uu_d$$

$$f^{n+1}(uu_d) := (\downarrow f)(f^n(uu_d))$$

Note that $\downarrow uu_{(d \overset{c}{\to} d)}\,\mathbf{lift} = uu_{(d \overset{c}{\to} d)}$ and therefore the above definition is sound.

6 Models

6.1 Interpretation of sort assertions

In this section we define the interpretation of well-formed terms. The interpretation of $(\chi, \Gamma, e, \tau) \in T_{\Sigma,\tau}(\chi, \Gamma)$ is defined inductively on the structure of the normal sort derivation $ND : \Gamma \rhd_\chi e :: \tau$. The technique used is again due to [15].

Definition 23 Satisfaction of a variable context.
Let $\Sigma = (\Omega, F, O)$ be a polymorphic signature with $\Omega = (K, \leq, SC)$ and let $\mathcal{A} = (\mathcal{SA}, \mathcal{F}, \mathcal{O})$ be a a polymorphic Σ–algebra.

If Γ is a variable context and

$$\nu = \{\nu_k : \chi_k \to k^{\mathcal{SA}}\}_{k\in K\setminus\{map\}} \quad \text{sort environment (order-sorted)}$$
$$\eta : \psi \to \bigcup_{d\in cpo^{\mathcal{SA}}} d \quad \text{object environment (unsorted)}$$

then η satisfies Γ in sort environment ν (in symbols $\eta \models_\nu \Gamma$) iff

$$\eta \models_\nu \Gamma \Leftrightarrow \text{for all } x{:}\tau \in \Gamma.\eta(x) \in \nu^\star(\tau)$$

Definition 24 Update of object environments.

$$\eta[a/x](y) := \begin{cases} a & \text{if } x = y \\ \eta(y) & \text{otherwise} \end{cases}$$

Now we define an order sorted meaning function $\mathcal{A}[\![\cdot]\!]_{\nu,\eta}$ that maps normal sort derivations $ND : \Gamma \rhd_\chi e :: \tau$ to elements in $\mathcal{A}$. Since normal sort derivations always exist and are unique this leads to a total meaning function $\mathcal{A}[\![\cdot]\!]_{\nu,\eta}$: $T_\Sigma(\chi, \Gamma) \to \mathcal{A}$.

Definition 25 Meaning of a sort derivation.
The meaning of a normal sort derivation $ND : \Gamma \rhd_\chi e :: \tau$ in a polymorphic algebra $\mathcal{A}$ in sort context ν and variable context Γ such that $\eta \models_\nu \Gamma$ is given by $\mathcal{A}[\![ND : \Gamma \rhd_\chi e :: \tau]\!]_{\nu,\eta}$ which is recursively defined on the structure of ND. The defining clauses are given below.

Base cases:

(var) $\mathcal{A}[\![x:\tau \rhd_\chi x :: \tau]\!]_{\nu,\eta} = \eta(x)$ (const) $\mathcal{A}[\![\emptyset \rhd_\chi c :: \tau]\!]_{\nu,\eta} = c^{\mathcal{A}}$

(Π-inst)

$$\mathcal{A}[\![\emptyset \rhd_\chi f[\tau_1,\ldots,\tau_n] :: \tau]\!]_{\nu,\eta} = f^{\mathcal{A}}(\nu^\star(\tau_1),\ldots,\nu^\star(\tau_n))$$

Inductive cases:
(weak)

$$\mathcal{A}[\![\Gamma \cup \{x_1:\tau_1,\ldots,x_n:\tau_n\} \rhd_\chi e :: \tau]\!]_{\nu,\eta} = \mathcal{A}[\![\Gamma \rhd_\chi e :: \tau]\!]_{\nu,\eta}$$

(map-appl)

$$\mathcal{A}[\![\Gamma \rhd_\chi oe :: \tau_2]\!]_{\nu,\eta} = o^{\mathcal{A}}(\mathcal{A}[\![\Gamma \rhd_\chi e :: \tau_1]\!]_{\nu,\eta})$$

(Πmap-appl)

$$\mathcal{A}[\![\Gamma \rhd_\chi o[\tau_1,\ldots,\tau_n]e :: \sigma_2]\!]_{\nu,\eta} = o^{\mathcal{A}}(\nu^\star(\tau_1),\ldots,\nu^\star(\tau_n))(\mathcal{A}[\![\Gamma \rhd_\chi e :: \sigma_1]\!]_{\nu,\eta})$$

(tuple)

$$\mathcal{A}[\![\Gamma \rhd_\chi \langle e_1,\ldots,e_n\rangle :: \tau_1\times\ldots\times\tau_n]\!]_{\nu,\eta} = (\mathcal{A}[\![\Gamma \rhd_\chi e_1 :: \tau_1]\!]_{\nu,\eta},\ldots,\mathcal{A}[\![\Gamma \rhd_\chi e_n :: \tau_n]\!]_{\nu,\eta})$$

(abstr)[8]

$$\mathcal{A}[\![\Gamma \rhd_\chi \lambda x:\tau_1.e :: \tau_1\rightarrow\tau_2]\!]_{\nu,\eta} =$$
the <u>unique</u> pair $(f,0) \in \nu^\star(\tau_1\rightarrow\tau_2)$ with
$\forall a \in \nu^\star(\tau_1). f(a) = \mathcal{A}[\![\Gamma, x:\tau_1 \rhd_\chi e :: \tau_2]\!]_{\nu,\eta[a/x]}$

(patt-abstr)

$$\mathcal{A}[\![\Gamma \rhd_\chi \lambda\langle x_1:\tau_1,\ldots,x_n:\tau_n\rangle.e :: \tau_1\times\ldots\times\tau_n\rightarrow\tau]\!]_{\nu,\eta} =$$
the <u>unique</u> pair $(f,0) \in \nu^\star(\tau_1\times\ldots\times\tau_n\rightarrow\tau)$ with
$\forall a_1 \in \nu^\star(\tau_1),\ldots,a_n \in \nu^\star(\tau_n). f((a_1,\ldots,a_n)) = \mathcal{A}[\![\Gamma, x_1:\tau_1,\ldots,x_n:\tau_n \rhd_\chi e :: \tau]\!]_{\nu,\eta[a_1/x_1,\ldots,a_n/x_n]}$

---

[8] the †-test ensures that the clauses for (abstr) and (patt-abstr) are well defined.

(appl)

$$\mathcal{A}[\Gamma \rhd_\chi e_1 e_2 :: \tau_2]_{\nu,\eta} =\downarrow (\mathcal{A}[\Gamma \rhd_\chi e_1 :: \tau_1 \rightarrow \tau_2]_{\nu,\eta})(\mathcal{A}[\Gamma \rhd_\chi e_2 :: \tau_1]_{\nu,\eta})$$

(universal quantifier)

$$\mathcal{A}[\Gamma \rhd_\chi \forall^\perp x{:}\tau.e :: \mathsf{Bool}]_{\nu,\eta} =$$

$$= \begin{cases} tt \text{ if } \forall a \in \nu^\star(\tau).(\mathcal{A}[\Gamma, x{:}\tau \rhd_\chi e :: \mathsf{Bool}]_{\nu,\eta[a/x]} = tt) \\ ff \text{ if } \exists a \in \nu^\star(\tau).(\mathcal{A}[\Gamma, x{:}\tau \rhd_\chi e :: \mathsf{Bool}]_{\nu,\eta[a/x]} = ff) \\ uu \quad \text{otherwise} \end{cases}$$

(existential quantifier)

$$\mathcal{A}[\Gamma \rhd_\chi \exists^\perp x{:}\tau.e :: \mathsf{Bool}]_{\nu,\eta} =$$

$$= \begin{cases} tt \text{ if } \exists a \in \nu^\star(\tau).(\mathcal{A}[\Gamma, x{:}\tau \rhd_\chi e :: \mathsf{Bool}]_{\nu,\eta[a/x]} = tt) \\ ff \text{ if } \forall a \in \nu^\star(\tau).(\mathcal{A}[\Gamma, x{:}\tau \rhd_\chi e :: \mathsf{Bool}]_{\nu,\eta[a/x]} = ff) \\ uu \quad \text{otherwise} \end{cases}$$

6.2 Satisfaction and Models

In this subsection we define the satisfaction relation for boolean terms and sentences (closed boolean terms) and also the notion of a model.

Definition 26 Satisfaction.
Let

$\mathcal{A} = (\mathcal{SA}, \mathcal{F}, \mathcal{O})$	Σ-Algebra
$\nu = \{\nu_k : \chi_k \rightarrow k^{\mathcal{SA}}\}_{k \in K \setminus \{map\}}$	sort environment (order-sorted)
$\eta : \psi \rightarrow \bigcup_{d \in cpo^{\mathcal{SA}}} d$	object environment (unsorted)

and Γ a variable context with $\eta \models_\nu \Gamma$ then:
$\mathcal{A}$ satisfies $(\chi, \Gamma, e, \mathsf{Bool}) \in \mathbf{Form}(\Sigma, \chi, \Gamma)$ wrt. sort environment ν and object environment η (in symbols $\mathcal{A} \models_{\nu,\eta} (\chi, \Gamma, e, \mathsf{Bool})$) iff

$$\mathcal{A} \models_{\nu,\eta} (\chi, \Gamma, e, \mathsf{Bool}) \Leftrightarrow \mathcal{A}[\Gamma \rhd_\chi e :: \mathsf{Bool}]_{\nu,\eta} = tt$$

A special case of the above definition is the satisfaction of sentences. Let $(\chi, \emptyset, e, \mathsf{Bool}) \in \mathbf{Sen}(\Sigma, \chi)$ and η_0 an arbitrary environment, then:

$$\mathcal{A} \models_\nu (\chi, \emptyset, e, \mathsf{Bool}) \Leftrightarrow \mathcal{A}[\emptyset \rhd_\chi e :: \mathsf{Bool}]_{\nu,\eta_0} = tt$$

$$\mathcal{A} \models (\chi, \emptyset, e, \mathsf{Bool}) \Leftrightarrow \mathcal{A} \models_\nu (\chi, \emptyset, e, \mathsf{Bool}) \text{ for every } \nu$$

Now we are able to define models $\mathcal{A}$ of specifications $S = (\Sigma, E)$.

Definition 27 Models.
Let $S = (\Sigma, E)$ be a specification. A polymorphic Σ-algebra $\mathcal{A}$ is a model of S (in symbols $\mathcal{A} \models S$) iff

$$\mathcal{A} \models S \Leftrightarrow \forall p \in E.\ \mathcal{A} \models p$$

7 Conclusions

We have presented the semantics of the kernel part of the SPECTRUM language. Our work differs in many respects from other approaches. In contrast to LCF we allow the use of type classes. Moreover arbitrary non–continuous functions can be used for specification purposes. This also permits to handle predicates and boolean functions in a uniform manner. In contrast with other semantics for polymorphic lambda calculus (e.g. [15]) we did not provide an explicit type binding operator on the object level. This is not a restriction for languages having an ML–like polymorphism but allows a more simple treatment of the sort language. More precisely we used order sorted algebras instead of the more complex applicative structures. Order sorted algebras were also essential in the description of type classes.

8 Acknowledgment

For comments on draft versions and stimulating discussions we like to thank M. Löwe, B. Möller, F. Nickl, B. Reus, D. Sannella, T. Streicher, A. Tarlecki, M. Wirsing and U. Wolter.
Special thanks go also to our colleagues H. Hussmann, C. Facchi, R. Hettler and D. Nazareth to Tobias Nipkow whose work inspired our treatment of type classes and to Manfred Broy whose role was decisive in the design of the SPECTRUM language.

References

1. M. Broy. Requirement and Design Specification for Distributed Systems. *LNCS*, 335:33–62, 1988.
2. M. Broy, C. Facchi, R. Grosu, R. Hettler, H. Hussmann, D. Nazareth, F. Regensburger, O. Slotosch, and K. Stølen. The Requirement and Design Secification Language SPECTRUM. An Informal Introduction. Version 1.0. Part I. Technical Report TUM-I9311, Technische Universität München. Institut für Informatik, May 1993.
3. M. Broy, C. Facchi, R. Grosu, R. Hettler, H. Hussmann, D. Nazareth, F. Regensburger, O. Slotosch, and K. Stølen. The Requirement and Design Secification Language SPECTRUM. An Informal Introduction. Version 1.0. Part II. Technical Report TUM-I9312, Technische Universität München. Institut für Informatik, May 1993.
4. R. Milner C Wadsworth M. Gordon. *Edinburgh LCF: A Mechanised Logic of Computation*, volume 78 of *LNCS*. Springer, 1979.
5. J. Camilleri. The HOL System Description, Version 1 for HOL 88.1.10. Technical report, Cambridge Research Center, 1989.
6. L. Cardelli and P. Wegner. On Understanding Types, Data Abstraction, and Polymorphism. *ACM Computing Surveys*, 17(4):471–523, December 1985.
7. M.-C. Gaudel. Towards Structured Algebraic Specifications. *ESPRIT '85', Status Report of Continuing Work (North-Holland)*, pages 493–510, 1986.
8. M. Gogolla. Partially Ordered Sorts in Algebraic Specifications. In B. Courcelle, editor, *Proc. 9th CAAP 1984, Bordeaux*. Cambridge University Press, 1976.

9. J.A. Goguen and J. Meseguer. Order–Sorted Algebra Solves the Constructor–Selector, Multiple Representation and Coercion Problems. In *Logic in Computer Science*, IEEE, 1987.
10. R. Grosu and F. Regensburger. The Logical Framework of SPECTRUM. Technical Report TUM-I9402, Institut für Informatik, Technische-Universität München, 1994.
11. C. A. Gunter. *Semantics of Programming Languages: Structures and Techniques.* MIT Press, 1992.
12. J.V. Guttag, J.J. Horning, and J.M. Wing. Larch in Five Easy Pieces. Technical report, Digital, Systems Research Center, Paolo Alto, California, 1985.
13. P. Hudak, S. Peyton Jones, and P. Wadler, editors. *Report on the Programming Language Haskell, A Non-strict Purely Functional Language (Version 1.2).* ACM SIGPLAN Notices, May 1992.
14. J. C. Mitchell. *Introduction to Programming Language Theory.* MIT Press, 1993.
15. J.C. Mitchell. Type Systems for Programming Languages. In *Handbook of Theoretical Computer Science*, chapter 8, pages 365–458. Elsevier Science Publisher, 1990.
16. T. Nipkow. Order-Sorted Polymorphism in Isabelle. In G. Huet and G. Plotkin, editors, *Logical Environments*, pages 164–188. CUP, 1993.
17. Tobias Nipkow and Christian Prehofer. Type checking type classes. In *Proc. 20th ACM Symp. Principles of Programming Languages*, pages 409–418, 1993.
18. F. Regensburger. *HOLCF: Eine konservative Erweiterung von HOL durch LCF.* PhD thesis, Technische Universität München, 1994. to appear.
19. D. Sannella and M. Wirsing. A Kernel Language for Algebraic Specification and Implementation. Technical Report CSR-131-83, University of Edinburgh, Edinburgh EH9 3JZ, September 1983.
20. G. Smolka, W. Nutt, J. Goguen, and J. Meseguer. Order-Sorted Equational Computation. In *Resolution of Equations in Algebraic Structures.* Academic Press, 1989.
21. C. Strachey. Fundamental Concepts in Programming Languages. In *Lecture Notes for International Summer School in Computer Programming*, Copenhagen, 1967.
22. P. Wadler and S. Blott. How to Make Ad-hoc Polymorphism Less Ad hoc. In *16th ACM Symposium on Principles of Programming Languages*, pages 60–76, 1989.

ATLAS: A TYPED LANGUAGE FOR ALGEBRAIC SPECIFICATION

B.M. Hearn and K. Meinke

Department of Computer Science
University College of Swansea
Singleton Park
Swansea SA2 8PP
Great Britain

Abstract. We introduce an implementation of rewriting for type and combinator terms called *ATLAS*. This system implements the algebraic and term rewriting theory for abstract types and combinators developed in Meinke [1991, 1992b]. The system is intended to support the execution of equational specifications of abstract types and combinators. The type checking algorithms of the system also allow it to function as a framework for defining logics and proof checking. We present a short tutorial introduction to *ATLAS* by means of examples taken from first and higher order algebraic specifications and logics.

1. INTRODUCTION.

Given a many–sorted Σ algebra A it is natural, and often useful for purposes such as algebraic specification, to impose some mathematical structure on the sort set S. One possibility is to impose a poset structure on S. This leads to the theory of order–sorted algebras introduced in Goguen [1978]. Another possibility is to impose a many–sorted universal algebraic structure on S. This leads to the theory of *algebras of types and combinators.* Basic results in the theory of algebras of types and combinators, including results on soundness and completeness of equation and type assignment calculi, and an initial model semantics for equational type specifications were presented in Meinke [1991]. The equivalence of typed equational logic and term rewriting of type and combinator expressions was established in Meinke [1992b].

In this paper we present an implementation of rewriting for type and combinator terms called *ATLAS* (A Typed Language for Algebraic Specifications). This system is primarily intended to support execution of equational specifications of abstract types and combinators. However, as we will show the system also provides a logical framework for defining logics and proof checking. We will give a short tutorial introduction to *ATLAS* by means of examples which attempt to convey the wide range of applications of this formalism. These include: first and higher order data type specifications, parameterised specifications, equationally specified type disciplines for functional languages, and a framework for defining logics (with Hilbert style proof systems) and proof checking. Although we describe the formal syntax of *ATLAS* specifications and some aspects of the software interface, we do not discuss the architecture or algorithms of the *ATLAS* system which will be presented elsewhere.

Although the paper is largely self-contained background definitions and results for algebraic specifications may be found in Wirsing [1990]; for basic universal algebra the reader may consult Meinke and Tucker [1992].

The structure of the paper is as follows. In Section 2 we briefly review the mathematical theory of rewriting terms and types. In Section 3 we describe the concrete syntax of *ATLAS* specifications. In Section 4 we consider case studies of *ATLAS* specifications taken from abstract data type theory, the theory of hardware algorithms and the metamathematics of proof systems and proof checking.

2. REWRITING TERMS AND TYPES.

We shall briefly review the theory of rewriting terms and types presented in Meinke [1992b] which provides the mathematical foundation for *ATLAS*. In particular this theory provides an initial model semantics for *ATLAS* specifications.

2.1. Definition. By a *type signature* we mean a K–sorted signature Γ for some non–empty set K. In this context we term K a set of *kinds* and Γ a *K-kinded type signature.* □

Let Γ be a K–kinded type signature and $Y = \langle Y_k \mid k \in K \rangle$ be a family of sets of variable symbols. We have the usual notion of *term algebra* $T(\Gamma, Y)$ over Γ and Y. The elements of $T(\Gamma, Y)_k$ are termed *type terms* of kind k. As usual for each type term $\tau \in T(\Gamma, Y)_k$ we may define the set $Occ(\tau) \subseteq \mathbf{N}^*$ of *occurrences* in τ together with the map $Knd_\tau : Occ(\tau) \to K$ which assigns to each occurrence in τ its kind, inductively as usual. Furthermore for any occurrence $\bar{i} \in Occ(\tau)$ of kind $k' \in K$ and any type term $\sigma \in T(\Gamma, Y)_{k'}$ we may inductively define the *subterm of* τ *at* $\bar{i}$ denoted by $\tau(\bar{i})$ and the substitution of σ into τ at $\bar{i}$ denoted by $\tau(\bar{i}/\sigma)$. A *type rewrite rule* of kind $k \in K$ is a pair (t, t'), usually written $t \to t'$, where $t, t' \in T(\Gamma, Y)_k$.

Given a K–kinded type signature Γ and a subset $K' \subseteq K$, the K'–kinded ground type terms $\tau \in T(\Gamma)_k$ for $k \in K'$ may be used as sorts in a many sorted signature of typed combinators.

2.2. Definition. Let Γ be a K–kinded type signature, and $K' \subseteq K$ be any subset. By a *combinator signature* Σ over Γ and K' we mean a $\cup_{k \in K'} T(\Gamma)_k$–sorted signature Σ. □

The notions of *term, occurrence, substitution* and *rewrite rule* for a combinator signature Σ over a type signature Γ are complicated by the parametric role of Γ. In the sequel we consider some fixed, but arbitrarily chosen family $X = \langle X_\tau \mid k \in K'$ and $\tau \in T(\Gamma)_k \rangle$ of disjoint sets of combinator variables. The standard definition of term formation over Σ and X is too restrictive in our context, we use instead a more liberal notion of *preterm* over Σ and X. Crucially the notion of preterm allows term formation using a function symbol f of domain type $\tau(1)\ldots\tau(n)$ where the ith subterm has a nominal type syntactically different from, but semantically identical to (by virtue of type equations) $\tau(i)$.

2.3. Definition. Define the K' indexed family $T_{pre}(\Sigma, X) = \langle\, T_{pre}(\Sigma, X)_k \mid k \in K' \,\rangle$ of sets $T_{pre}(\Sigma, X)_k$ of *preterms of kind* k over Σ and X, and for each preterm $t \in T_{pre}(\Sigma, X)_k$ its *nominal type* $\tau_t \in T(\Gamma)_k$, inductively.

(i) For any kind $k \in K'$, any types $\sigma, \tau \in T(\Gamma)_k$ and any variable $x \in X_\sigma$,

$$x^\tau \in T_{pre}(\Sigma, X)_k,$$

and has nominal type τ.

(ii) For any kind $k \in K'$, any types $\sigma, \tau \in T(\Gamma)_k$ and any constant symbol $c \in \Sigma_{\lambda, \sigma}$,

$$c^\tau \in T_{pre}(\Sigma, X)_k,$$

and has nominal type τ.

(iii) For any $n \geq 1$, any kinds $k_0, \ldots, k_n \in K'$, any types $\tau_i \in T(\Gamma)_{k_i}$ for $1 \leq i \leq n$ and $\sigma, \tau \in T(\Gamma)_{k_0}$, any function symbol $f \in \Sigma_{\tau_1 \ldots \tau_n, \sigma}$ and any preterms $t_i \in T_{pre}(\Sigma, X)_{k_i}$ for $1 \leq i \leq n$,

$$f^\tau(t_1, \ldots, t_n) \in T_{pre}(\Sigma, X)_{k_0}$$

and has nominal type τ.

We let $T_{pre}(\Sigma)_k$ denote the set of all ground preterms over Σ of kind k. □

Clearly combinator preterms provide the most liberal notion of term formation consistent with the sort discipline provided by the kind set K. A sound and complete type assignment calculus then provides a means of identifying the meaningful (i.e. typable) combinator terms in the context of a particular equational theory of types (cf. Meinke [1991]).

Since each combinator preterm t is explicitly annotated throughout by type terms, each node in the parse tree $P(t)$ of t is labelled with the parse tree $P(\tau)$ of a type term τ. This two level structure of the parse tree $P(t)$ allows us to perform rewriting on either level, and is reflected in the definition of *occurrences* and *substitution* for t.

2.4. Definition. For any kind $k \in K'$ and any combinator preterm $t \in T_{pre}(\Sigma, X)_k$ we define the sets $Occ^{com}(t) \subseteq \mathbf{N}^*$ and $Occ^{typ}(t) \subseteq \mathbf{N}^* \times \mathbf{N}^*$ of all *combinator occurrences* and *type occurrences* respectively in t. We also define the maps

$$Knd_t^{com} : Occ^{com}(t) \to K, \quad Knd_t^{typ} : Occ^{typ}(t) \to K$$

which assign kinds to combinator occurrences and type occurrences. These sets and mappings are defined inductively.

(i) For any kind $k \in K'$, any type term $\tau \in T(\Gamma)_k$ and any preterm $t \in T_{pre}(\Sigma, X)_k$ with nominal type τ,

$$\lambda \in Occ^{com}(t)$$

and

$$Knd_t^{com}(\lambda) = k.$$

Also for any $\bar{i} \in Occ(\tau)$,

$$(\lambda, \bar{i}) \in Occ^{typ}(t)$$

and

$$Knd_t^{typ}(\lambda, \bar{i}) = Knd_\tau(\bar{i}).$$

(ii) For any $n \geq 1$, any kinds $k_0, \ldots, k_n \in K'$, any types $\tau_i \in T(\Gamma)_{k_i}$ for $1 \leq i \leq n$ and $\sigma, \tau \in T(\Gamma)_{k_0}$, any function symbol $f \in \Sigma_{\tau_1 \ldots \tau_n, \sigma}$ and any preterms $t_i \in T_{pre}(\Sigma, X)_{k_i}$ for $1 \leq i \leq n$, let t denote the preterm $f^\tau(t_1, \ldots, t_n)$. For each $1 \leq m \leq n$, if $\bar{i} \in Occ^{com}(t_m)$ then

$$m.\bar{i} \in Occ^{com}(t)$$

and

$$Knd_t^{com}(m.\bar{i}) = Knd_{t_m}^{com}(\bar{i}).$$

Also for any $(\bar{i}, \bar{j}) \in Occ^{typ}(t_m)$,

$$(m.\bar{i}, \bar{j}) \in Occ^{typ}(t)$$

and

$$Knd_t^{typ}(m.\bar{i}, \bar{j}) = Knd_{t_m}^{typ}(\bar{i}, \bar{j}).$$

□

To rewrite a combinator subexpression in a combinator preterm $t \in T_{pre}(\Sigma, X)_k$ we require substitution at a combinator occurrence $\bar{i}$ of t. To rewrite a type subexpression in t we require substitution at a type occurrence $(\bar{i}, \bar{j})$ of t. We define both substitutions by induction on the complexity of occurrences.

2.5. Definition. For any kind $k \in K'$, any preterm $t \in T_{pre}(\Sigma, X)_k$, any combinator occurrence $\bar{i} \in Occ^{com}(t)$ of kind $k' \in K'$ and any preterm $t' \in T_{pre}(\Sigma, X)_{k'}$ we define the *combinator subterm* of t at $\bar{i}$, denoted by $t(\bar{i})$, and the *substitution of* t' *into* t *at* $\bar{i}$ denoted by $t(\bar{i}/t')$ inductively.

(i) For any kind $k \in K'$ and any preterm $t \in T_{pre}(\Sigma, X)_k$,

$$t(\lambda) = t,$$

and for any preterm $t' \in T_{pre}(\Sigma, X)_k$,

$$t(\lambda/t') = t'.$$

(ii) For any $n \geq 1$, any kinds $k_0, \ldots, k_n \in K'$, any types $\tau_i \in T(\Gamma)_{k_i}$ for $1 \leq i \leq n$, and $\sigma, \tau \in T(\Gamma)_{k_0}$, any function symbol $f \in \Sigma_{\tau_1 \ldots \tau_n, \sigma}$, any preterms $t_i \in T_{pre}(\Sigma, X)_{k_i}$ for $1 \leq i \leq n$, any $1 \leq m \leq n$ and any combinator occurrence $\bar{i} \in Occ^{com}(t_m)$,

$$f^\tau(t_1, \ldots, t_n)(m.\bar{i}) = t_m(\bar{i}),$$

and if $Knd^{com}_{t_m}(\bar{\imath}) = k'$ then for any preterm $t' \in T_{pre}(\Sigma, X)_{k'}$,

$$f^{\tau}(t_1, \ldots, t_n)(m.\bar{\imath}/t') = f^{\tau}(t_1, \ldots, t_{m-1}, t_m(\bar{\imath}/t'), t_{m+1}, \ldots, t_n).$$

For any type occurrence $(\bar{\imath}, \bar{\jmath}) \in Occ^{typ}(t)$ of kind $k' \in K'$ and any type term $\tau \in T(\Gamma)_{k'}$ we define the *type subterm* of t at $(\bar{\imath}, \bar{\jmath})$, denoted by $t(\bar{\imath}, \bar{\jmath})$ and the *substitution of* τ *into* t *at* $(\bar{\imath}, \bar{\jmath})$, denoted by $t(\bar{\imath}, \bar{\jmath}/\tau)$ inductively.

(iii) For any kind $k \in K'$, any type terms $\sigma, \tau \in T(\Gamma)_k$, any combinator constant symbol $c \in \Sigma_{\lambda, \sigma}$ and any occurrence $\bar{\jmath} \in Occ(\tau)$,

$$c^{\tau}(\lambda, \bar{\jmath}) = \tau(\bar{\jmath})$$

and if $Knd_{\tau}(\bar{\jmath}) = k'$ then for any type term $\tau' \in T(\Gamma)_{k'}$,

$$c^{\tau}(\lambda, \bar{\jmath}/\tau') = c^{\tau(\bar{\jmath}/\tau')}.$$

(iv) For any $n \geq 1$, any kinds $k_0, \ldots, k_n \in K'$, any type terms $\tau_i \in T(\Gamma)_{k_i}$ for $1 \leq i \leq n$ and $\sigma, \tau \in T(\Gamma)_{k_0}$, any combinator function symbol $f \in \Sigma_{\tau_1 \ldots \tau_n, \sigma}$, any preterms $t_i \in T_{pre}(\Sigma, X)_{k_i}$ for $1 \leq i \leq n$ and for any occurrence $\bar{\jmath} \in Occ(\tau)$,

$$f^{\tau}(t_1, \ldots, t_n)(\lambda, \bar{\jmath}) = \tau(\bar{\jmath})$$

and if $Knd_{\tau}(\bar{\jmath}) = k'$ then for any type term $\tau' \in T(\Gamma)_{k'}$

$$f^{\tau}(t_1, \ldots, t_n)(\lambda, \bar{\jmath}/\tau') = f^{\tau(\bar{\jmath}/\tau')}(t_1, \ldots, t_n).$$

For any $1 \leq m \leq n$ and any type occurrence $(\bar{\imath}, \bar{\jmath}) \in Occ^{typ}(t_m)$,

$$f^{\tau}(t_1, \ldots, t_n)(m.\bar{\imath}, \bar{\jmath}) = t_m(\bar{\imath}, \bar{\jmath})$$

and if $Knd^{typ}_{t_m}(\bar{\imath}, \bar{\jmath}) = k'$ then for any type term $\tau' \in T(\Gamma)_{k'}$,

$$f^{\tau}(t_1, \ldots, t_n)(m.\bar{\imath}, \bar{\jmath}/\tau') = f^{\tau}(t_1, \ldots, t_{m-1}, t_m(\bar{\imath}, \bar{\jmath}/\tau'), t_{m+1}, \ldots, t_n).$$

□

To rewrite combinator expressions over Σ we introduce *combinator rewrite rules*.

2.6. Definition. A *combinator rewrite rule* r of kind $k \in K'$ is an ordered pair of combinator preterms (t_1, t_2) with $t_1, t_2 \in T_{pre}(\Sigma, X)_k$. We write $t_1 \to t_2$ to denote the rule (t_1, t_2). □

2.7. Definition. Let $k \in K'$ be any kind and $t, t' \in T_{pre}(\Sigma, X)_k$ be any combinator preterms of kind $k \in K'$.

(i) We say that t *rewrites* to t' *by a type rewrite rule* $\tau_1 \to \tau_2$ of kind $k' \in K$ at a type occurrence $(\bar{\imath}, \bar{\jmath}) \in Occ^{typ}(t)$ of kind k' if for some substitution $\theta : Y \to T(\Gamma)$,

$$t(\bar{\imath}, \bar{\jmath}) = \overline{\theta}(\tau_1)$$

and

$$t' = t(\bar{\imath}, \bar{\jmath}/\overline{\theta}(\tau_2)).$$

If t rewrites to t' by a type rewrite rule ρ at a type occurrence $(\bar{i}, \bar{j}) \in Occ^{typ}(t)$ then we write

$$t \xrightarrow{(\bar{i}, \bar{j}), \rho} t'$$

and when the occurrence $(\bar{i}, \bar{j})$ is known or irrelevant we may simply write $t \xrightarrow{\rho} t'$.

(ii) We say that t *rewrites* to t' *by a combinator rewrite rule* $t_1 \rightarrow t_2$ of kind $k' \in K'$ at a combinator occurrence $\bar{i} \in Occ^{com}(t)$ of kind k' if for some substitution $\theta : X \rightarrow T_{pre}(\Sigma, X)$,

$$t(\bar{i}) = \bar{\theta}(t_1)$$

and

$$t' = t(\,\bar{i} / \bar{\theta}(t_2)\,).$$

If t rewrites to t' by a combinator rewrite rule r at a combinator occurrence $\bar{i} \in Occ^{com}(t)$ then we shall write

$$t \xrightarrow{\bar{i}, r} t'$$

and when the occurrence $\bar{i}$ is known or irrelevant we may simply write $t \xrightarrow{r} t'$. □

Finally we can define the rewrite relation $\xrightarrow{R}$ on combinator preterms induced by a *type rewrite system* R on preterms.

2.8. Definition. By a *type rewrite system* R we mean an ordered pair $R = (R_{typ}, R_{com})$, where R_{typ} is a set of type rewrite rules and R_{com} is a set of combinator rewrite rules.

Define the K'–indexed family $\xrightarrow{R} = \langle \xrightarrow{R}_k \mid k \in K' \rangle$ of binary relations on $T_{pre}(\Sigma, X)$ by

$$t \xrightarrow{R}_k t' \;\Leftrightarrow\; t \xrightarrow{\rho} t' \text{ for some } \rho \in R_{typ} \text{ or } t \xrightarrow{r} t' \text{ for some } r \in R_{com},$$

for any $k \in K'$ and $t, t' \in T_{pre}(\Sigma, X)_k$. Let $\xrightarrow{R}^*$ denote the reflexive, transitive closure of $\xrightarrow{R}$. □

Rewriting types and combinators provides a natural computational method to support equational reasoning for types and combinators.

2.9. Definition. By an *equational type specification* we mean a pair

$$Spec = (\,(\Gamma, \varepsilon), (\Sigma, E)\,)$$

where Γ is a K–kinded type signature, ε is an equational theory over Γ and some family Y of sets of variables, Σ is a combinator signature over Γ and $K' \subseteq K$, and E is a set of equations in preterms over Σ and some family X of sets of variables, i.e. formulas of the form $t = t'$ where $t, t' \in T_{pre}(\Sigma, X)_k$. □

In Meinke [1991] a semantics for validity of equations in preterms is given with which the model category $Mod(Spec)$ of an equational type specification $Spec$ is defined as a cofibered category of algebras. It is shown that the category

Mod(*Spec*) admits an initial object $I(Spec)$ which can be concretely constructed as the quotient of the algebra of typable ground combinator preterms factored by provable equivalence. The combinator algebra $I(Spec)$ provides an initial model semantics for *ATLAS* specifications. In Meinke [1992b] it is shown that rewriting types and combinators, equational reasoning with types and combinators and calculation in the initial model $I(Spec)$ are equivalent in the following sense.

2.10. Definition. For any equational type specification $Spec = ((\Gamma, \varepsilon), (\Sigma, E))$ we let $R(\varepsilon, E)$ denote the type rewrite system with

$$R(\varepsilon, E)_{typ} = \{ \tau \to \tau', \tau' \to \tau \mid \tau = \tau' \in \varepsilon \}$$

and

$$R(\varepsilon, E)_{com} = \{ t \to t', t' \to t \mid t = t' \in E \}.$$

□

2.11. Theorem. *Let* $Spec = ((\Gamma, \varepsilon), (\Sigma, E))$ *be an equational type specification. Suppose that each equation* $e \in E$ *is well typed with respect to* $Alg(\Gamma, \varepsilon)$. *For any* $k \in K'$ *and any ground preterms* $t, t' \in T_{pre}(\Sigma)_k$, *if* t *and* t' *are well typed with respect to* $Alg(\Gamma, \varepsilon)$ *then the following are equivalent:*

(i) $(\varepsilon, E) \vdash t = t'$;
(ii) $I(Spec) \models t = t'$;
(iii) $t \stackrel{R(\varepsilon, E)^*}{\longrightarrow} t'$.

Proof. See Meinke [1992b]. □

Note in Theorem 2.11 that an equation $t = t'$ in combinator preterms $t, t' \in T_{pre}(\Sigma, X)_k$ is well typed with respect to $Alg(\Gamma, \varepsilon)$ if, and only if, $Alg(\Gamma, \varepsilon) \models \tau_t = \tau_{t'}$, where τ_t and $\tau_{t'}$ are the nominal types of t and t' respectively.

3. TYPE SPECIFICATIONS IN *ATLAS*.

In this section we describe the concrete syntax of type specifications in *ATLAS*. The definition of *ATLAS* specification syntax is divided into context-free and context-sensitive components and *ATLAS* specifications are processed in two stages based on this separation. We formally define the context-free syntax of *ATLAS* but only briefly describe the context-sensitive syntax by means of examples. A full account of both context-free and context-sensitive definitions will appear in Hearn[1994].

In Section 3.1. we describe the hierarchical organisation of *ATLAS* specifications into three levels for kind, type and combinator specifications. In Sections 3.2 to 3.4 we describe each of these levels in detail. In Appendix 1 we include the syntactic definitions of kind, type and combinator terms in *ATLAS*.

3.1. The Hierarchical Structure of *ATLAS* Specifications. We define the context-free syntax of *ATLAS* specifications using BNF production rules.

The outermost production for an *ATLAS* specification declares that the specification is hierarchically organised into three levels for *kind*, *type* and *combinator* specifications.

$$< Atlas\_spec > ::= \underline{\texttt{Atlas Specification}} < identifier > \underline{\texttt{is}}$$
$$< kind\_spec >< type\_spec >< comb\_spec > \underline{\texttt{end.}}$$

In order to discuss the syntax of these three levels it is helpful to see a small but complete *ATLAS* specification. Below we show how simple typed combinatory logic may be encoded into *ATLAS*.

```
Atlas Specification Typed_Combinatory_Logic is
(* An encoding of typed combinatory logic *)

Kind Specification Typed_Combinatory_Logic is
  Constants
    TYPE : kind;
  Operations;
  Variables;

Type Specification Typed_Combinatory_Logic is
  Constants;
  Operations
    arrow : TYPE * TYPE -> TYPE;
  Variables
    x,y,z : TYPE;
  Rules;

Combinator Specification Typed_Combinatory_Logic is
  Using kinds TYPE;
  Constants
    K : (x arrow (y arrow x)),
    S : ((x arrow (y arrow z)) arrow ((x arrow y)
        arrow (x  arrow z))),
    I : (x arrow x);
  Operations
    eval : (x arrow y) * x -> y {implicit};
  Variables
    X:x, Y:y, Z:z;
  Rules
```

```
   (((S  X)  Y)  Z) -> ((X Z) (Y Z)),
   ((K  X) Y) -> X,
   (I X) -> X;
end.
```

3.2. Kind Specifications. The first level of an *ATLAS* specification defines a signature for kinds. The productions for a kind specification are as follows.

```
     < kind_spec > ::= < kind_head >< kind_const_id >
                       < kind_op_id >< kind_var_id >
     < kind_head > ::= Kind Specification < identifier > is
 < kind_const_id > ::= Constants < kind_consts > ; | Constants ;
   < kind_consts > ::= < kind_const > | < kind_const > , < kind_consts >
    < kind_const > ::= < identifier_list > : kind
    < kind_op_id > ::= Operations < kind_ops > ; | Operations ;
      < kind_ops > ::= < kind_op > | < kind_op > , < kind_ops >
       < kind_op > ::= < identifier_list > : < kind_op_domain > -> kind
                       | < identifier > : kind * kind - >
                       kind {implicit}
< kind_op_domain > ::= kind | kind * < kind_op_domain >
   < kind_var_id > ::= Variables < kind_vars > ; | Variables;
     < kind_vars > ::= < kind_var > | < kind_var > , < kind_vars >
      < kind_var > ::= < identifier_list > : kind
```

A kind specification declares constant, operation and variable symbols for a single(`kind`)-sorted signature. Note that binary operation symbols can be declared to be *implicit.* Implicit operation symbols may be omitted from the construction of a term and will be explicitly reinstated at compile time. *ATLAS* does not support the declaration of rewrite rules for kinds.

Notice that since rewrite rules for kinds are omitted from *ATLAS* specifications the initial model semantics of a three level *ATLAS* specification may still be defined using the methods of Section 2 by taking the kind sort set to be the set of all ground kind terms.

3.3. Type Specifications. The second level of an *ATLAS* specification defines a signature and rewrite system for types. The productions for a type specification are as follows.

```
< type_spec > ::= < type_head >< type_const_id >< type_op_id >
                  < type_var_id >< type_rule_id >
< type_head > ::= Type Specification < identifier > is
< type_const_id > ::= Constants < type_consts > ; | Constants ;
< type_consts > ::= < type_const > | < type_const > , < type_consts >
< type_const > ::= < identifier_list > : < kind_term >
< type_op_id > ::= Operations < type_ops > ; | Operations ;
< type_ops > ::= < type_op > | < type_op > , < type_ops >
< type_op > ::= < identifier_list > : < type_dom > -> < kind_term >
              | < identifier > : < kind_term > * < kind_term >
                -> < kind_term > {implicit}
< type_dom > ::= < kind_term > | < kind_term > * < type_dom >
< type_var_id > ::= Variables < type_vars > ; | Variables ;
< type_vars > ::= < type_var > | < type_var > , < type_vars >
< type_var > ::= < identifier_list > : < kind_term >
< type_rule_id > ::= Rules < type_rules > ; | Rules ;
< type_rules > ::= < type_rule > | < type_rule > , < type_rules >
< type_rule > ::= < type_term > -> < type_term >
```

A type specification declares constant, operation and variable symbols for a many-sorted signature where the sorts are taken to be terms over the previously defined kind signature. Thus the signature may be infinitely sorted. Again binary type operations may be declared to be implicit.

3.4. Combinator Specifications. The third level of an *ATLAS* specification gives a signature and rewrite system for typed combinators. The productions for a combinator specification are as follows.

```
< comb_spec > ::= < comb_head >< kind_usage >< comb_const_id >
                  < comb_op_id >< comb_var_id >< comb_rule_id >
< comb_head > ::= Combinator Specification < identifier > is
< kind_usage > ::= Using kinds < kinds > ;
< kinds > ::= < identifier > | < identifier > , < identifiers >
< comb_const_id > ::= Constants < comb_consts > ; | Constants ;
```

```
< comb_consts > ::= < comb_const >
                  | < comb_const > , < comb_consts >
 < comb_const > ::= < identifier_list > : < type_term >
 < comb_op_id > ::= Operations < comb_ops > ; | Operations ;
   < comb_ops > ::= < comb_op > | < comb_op > , < comb_ops >
    < comb_op > ::= < identifier_list > : < comb_op_domain > - >
                    < type_term >
                  | < identifier > : < type_term > *
                    < type_term > - > < type_term > {implicit}
< comb_op_domain > ::= < type_term >
                  | < type_term > * < comb_op_domain >
< comb_var_id > ::= Variables < comb_vars > ; | Variables ;
  < comb_vars > ::= < comb_var > | < comb_var > , < comb_vars >
   < comb_var > ::= < identifier_list > : < type_term >
< comb_rule_id > ::= Rules < comb_rules > ; | Rules ;
 < comb_rules > ::= < comb_rule > | < comb_rule > , < comb_rules >
  < comb_rule > ::= < comb_term > - > < comb_term >
```

A combinator specification declares constant, operation and variable symbols for a many-sorted signature where the sorts are taken to be type terms over the previously defined type signature (itself sorted by the associated kind signature). Thus, again the signature can be infinitely sorted. Type terms appearing in combinator declarations may only have kinds declared after the statement `Using kinds`.

4. A SPECIFICATION TUTORIAL.

In this section we provide a brief tutorial in constructing *ATLAS* specifications by means of a survey of specification case studies. Our case studies are chosen from:

(i) first order and parameterised algebraic specifications,
(ii) higher order algebraic specifications, and
(iii) propositional logics.

4.1. First Order Algebraic Specifications. We shall consider the classical data type stack which is commonly used as a benchmark for comparing algebraic specification formalisms. A recent survey of stack specifications may be found in Bergstra and Tucker [1990]. We begin with an *ATLAS* specification of the simple stack which we shall then refine in two successive stages to demonstrate the type formalism of *ATLAS*.

```
Atlas Specification Simple_Stack is

Kind Specification Simple_Stack is
  Constants
    TYPE : kind;
  Operations;
  Variables ;

Type Specification Simple_Stack is
  Constants
    nat : TYPE,
    stack : TYPE;
  Operations;
  Variables ;
  Rules ;

Combinator Specification Simple_Stack is
  Using kinds TYPE;
  Constants
    error : nat,
    0 : nat,
    empty : stack;
  Operations
    succ : nat -> nat,
    push : nat * stack -> stack,
    pop  : stack -> stack,
    top  : stack -> nat;
  Variables
    X : nat,
    Y : stack;
  Rules
    pop(push(X,Y)) -> Y,
    top(push(X,Y)) -> X,
    pop(empty) -> empty,
    top(empty) -> error;
end.
```

The specification `Simple_Stack` gives a paradigm for encoding any many-sorted algebraic specification (Σ, E) into *ATLAS*. Given an S-sorted signature Σ we encode each sort $s \in S$ as a type constant **s** : TYPE. Each symbol $c \in \Sigma_{\lambda,s}$ is encoded as a combinator constant symbol **c** : **s**. Each operation symbol $f \in \Sigma_{w,s}$

for $w = s(1) \ldots s(n) \in S^+$ is encoded as a combinator function symbol `f : s(1) * ... * s(n) -> s`. Each equation $t = t' \in E$ is encoded as an appropriately oriented combinator rewrite rule `t -> t'` or `t' -> t`. (In practise we may need to apply completion to the equations of E according to some termination ordering in order to obtain a confluent and terminating set of combinator rewrite rules from E. The current implementation of *ATLAS* does not support completion.)

The specification `Simple_Stack` specifies stacks of one particular fixed type namely natural numbers. As is well known the stack may be defined to be parametric in type. Our first refinement of `Simple_Stack` encodes the parametric stack as a *type construction.* We replace the type constant `stack : TYPE` by the type forming operation `stack : TYPE -> TYPE`.

```
Atlas Specification Parametric_Stack is

Kind Specification Parametric_Stack is
  Constants
    TYPE : kind;
  Operations;
  Variables ;

Type Specification Parametric_Stack is
  Constants
    nat : TYPE;
  Operations
    stack : TYPE -> TYPE;
  Variables
    x : TYPE;
  Rules ;

Combinator Specification Parametric_Stack is
  Using kinds TYPE;
  Constants
    error : x,
    empty : stack(x);

  Operations
    push : x * stack(x) -> stack(x),
    pop  : stack(x) -> stack(x),
    top  : stack(x) -> x;
  Variables
    X : x,
    Y : stack(x);
  Rules
    pop(push(X,Y)) -> Y,
```

```
   top(push(X,Y)) -> X,
   pop(empty) -> empty,
   top(empty) -> error;
end.
```

In Parametric_Stack we can form type terms such as

$$\mathtt{nat} : \mathsf{TYPE}, \quad \mathtt{stack(nat)} : \mathsf{TYPE}, \quad \mathtt{stack(stack(nat))} : \mathsf{TYPE}.$$

Notice the use of type variables in the combinator specification to declare parametric constants `error` : `x` and `empty` : `stack(x)`, also operations `pop`, `top` and `push`. This type of parametric declaration is often termed *free variable polymorphism*. The type and combinator declarations together capture the parametric syntax of the stack.

The combinator rewrite rules in Parametric_Stack have all type and kind annotations omitted. To illustrate the kind and type inference mechanism of *ATLAS* we show the same rules after compilation.

```
Rules
pop(push(X:A:TYPE,Y:stack(A:TYPE):TYPE):stack(A:TYPE):TYPE)
:stack(A:TYPE):TYPE
 ->
Y:stack(A:TYPE):TYPE,

top(push(X:F:TYPE,Y:stack(F:TYPE):TYPE):stack(F:TYPE):TYPE):F:TYPE
   ->
X:F:TYPE,

pop(empty:stack(K:TYPE):TYPE):stack(K:TYPE):TYPE
   ->
empty:stack(K:TYPE):TYPE,

top(empty:stack(N:TYPE):TYPE):N:TYPE
   ->
error:N:TYPE;
```

Notice that type variables also feature in these rules, i.e. *ATLAS* supports free variable polymorphism in rewrite rules. Semantically all declarations featuring kind or type variables are equivalent to *schemes* of declarations, one for each instantiation by ground terms. This allows finite declarations of infinite collections of syntax and rewrite rules and is a powerful feature of *ATLAS* that is essential for applications such as proof checking (cf. Section 4.3).

The specification Parametric_Stack provides a paradigm for encoding certain kinds of parameterised first order algebraic specifications that can also be applied to lists, trees, arrays, queues etc. As is well known, Parametric_Stack is flawed by an inadequate treatment of error handling. In particular, the natural number type nat : TYPE is spoiled by the non–standard value **error : nat** which would invalidate the induction principle for natural numbers. A refinement of this specification which remedies the problem is the so called *stratified stack* which can be encoded in *ATLAS* as follows.

```
Atlas Specification Stratified_Stack is

Kind Specification Stratified_Stack is
  Constants
    TYPE : kind,
    NAT  : kind;
  Operations;
  Variables ;

Type Specification Stratified_Stack is
  Constants
    0 : NAT,
    error : TYPE;
  Operations
    succ : NAT -> NAT,
    stack : TYPE * NAT -> TYPE;
  Variables
    x : TYPE,
    y : NAT;
  Rules ;

Combinator Specification Stratified_Stack is
  Using kinds TYPE;
  Constants
    empty : stack(x,0),
    error_flag  : error;
  Operations
    push : x * stack(x,y) -> stack(x,succ(y)),
    pop  : stack(x,succ(y)) -> stack(x,y),
    pop  : stack(x,0) -> stack(x,0),
    top  : stack(x,succ(y))-> x,
    top  : stack(x,0) -> error;
  Variables
    X : x,
    Y : stack(x,y);
```

```
  Rules
    pop(push(X,Y)) -> Y,
    top(push(X,Y)) -> X,
    pop(empty) -> empty,
    top(empty) -> error_flag;
end.
```

In `Stratified_Stack` the type construction `stack : TYPE * NAT -> TYPE` is parametric in both type and height. The empty stack of type τ is represented as a value of type `stack`$(\tau, 0)$: TYPE. This allows us to modify the type declarations for `top` and `pop`, in particular we now define `top : stack(x, 0) -> error` where `error : TYPE` is a new type constant and values such as `flag : error` may be used for error messages and exception handling.

4.2. Higher Order Algebraic Specification. We shall consider higher type algebraic specifications in *ATLAS*. A basic theory for higher type algebraic specification methods is given in Möller [1987, 1994], Möller et al [1988] and Meinke [1992a, 1993]. Higher order specification methods are particularly useful in the design of hardware algorithms and programming languages. Both of these applications typically require second, and occasionally third order operations. An example of a rudimentary typed functional programming language defined using a higher (ω) order specification, namely typed combinatory logic, has already been given in 3.1. A case study of specification and verification of a systolic convolution algorithm over the ring **Z** of integers is given in Meinke and Steggles [1994]. This specification of convolution with respect to a fixed number of parameters as a second order *stream transformer* may be formalised in *ATLAS* as follows.

```
Atlas Specification Conv3 is
(* convolution for fixed data length 3*)

Kind Specification Conv3 is
  Constants
    TYPE : kind ;
  Operations;
  Variables;

Type Specification Conv3 is
  Constants
    nat, int : TYPE;
  Operations
    arrow : TYPE * TYPE -> TYPE;
  Variables;
```

```
  Rules ;

Combinator Specification Conv3 is
  Using kinds TYPE;
  Constants
    0, 1 : int;
  Operations
    succ : nat -> nat,
    minus : int -> int,
    add,mult : int * int -> int,
    eval : (nat arrow int) * nat -> int {implicit},
    Conv3 : int * int * int * (nat arrow int) ->
    (nat arrow int);
  Variables
    x,x1,y,y1,z,z1 : int,
    x2,x3 : int,
    t : nat,
    s : (nat arrow int);
  Rules
    add(x,add(y,z)) -> add(add(x,y),z),
    mult(x,mult(y,z)) -> mult(mult(x,y),z),
    add(x,0) -> x,
    add(x,minus(x)) -> 0,
    mult(x,1) -> x,
    mult(1,x) -> x,
    mult(x,add(y,z)) -> add(mult(x,y),mult(x,z)),
    mult(add(x,y),z) -> add(mult(x,z),mult(y,z)),
    (Conv3(x1,x2,x3,s) t) ->
    add(mult(x1,(s t)),add(mult(x2, (s succ(t))),
    mult(x3, (s succ(t)))));
end.
```

In Bergstra and Tucker [1980] it is shown that the second order algebra PR of primitive recursion has no recursively enumerable equational specification under initial algebra semantics. In Meinke [1994] it is shown that the algebra of primitive recursion has a recursive equational specification $(\Sigma^{\mathcal{PR}}, E^{\mathcal{PR}})$ under second order initial algebra semantics, thus higher order algebraic specification methods are strictly more powerful than first order initial algebra methods. By using type variables and schemes of declarations in *ATLAS* we can represent this countably infinite signature $\Sigma^{\mathcal{PR}}$ and the equations $E^{\mathcal{PR}}$, oriented as rewrite rules, as a finite *ATLAS* specification.

```
Atlas Specification Primitive_Recursion is
(* Second order algebra of primitive recursion *)

Kind Specification Primitive_Recursion is
  Constants
    TYPE : kind,
    NAT : kind;
  Operations;
  Variables;

Type Specification Primitive_Recursion is
  Constants
    0 : NAT,
    nat : TYPE;
  Operations
    cross,arrow : TYPE * TYPE -> TYPE,
    succ : NAT -> NAT,
    nat : NAT -> TYPE;
  Variables
    a,b,y : TYPE,
    x : NAT,
    k,l,j : NAT;
  Rules
    nat(0) -> nat,
    nat(succ(x)) -> (nat cross nat(x));

Combinator Specification Primitive_Recursion is
  Using kinds TYPE;
  Constants
    0 : nat;
  Operations
    succ : nat -> nat,
    o : (nat(k) arrow nat(j)) * (nat(l) arrow nat(k)) ->
        (nat(l) arrow nat(j)),
    | : (nat(k) arrow nat(j)) * (nat(k) arrow nat) ->
        (nat(k) arrow nat(succ(j))),
    pr : (nat(k) arrow nat) * (nat(succ(succ(k))) arrow nat)
         -> (nat(succ(k)) arrow nat),
    eval : (a arrow b) * a -> b {implicit},
    pair : nat(k) * nat -> nat(succ(k)),
    proj1 : nat(succ(k)) -> nat(k),
    proj2 : nat(succ(k)) -> nat;
  Variables
    W : ( nat( succ ( succ(l) ) ) arrow nat ),
```

```
    X : ( nat(k) arrow nat (j) ),
    Y : ( nat(l) arrow nat(k) ),
    Z : ( nat(l) arrow nat ),
    x : nat(l),
    y : nat;
  Rules
    ((X o Y) x) -> (X (Y x)),
    ((Y | Z) x) -> pair((Y x),(Z x)),
    (pr(Z, W) pair(x, 0)) -> (Z x),
    (pr(Z, W) pair(x, succ(y))) ->
    (W pair(pair(x, y),(pr(Z, W) pair(x, y)))),
    proj1(pair(x,y)) -> x,
    proj2(pair(x,y)) -> y;
end.
```

4.3. *ATLAS* as a Logical Framework. The well known *Curry Howard isomorphism* provides a paradigm to interpret logical propositions as type expressions and proofs as combinator expressions. Proof checking then becomes type checking of combinator terms. Since *ATLAS* can accept arbitrary type and combinator declarations it can accept arbitrary logical languages and proof rules. Thus *ATLAS* can be used as a logical framework in the sense of Huet and Plotkin [1991]. Since *ATLAS* computes with typed combinators rather than typed λ–terms then logics must be encoded into *ATLAS* via Hilbert style rather than natural deduction proof systems.

To briefly illustrate *ATLAS* as a logical framework let us consider some encodings of propositional logics. Intuitionistic propositional logic may be encoded as follows.

```
Atlas Specification Int_Prop_Logic is
(* Intuitionistic Propositional  Logic *)
Kind Specification Int_Prop_Logic is
  Constants
    PROP : kind;
  Operations;
  Variables;

Type Specification Int_Prop_Logic is
  Constants
    false : PROP;
  Operations
    not : PROP -> PROP,
    implies : PROP * PROP -> PROP,
    or,and,bi_implies : PROP * PROP -> PROP;
  Variables
```

```
    x,y,z : PROP;
  Rules
    not (x) -> (x implies false ),
    (x bi_implies y) -> (( x implies y) and (y implies x));

Combinator Specification Int_Prop_Logic is
Using kinds PROP;
  Constants
    K : (x implies  (y implies  x)),
    S : ((x implies  (y implies  z)) implies  ((x implies  y)
        implies (x  implies z))),
    I : (x implies x),
    CI : (x implies (y implies (x and y))),
    CEL : ((x and y) implies  x),
    CER : ((x and y) implies y),
    DIL : (x implies  (x or y)),
    DIR : (y implies  (x or y)),
    DE : ((x implies  z) implies  ((y implies  z) implies
         ((x or y) implies  z))),
    XF : (false implies z);
  Operations
    MP : (x implies y) * x -> y {implicit};
  Variables
    X:x, Y:y, Z:z;

  Rules
    (((S  X)  Y)  Z) -> ((X Z) (Y Z)),
    ((K  X) Y) -> X,
    (I X) -> X,
    (CEL  ((CI  X) Y)) -> X,
    (CER  ((CI  X)   Y)) -> Y,
    ((CI  (CEL  X))  (CER  X)) -> X,
    (CEL  (XF Y)) -> (XF Y),
    (CER (XF Y)) -> (XF Y),
    ((XF  Y)  X) -> (XF Y),
    (XF Y) -> Y;
end.
```

Notice the use of type variables in the combinator specification to formalise schemes of axioms and proof rules. Classical propositional logic may be obtained by extending `Int_Prop_Logic` in one of several ways. For example we may add the type rewrite rule `not(not(x)) -> x` or alternatively the combinator constant `XM : (x or (not(x)))`.

As well as type checking, the rewrite engine of *ATLAS* allows us to carry out proof normalisation as combinator rewriting. In intuitionistic logics, proofs encode algorithms and proof normalisation corresponds to algorithm execution. This subject is an extensive and active area of current research in logics of programs but for reasons of space it must fall outside the scope of an introductory tutorial to *ATLAS* .

Current work on *ATLAS* concerns the development of modularisation facilities suitable for specification in the large. Such modularisation primitives should allow each specification level (kind, type, combinator) to form a module and therefore generalise the formalism from three fixed levels to an arbitrary number. This work will be described in Hearn [1994].

We thank U. Berger, J.R. Hindley, J.V. Tucker and E.G. Wagner for helpful comments on this work. We also acknowledge the financial support of the Science and Engineering Research Council, the British Council and IBM T.J. Watson Research Center.

APPENDIX 1 : SYNTAX OF *ATLAS* TERMS.

We supplement Section 3 with the syntactic definitions of kind, type and combinator terms.

Let us consider in detail the syntax of *kind terms.* As we shall see, the syntax of type terms and combinator terms inherit the syntax of kind terms introduced as annotations (cf. Definition 2.3). The production rules for kind terms are as follows.

$$
\begin{array}{rcl}
< kind\_term > & ::= & < identifier > \\
 & & \mid < identifier > \underline{(} < kind\_term\_list > \underline{)} \\
 & & \mid \underline{(} < kind\_term >< identifier >< kind\_term > \underline{)} \\
 & & \mid \underline{(} < kind\_term >< kind\_term > \underline{)} \\
< kind\_term\_list > & ::= & < kind\_term > \\
 & & \mid < kind\_term > \underline{,} < kind\_term\_list >
\end{array}
$$

These productions state that kind terms are normally in prefix notation, although binary operations may be used in infix notation and may also be implicitly present.

The syntax of *type terms*, which appear in type rewrite rules, make use of the syntax of kind terms as annotations. We define the productions for type terms as follows.

< *type_term* > ::= < *identifier* >
| < *identifier* > : < *kind_term* >
| < *identifier* > (< *type_term_list* >)
| < *identifier* > (< *type_term_list* >) :
< *kind_term* >
| (< *type_term* > < *identifier* > < *type_term* >)
| (< *type_term* > < *type_term* >)
< *type_term_list* > ::= < *type_term* >
| < *type_term* > , < *type_term_list* >

Type terms are normally in prefix notation although binary type operationsIt is important to observe that type terms may include annotations in the form of kind terms. Such annotations can be implicit, for example the term production rule declares that an atomic type term may be an identifier or an identifier explicitly annotated with a kind term. Kind annotations may be omitted in which case a *kind inference* algorithm will be used to reinstate these where possible. Combinator terms have the following syntactic definition (cf. type and kind terms).

< *comb_term* > ::= < *identifier* >
| < *identifier* > : < *type_term* >
| < *identifier* > (< *comb_term_list* >)
| < *identifier* > (< *comb_term_list* >) :
< *type_term* >
| (< *comb_term* >< *identifier* >< *comb_term* >)
| (< *comb_term* >< *comb_term* >)
< *comb_term_list* > ::= < *comb_term* >
| < *comb_term* > , < *comb_term_list* >

Combinator terms are normally in prefix form although binary operations may also appear in infix form. Binary combinator operations may occur implicitly. Notice that combinator terms may include annotation in the form of type terms. This type annotation can also be implicit (recall kind annotation on type terms can be implicit). If type annotation is implicit then type and kind inference algorithms will be used to explicitly reinstate type annotations where possible.

REFERENCES.

J. A. Bergstra and J. V. Tucker, A natural data type with a finite equational final semantics but no effective equational initial semantics specification, Bull. EATCS 11, 23-33 (1980).

J. A. Bergstra and J. V. Tucker, The inescapable stack: an exercise in algebraic specification with total functions, second revised edition, P8804b, Programming Research Group, University of Amsterdam, 1990.

J. A. Goguen, Order–sorted algebra, semantics and theory of computation, Report 14, Computer Science Dept, UCLA, 1978.

B.M. Hearn, Implementation of a Typed Language for Algebraic Specifications, PhD. Thesis, Department of Computer Science, University College of Swansea, (to appear), 1994.

G. Huet and G. Plotkin (eds), Logical Frameworks, Cambridge University Press, Cambridge, 1991.

K. Meinke, Equational specification of abstract types and combinators, Report CSR 11–91, Dept. of Computer Science, University College Swansea, to appear in: G. Jäger (ed), Proc. Computer Science Logic '91, Lecture Notes in Computer Science 626, Springer Verlag, Berlin, 1991.

Meinke [1992a], K. Meinke, Universal algebra in higher types, Theoretical Computer Science, 100, (1992) 385–417.

Meinke [1992b], K. Meinke, Algebraic semantics of rewriting terms and types, in M. Rusinowitch and J-L. Remy (eds), Proc. Third Int. Conf. on Conditional Term Rewriting Systems, Lecture Notes in Computer Science 656, Springer Verlag, Berlin 1992.

K. Meinke, Subdirect representation of higher type algebras, Report CSR 14–90, Dept. of Computer Science, University College Swansea, in: K. Meinke and J.V. Tucker (eds), Many–Sorted Logic and its Applications, John Wiley, 1993.

K. Meinke, A recursive second order initial algebra specification of primitive recursion, to appear in Acta Informatica, 1994.

K. Meinke and L.J. Steggles, Specification and verification in higher order algebra: a case study of convolution, this conference proceedings, 1994.

K. Meinke and J.V. Tucker, Universal algebra, 189–411 in: S. Abramsky, D. Gabbay and T.S.E. Maibaum, (eds) Handbook of Logic in Computer Science, Oxford University Press, Oxford, 1992.

B. Möller, Higher–order algebraic specifications, Facultät für Mathematik und Informatik, Technische Universität München, Habilitationsschrift, 1987.

B. Möller, Ordered and continuous models of higher order algebraic specifications, this conference proceedings, 1994.

B. Möller, A. Tarlecki and M. Wirsing, Algebraic specifications of reachable higher–order algebras, in: D. Sannella and A. Tarlecki (eds), Recent Trends in Data Type Specification, Lecture Notes in Computer Science 332,(Springer Verlag, Berlin, 1988) 154–169.

M. Wirsing, Algebraic Specification, 675-788 in: J. Van Leeuwen (ed.), Handbook of Theoretical Computer Science, Volume B - Formal Models and Semantics, Elsevier, 1990.

Compilation of Combinatory Reduction Systems

Stefan Kahrs*

University of Edinburgh
Laboratory for Foundations of Computer Science

Abstract. Combinatory Reduction Systems generalise Term Rewriting Systems. They are powerful enough to express β-reduction of λ-calculus as a single rewrite rule. The additional expressive power has its price — CRSs are much harder to implement than ordinary TRSs.
We propose an abstract machine suitable for executing CRSs. We define what it means to execute an instruction, and give a translation from CRS rules into sequences of instructions. Applying a rewrite rule to a term is realised by initialising the machine with this term, and then successively executing the instructions of the compiled rule.

1 Introduction

Combinatory Reduction Systems were introduced by Klop in 1980 [9]. CRSs in their original form generalise *applicative* TRSs [10]. We shall concentrate here on *functional* CRSs, as defined by Kennaway in [8]; they generalise ordinary TRSs. The techniques of this paper can easily be adapted to applicative CRSs.

Functional CRSs extend TRSs in two respects. Firstly, they support a notion of variable binding. Substitution has to respect variable bindings: it is not allowed to capture bound variables. Thus, if a rewrite rule contains a subterm $\lambda x.y$, where y is a free variable, then no instance of the rewrite rule can substitute anything for y which contains x freely. However, this excludes rules such as the S-combinator introduction rule in the translation from λ-calculus into Combinatory Logic:

$$\mathrm{Lambda}([x]\mathrm{App}(y,z)) \;\rightarrow\; \mathrm{App}(\mathrm{App}(\mathrm{S},\mathrm{Lambda}([x]y)),\mathrm{Lambda}([x]z))$$

For the S-combinator introduction rule, it is necessary that the terms substituted for y and z may contain x freely, but this is not possible with ordinary first-order substitutions. To overcome this restriction, CRSs have another extension: they support second-order variables and second-order substitutions. For first-order terms, the application of a substitution to a term and the computation of a matching substitution are rather simple operations. Thus, a rewriting step itself is a rather simple operation. This is not longer true in the second-order world, i.e. if the terms contain second-order variables; see [4] where Huet and Lang present an algorithm to compute all principal matches for the second-order case. For

* The research reported here was partially supported by SERC grant GR/J07303.

second-order variables, substitution application incorporates β-reduction, and consequently matching may involve β-expansion.

Combinatory Reduction Systems support second-order rewriting only in a restricted way that uniquely determines the necessary β-expansions during matching. Nevertheless, β-expansion and β-reduction remain rather expensive operations — it would be nice to compile CRSs and to statically perform as much of the expansion/reduction process as possible. This is our goal. We define an abstract machine, suitable for executing CRSs, and we give a translation from rewrite rules into instruction sequences for this machine.

We do not address the problems arising from considering CRSs as a programming language; we look at rewriting in the (very) small, not in the large. Apart from the problem of sheer size, the main reason for concentrating on this issue is the close similarity to corresponding problems in related systems. A CRS operates by performing the same set of elementary actions *when checking the applicability of a rule* as a first-order TRS with some notion of variable binding does. There is only one elementary action that does not exist in the first-order case, the equality check corresponding to a non-initial occurrence of a second-order variable, but even this is not too different.

I wrote the implementation originally in C (for applicative CRSs), but for the sake of presentation I use Standard ML in this paper, see [16, 17]. To shorten the code I shall treat several functions as predefined, i.e. I use — without further explanation — several functions which are not primitive in Standard ML, but are standard in the Functional Programming community. Definitions for all these functions can be found in [17], appendix 2. However, I redefine `select n` to be `hd o (drop n)`, i.e. I index lists starting from 0 (instead of from 1). The ML code for these functions can be obtained by anonymous ftp from `ftp.dcs.ed.ac.uk`, directory `pub/smk/CRS`, file `reade.ml`; the file `compile.ml` contains an extended version of the ML implementation presented in this paper.

2 Preliminaries: Combinatory Reduction Systems

First, we recall the basic definitions for Combinatory Reductions Systems. Instead of giving the usual "mathematical textbook" style definitions, we express these definitions in ML. Another difference from the usual presentation [8] is that we consider CRS terms in "de Bruijn style" [2] with variable names being replaced by natural numbers.

The *alphabet* of a CRS is determined by two types `mvar` and `symbol`, the elements of which we call *metavariables* and *symbols*, respectively. For the purposes of this paper, we take them to be strings of characters, i.e. we fix a (sufficiently large) CRS alphabet.

```
type mvar = string and symbol = string
```

Given a CRS alphabet `A`, we can define the *terms* over it as the values of type `term`:

```
datatype term =
      Var of int | Sym of symbol * term list |
      Meta of mvar * term list | Abst of term;
```

Usually, symbols and metavariables come equipped with an arity function, such that a term of the form Sym(s,ts) is only well-formed if the arity of s is equal to the length of ts. We do not impose this restriction here and allow overloading of symbols and metavariables. CRS theory traditionally [9, 8] distinguishes a subclass of terms not containing the constructor Meta, restricting the word "terms" for them and calling terms which may contain metavariables "metaterms". This distinction is not essential and we avoid it here.

The following notion of substitution supports replacement of more than one variable at a time. A substitution is determined by a function from natural numbers (de Bruijn indices) to terms.

```
fun subst f (Var x) = f x
 |  subst f (Sym(s,ts)) = Sym(s,map (subst f) ts)
 |  subst f (Meta(s,ts)) = Meta(s,map (subst f) ts)
 |  subst f (Abst t) = Abst (subst (fn n =>
      if n=0 then Var 0 else lift (f(n-1))) t)
and lift t = subst (fn x => Var(x+1)) t
```

This definition is slightly unusual and also rather inefficient, but it is more general than the usual definition (see [3], for example) which uses two recursive functions instead of one. The added generality enables us to easily state the substitution lemma of λ-calculus [1]:

Proposition 1. *Let* t *be a term and* f *and* g *functions of type* int->term. *Then:*

$$\texttt{subst f (subst g t)} = \texttt{subst ((subst f) o g) t}$$

This proposition does not even require totality of f and g, in the sense that a = b holds iff either the expressions a and b are both undefined or both are defined and denote equal values. Based on substitution, β-reduction and n-fold β-reduction are defined as follows:

```
fun beta arg (Abst b) = subst (fn 0 => arg | k => Var (k-1)) b
 |  beta _ _ = error "no beta-redex"
val betas = foldright beta
```

Using betas for multiple argument β-reduction is rather inefficient, as each argument requires its own β-reduction including (repeated) adjustments of de Bruijn indices. We can characterise the effect of n-fold β-reduction more concisely:

Proposition 2. *Let* t *be a term and* ts *be a list of terms. If* k=length ts *then:*

```
betas (repeat Abst k t) ts  =
subst (fn x => if x<k then select x ts else Var(x-k)) t
```

The functional `repeat` is function iteration: `repeat f n` is the function `f` composed with itself `n` times. Proposition 2 is typical for the kind of benefit we can expect from a compilation of CRSs: the expression on the right-hand side of the equation uses much less adjustments of variable indices, e.g. the terms from `ts` are not adjusted at all.

We have a second notion of substitution, the substitution of metavariables. A metavariable substitution is determined by a *valuation*, a function from pairs of metavariables and their arities to terms. Notice that a valuation is defined here on pairs, because we allow overloaded metavariables. A valuation has to satisfy a further condition: each pair `(z,n)` has to be mapped to a term (its *substitute*) of the form `(repeat Abst n t)` for some `t`, i.e. it "starts" with at least `n` abstractions. The reason for this restriction is the definition of `betas`: an application `betas t xs` is only well-defined if `t` starts with at least as many abstractions as the list `xs` is long.

```
fun metasubst v (Var x) = Var x
  | metasubst v (Sym(s,ts)) = Sym(s,map (metasubst v) ts)
  | metasubst v (Meta(z,ts)) =
        betas (v (z,length ts)) (map (metasubst v) ts)
  | metasubst v (Abst t) = Abst (metasubst (lift o v) t)
```

The definitions of metavariable substitution in the literature [9, 8] do not mention that there is a danger of name capture for metavariable substitution — but there is one; in the above version, this is reflected by the presence of `lift` in the last clause.

For the definition of "CRS rule" we need several auxiliary functions, e.g. for extracting the free/meta- variables of a term.

```
val freevars =
    let fun fv n (Var k) = if k<n then [] else [k-n]
          | fv n (Sym(_,ts)) = foldright (append o fv n) [] ts
          | fv n (Meta(_,ts)) = foldright (append o fv n) [] ts
          | fv n (Abst t) = fv (n+1) t
    in fv 0 end;
fun metas (Var _) = []
  | metas (Sym(_,ts)) = foldright (append o metas) [] ts
  | metas (Meta(z,ts)) =
        (z,length ts)::foldright (append o metas) [] ts
  | metas (Abst t) = metas t;
```

Left-hand sides of CRS rules obey a strong syntactic restriction, the so-called *simplicity condition*. Dale Miller coined this name [12] for arbitrary terms in λ^τ, but the corresponding condition for CRS rules was already present in Klop's thesis [9]; Nipkow [14] calls terms satisfying this property "higher-order patterns". The absence of third-order variables from CRSs simplifies the definition of simplicity for CRS terms. Second-order unification of simple terms is decidable and has most general solutions [12] and so has second-order matching of arbitrary terms against simple terms.

```
fun simple (Var _) = true
  | simple (Sym(_,ts)) = all simple ts
  | simple (Meta(_,ts)) =
      all (fn Var _ => true | _ => false) ts andalso
      let fun isset [] = true
            | isset (x::xs) = not(member xs x) andalso isset xs
      in isset (foldright (append o freevars) [] ts) end
  | simple (Abst t) = simple t;
```

A *CRS rule* is a pair of terms p satisfying the predicate crs_rule, i.e. such that crs_rule p evaluates to true.

```
fun crs_rule (left,right) =
    contains (metas right) (metas left) andalso
    (fn Sym _ => true | _ => false) left andalso
    freevars left = [] andalso freevars right = [] andalso
    simple left;
```

The conditions in the first two lines (of crs_rule) are typical restrictions for rewrite systems. CRS rules are required to contain no free variables. This restriction is a consequence of the slightly artificial distinction between metavariables and variables. We could indeed identify variables and metavariables in terms, using de Bruijn indices instead of mvar for metavariables etc. For the purposes of this paper, it is useful to keep the distinction.

Assume a fixed CRS alphabet and terms over this alphabet. A *Combinatory Reduction System* consists of a set R of CRS rules. The rewrite relation associated with a CRS is (as usual for rewrite systems) obtained from the set of rules, by interpreting the set of rules as a relation (set of pairs) and closing it under certain properties.

Definition 3. A binary relation $\rightarrow$ on terms is called a *rewrite relation*, if we have the following (for arbitrary t, z, k etc.):

$$\texttt{metas t} = \texttt{[(z,k)]} \wedge \texttt{a} \rightarrow \texttt{b} \wedge \\ \texttt{context} = \texttt{fn x => metasubst (fn \_ => repeat Abst k x) t} \Rightarrow \\ \texttt{context a} \rightarrow \texttt{context b}$$

Notice that any rewrite relation is substitutive and compatible (in the usual sense). Notice also that "rewrite relation" only refers to substitution of *variables*, while CRS rules also contain *metavariables.* We therefore need another property:

Definition 4. A binary relation $\rightarrow$ on terms is called *meta-substitutive*, if for arbitrary total functions g (of the right type) we have:

$$\texttt{a} \rightarrow \texttt{b} \Rightarrow \texttt{metasubst g a} \rightarrow \texttt{metasubst g b}$$

The rewrite relation associated with a CRS R is the smallest rewrite relation containing the meta-substitutive closure of R.

3 Compilation of a CRS

In the following, we restrict our efforts on implementing the meta-substitutive closure of a single rule, i.e. applying a CRS rule at the root of a term. The reasons for this limited effort are:

- A naïve, non-efficient extension to the general case is straightforward (see the `ftp` source). Working with de Bruijn indices makes it unnecessary to treat rewriting within abstractions in any special way; there is no need to "freeze" variables or to adjust indices.
- Eliminating most the inefficiencies of the naïve approach requires techniques known from the implementation of first-order TRSs. However, when I applied such techniques in an implementation of λ-rewriting (see [5]; λ-rewriting systems also require the right-hand sides of rules to be simple), the second-order patterns had no impact on the approach at all. This observation generalises to arbitrary CRSs.
- We are interested in the rewrite relation of a CRS as such, not just in the relation that relates terms to their normal forms. Thinking of CRSs as the kernel of a programming language would be a slightly different undertaking, e.g. it would be worthwhile then to allow conditional CRS rules and to require a constructor discipline.

The implementation of a CRS rule we shall develop is based on the *compilation* of the rule. The CRS rule (the pair of terms) is translated into a different representation, which is more suitable for the computation of CRS matching and metavariable substitution. This representation is an instruction sequence for modifying the state of an abstract machine; the abstract machine is especially designed for the execution of CRS rules.

4 Abstract Machine

The chosen abstract machine for the execution of compiled CRS rules is a stack-based machine, very much in the spirit of abstract machines for term rewriting, or Landin's SECD machine [11] for the λ-calculus.

The most significant difference to the SECD machine is the omission of a *dump*, a stack of states of the machine. The dump is superfluous, because rule application does not depend on other rule applications. Even if we had such a dependency (e.g. conditional CRS rules) and needed a dump, it would not be necessary to store the entire state, as most components are stack-like.

4.1 Machine Components

Instead of carrying the state of the Abstract Machine (the tuple of its components) around, we use the Standard ML state for this purpose. Therefore, all components of the abstract machine are given as references, that is: as updateable pointers to values.

The components of the machine are the following:

– The stack of *current* terms, which contains (subterms of) the term the rule is applied to. We call the top of this stack "the current term".

```
val CURR = ref ([]:term list);
```

– The environment, which represents the matching valuation.

```
val ENV = ref ([]:term list);
```

Notice that the environment is not an association list; the compilation replaces metavariables by relative addresses w.r.t. this environment. Also, entries in the environment are not (necessarily) k-fold abstractions when associated with k-ary metavariables. For the precise correspondence between the environment and the valuation it represents, see section 5.

– The stack which is used to create the instance of the right-hand side of the rule. After successfully executing all instructions corresponding to a rule, the stack contains only one term, the instance of the right-hand side.

```
val STACK = ref ([]:term list);
```

– An offset to adjust de Bruijn indices for free variables.

```
val OFFSET = ref 0;
```

– A stack of numbers to adjust de Bruijn indices for bound variables of the right-hand side of a rule. This is a bit delicate when they occur as arguments of a metavariable.

```
val NSTACK = ref ([]:int list);
```

The rôles of OFFSET and in particular of NSTACK are rather subtle. The idea behind these components is to allow nested substitutions like $t[u[s/y]/x]$ to be computed outside-in rather than inside-out. This minimises adjustments of variable indices and term traversals. However, there is a problem with implementing outside-in: s may contain a bound variable whose λ-binding is in t and the new index for this variable depends on the occurrences of x and y in t and u. NSTACK is used to keep track of such occurrences.

OFFSET and NSTACK represent a substitution, a certain variable lifting. In any given state of the abstract machine, we can retrieve it as follows:

```
fun offset () =
    let val off = !OFFSET and nst = !NSTACK
    in  subst (fn x=>
                if x < length nst then Var(x+select x nst)
                else Var(x+off))
    end
```

Initially, OFFSET is set to 0 and NSTACK is empty, making offset() the identity on terms.

The abstract machine also has an implicit component: the *control*, the list of instructions that have yet to be executed. The instructions do not modify themselves, thus it is not necessary to include the control as an explicit component of the machine.

The failure of matching is expressed using the exception mechanism of Standard ML. The function `test` is used to check Boolean expressions that have to be true when matching succeeds.

```
exception failure;
fun test p = if p then () else raise failure;
```

4.2 Machine Instructions

The abstract machine has instructions for matching a term against the left-hand side of a rule and instructions for applying the resulting valuation to the right-hand side. The instructions are the following:

```
datatype instruction =
   IS of string*int | NEXT | ISABST | CHECK of int list | SET |
   EQVAR of int | EQI of int*int*((int*int)list) | EQIMM of int |
   PUSHI of int*int*((int*(instruction list)) list) |
   PUSHIMM of int | PUSHVAR of int | CELL of string*int |
   LAMBDA | ADBMAL
```

If `s` is an instruction then `exec s` changes the state of the abstract machine. Because of its size (14 alternatives), the definition of `exec` below is divided into several parts. Execution and compilation are defined independently.

Matching Instructions The instructions in the first two lines of the definition of type `instruction` are used for computing the matching valuation. Matching manipulates two objects: the stack of current terms and the environment (valuation).

```
fun exec (IS(s,n)) =
      (case !CURR of Sym(s',bs)::ps =>
           (test (s=s' andalso length bs = n); CURR := bs @ ps)
       | _ => raise failure)
 | exec ISABST =
      (case !CURR of Abst t::ps => CURR:=t::ps
       | _ => raise failure)
 | exec NEXT = CURR := tl (!CURR)
```

`IS(s,n)` checks that the current term is an application of the symbol `s` to `n` arguments and replaces it by its argument list. `ISABST` is the corresponding instruction for abstractions. Matching fails if the current term does not have the required form. `NEXT` is used to select the next argument. In Higher-Order Rewrite

Systems [13, 6], an ISABST instruction would not have to check anything, as the type system already guarantees that the current term is an abstraction. In a certain sense, ISABST checks for the presence of the symbol Λ introduced by the translation from CRSs into HRSs, see [6, 15].

```
| exec (CHECK vs) =
  let fun check n (Var k) = k<n orelse not(member vs (k-n))
        | check n (Sym(_,bs)) = all (check n) bs
        | check n (Abst t) = check (n+1) t
        | check n (Meta(_,bs)) = all (check n) bs
  in test (check 0 (hd(!CURR))) end
```

The instruction CHECK vs checks that the current term does not contain any of the variable indices in vs. These checks are necessary because of the presence of bound variables, *not* because of the presence of higher-order variables.

```
| exec SET = ENV := hd(!CURR) :: !ENV
```

SET stores the current term on top of the environment. Since the current term is always a subterm of the original current term, the environment does not directly contain the substitutes of k-ary metavariables, but rather the substitutes without the k abstractions and modulo some change of bound variables.

```
| exec (EQVAR n) = (case hd(!CURR) of Var m => test (m=n)
                                     | _ => raise failure)
```

The instruction (EQVAR n) corresponds to a "positive" occurrence of a bound variable with index n in the left-hand side of a rule. Only bound variables outside metavariable applications correspond to EQVAR instructions.

```
| exec (EQI(n,d,vvs)) =
  let fun equal k (Var x, Var y)=
            if x<k then x=y
            else assoc vvs (fn a=>a+d) (x-k) = y-k
        | equal k (Sym(s,bs), Sym(t,at)) =
            s=t andalso eqlist k bs at
        | equal k (Abst a, Abst b) = equal (k+1) (a,b)
        | equal k (Meta(z,bs), Meta(y,at)) =
            z=y andalso eqlist k bs at
        | equal _ _ = false
      and eqlist k xs ys = length xs=length ys andalso
            all(equal k)(pairlists xs ys)
  in test (equal 0 (select n (!ENV),hd(!CURR))) end
| exec (EQIMM n) = test (hd(!CURR) = select n (!ENV))
```

EQI instructions are used for non-left-linear CRS rules to match non-initial occurrences of metavariables. EQIMM n is a cheap version that compares the current term with the term stored earlier at the n-th place in the environment.

EQI(n,d,xs) does the same, but in a more difficult setting when the variables cannot be compared one-to-one; xs is an association list for translating bound variables, and d has to be added to the indices corresponding to free variables.

The instruction EQIMM is redundant in the sense that all that it does can be done with EQI as well. But it cannot be done quite so well, e.g. if terms are uniquely represented (using a cache; see [7]) then the execution of the instruction EQIMM takes constant time, while EQI is linear in the size of the term to which the rule is applied.

Instantiating the Right-Hand Side The instructions in the last three lines of the definition of type instruction are used for generating the instance of the right-hand side of the rule. In particular, they manipulate the STACK which will finally contain this instance.

```
| exec (CELL (s,n)) =
      let val (args,rest) = split n (!STACK)
      in STACK := Sym(s,rev args)::rest end
```

CELL(s,n) pops n elements $t_1, \cdots, t_n$ from the stack (n may be 0) and replaces them by the term $s(t_1, \cdots, t_n)$.

```
| exec LAMBDA = NSTACK := 0 :: !NSTACK
| exec ADBMAL = let val t::ts = !STACK in
      STACK := Abst t :: ts; NSTACK := tl(!NSTACK) end
```

The instructions LAMBDA and ADBMAL are used to create abstractions. The changes to NSTACK are very often insignificant, but they do matter if the code executed between a LAMBDA and an ADBMAL involves PUSHVAR instructions.

```
| exec (PUSHI(n,d,acs)) = push (select n (!ENV)) d acs
| exec (PUSHIMM n) = STACK := select n (!ENV) :: !STACK
```

PUSHI(n,d,acs) pushes the term stored at the n-th place of the environment onto the stack, or more precisely: a substitution instance of this term. The substitution is the one corresponding to the k-fold β-reduction that occurs when a k-ary metavariable is meta-substituted; the function push, which creates the substitution instance is defined below. For each occurrence of a free variable, d has to be added to its index. Each occurrence of a (non-local) bound variable invokes an instruction sequence from acs producing a term, i.e. acs associates bound variables to instruction sequences. PUSHIMM is a cheap version of PUSHI, analogous to EQIMM for EQI.

```
| exec (PUSHVAR x) =
      STACK := Var(x+ select x (!NSTACK)) :: !STACK
```

PUSHVAR(x) pushes the variable with index x onto the stack. We have to add the x-th component of !NSTACK to that index to adapt it to the context in which it occurs. This will always be 0 if no metavariable intervenes between binding and using occurrence of that variable. For the same reason, the corresponding matching instruction `EQVAR` does not have to bother about `NSTACK` — `EQVAR` is only generated for variables outside metavariable applications.

The evaluation of `push t d acs` pushes a term `subst f t` to the stack, where the substitution function `f` is determined by `d`, `acs` and the state of the abstract machine. The state of the abstract machine is important, because `push` may execute instructions (`push` and `exec` are mutually recursive) from `acs`, including accesses to the environment. This is the above mentioned outside-in strategy of computing nested substitutions.

```
and push t d acs = let
    fun push' n (t as Var x) =
          if x<n then STACK := t::(!STACK)
          else let val bs=(assoc acs (fn k =>[])(x-n));
               in if bs=[] then
                      STACK:= Var(x+d+ !OFFSET)::(!STACK)
                  else (NSTACK := map (fn y=>y+n) (!NSTACK);
                       OFFSET := !OFFSET + n;
                       map exec bs;
                       OFFSET:= !OFFSET - n;
                       NSTACK := map (fn y=>y-n) (!NSTACK))
               end
      | push' n (Abst t) = (push'(n+1) t;
            let val b::rs= !STACK in STACK:=(Abst b)::rs end)
      | push' n (Sym(s,bs)) =
            (map (push' n) bs; exec(CELL(s,length bs)))
      | push' n (Meta(z,bs)) = (map (push' n) bs;
             let val (args,rest) = split (length bs) (!STACK)
             in STACK:= Meta(z,rev args) :: rest end)
    in push' 0 t end;
```

The most interesting part of `push` is the treatment of variables. Local bound variables (`x<n`) are pushed unchanged. Free variables (not found in the association list) are pushed with a slight change, their variable index is increased by `d`. This is the statically known difference in number of surrounding abstractions of the first and the current occurrence of the metavariable corresponding to this call of `push`. Other bound variables are associated with an instruction sequence `bs` from `acs`. For the execution of such an instruction sequence, all entries of `NSTACK` have to be increased by the number of abstractions that surround the occurrence of this bound variable in `t`.

A CRS rule corresponds to a sequence of instructions. To apply a CRS rule to a term we push the term onto `CURR` and execute the instructions:

```
fun run t cs = (CURR := t :: !CURR; map exec cs;
                case !STACK of r::rs => (STACK:=rs; [r]))
    handle failure => [];
```

The function run returns a singleton list containing the rewrite result if rule application succeeds and the empty list otherwise — the exception failure indicates the non-existence of a matching substitution.

It is not always meaningful to apply run t to an instruction sequence, because several instructions (e.g. CELL(s,n)) assume the abstract machine to be in a certain state. However, execution and compilation should fit together in the sense that they implement the meta-substitutive closure of a CRS rule. For the compilation function compile we are going to define, the following proposition should hold: Let l → r be a CRS rule, t and u be CRS terms, then

$$\exists \texttt{f}.\ \texttt{metasubst f l} = \texttt{t} \wedge \texttt{metasubst f r} = \texttt{u} \quad \Rightarrow \quad \texttt{run t (compile(l,r))} = \texttt{[u]}.$$

This is the completeness of compilation, i.e. the ability to rewrite each redex. We also want a soundness property, the non-ability to rewrite any non-redex:

$$\neg\exists \texttt{f}.\ \texttt{metasubst f l} = \texttt{t} \quad \Rightarrow \quad \texttt{run t (compile(l,r))} = \texttt{[]}$$

Complete code is almost by default sound; the only likely sources of a soundness violation are overly weak CHECK instructions or a wrong treatment of non-left-linear rules, and indeed for certain CRS rules completeness implies soundness.

5 Symbol Table

The compilation function presented later uses a symbol table for the metavariables of a rule. It has a similar purpose as the symbol table (for identifiers) used in an implementation of a programming language. The symbol table is a list of entries, each entry having the following form:

```
type entry = { MV: mvar*int, BV: int,
               LOC: int ref, ARG: int list };
fun lookup z (e::es : entry list) =
      if #MV(e)=z then e else lookup z es
  | lookup z [] = error "internal error"
```

Type entry is an SML record type; each entry corresponds to a metavariable in a CRS rule. MV is its name plus arity, BV is the number of abstractions at its first occurrence, which is essential because of the representation of variables as de Bruijn indices. LOC is the location of the metavariable in the environment. ARG represents the list of arguments of the first occurrence — it can be given as an integer list, because the restrictions for CRS rules ensure that it is a list of bound variables, hence a list of de Bruijn indices. For each element e of type entry we assume that the arity and the length of the argument list are equal, i.e. #2(#MV e) = length(#ARG e).

```
val unvar = map (fn (Var x) => x | _ => error "lhs not simple");
```

```
fun update n t (Var x) = t:entry list
  | update n t (Sym(s,ts)) = foldleft (update n) t ts
  | update n t (Abst m) = update (n+1) t m
  | update n t (Meta(z,ts)) =
        let val mv = (z,length ts)
            fun enter [] =
                    [ {LOC=ref 0, MV=mv, BV=n, ARG=unvar ts} ]
              | enter (tab as (e as {MV=y, ...})::es) =
                    if mv=y then tab else e::enter es
        in foldleft (update n) (enter t) ts end;
```

The function update traverses a simple term t and updates the symbol table, storing all initial occurrences of metavariables in t that are not already in the table. If the local function enter is applied to an empty symbol table, then we have an initial occurrence of a metavariable and we create a new entry. By the restrictions on CRS rules it is guaranteed that the argument list ts of an initial occurrence is always a list of variables, i.e. the error in unvar does not arise.

```
fun create_table l =
    let fun locations rn ({LOC=rl, ...}:entry) =
                     (rl:= !rn; rn:= !rn+1);
        val tab = update 0 [] l
        val no = ref 0
    in map (locations no) (rev tab); (tab,no) end
```

The function create_table generates the symbol table for a CRS rule l → r. It traverses the left-hand side of the rule, and then assigns a relative address to each metavariable. A symbol table for a CRS rule l → r contains all metavariables occurring in the rule, because they all have to occur in the left-hand side l; each metavariable in a symbol table of length n is associated with a unique location between 0 and $n-1$. This location is the relative address (in !ENV) of the substitute of the metavariable after matching has succeeded.

The combination of a symbol table, which uses locations between 0 and $k-1$, and an environment !ENV of length at least k represents a valuation defined on the entries of the symbol table.

```
fun getval tab = fn z =>
        let val e = lookup z tab
            val k = length(#ARG e)
            val su = assoc
                     (pairlists (#ARG e)(map Var (0 upto (k-1))))
                     (fn m => Var(m + k - #BV e))
        in  repeat Abst k (subst su (select (!(#LOC e)) (!ENV)))
        end;
```

The environment always carries subterms of the initial current term. This is simpler than to replace them by the proper substitutes of metavariables. But this means that we always make an assumption about the connection between symbol table and environment:

Definition 5. Let `tab` be a symbol table. An environment `ENV` *covers* a pair `(mv,n)` of a metavariable `mv` with its arity `n` iff

1. `(lookup (mv,n) tab)` is defined (call it `e`);
2. `(select (!(#LOC e))(!ENV))` is defined (call it `a`) and
3. the free variables of `a` are all either greater or equal than `#BV e` or occur in `#ARG e`.

We say that an environment *covers a term* if it covers all metavariables occurring in it.

Similarly as in proposition 2, we can characterise the n-fold β-reduction that corresponds to substitution of metavariables:

Lemma 6. *Let* `tab` *be a symbol table,* `mv` *be a metavariable, and* `ts` *be a list of terms, such that* `(mv,length ts)` *is covered by* `ENV`*. For arbitrary natural numbers* `n` *we have then:*

```
metasubst (repeat lift n o getval tab) (Meta(mv,ts))  =
subst (assoc(pairlists(#ARG e)ts')
      (fn m=>Var(m + n - #BV e))) (select(!(#LOC e))(!ENV))
```

where `ts'` *abbreviates* `map (metasubst(repeat lift n o getval tab)) ts`.

Proof. `metasubst (repeat lift n o getval tab) (Meta(mv,ts))`

```
= betas (repeat lift n (repeat Abst k
                (subst su (select(!(#LOC e))(!ENV))))) ts'
= { fun lift' n k = subst (fn x=>if x<k then Var x else Var(x+n)) }
betas (repeat Abst k (lift' n k
                (subst su (select(!(#LOC e))(!ENV))))) ts'
= subst (fn x=>if x<k then select x ts' else Var(x-k))
        (lift' n k (subst su (select (!(#LOC e))(!ENV))))
= subst(fn x=>if x<k then select x ts' else Var(x+n-k))
        (subst su (select (!(#LOC e))(!ENV)))
= { fun f' x = if x<k then select x ts' else Var(x+n-k) }
subst ((subst f') o su) (select(!(#LOC e))(!ENV))
```

We now simplify the substitution function:

```
subst f' o su
= assoc(pairlists(#ARG e)(map (subst f')(map Var(0 upto(k-1)))))
        (subst f' o (fn m=>Var(m+k-#BV e)))
= assoc(pairlists(#ARG e)(map f' (0 upto (k-1)))))
        (fn m=>f'(m+k-#BV e))
= assoc(pairlists(#ARG e)ts')(fn m=>f'(m+k-#BV e))
```

We can restrict any substitution function to the free variables of the term it is applied to. By the "cover" assumption about the environment, we know that `m` is greater or equal than `#BV e` whenever `assoc` uses its default function. This allows us to simplify `f'(m+k-#BV e)` to `Var(m+n-#BV e)`. □

The natural number n in `repeat lift n` corresponds to the number of abstractions that surround a metavariable occurrence. This number is statically known: the component `d` in an instruction `PUSHI(k,d,acs)` is the difference `(n-#BV e)`. This static information is useful to detect the case in which the substitution the lemma describes is the identity substitution; see the section on optimisation.

The components `OFFSET` and `NSTACK` of the abstract machine are used for a different variable lifting, the offsets of which are not statically known. The idea is not to compute the list `ts'` in advance and then to lift its variable indices whenever the substitution of the lemma is applied to an abstraction, but rather to collect all lifting information before instantiating `ts`.

6 Generating Code

Code generation for left-hand and right-hand sides of a rule are fairly independent — the only dependency between the two is the symbol table, since it gives relative addresses of metavariable substitutes in the environment.

6.1 Generating Match Code

The generated match code is very similar to how an interpretative implementation of CRS rewriting would operate.

```
fun bvset n ts = filter (not o member (unvar ts)) (0 upto (n-1));
fun lhs tab n k =
    let fun lhs' (Sym(s,ts)) = IS(s,length ts) ::
                foldleft(fn cs=>fn t => cs @ lhs' t) [] ts
        | lhs' (Var x) = [EQVAR x, NEXT]
        | lhs' (Abst t) = ISABST::lhs tab (n+1) k t
        | lhs' (Meta(z,ts)) =
                let val {LOC=ref l, ARG=ns, BV=no, ... } =
                    lookup (z,length ts) tab
                in if !k>l then
                       (k:= !k-1; [CHECK(bvset n ts),SET,NEXT])
                   else [EQI(l - !k,n-no,
                         pairlists ns (unvar ts)),NEXT]
                end
    in lhs' end;
```

The code for the initial occurrence of a metavariable checks for illegal name capture (`CHECK`) and stores the current term in the environment (`SET`); the `CHECK` establishes the precondition for environment components used in lemma 6. At a non-initial occurrence, the earlier stored term has to be compared with the current one. Which occurrence is considered to be the initial one depends on the term traversal — we traverse the term from left to right (for no good reason). Executing (successfully) the code of `lhs tab n k` means to remove one term from the stack of current terms.

Lemma 7. *Let* `t` *be a simple and closed term.*
Let `(tab,rn)=create_table t`*. Let* `f` *be a valuation defined on the metavariables in* `t`*. Then* `f` *is pointwise the same as*

```
(CURR:=metasubst f t :: !CURR;
map exec (lhs tab 0 rn t);
getval tab)
```

and the environment `ENV` *covers* `t` *w.r.t.* `tab` *after the evaluation of this expression.*

Proof. By induction on the structure of `t`. Sketch: we have to prove a more general lemma, because most of the required properties are not closed under taking subterms. In particular: `t` is simple and does not contain free variables greater or equal than `n`; `tab` is a symbol table containing all metavariables of `t`; `rn` refers to a natural number, such that for all entries `e` in `tab` of metavariables occurring in `t` the following holds: either `!(#LOC e)<!rn`, or the environment covers `#MV e`. To get the right valuation using `getval`, we have to compose `getval tab` with `repeat lift n`, and we also have to place `!rn` arbitrary terms on top of the environment after executing the code from `(lhs tab n rn t)`; after evaluating this expression to an instruction sequence containing k SET instructions, `!rn` is reduced by k. □

Lemma 7 states the completeness of our matching procedure. We can also claim a soundness property (with a very similar proof), i.e. that execution raises the exception `failure`, if the current term is not a valuation instance of `t`.

6.2 Generating Code for the Right-hand Side

Code generation for the right-hand side is slightly simpler, because we do not have to distinguish between initial and non-initial occurrences of metavariables. The complications concerning possible nesting of metavariable applications arise at run-time, not at compile-time.

```
fun rhs tab n =
   let fun rhs' (Sym(s,ts)) =
                 foldright append [CELL(s,length ts)] (map rhs' ts)
         | rhs' (Var x) = [PUSHVAR x]
         | rhs' (Abst t) = [LAMBDA] @ rhs tab (n+1) t @ [ADBMAL]
         | rhs' (Meta(z,ts)) =
                 let val {LOC = ref l, ARG=ns, BV=m, ...}
                        = lookup (z,length ts) tab
                 in  [PUSHI(l,n-m,pairlists ns (map rhs' ts))]
                 end
   in rhs' end
```

For code generation with `rhs`, we can make the following claim:

Lemma 8. *Let* `r` *be a term and* `tab` *be a symbol table such that the environment covers* `r`*. Let* `n` *be the length of* `!NSTACK`*. If all free variables in* `r` *are smaller than* `n`*, then the evaluation of* `map exec (rhs tab n r)` *puts a term* `r'` *on top of* `STACK` *and leaves all other components of the abstract machine unchanged, where* `r'` *is equal to*

```
offset()(metasubst (repeat shift n o getval tab) r).
```

Proof. By induction over the term structure of `r`. Sketch: Symbol applications and variables are trivial; for abstractions notice that `offset()(Abst t)` is equivalent to `Abst(#2(exec LAMBDA, lift(offset() t), exec ABDMAL))`, see the definition of `subst`. For metavariables we can use lemma 6 and proposition 1 to characterise the `metasubst` application and its composition with `offset()` as a single substitution. It then remains to show that the evaluation of the expression `(push t d (pairlists ns (map (rhs tab n) ts)))` stores the term we obtain from `(subst(assoc(pairlists ns us)(fn m=>Var(m+d+!OFFSET)))t)` on top of `STACK`; in this expression, `us` is shorthand for pointwise applying `offset() o (metasubst ...)` to the list `ts`. This requires again an inductive proof on the term structure of `t`; notice here that the evaluation of

```
(OFFSET:=!OFFSET+k; NSTACK:=map(fn x=>x+k)(!NSTACK);
map exec (rhs tab n u);
OFFSET:=!OFFSET-k; NSTACK:=map(fn x=>x-k)(!NSTACK))
```

is equivalent to

```
(map exec (rhs tab n u);
STACK:=repeat lift k(hd(!STACK)) :: tl(!STACK))
```

Notice further that `rhs` never returns an empty list; therefore, using `(fn x=>[])` as default function for `assoc` (see the definition of `push`) is a proper way to distinguish free variables. □

The proof requires two nested inductions, because I have chosen a lazy compilation scheme for outside-in computation of nested substitutions. An eager compilation scheme for computing them inside-out would generate the instances of the argument list `ts` of a metavariable on the stack.

6.3 Full Compilation

Compiling a rule simply involves of compiling its parts and concatenating the code afterwards.

```
fun compile (l,r) =
    let val (tab,rn) = create_table l
    in  lhs tab 0 rn l @ rhs tab 0 r
    end;
```

Completeness of compilation is now easily established:

Theorem 9. *Let* (l,r) *be a CRS rule and* f *be a valuation defined on the metavariables in* l. *Let* !NSTACK=[] *and* !OFFSET=0. *Then:*

run(metasubst f l)(compile(l,r)) = [metasubst f r].

Proof. Immediate by distributivity of map over @ and lemmas 7 and 8, taking n = 0. □

Similarly, soundness of compilation follows from the soundness of matching.

6.4 An Example

To get a feeling for the code the compilation produces, let us look at an example. Here is the chain rule of symbolic derivation, for the sake of legibility represented as a nameful CRS rule, i.e. with variable names instead of de Bruijn indices.

$$D([x]App(f, g(x))) \;\to\; M(D([x]g(x)), B(D(f), [x]g(x)))$$

Explanation: x is the only variable, f and g are metavariables, and the symbols D, B, *App*, and M can be interpreted as follows: D is the derivation operator, B is function composition, *App* is function application, and M is multiplication of functions (pointwise). The substitute for f cannot contain x freely; this is the usual side-condition of the chain rule, which is here implicit, as f is a nullary metavariable. Applying the function compile to the above rule (after converting it to a pair of CRS terms, of course) we obtain the code in figure 1.

1 IS(D,1)	10 LAMBDA
2 ISABST	11 PUSHI(0,0,[(0,[PUSHVAR 0])])
3 IS(App,2)	12 ADBMAL
4 CHECK [0]	13 CELL(D,1)
5 SET	14 PUSHI(1,-1,[])
6 NEXT	15 CELL(D,1)
7 CHECK []	16 LAMBDA
8 SET	17 PUSHI(0,0,[(0,[PUSHVAR 0])])
9 NEXT	18 ADBMAL
	19 CELL(B,2)
	20 CELL(M,2)

Fig. 1. Code of a compiled rule

The first 9 instructions (left column) were generated by lhs, i.e. their purpose is to match the left-hand side of the rule; the remaining 11 instructions are for the creation of an instance of the right-hand side. The code mimics exactly what an interpreted rule application would do, only the term structure of the left-hand and right-hand sides of the rule has been flattened into a list of instructions and the names of metavariables have been replaced by relative addresses.

7 Optimisation

The code generated by the naïve compilation can be improved in many ways. Most of these improvements are minor, they allow to compactify the code, sometimes requiring extensions to the instruction set to make better use of the resources of the abstract machine. We do not have the space here to elaborate on that; several optimisations can be found in the `ftp` source.

One kind of optimisation has a major effect: it would be nice to "compile away" as many second-order substitutions as possible. An important observation here is that the (named) term $(\lambda x.t)x$ can be β-reduced to t, but that it is a waste of resources to perform this β-reduction in the standard way. This waste of resources is the difference between the complexities $\mathcal{O}(n)$ and $\mathcal{O}(1)$ of executing a `PUSHI` and `PUSHIMM` instruction, respectively.

Such wasteful β-reductions typically occur in λ-rewriting systems, i.e. in CRSs which only move bound variables around rather than replacing them, e.g. symbolic derivation, translation of λ-calculus into Combinatory Logic, etc. Similarly to the way in which any HRS can be translated into a CRS plus β-reduction (see [15]), any CRS can be translated into a λ-rewriting system plus β-reduction; however, β-reduction is expressible within CRSs but not within λ-rewriting systems.

```
fun single (c as EQI(n,0,xs)) =
        if all (op =) xs then EQIMM n else c
  | single (c as PUSHI(n,0,acs)) =
        if all (fn (m,xs)=> xs=[PUSHVAR m]) acs
        then PUSHIMM n else c
  | single c = c
```

The function `single` maps an instruction to an equivalent instruction, being exactly the mentioned detection of trivial β-redexes. Notice that the `d` component, the difference in surrounding abstractions between the current and initial occurrence of the metavariable these instructions correspond to, is required to be 0 — otherwise we need a term traversal to adjust indices of free variables. But even with $\mathtt{d} = 0$ we also need `OFFSET` to be set to 0 for the same reason, and `NSTACK` to contain only zeros for a similar one. We can say that `exec c` and `exec(single c)` have the same effect on the state of the abstract machine, provided `!OFFSET=0` and `all(fn x=>x=0)(!NSTACK)`. Initially, `OFFSET` and `NSTACK` satisfy these conditions; moreover they are only ever locally violated within the function `push`. Thus, we can apply `single` to all instructions on the outermost level, i.e. to those that do not occur within a `PUSHI` instruction.

This optimisation is applicable to the example in figure 1; instructions 11 and 17 can both be replaced by `PUSHIMM 0`. These instructions correspond to right-hand side occurrences of the second-order metavariable g. The corresponding optimisation for f (instruction 14) is not possible, for we have to decrease de Bruijn indices of free variables in its substitute by 1.

8 Conclusion

We have defined an abstract machine for executing Combinatory Reduction Systems and a compiler for translating CRS rules into instructions of that machine. The correctness of this translation has been established. Nearly all important actions of the system are performed on the low level of instructions — the only exception being the function **push** which provides an interaction between an interpretative term traversal and the execution of code. The code can be seen as a linearisation of the actions one would perform in a similar way when *interpreting* the rewrite rule.

References

1. Hendrik P. Barendregt. *The Lambda-Calculus, its Syntax and Semantics.* North-Holland, 1984.
2. N. G. de Bruijn. Lambda calculus notation with nameless dummies, a tool for automatic formula manipulation. *Indagationes Mathematicae*, 34:381–392, 1972.
3. T. Hardin. How to get confluence for explicit substitutions. In M.R. Sleep, M.J. Plasmeijer, and M.C.J.D. van Eekelen, editors, *Term Graph Rewriting*, chapter 3, pages 31–45. John Wiley & Sons, 1993.
4. Gérard Huet and Bernard Lang. Proving and applying program transformations expressed with second-order patterns. *Acta Informatica*, 11:31–55, 1978.
5. Stefan Kahrs. *λ-rewriting.* PhD thesis, Universität Bremen, 1991. (in German).
6. Stefan Kahrs. Context rewriting. CTRS'92, pages 21–35. LNCS 656.
7. Stefan Kahrs. Unlimp – uniqueness as a leitmotiv for implementation. PLILP'92, pages 115–129. LNCS 631.
8. J.R. Kennaway. Sequential evaluation strategies for parallel-or and related reduction systems. *Annals of Pure and Applied Logic*, 43:31–56, 1989.
9. Jan Willem Klop. *Combinatory Reduction Systems.* PhD thesis, Centrum voor Wiskunde en Informatica, 1980.
10. Jan Willem Klop. Term rewriting systems. In S. Abramsky, D. M. Gabbai, and T. S. E. Maibaum, editors, *Handbook of Logic in Computer Science, Volume 2*, pages 1–116. Oxford University Press, 1992.
11. P. J. Landin. The mechanical evaluation of expressions. *Computer Journal*, 6:308–320, 1964.
12. Dale Miller. A logic programming language with lambda-abstraction, function variables, and simple unification. *Extensions of Logic Programming*, pages 253–281, 1991. LNCS 475.
13. Tobias Nipkow. Higher order critical pairs. LICS'91, pages 342–349.
14. Tobias Nipkow. Functional unification of higher-order patterns. LICS'93.
15. Vincent van Oostrom and Femke van Raamsdonk. Comparing combinatory reduction systems and higher-order rewrite systems. HOA'93. (This volume).
16. Laurence C. Paulson. *ML for the Working Programmer.* Cambridge University Press, 1991.
17. Chris Reade. *Elements of Functional Programming.* Addison-Wesley, 1989.

SPECIFICATION AND VERIFICATION IN HIGHER ORDER ALGEBRA: A CASE STUDY OF CONVOLUTION

K. Meinke and L.J. Steggles

Department of Computer Science,
University College of Swansea,
Singleton Park,
Swansea SA2 8PP,
Great Britain.

Abstract. We present a case study of higher order algebraic methods applied to the specification of convolution as a second order transformation on streams. Two systolic synchronous concurrent algorithms (SCAs) for convolution are formally specified and verified using higher order equational logic. We then study the metamathematics of these verification proofs by means of non–standard models.

1 Introduction.

Many–sorted first order algebraic methods are widely recognised as a practical and theoretically well founded formalism that can support many aspects of algorithm design such as specification, verification, transformation and refinement. The literature on these methods is extensive: recent surveys are Wirsing [1990], Ehrich [1992] and Meinke and Tucker [1992], while an introduction to the methods can be found in Ehrig and Mahr [1985]. However many algebras arising in practise, for example *algebras of streams* used in hardware specification, are best modelled as *higher order algebras.* By a higher order algebra we mean a many–sorted universal algebra in which the carrier sets may contain higher order objects such as functions, functions acting on functions, etc. Theoretical results on the scope and limits of first order algebraic methods (see for example the surveys Bergstra and Tucker [1987] or Goguen and Meseguer [1985]) show that many higher order algebras fail to possess a recursively enumerable equational specification under first order initial algebra semantics. However, some of these algebras can be shown to possess a recursive equational specification under higher order initial algebra semantics (see Meinke [1994a]). Thus there arises both a practical motivation and a need on fundamental theoretical grounds to consider *higher order algebraic methods* as a formalism for algorithm design.

Early contributions to the theory of higher order algebraic methods include Maibaum and Lucena [1980], Poigné [1986], Broy [1987], Möller [1987a, 1987b, 1994] and Möller et al [1988]. A comprehensive model theory for higher order algebras is presented in Meinke [1992a, 1992b]. This work includes results on existence and construction of free and initial algebras, completeness of calculi for higher order equations, model theoretic characterisations of axiomatically defined classes, and subdirect representations.

Few case studies of higher order algebraic specifications have appeared in the literature to date. Thus little is known about the use of these methods in practise. In this paper our main aim is to illustrate the use of higher order algebraic methods by means of a detailed case study in the specification of convolution as a second order transformation on streams. We then formally specify and verify two systolic synchronous concurrent algorithms (SCAs) for convolution using higher order equational logic.

A mathematical theory of SCAs, to which this paper also contributes, has been developed including work on theoretical foundations and models of computation (Thompson [1987]) specification, verification and top down design (Harman and Tucker [1988a, 1988b], Eker and Tucker [1988, 1989], Hobley et al [1988]) and software tools (Martin and Tucker [1987]). Our convolution algorithms have been adapted from the SCA specification case study Hobley et al [1988], which uses purely semantical techniques from many-sorted first order universal algebra to carry out specification and verification. By contrast, our own case study is fully formalised within higher order equational logic. This formal approach supports metamathematical investigations into the logic of SCAs, in particular into the role of *higher order constructs* and the perplexing role of *induction principles*. Using non-standard models for higher order equational logic we prove some simple independence results in the metamathematics of SCA verification.

The structure of this paper is as follows. In section 2 we provide some essential definitions and results in higher order algebra necessary for our case study. In section 3 we give a second order initial algebra specification of convolution as a transformation on streams over the ring of integers. In section 4 we present second order equational specifications of two systolic SCAs for convolution. These algorithms differ only slightly in the programmability of weights, yet turn out to require quite different proof theoretic strength for their verification. In section 5 we carry out a formal verification of correctness for the non-programmable convolver in pure higher order equational logic. We then give a formal verification of correctness for the programmable convolver in higher order equational logic augmented with the proof rule of *free variable induction* (for equations). We show that free variable induction is an essential rule in the latter proof of correctness by constructing a non-standard model of the programmable convolver specification (based on a nonstandard model of time) where both the induction principle and the correctness formula are invalid. Thus the correctness formula cannot be proved in pure higher order equational logic without induction. Finally, in section 6, we conclude by discussing the implications of our results with regard to algebraic specification techniques and also for the logic of SCAs.

We have attempted to make the paper self contained. Further background material on universal algebra may be found in Cohn [1965], Wechler [1992] and Meinke and Tucker [1992], an overview of the theory of synchronous concurrent algorithms is Thompson and Tucker [1991].

2 Higher Order Algebraic Methods.

In this section we present the basic definitions and fundamental theoretical results of higher order algebra necessary for our case study. A full account of the theory of higher order algebra is Meinke [1992a]. We begin by fixing our notation for many–sorted first order universal algebra which is taken from Meinke and Tucker [1992].

A *many–sorted signature* Σ consists of a non–empty set S, the elements of which are *sorts*, and an $S^* \times S$–indexed family $\langle \Sigma_{w,\,s} \mid w \in S^*,\ s \in S \rangle$ of sets, where S^* denotes the set of all words over S. For the empty word $\lambda \in S^*$ and any sort $s \in S$, each element $c \in \Sigma_{\lambda,\,s}$ is a *constant symbol* of sort s; for each non–empty word $w = s(1) \ldots s(n) \in S^+$ and any sort $s \in S$, each element $f \in \Sigma_{w,\,s}$ is a *function symbol* of *domain type* w, *codomain type* s and *arity* n. Thus we can define Σ to be the pair $(S, \langle \Sigma_{w,\,s} \mid w \in S^*,\ s \in S \rangle)$.

Let Σ be an S–sorted signature. An *S–sorted Σ algebra* is an ordered pair (A, Σ^A), consisting of an S–indexed family $A = \langle A_s \mid s \in S \rangle$ of *carrier sets* A_s and an $S^* \times S$–indexed family $\Sigma^A = \langle \Sigma^A_{w,s} \mid w \in S^*,\ s \in S \rangle$ of sets of constants and algebraic operations. For each sort $s \in S$, $\Sigma^A_{\lambda,s} = \{\ c_A \mid c \in \Sigma_{\lambda,s}\ \}$, where $c_A \in A_s$ is a *constant* that interprets c. For each $w = s(1) \ldots s(n) \in S^+$ and each sort $s \in S$, $\Sigma^A_{w,s} = \{\ f_A \mid f \in \Sigma_{w,s}\ \}$, where $f_A : A^w \rightarrow A_s$ is an *operation* with domain $A^w = A_{s(1)} \times \ldots \times A_{s(n)}$ and codomain A_s which interprets f. As usual, we allow A to denote both a Σ algebra and its S–indexed family of carrier sets.

We assume that the reader is familiar with basic universal algebraic constructions and results, such as congruence and quotient construction; the construction of an initial model for an equational class (variety) of Σ algebras as a quotient of the ground term algebra $T(\Sigma)$, and the soundness and completeness of many–sorted first order equational logic.

The theory of higher order universal algebra can be developed within the framework of many–sorted first order universal algebra. We begin by defining notations for higher order types.

2.1 Definition. Let B be any non–empty set, the members of which will be termed *basic types*, the set B being termed a *type basis*. The *type hierarchy* $H(B)$ generated by B is the set $H(B) = \bigcup_{n \in \omega} H_n(B)$ of formal expressions built up inductively by

$$H_0(B) = B,$$

and

$$H_{n+1}(B) = H_n(B) \cup \{\ (\sigma \times \tau),\ \ (\sigma \rightarrow \tau)\ \mid \sigma, \tau \in H_n(B)\ \}.$$

Each element $(\sigma \times \tau) \in H(B)$ is termed a *product type* and each element $(\sigma \rightarrow \tau) \in H(B)$ is termed a *function type*.

We can assign an *order* to each type $\sigma \in H(B)$ as follows. Each basic type $\sigma \in B$ has order 0. If $\sigma, \tau \in H(B)$ have order m and n respectively then $(\sigma \times \tau)$ has order $sup\{m, n\}$ and $(\sigma \rightarrow \tau)$ has order $sup\{m, n\} + 1$.

A *type structure* S over a type basis B is a subset $S \subseteq H(B)$, which is closed under subtypes in the sense that for any $\sigma, \tau \in H(B)$, if $(\sigma \to \tau) \in S$ or $(\sigma \times \tau) \in S$ then both $\sigma \in S$ and $\tau \in S$. We say that S is a *basic type structure* over B if, and only if, $S \subseteq B$. A type structure S over a basis B is said to be an *nth-order type structure* if, and only if, the order of each type $\tau \in S$ is strictly less than n. □

In a *higher order signature* Σ we take a type structure S as the sort set and include distinguished evaluation and projection operation symbols for all function and product types as follows.

2.2 Definition. Let S be a type structure over a type basis B. An *S–typed signature* Σ is an S–sorted signature such that for each product type $(\sigma \times \tau) \in S$ we have two unary *projection operation symbols*

$$proj^{(\sigma\times\tau),\sigma} \in \Sigma_{(\sigma\times\tau),\sigma} \quad proj^{(\sigma\times\tau),\tau} \in \Sigma_{(\sigma\times\tau),\tau}.$$

Also for each function type $(\sigma \to \tau) \in S$ we have a binary *evaluation operation symbol*

$$eval^{(\sigma\to\tau)} \in \Sigma_{(\sigma\to\tau)\,\sigma,\tau}\,.$$

□

When the types σ and τ are clear, we let $proj^1$ and $proj^2$ denote the projection operation symbols $proj^{(\sigma\times\tau),\sigma}$ and $proj^{(\sigma\times\tau),\tau}$ respectively. Next we introduce the intended interpretations of a higher order signature Σ.

2.3 Definition. Let S be a type structure over a type basis B. Let Σ be an S–typed signature and A be an S–sorted Σ algebra. We say that A is an *S–typed Σ algebra* if, and only if, for each product type $(\sigma \times \tau) \in S$ we have $A_{(\sigma\times\tau)} \subseteq A_\sigma \times A_\tau$, and for each function type $(\sigma \to \tau) \in S$ we have $A_{(\sigma\to\tau)} \subseteq [A_\sigma \to A_\tau]$, i.e. $A_{(\sigma\to\tau)}$ is a subset of the set of all (total) functions from A_σ to A_τ. Furthermore, for each product type $(\sigma \times \tau) \in S$ the operations

$$proj^1_A : A_{(\sigma\times\tau)} \to A_\sigma, \quad proj^2_A : A_{(\sigma\times\tau)} \to A_\tau$$

are the first and second *projection mappings* defined on each $a = (a_1, a_2) \in A_{(\sigma\times\tau)}$ by

$$proj^1_A(a) = a_1, \quad proj^2_A(a) = a_2;$$

also, for each function type $(\sigma \to \tau) \in S$, $eval^{(\sigma\to\tau)}_A : A_{(\sigma\to\tau)} \times A_\sigma \to A_\tau$ is the *evaluation operation* on the function space $A_{(\sigma\to\tau)}$ defined by

$$eval^{(\sigma\to\tau)}_A(a, n) = a(n)$$

for each $a \in A_{(\sigma\to\tau)}$ and $n \in A_\sigma$. If A is an S–typed Σ algebra we say that A is *complete* if, and only if, for each product type $(\sigma \times \tau) \in S$ we have $A_{(\sigma\times\tau)} = A_\sigma \times A_\tau$, and for each function type $(\sigma \to \tau) \in S$ we have

$A_{(\sigma\to\tau)} = [A_\sigma \to A_\tau].$ □

From the viewpoint both of algebra and specification theory we are mainly concerned with the structure of higher order algebras up to isomorphism. This structure can be characterised by first order sentences as follows. Let B be a type basis, $S \subseteq H(B)$ be a type structure over B, Σ be an S–typed signature and X be an S–indexed family of infinite sets of variables. The set $Ext = Ext_\Sigma$ of *extensionality sentences* over Σ is the set of all Σ sentences of the form

$$\forall x \forall y \Big(\forall z \Big(eval^{(\sigma\to\tau)}(x,z) = eval^{(\sigma\to\tau)}(y,z) \Big) \Rightarrow x = y \Big),$$

for each function type $(\sigma \to \tau) \in S$, where $x, y \in X_{(\sigma\to\tau)}$, $z \in X_\sigma$, and

$$\forall x \forall y \Big(proj^1(x) = proj^1(y) \ \wedge \ proj^2(x) = proj^2(y) \Rightarrow x = y \Big),$$

for each product type $(\sigma \times \tau) \in S$, where $x, y \in X_{(\sigma\times\tau)}$. A Σ algebra A is *extensional* if, and only if, $A \models Ext$. An extensional Σ algebra can be isomorphically collapsed to an S–typed Σ algebra by a construction known as the *Shepherdson Mostowski collapse* which gives the following representation theorem for extensional Σ algebras.

2.4 Collapsing Theorem. (Shepherdson, Mostowski) *Let A be an S–sorted Σ algebra. Then A is isomorphic to an S–typed Σ algebra if, and only if, A is extensional.*

Proof. See Meinke [1992a]. □

We are interested in specifying higher order algebras or classes of higher order algebras by means of higher order formulas, i.e. many-sorted first order formulas over a higher order signature Σ. In this paper we confine our attention to specification methods based on *higher order equations* over Σ. By a higher order equation over an S–typed signature Σ and S–indexed family X of sets of variables we mean a formula of the form

$$t = t',$$

where t, $t' \in T(\Sigma, X)_\tau$ are terms over the term algebra $T(\Sigma, X)$ of the same type $\tau \in S$. We let $Eqn(\Sigma, X)$ denote the set of all higher order equations over Σ and X. Given any Σ algebra A, we have the usual notion of truth for an equation under an assignment $\alpha : X \to A$, and the usual validity relation $\models$ on an equation or set of equations.

Let E be any set of (higher order) equations over Σ and X, then the *extensional equational class*

$$Alg_{Ext}(\Sigma, E) = \{ \ A \in Alg(\Sigma) \ | \ A \models E \cup Ext \ \}$$

can be shown to be an *extensional variety*, i.e. a class of extensional Σ algebras closed under the formation of extensional homomorphic images, extensional subalgebras and direct products. In general $Alg_{Ext}(\Sigma, E)$ does not admit an initial algebra. However, by a basic result of higher order universal algebra (see Meinke [1992a]), $Alg_{Ext}(\Sigma, E)$ admits an algebra $I_{Ext}(\Sigma, E)$ which is initial in the subclass $Min_{Ext}(\Sigma, E)$ of all *minimal*, extensional Σ algebras which are models of E. Thus $I_{Ext}(\Sigma, E)$ is initial in a weaker, but nontrivial sense and unique up to isomorphism. We call this algebra the *initial extensional model* or *higher order initial model.* Since $I_{Ext}(\Sigma, E)$ is extensional, it can be isomorphically collapsed to an S–typed Σ algebra and thus serves as the appropriate higher order initial algebra semantics of the pair (Σ, E) viewed as a higher order equational specification.

Higher order initial models can be concretely constructed from syntax using a *higher order equational calculus.* This calculus extends the many–sorted first order equational calculus with additional inference rules for higher types.

2.5 Definition. The (ordinary) inference rules of *higher order equational logic* are the following.

(i) For any type $\tau \in S$ and any term $t \in T(\Sigma, X)_\tau$,

$$\frac{}{t = t}$$

is a *reflexivity* rule.

(ii) For any type $\tau \in S$ and any terms $t_0, t_1 \in T(\Sigma, X)_\tau$,

$$\frac{t_0 = t_1}{t_1 = t_0}$$

is a *symmetry* rule.

(iii) For any type $\tau \in S$ and any terms $t_0, t_1, t_2 \in T(\Sigma, X)_\tau$,

$$\frac{t_0 = t_1, \quad t_1 = t_2}{t_0 = t_2}$$

is a *transitivity* rule.

(iv) For each type $\sigma \in S$, any terms $t, t' \in T(\Sigma, X)_\sigma$, any type $\tau \in S$, any variable symbol $x \in X_\tau$ and any terms $t_0, t_1 \in T(\Sigma, X)_\tau$,

$$\frac{t = t', \quad t_0 = t_1}{t[x/t_0] = t'[x/t_1]}$$

is a *substitution* rule.

(v) For each product type $(\sigma \times \tau) \in S$ and any terms $t_0, t_1 \in T(\Sigma, X)_{(\sigma \times \tau)}$,

$$\frac{proj^1(t_0) = proj^1(t_1), \quad proj^2(t_0) = proj^2(t_1)}{t_0 = t_1}$$

is a *projection* rule.

(vi) For each function type $(\sigma \to \tau) \in S$, any terms $t_0, t_1 \in T(\Sigma, X)_{(\sigma\to\tau)}$ and any variable symbol $x \in X_\sigma$ not occurring in t_0 or t_1,

$$\frac{eval^{(\sigma\to\tau)}(t_0, x) = eval^{(\sigma\to\tau)}(t_1, x)}{t_0 = t_1}$$

is an *evaluation* rule. □

Let $\vdash$ denote the inference relation between equational theories $E \subseteq Eqn(\Sigma, X)$ and equations $e \in Eqn(\Sigma, X)$, defined by $E \vdash e$ if, and only if, there exists a proof of e from E using the inference rules of higher order equational logic. For the higher order equational calculus we have a completeness theorem with respect to extensional models. For the usual technical reasons (see for example Goguen and Meseguer [1982]) we impose the assumption of non–voidness on signatures; an S–sorted signature Σ is *non–void* if for each sort $s \in S$, there exists a ground term $t \in T(\Sigma)_s$.

2.6 Completeness Theorem. *Let Σ be a non–void S–typed signature and $E \subseteq Eqn(\Sigma, X)$ be any equational theory. For any equation $e \in Eqn(\Sigma, X)$,*

$$E \vdash e \Leftrightarrow Alg_{Ext}(\Sigma, E) \models e.$$

Proof. See Meinke [1992a]. □

Given an S–typed signature Σ and a set E of equations over Σ we may define the congruence $\equiv^E$ on the ground term algebra $T(\Sigma)$ by $t \equiv^E_\tau t'$ if, and only if, $E \vdash t = t'$, for any $\tau \in S$ and $t, t' \in T(\Sigma)_\tau$. However, in general the quotient algebra $T(\Sigma)/\equiv^E$ is not extensional. To construct a quotient algebra of $T(\Sigma)$ which is both a model of E and extensional, we must add an infinitary inference rule to the higher order equational calculus.

2.7 Definition. Let S be a type structure over a type basis B. Let Σ be an S–typed signature. For each function type $(\sigma \to \tau) \in S$ and any terms $t_0, t_1 \in T(\Sigma, X)_{(\sigma\to\tau)}$,

$$\frac{\langle eval^{(\sigma\to\tau)}(t_0, t) = eval^{(\sigma\to\tau)}(t_1, t) \mid t \in T(\Sigma)_\sigma \rangle}{t_0 = t_1}$$

is an (infinitary) *ω–evaluation rule.* □

Let $\vdash_\omega$ denote the inference relation between equational theories $E \subseteq Eqn(\Sigma, X)$ and equations $e \in Eqn(\Sigma, X)$, defined by $E \vdash_\omega e$ if, and only if, there exists an infinitary proof of e from E using the inference rules of higher order equational logic and the ω–evaluation rules. Define the extensional congruence $\equiv^{E,\omega} = \langle \equiv^{E,\omega}_\tau \mid \tau \in S \rangle$ on the term algebra $T(\Sigma)$ by

$$t \equiv^{E,\omega}_\tau t' \Leftrightarrow E \vdash_\omega t = t'$$

for each type $\tau \in S$ and any terms $t, t' \in T(\Sigma)_\tau$. Factoring $T(\Sigma)$ by the congruence $\equiv^{E,\omega}$ gives a concrete construction of the initial extensional model $I_{Ext}(\Sigma, E)$. Recalling that a Σ algebra A is minimal if, and only if, A has no proper subalgebra we have the following theorem.

2.8 Theorem. *Let Σ be an S–typed signature. Let E be any equational theory over Σ. Then*

$$T(\Sigma)/\equiv^{E,\omega} \cong I_{Ext}(\Sigma, E).$$

Thus $T(\Sigma)/\equiv^{E,\omega}$ is initial in the class $Min_{Ext}(\Sigma, E)$ of all minimal extensional models of E.

Proof. See Meinke [1992a]. □

This concrete construction of the initial extensional model of a higher order equational theory can be used to verify the correctness of specifications.

2.9 Definition. By a *higher order equational specification* we mean a pair

$$Spec = (\Sigma, E),$$

where Σ is an S–typed signature for some type structure S over a type basis B and E is a set of equations over Σ and some S–indexed family X of sets of variables.

Let A be any Σ algebra. We say that a higher order equational specification $Spec = (\Sigma, E)$ is a *correct specification of A under higher order initial algebra semantics* if, and only if,

$$I_{Ext}(\Sigma, E) \cong A.$$

□

By Theorem 2.8, to establish that a specification (Σ, E) is correct for a Σ algebra A, it suffices to check that $T(\Sigma)/\equiv^{E,\omega} \cong A$. This method will be used in Sections 3 and 4.

3 A Second Order Algebra of Convolution.

In this section we introduce a second order algebra $Conv^n$ of convolution for sample size $n \in \mathbf{N}^+$ over streams of integers from the ring $\mathbf{Z}$. We give a second order equational specification of $Conv^n$ and prove that our specification is correct under higher order initial algebra semantics. Throughout our specification and verification studies the sample size n is a fixed but arbitrarily chosen parameter of the metatheory. As well as requiring first order types T for time and *ring* for integer ring elements, $Conv^n$ requires one second order type $(T \rightarrow ring)$ for streams of integer ring elements.

We begin by recalling the usual first order equational theory of rings with unity (cf. Cohn [1982]).

3.1 Definition. Let $\Sigma = \Sigma^{Ring}$ be a single-sorted first order signature for rings with unity defined as follows. Let S = $\{ring\}$ be the sort set and let Σ be the $S^* \times S$-indexed family of sets defined by:

$$\begin{aligned} \Sigma_{\lambda,ring} &= \{0,1\}, \\ \Sigma_{ring,ring} &= \{-\}, \\ \Sigma_{ring\ ring,ring} &= \{+,\times\}, \end{aligned}$$

and for all other $w \in S^*$ and $s \in S$, $\Sigma_{w,s} = \emptyset$. □

3.2 Definition. Let X be an infinite set of variables. The first order equational theory E^{Ring} of rings with unity consists of the following equations over Σ^{Ring} and X:

$$x + y = y + x, \tag{1}$$
$$x + (y + z) = (x + y) + z, \tag{2}$$
$$x \times (y \times z) = (x \times y) \times z, \tag{3}$$
$$x + 0 = x, \tag{4}$$
$$x + (-x) = 0, \tag{5}$$
$$x \times 1 = x, \tag{6}$$
$$1 \times x = x, \tag{7}$$
$$x \times (y + z) = (x \times y) + (x \times z), \tag{8}$$
$$(x + y) \times z = (x \times z) + (y \times z). \tag{9}$$

□

The class $Alg(\Sigma^{Ring}, E^{Ring})$ of all Σ^{Ring} algebras satisfying the set E^{Ring} of ring equations forms an equational class or variety. Therefore by a basic result of universal algebra, $Alg(\Sigma^{Ring}, E^{Ring})$ admits an initial model $I(\Sigma^{Ring}, E^{Ring})$ which is unique up to isomorphism. This initial model can be concretely constructed as the quotient algebra $T(\Sigma^{Ring})/\equiv^{E^{Ring}}$ where $t \equiv^{E^{Ring}} t' \Leftrightarrow E^{Ring} \vdash t = t'$ and $\vdash$ denotes the provability relation in many-sorted first order equational logic. The structure of this initial model can be easily described. Let

$$\mathbf{Z} = (\mathbf{Z}; 0_{\mathbf{Z}}, 1_{\mathbf{Z}}; -_{\mathbf{Z}}, +_{\mathbf{Z}}, \times_{\mathbf{Z}})$$

denote the ring of integers.

3.3 Proposition. $I(\Sigma^{Ring}, E^{Ring}) \cong \mathbf{Z}$.

Proof. Exercise. □

Thus, we may take the pair $Spec^{Ring} = (\Sigma^{Ring}, E^{Ring})$ as a correct equational specification of the ring $\mathbf{Z}$ of integers under first order initial algebra semantics. For the purposes of specification it is useful to fix a numeral representation of each integer $x \in \mathbf{Z}$ as a Σ^{Ring} term.

3.4 Definition. For each $x \in \mathbf{Z}$ define $\overline{x} \in T(\Sigma^{Ring})$ as follows:

$$\overline{x} = \begin{cases} 0, & \text{if } x = 0; \\ 1, & \text{if } x = 1; \\ 1 + \overline{(x-1)}, & \text{if } x > 1; \\ -\overline{(-x)}, & \text{if } x < 0. \end{cases}$$

□

Simple facts about this numeral representation are:

3.5 Proposition. *For all $a, b \in \mathbf{Z}$:*
(i) $\overline{a}_{\mathbf{Z}} = a$;
(ii) $E^{Ring} \vdash \overline{(a+b)} = (\overline{a}) + (\overline{b})$;
(iii) $E^{Ring} \vdash \overline{(a \times b)} = (\overline{a}) \times (\overline{b})$.

Proof.
(i) A simple induction.
(ii) and (iii) By (i) we have $\mathbf{Z} \models \overline{(a+b)} = \overline{a} + \overline{b}$ and $\mathbf{Z} \models \overline{(a \times b)} = \overline{a} \times \overline{b}$ and so the results follow by Proposition 3.3 and initiality. □

To define the second order algebra $Conv^n$ of convolution we extend the signature Σ^{Ring} to a second order signature Σ^{Conv^n}. We add a basic type T for time, a second order type $(T \to ring)$ for streams of ring elements, a constant symbol $\hat{a}$ for each stream $a : \mathbf{N} \to \mathbf{Z}$ of integers and an operation symbol $conv^n$ for convolution of sequences of n elements from a stream of integers.

3.6 Definition. Let $\Sigma = \Sigma^{Conv^n}$ be a second order signature defined as follows. Let $S \subseteq H(B)$ be the second order type structure $S = \{T, ring, (T \to ring)\}$ defined over the type basis $B = \{T, ring\}$. Let Σ be the $S^* \times S$-indexed family of sets defined by:

$$\Sigma_{\lambda,T} = \{0\},$$
$$\Sigma_{\lambda,ring} = \Sigma^{Ring}_{\lambda,ring},$$
$$\Sigma_{\lambda,(T\to ring)} = \{\hat{a} \mid a : \mathbf{N} \to \mathbf{Z}\},$$
$$\Sigma_{T,T} = \{succ\},$$
$$\Sigma_{ring,ring} = \Sigma^{Ring}_{ring,ring},$$
$$\Sigma_{ring\ ring,ring} = \Sigma^{Ring}_{ring\ ring,ring},$$
$$\Sigma_{(T\to ring)\ T,ring} = \{eval^{(T\to ring)}\},$$
$$\Sigma_{ring^n\ (T\to ring),(T\to ring)} = \{conv^n\},$$

(where $ring^1 = ring$ and $ring^{n+1} = ring\ ring^n$) and for all other $w \in S^*$ and $s \in S$, $\Sigma_{w,s} = \emptyset$.

□

Let $X = \langle X_s \mid s \in S \rangle$ denote an S-indexed family of infinite sets of variables. We may write $t(t')$ as an abbreviation for the term $eval^{(T \to ring)}(t, t')$ for any $t \in T(\Sigma, X)_{(T \to ring)}$ and $t' \in T(\Sigma, X)_T$.

We can now define the second order algebra $Conv^n$ of convolution over the ring $\mathbf{Z}$ of integers.

3.7 Definition. Let $C = Conv^n$ be the S-typed Σ^{Conv^n} algebra with $C_T = \mathbf{N}, C_{ring} = \mathbf{Z}$ and $C_{(T \to ring)} = [\mathbf{N} \to \mathbf{Z}]$. We define the constants $0_C = 0 \in \mathbf{N}$, for $0 \in \Sigma^{Conv^n}_{\lambda,T}$ and $0_C = 0 \in \mathbf{Z}$, for $0 \in \Sigma^{Conv^n}_{\lambda,ring}$, $1_C = 1 \in \mathbf{Z}$ and $\hat{a}_C = a$ for each $a : \mathbf{N} \to \mathbf{Z}$. Define

$$succ_C : C_T \to C_T, \qquad succ_C(x) = x + 1.$$

The operations $\dot{-}_C, +_C$ and $\times_C$ are the corresponding ring operations of $\mathbf{Z}$. Define

$$eval_C^{(T \to ring)} : C_{(T \to ring)} \times C_T \to C_{ring}, \qquad eval_C^{(T \to ring)}(a, t) = a(t).$$

(We may write $a(t)$ as an abbreviation for $eval_C^{(T \to ring)}(a, t)$, for any $a \in C_{(T \to ring)}$ and $t \in C_T$.) Finally, define

$$conv_C^n : (C_{ring})^n \times C_{(T \to ring)} \to C_{(T \to ring)}$$

by

$$conv_C^n(w_1, \ldots, w_n, a)(t) = (w_n \times a(t + (n-1))) + \cdots + (w_1 \times a(t)).$$

□

We can give a second order equational specification of the algebra $Conv^n$ as follows.

3.8 Definition. Define the second order equational specification $Spec^{Conv^n} = (\Sigma^{Conv^n}, E^{Conv^n})$, where the equational theory E^{Conv^n} consists of the equations (1), ..., (9) of the equational theory E^{Ring} of rings with unity together with the following second order equations that axiomatise all streams and the operation $conv^n$.

For each $a : \mathbf{N} \to \mathbf{Z}$ and each $m \in \mathbf{N}$ we have (recalling the numeral representation of integers) the equation

$$\hat{a}(succ^m(0)) = \overline{a(m)}. \tag{10}$$

For $w_1, \ldots, w_n \in X_{ring}$, $s \in X_{(T \to ring)}$ and $x \in X_T$ we have the equation

$$conv^n(w_1, \ldots, w_n, s)(x) = (w_n \times s(succ^{(n-1)}(x))) + \cdots + (w_1 \times s(x)). \tag{11}$$

□

It remains to show that $Spec^{Conv^n}$ is a correct equational specification of $Conv^n$ under higher order initial algebra semantics.

3.9 Correctness Theorem. $I_{Ext}(Spec^{Conv^n}) \cong Conv^n$.

Proof. Let $\Sigma = \Sigma^{Conv^n}$, $E = E^{Conv^n}$ and $C = Conv^n$. Since C and $I_{Ext}(\Sigma, E)$ are both minimal, extensional algebras and as $I_{Ext}(\Sigma, E)$ is initial in the class of all minimal, extensional Σ, E algebras it suffices to construct a homomorphism $\phi : C \to I_{Ext}(\Sigma, E)$. We define a family of mappings $\phi = \langle \phi_s : C_s \to I_{Ext}(\Sigma, E)_s \mid s \in S \rangle$ and prove that ϕ is a homomorphism.

Define

$$\phi_T(m) = [succ^m(0)],$$

for each $m \in C_T$, and

$$\phi_{ring}(x) = [\overline{x}],$$

for each $x \in C_{ring}$ and where for each $x \in \mathbf{Z}$ the numeral $\overline{x} \in T(\Sigma)_{ring}$ is given by Definition 3.4.

Finally define

$$\phi_{(T\to ring)}(a) = [\hat{a}]$$

for each $a \in C_{(T\to ring)}$.

We shall only verify the homomorphism condition for ϕ with respect to the second order operation $conv^n_C$. Checking this condition for all other constants and operations of the algebra is routine. Suppose $w_1, \ldots, w_n \in C_{ring}$ and $a \in C_{(T\to ring)}$ then we must show that

$$\begin{aligned} &\phi_{(T\to ring)}(conv^n_C(w_1, \ldots, w_n, a)) \;= \\ &conv^n_{I_{Ext}(\Sigma,E)}(\phi_{ring}(w_1), \ldots, \phi_{ring}(w_n), \phi_{(T\to ring)}(a)). \end{aligned} \tag{1}$$

We prove this by first showing that the following relationship holds for all $m \in \mathbf{N}$

$$\begin{aligned} &\phi_{(T\to ring)}(conv^n_C(w_1, \ldots, w_n, a))([succ^m(0)]) = \\ &conv^n_{I_{Ext}(\Sigma,E)}(\phi_{ring}(w_1), \ldots, \phi_{ring}(w_n), \phi_{(T\to ring)}(a))([succ^m(0)]). \end{aligned} \tag{2}$$

For any $m \in \mathbf{N}$ and $t = succ^m(0)$ we have

$$\phi_{(T\to ring)}(conv^n_C(w_1, \ldots, w_n, a))([t]) = \phi_{(T\to ring)}(\beta)([t])$$

where $conv^n_C(w_1, \ldots, w_n, a) = \beta$,

$$= [\hat{\beta}]([t])$$

by definition of ϕ,

$$= [\hat{\beta}(t)]$$

by definition of $I_{Ext}(\Sigma, E)$,

$$= [\overline{\beta(m)}]$$

by equation 3.8.(10),

$$= [\overline{(w_n \times a(m + (n-1))) + \cdots + (w_1 \times a(m))}]$$

by definition of β and $conv_C^n$,

$$= [(\overline{w_n} \times \overline{a(m + (n-1))}) + \cdots + (\overline{w_1} \times \overline{a(m)})]$$

by Proposition 3.5.(ii) and 3.5.(iii),

$$= [(\overline{w_n} \times \hat{a}(succ^{m+(n-1)}(0))) + \cdots + (\overline{w_1} \times \hat{a}(succ^m(0)))]$$

by equation 3.8.(10),

$$= [conv^n(\overline{w_1}, \ldots, \overline{w_n}, \hat{a})(t)]$$

by equation 3.8.(11),

$$= conv^n_{I_{Ext}(\Sigma,E)}([\overline{w_1}], \ldots, [\overline{w_n}], [\hat{a}])([t])$$

by definition of $I_{Ext}(\Sigma, E)$,

$$= conv^n_{I_{Ext}(\Sigma,E)}(\phi_{ring}(w_1), \ldots, \phi_{ring}(w_n), \phi_{(T \to ring)}(a))([t])$$

by definition of ϕ. Thus, letting

$$\phi_{(T \to ring)}(conv_C^n(w_1, \ldots, w_n, a)) = [t_0],$$

$$conv^n_{I_{Ext}(\Sigma,E)}(\phi_{ring}(w_1), \ldots, \phi_{ring}(w_n), \phi_{(T \to ring)}(a)) = [t_1],$$

we have for all $m \in \mathbf{N}$, by (2) above that

$$[eval^{(T \to ring)}(t_0, succ^m(0))] = [eval^{(T \to ring)}(t_1, succ^m(0))].$$

Thus

$$E^{Conv^n} \vdash_{\omega} eval^{(T \to ring)}(t_0, succ^m(0)) = eval^{(T \to ring)}(t_1, succ^m(0)),$$

so by the ω-evaluation rule and the fact that each term of sort T has the form $succ^m(0)$ for $m \in \mathbf{N}$ we have

$$E^{Conv^n} \vdash_{\omega} t_0 = t_1$$

and so $[t_0] = [t_1]$ which establishes (1). □

4 Implementation of Convolution.

In this section we present two systolic algorithms, first introduced in Kung [1982], which implement convolution of sample size n. We formalise both algorithms as second order algebras and specify each by means of second order equations. The first algorithm we consider has fixed or *non-programmable* weights which restrict it to implementing a fixed convolution function. By contrast, the second algorithm has *programmable* weights allowing it to implement different convolution functions. Both algorithms are examples of a *synchronous concurrent algorithm* (SCA) and we use the notation and terminology of SCA theory (see for example Thompson and Tucker [1991]) in presenting the algorithms. We begin with an introduction to the SCA model of computation.

A synchronous concurrent algorithm (SCA) consists of a network of processing elements, called *modules*, which communicate via interconnections called *channels*. We refer to the network of an SCA as its *architecture*. All modules compute and communicate in parallel and these actions are synchronised to a global clock $T = \{0, 1, 2, \ldots\}$. The network computes over a data type modelled as an algebra A. Each *processor module* M_i performs a single computation which we specify by means of a total function $f_i : A^{\lambda(i)} \to A$ over the algebra A, where module M_i has $\lambda(i)$ input channels and one output channel. Communication channels are allowed to finitely branch but they may not merge. Data enters the network at distinguished modules called *sources* and leaves at distinguished modules called *sinks*. In the case that A is a many-sorted algebra the modules and channels of the network are strongly typed by the sort set S.

The operational semantics of an SCA can be described for each clock cycle $t \in T$ as follows. Initially, for $t = 0$, each module places a predefined *initial value* on its output channel. If the correct operation of an SCA depends upon the initial value of certain modules then the SCA is termed *programmable*. At each subsequent clock cycle, $t + 1$, each processor module M_i applies its associated function f_i to the data $a_1, \ldots, a_{\lambda(i)}$ on its input channels at time t and places the result $f_i(a_1, \ldots, a_{\lambda(i)})$ on its output channel at time $t + 1$. Each source module reads in a single input datum $\alpha(t) \in A$ from an *input stream* $\alpha : T \to A$. Each sink module delivers a single output datum $\beta(t) \in A$ to an *output stream* $\beta : T \to A$. Thus, the behaviour of the entire network may be described by a system of simultaneous primitive recursive functions.

An SCA processes infinite streams of input data and produces infinite streams of output data. Thus, we may describe its *black box behaviour* in terms of a second order transformation on streams $NET : A^k \times [T \to A]^p \to [T \to A]^q$ called the *network function*, where k is the total number of modules, p is the number of source modules and q the number of sink modules. Given an initial state $\overline{a} \in A^k$ for the entire network and p input streams $\alpha_1, \ldots, \alpha_p : T \to A$ then $NET(\overline{a}, \alpha_1, \ldots, \alpha_p)$ defines the q output streams $\beta_1, \ldots, \beta_q : T \to A$ produced by the network.

To formalise the interaction of an SCA with its external environment we must specify *input and output scheduling functions*. These reconcile the system clock T with some external environment clock by specifying when data is to be entered and retrieved from the network. The complete description of an SCA, in full detail, is obtained by combining the scheduling functions with the network specification.

This brief introduction to the SCA model of computation will become clearer upon consideration of our two convolution algorithms. The first of these, referred to in the sequel as the *non-programmable convolver*, we describe in an informal way as follows.

4.1 Definition. The architecture of the non-programmable convolver is depicted in Figure 1. The function performed by each module $M_{i,j}$ is defined as follows.

(i) Module $M_{1,j}$, for $j = 1, \ldots, n$, computes the identity function $Id_{\mathbf{Z}} : \mathbf{Z} \to \mathbf{Z}$, where $Id_{\mathbf{Z}}(x) = x$.
(ii) Module $M_{3,j}$, for $j = 1, \ldots, n$, returns a constant weight value $w_{(n+1)-j} \in \mathbf{Z}$.
(iii) Module $M_{2,j}$, for $j = 1, \ldots, n-1$, computes the ring inner product function $IP_{\mathbf{Z}} : \mathbf{Z} \times \mathbf{Z} \times \mathbf{Z} \to \mathbf{Z}$, where $IP_{\mathbf{Z}}(x, y, z) = (x \times_{\mathbf{Z}} y) +_{\mathbf{Z}} z$.
(iv) Module $M_{2,n}$ computes the ring multiplication operation $\times_{\mathbf{Z}}$.

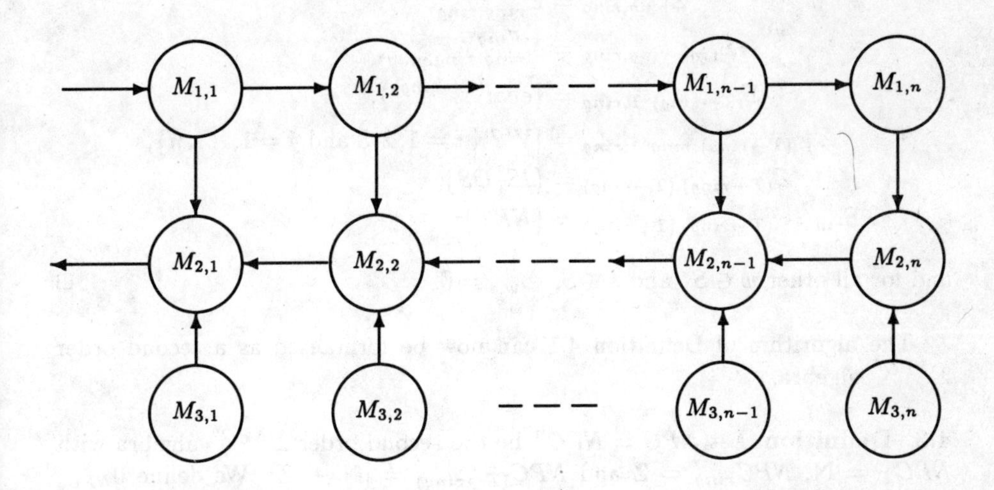

Fig. 1. Architecture of the non-programmable convolver.

The input schedule *ISCHED* for this architecture loads data into the network during alternate clock cycles and is defined by

$$ISCHED : [\mathbf{N} \to \mathbf{Z}] \to [\mathbf{N} \to \mathbf{Z}], \quad ISCHED(a)(t) = \begin{cases} 0, & \text{if } t \text{ is odd,} \\ a(t/2), & \text{otherwise,} \end{cases}$$

for $a : \mathbf{N} \to \mathbf{Z}$ and $t \in \mathbf{N}$. The output schedule *OSCHED* takes into account the initialisation period of the network, which is $2n$ clock cycles, and then retrieves data during alternate clock cycles. It is defined by

$$OSCHED : [\mathbf{N} \to \mathbf{Z}] \to [\mathbf{N} \to \mathbf{Z}], \quad OSCHED(a)(t) = a(2(n+t)),$$

for $a : \mathbf{N} \to \mathbf{Z}$ and $t \in \mathbf{N}$. □

We will formalise this SCA as a second order algebra NPC^n. We first define a second order signature Σ^{NPC^n} to name all algorithm components.

4.2 Definition. Let $\Sigma = \Sigma^{NPC^n}$ be a second order signature defined as follows. Let $S \subseteq H(B)$ be the second order type structure $S = \{T, ring, (T \to ring)\}$ defined over the type basis $B = \{T, ring\}$. Let Σ be the $S^* \times S$-indexed family of sets defined by

$$\begin{aligned}
\Sigma_{\lambda,T} &= \{0\},\\
\Sigma_{\lambda,ring} &= \Sigma^{Ring}_{\lambda,ring},\\
\Sigma_{\lambda,(T\to ring)} &= \{\hat{a} \mid a : \mathbf{N} \to \mathbf{Z}\},\\
\Sigma_{T,T} &= \{succ, twice\},\\
\Sigma_{ring,ring} &= \Sigma^{Ring}_{ring,ring},\\
\Sigma_{ring\ ring,ring} &= \Sigma^{Ring}_{ring\ ring,ring},\\
\Sigma_{(T\to ring)\ T,ring} &= \{eval^{(T\to ring)}\},\\
\Sigma_{T\ (T\to ring)\ ring^{3n},ring} &= \{V^{i,j} \mid i = 1,2,3 \text{ and } j = 1,\ldots,n\},\\
\Sigma_{(T\to ring),(T\to ring)} &= \{IS, OS\},\\
\Sigma_{ring^{3n}\ (T\to ring),(T\to ring)} &= \{NET\},
\end{aligned}$$

and for all other $w \in S^*$ and $s \in S$, $\Sigma_{w,s} = \emptyset$. □

The algorithm of Definition 4.1 can now be formalised as a second order Σ^{NPC^n} algebra.

4.3 Definition. Let $NPC = NPC^n$ be the second order Σ^{NPC^n} algebra with $NPC_T = \mathbf{N}$, $NPC_{ring} = \mathbf{Z}$ and $NPC_{(T\to ring)} = [\mathbf{N} \to \mathbf{Z}]$. We define 0_{NPC}, 1_{NPC}, $-_{NPC}$, $+_{NPC}$ and $\times_{NPC}$ to be the corresponding constants and operations of the ring $\mathbf{Z}$; the constants $0_{NPC}, \hat{a}_{NPC}$ for each $a \in [\mathbf{N} \to \mathbf{Z}]$ and the operations $succ_{NPC}$ and $eval^{(T\to ring)}_{NPC}$ are defined as before in the second order algebra $Conv^n$ (Definition 3.7).

For each module $M_{i,j}$, for $i = 1,2,3$ and $j = 1,\ldots,n$, we define a *value function*

$$V^{i,j}_{NPC} : \mathbf{N} \times [\mathbf{N} \to \mathbf{Z}] \times \mathbf{Z}^{3n} \to \mathbf{Z},$$

formalising the operation of $M_{i,j}$. Let $a : \mathbf{N} \to \mathbf{Z}$ and $w_1, \ldots, w_n \in \mathbf{Z}$ be the fixed weights used in convolution, $t \in \mathbf{N}$ and $\mathbf{x} = x_{1,1}, \ldots, x_{3,n} \in \mathbf{Z}^{3n}$, where $x_{i,j}$ is the initial value for the module $M_{i,j}$. Then $V^{i,j}_{NPC}(t, a, \mathbf{x})$ is the value computed by module $M_{i,j}$ at time t starting from the initial values $\mathbf{x}$ using the input stream a. For $i = 1, 2, 3$ and $j = 1, \ldots, n$ we define

$$V_{NPC}^{i,j}(0, a, \mathbf{x}) = x_{i,j}.$$

For $j = 2, \ldots, n$ define

$$V_{NPC}^{1,1}(t+1, a, \mathbf{x}) = a(t),$$
$$V_{NPC}^{1,j}(t+1, a, \mathbf{x}) = V_{NPC}^{1,j-1}(t, a, \mathbf{x}).$$

For $j = 1, \ldots, n$ define

$$V_{NPC}^{3,j}(t+1, a, \mathbf{x}) = w_{(n+1)-j}.$$

For $j = 1, \ldots, n-1$ define

$$V_{NPC}^{2,j}(t+1, a, \mathbf{x}) = (V_{NPC}^{3,j}(t, a, \mathbf{x}) \times V_{NPC}^{1,j}(t, a, \mathbf{x})) + V_{NPC}^{2,j+1}(t, a, \mathbf{x}),$$

and

$$V_{NPC}^{2,n}(t+1, a, \mathbf{x}) = V_{NPC}^{3,n}(t, a, \mathbf{x}) \times V_{NPC}^{1,n}(t, a, \mathbf{x}).$$

We define the network function for the architecture by

$$NET_{NPC} : \mathbf{Z}^{3n} \times [\mathbf{N} \to \mathbf{Z}] \to [\mathbf{N} \to \mathbf{Z}], \quad NET_{NPC}(\mathbf{x}, a)(t) = V_{NPC}^{2,1}(t, a, \mathbf{x}).$$

We define the scheduling functions using the following retiming function on the clock $\mathbf{N}$

$$twice_{NPC} : \mathbf{N} \to \mathbf{N}, \quad twice_{NPC}(x) = 2x.$$

Finally, we define the scheduling functions

$$IS_{NPC} : [\mathbf{N} \to \mathbf{Z}] \to [\mathbf{N} \to \mathbf{Z}],$$

$$IS_{NPC}(a)(2t+1) = 0,$$
$$IS_{NPC}(a)(2t) = a(t),$$

and

$$OS_{NPC} : [\mathbf{N} \to \mathbf{Z}] \to [\mathbf{N} \to \mathbf{Z}], \quad OS_{NPC}(a)(t) = a(2(n+t)).$$

□

We may now give a second order equational specification $Spec^{NPC^n(w_1,\ldots,w_n)}$ of the non- programmable convolver by formalising the semantical equations above.

4.4 Definition. We define the second order equational specification $Spec^{NPC^n}$ = $Spec^{NPC^n(w_1,\ldots,w_n)}$, where $w_1, \ldots, w_n \in \mathbf{Z}_{ring}$ are the weights to be used in convolution and $Spec^{NPC^n} = (\Sigma^{NPC^n}, E^{NPC^n})$. The second order equational theory E^{NPC^n} consists of the first order equations $(1), \ldots, (9)$ of the equational theory E^{Ring} of rings with unity (Definition 3.2) together with the following second order equations which axiomatise all streams, all value functions $V^{i,j}$, the network function NET, the retiming function *twice* and the input and output scheduling functions IS, OS. For each $a : \mathbf{N} \to \mathbf{Z}$ and each $m \in \mathbf{N}$ we have the equation

$$\hat{a}(succ^m(0)) = \overline{a(m)}. \tag{10}$$

Let $s \in X_{(T \to ring)}$, $t \in X_T$ and $\mathbf{x} = x_{1,1}, \ldots, x_{3,n} \in X^{3n}_{ring}$. We define equations for the value functions as follows. For $i = 1,2,3$ and $j = 1, \ldots, n$ we have equations

$$V^{i,j}(0, s, \mathbf{x}) = x_{i,j}, \tag{11}$$

$$V^{3,j}(succ(t), s, \mathbf{x}) = \overline{w_{(n+1)-j}}. \tag{12a}$$

For $j = 2, \ldots, n$ we have equations

$$V^{1,1}(succ(t), s, \mathbf{x}) = s(t), \tag{13}$$
$$V^{1,j}(succ(t), s, \mathbf{x}) = V^{1,j-1}(t, s, \mathbf{x}). \tag{14}$$

For $j = 1, \ldots, n-1$ we have equations

$$V^{2,j}(succ(t), s, \mathbf{x}) = (V^{3,j}(t, s, \mathbf{x}) \times V^{1,j}(t, s, \mathbf{x})) + V^{2,j+1}(t, s, \mathbf{x}), \tag{15}$$

and

$$V^{2,n}(succ(t), s, \mathbf{x}) = V^{3,n}(t, s, \mathbf{x}) \times V^{1,n}(t, s, \mathbf{x}). \tag{16}$$

Finally, we axiomatise the network, retiming and scheduling functions:

$$NET(\mathbf{x}, s)(t) = V^{2,1}(t, s, \mathbf{x}), \tag{17}$$
$$twice(0) = 0, \tag{18}$$
$$twice(succ(t)) = succ(succ(twice(t))), \tag{19}$$
$$IS(s)(succ(twice(t))) = 0, \tag{20}$$
$$IS(s)(twice(t)) = s(t), \tag{21}$$
$$OS(s)(t) = s(twice(succ^n(t))). \tag{22}$$

□

We can now assert the correctness of our specification $Spec^{NPC^n}$ with respect to the second order algebra NPC^n under higher order initial algebra semantics.

4.5 Correctness Theorem. $I_{Ext}(Spec^{NPC^n}) \cong NPC^n$.

Proof. Exercise, following the proof method of Theorem 3.9. □

Next we present a slightly different implementation of convolution as an SCA. This implementation is programmable and allows the user to load in the weights $w_1, \ldots, w_n \in \mathbf{Z}$ required for convolution as the initial values of modules $M_{3,n}, \ldots, M_{3,1}$ respectively. The weight value w_i is then restored by a feedback channel to module $M_{3,(n+1)-i}$ and remains constant throughout the computation.

Again we begin with an informal description of the architecture of this SCA, referred to in the sequel as the *programmable convolver*.

4.6 Definition. The architecture of the programmable convolver is depicted in Figure 2. The input and output schedules, *ISCHED* and *OSCHED*, and the

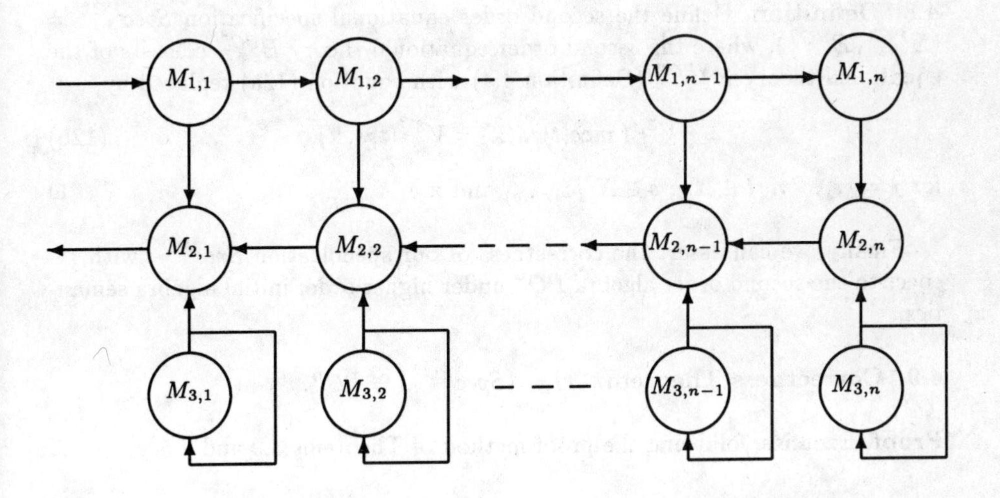

Fig. 2. Architecture of the programmable convolver.

function computed by each module $M_{i,j}$ are the same as in Definition 4.1 except for the functions computed by modules $M_{3,j}$, for $j = 1, \ldots, n$. These are redefined as follows. Module $M_{3,j}$, for $j = 1, \ldots, n$, computes the identity function $Id_{\mathbf{Z}} : \mathbf{Z} \to \mathbf{Z}$, where $Id_{\mathbf{Z}}(x) = x$. □

To formalise the programmable convolver we require a second order signature Σ^{PC^n} which we may take to be the same as Σ^{NPC^n}. Thus, we define

$$\Sigma^{PC^n} = \Sigma^{NPC^n}.$$

We may then formalise the programmable convolver as a second order Σ^{PC^n} algebra PC^n as follows.

4.7 Definition. Let $PC = PC^n$ be the second order Σ^{PC^n} algebra whose carrier sets, constants and functions are defined as in the second order algebra NPC^n (Definition 4.3) except that the value functions $V_{PC}^{3,j} : \mathbf{N} \times [\mathbf{N} \to \mathbf{Z}] \times \mathbf{Z}^{3n} \to \mathbf{Z}$, for $j = 1, \ldots, n$, are defined by

$$V_{PC}^{3,j}(0, a, \mathbf{x}) = x_{3,j},$$
$$V_{PC}^{3,j}(t+1, a, \mathbf{x}) = V_{PC}^{3,j}(t, a, \mathbf{x}),$$

for $t \in PC_T, a \in PC_{(T \to ring)}$ and $\mathbf{x} = x_{1,1}, \ldots, x_{3,n} \in PC_{ring}^{3n}$. □

We now give a second order equational specification $Spec^{PC^n}$ of the programmable convolver. Note that in this specification we do not axiomatise the weight values to be used in convolution since these are programmable.

4.8 Definition. Define the second order equational specification $Spec^{PC^n} = (\Sigma^{PC^n}, E^{PC^n})$, where the second order equational theory E^{PC^n} consists of the equational theory E^{NPC^n} (Definition 4.4) with equation (12a) replaced by

$$V^{3,j}(succ(t), s, \mathbf{x}) = V^{3,j}(t, s, \mathbf{x}), \tag{12b}$$

for $j = 1, \ldots, n$, $t \in X_T$, $s \in X_{(T \to ring)}$ and $\mathbf{x} \in X_{ring}^{3n}$. □

Finally, we can assert the correctness of our specification $Spec^{PC^n}$ with respect to the second order algebra PC^n under higher order initial algebra semantics.

4.9 Correctness Theorem. $I_{Ext}(Spec^{PC^n}) \cong PC^n$.

Proof. Exercise, following the proof method of Theorems 3.9 and 4.5. □

5 Formal Verification of Algorithm Correctness.

In the preceding sections we have formalised a user specification for convolution and two systolic implementations of convolution as second order equational specifications. In this section we will formally verify the correctness of both convolution algorithms with respect to the user specification. We first consider the non-programmable convolver and give a formal verification of its correctness in pure higher order equational logic. We then formally verify the correctness of the programmable convolver. To verify the latter algorithm we augment the higher order equational calculus with the additional proof rule of *free variable induction* (for equations). We show that the additional proof theoretic strength given by this rule is necessary for a formal proof of correctness by considering non-standard models of the algorithm specification.

We begin by defining a second order signature Σ^{Ver} and an equational theory E^{Ver} in which the formal verification of correctness of the non-programmable convolver can take place.

5.1 Definition. Let $S \subseteq H(B)$ be the second order type structure $S = \{T, ring, (T \to ring)\}$ defined over the type basis $B = \{T, ring\}$. Recalling $Spec^{Conv^n}$ (Definition 3.8) and $Spec^{NPC^n} = Spec^{NPC^n(w_1,\ldots,w_n)}$ (Definition 4.4) define the S-typed signature Σ^{Ver} by

$$\Sigma^{Ver} = \Sigma^{Conv^n} \cup \Sigma^{NPC^n}.$$

Let E^{Ver} be the second order equational theory defined over Σ^{Ver} by

$$E^{Ver} = E^{Conv^n} \cup E^{NPC^n}.$$

□

We present some basic technical facts needed in the proof of correctness of the non-programmable convolver.

5.2 Lemma. *Let* $m \in \mathbf{N}$, $s \in X_{(T\to ring)}, t \in X_T$ *and* $\mathbf{x} = x_{1,1}, \ldots, x_{3,n} \in X^{3n}_{ring}$, *then:*

(i) $E^{Ver} \vdash twice(succ^m(t)) = succ^{2m}(twice(t))$;

(ii) $E^{Ver} \vdash V^{1,i}(succ^i(t), s, \mathbf{x}) = s(t)$, *for* $i = 1, \ldots, n$;

(iii) $E^{Ver} \vdash V^{2,1}(succ^m(t), s, \mathbf{x}) = (V^{3,1}(succ^{m-1}(t), s, \mathbf{x}) \times$
$V^{1,1}(succ^{m-1}(t), s, \mathbf{x})) + \cdots + (V^{3,n-1}(succ^{m-(n-1)}(t), s, \mathbf{x}) \times$
$V^{1,n-1}(succ^{m-(n-1)}(t), s, \mathbf{x})) \ + \ V^{2,n}(succ^{m-(n-1)}(t), s, \mathbf{x})$,

where $m \geq n-1$.

Proof. Routine calculation. □

Intuitively, the concept of correctness for the non-programmable convolver should mean that the stream transformer computed by this algorithm, according to the specified input and output schedules and the network function, agrees with the user specification of convolution as a stream transformer. Formally, this concept of correctness can be expressed by the second order equation

$$OS(NET(\mathbf{x}, IS(s))) = conv^n(\overline{w_1}, \ldots, \overline{w_n}, s), \qquad \text{(I)}$$

where $w_1, \ldots, w_n \in \mathbf{Z}$ are the fixed weights used by the non-programmable convolver. We can now give a formal derivation of the correctness equation (I) in higher order equational logic using the axioms of the verification theory E^{Ver}. This verifies the correctness of the non-programmable convolver.

5.3 Correctness Theorem. (Non-programmable convolver).

$$E^{Ver} \vdash OS(NET(\mathbf{x}, IS(s))) = conv^n(\overline{w_1}, \ldots, \overline{w_n}, s),$$

for $\mathbf{x} = x_{1,1}, \ldots, x_{3,n} \in X^{3n}_{ring}$ *and* $s \in X_{(T \to ring)}$.

Proof. It suffices to show that

$$E^{Ver} \vdash OS(NET(\mathbf{x}, IS(s)))(t) = conv^n(\overline{w_1}, \ldots, \overline{w_n}, s)(t), \qquad (*)$$

where $t \in X_T$ and then to apply the evaluation rule. Now we know that

$$E^{Ver} \vdash OS(s)(t) = s(twice(succ^n(t))), \qquad (1)$$

by equation 4.4.(22), and

$$E^{Ver} \vdash s(twice(succ^n(t))) = s(succ^{2n}(twice(t))), \qquad (2)$$

by reflexivity, Lemma 5.2.(i) and substitution. Thus

$$E^{Ver} \vdash OS(s)(t) = s(succ^{2n}(twice(t))), \qquad (3)$$

by transitivity, (1) and (2), and so

$$E^{Ver} \vdash OS(NET(\mathbf{x}, IS(s)))(t) = NET(\mathbf{x}, IS(s))(succ^{2n}(twice(t))), \qquad (4)$$

by (3), reflexivity and substitution. Then

$$E^{Ver} \vdash NET(\mathbf{x}, IS(s))(succ^{2n}(twice(t))) = V^{2,1}(succ^{2n}(twice(t)), IS(s), \mathbf{x}), \qquad (5)$$

by equation 4.4.(17), reflexivity and substitution. We now "unwind" the recursive definition of the value functions. By Lemma 5.2.(iii), reflexivity and substitution

$$\begin{aligned} E^{Ver} \vdash V^{2,1}(succ^{2n}(twice(t)), IS(s), \mathbf{x}) \ = \ & (V^{3,1}(succ^{2n-1}(twice(t)), \\ IS(s), \mathbf{x}) \ \times \ & V^{1,1}(succ^{2n-1}(twice(t)), IS(s), \mathbf{x})) \ + \cdots + \\ (V^{3,n-1}(succ^{2n-(n-1)}(twice(t)), IS(s), \mathbf{x}) \ \times \ & V^{1,n-1}(succ^{2n-(n-1)}(twice(t)), \\ IS(s), \mathbf{x})) + & V^{2,n}(succ^{2n-(n-1)}(twice(t)), IS(s), \mathbf{x}), \end{aligned} \qquad (6)$$

and

$$\begin{aligned} E^{Ver} \vdash V^{2,n}(succ^{2n-(n-1)}(twice(t)), IS(s), \mathbf{x}) \ = \\ (V^{3,n}(succ^{2n-n}(twice(t)), IS(s), \mathbf{x}) \ \times V^{1,n}(succ^{2n-n}(twice(t)), IS(s), \mathbf{x})), \end{aligned} \qquad (7)$$

by equation 4.4.(16), reflexivity and substitution. Then

$$\begin{aligned} E^{Ver} \vdash (V^{3,1}(succ^{2n-1}(twice(t)), IS(s), \mathbf{x}) \ \times \ V^{1,1}(succ^{2n-1}(twice(t)), \\ IS(s), \mathbf{x})) \ + \cdots + (V^{3,n-1}(succ^{2n-(n-1)}(twice(t)), IS(s), \mathbf{x}) \times \\ V^{1,n-1}(succ^{2n-(n-1)}(twice(t)), IS(s), \mathbf{x})) + \\ V^{2,n}(succ^{2n-(n-1)}(twice(t)), IS(s), \mathbf{x}) = (V^{3,1}(succ^{2n-1}(twice(t)), IS(s), \mathbf{x}) \\ \times V^{1,1}(succ^{2n-1}(twice(t)), IS(s), \mathbf{x})) + \cdots + \\ (V^{3,n}(succ^{2n-n}(twice(t)), IS(s), \mathbf{x}) \ \times \ V^{1,n}(succ^{2n-n}(twice(t)), IS(s), \mathbf{x})), \end{aligned} \qquad (8)$$

by reflexivity and substitution using (7). We next "evaluate" the value functions. By reflexivity, Lemma 5.2.(ii), equation 4.4.(12a) and substitution

$$\begin{aligned} E^{Ver} \vdash (V^{3,1}(succ^{2n-1}(twice(t)), IS(s), \mathbf{x}) \ \times \ V^{1,1}(succ^{2n-1}(twice(t)), \\ IS(s), \mathbf{x})) + \cdots + \ (V^{3,n}(succ^{2n-n}(twice(t)), IS(s), \mathbf{x}) \ \times \\ V^{1,n}(succ^{2n-n}(twice(t)), IS(s), \mathbf{x})) \ = \ (\overline{w_n} \ \times \ IS(s)(succ^{2n-2}(twice(t)))) \\ + \cdots + \ (\overline{w_1} \ \times \ IS(s)(succ^{2n-2n}(twice(t)))), \end{aligned} \tag{9}$$

and

$$\begin{aligned} E^{Ver} \vdash (\overline{w_n} \ \times \ IS(s)(succ^{2n-2}(twice(t)))) \ + \cdots + \ (\overline{w_1} \ \times \\ IS(s)(succ^{2n-2n}(twice(t)))) \ = \ (\overline{w_n} \ \times \ IS(s)(twice(succ^{n-1}(t)))) \\ + \cdots + \ (\overline{w_1} \ \times \ IS(s)(twice(t))), \end{aligned} \tag{10}$$

by reflexivity, Lemma 5.2.(i), symmetry and substitution. Also

$$\begin{aligned} E^{Ver} \vdash (\overline{w_n} \times IS(s)(twice(succ^{n-1}(t)))) \ + \cdots + \ (\overline{w_1} \times IS(s)(twice(t))) \\ = \ (\overline{w_n} \ \times \ s(succ^{n-1}(t))) \ + \cdots + \ (\overline{w_1} \ \times \ s(t)), \end{aligned} \tag{11}$$

by reflexivity, equation 4.4.(21) and substitution, and

$$E^{Ver} \vdash (\overline{w_n} \times s(succ^{n-1}(t))) + \cdots + (\overline{w_1} \times s(t)) = conv^n(\overline{w_1}, \ldots, \overline{w_n}, s)(t) \tag{12}$$

by equation 3.8.(11), symmetry, reflexivity and substitution. Thus

$$E^{Ver} \vdash OS(NET(\mathbf{x}, IS(s)))(t) = conv^n(\overline{w_1}, \ldots, \overline{w_n}, s)(t), \tag{13}$$

by transitivity, and (4), (5), (6), (8), ...,(12), which establishes (*).

Applying the evaluation rule to (*) we derive the correctness equation (I), that is

$$E^{Ver} \vdash OS(NET(\mathbf{x}, IS(s))) = conv^n(\overline{w_1}, \ldots, \overline{w_n}, s). \tag{14}$$

□

Next we verify the correctness of the programmable convolver using higher order equational logic augmented with the proof rule of *free variable induction* (for equations). This proof rule is defined as follows.

5.4 Definition. Let S be a type structure over a type basis B such that $T \in S$ and let Σ be an S-typed signature such that $0 \in \Sigma_{\lambda,T}$ and $succ \in \Sigma_{T,T}$. Let X be an S-indexed family of infinite sets of variables and $E \subseteq Eqn(\Sigma, X)$. Then for $s \in S$, $t, t' \in T(\Sigma, X)_s$, $x \in X_T$ and c a new constant symbol of sort T not occurring in $\Sigma_{\lambda,T}$ we define the rule of *free variable induction* (for equations) as follows.

$$\frac{E \vdash t[x/0] = t'[x/0] \quad E \cup \{t[x/c] = t'[x/c]\} \vdash t[x/succ(c)] = t'[x/succ(c)]}{E \vdash t = t'}$$

It is understood that with each application of this proof rule a fresh constant c of sort T is introduced. Let $\vdash_{ind}$ denote the *inference relation* between equational theories $E \subseteq Eqn(\Sigma, X)$ and equations $e \in Eqn(\Sigma, X)$, defined by $E \vdash_{ind} e$ if, and only if, there exists a proof of e from E using the deduction rules of higher order equational logic and the rule of free variable induction. □

Note that all the formulas involved in the antecedents and succedent of this rule are equations. We leave it to the reader as an instructive exercise to check the conditions on Σ and E under which free variable induction is a sound inference rule with respect to an initial extensional model. We observe however that higher order equational logic together with the rule of free variable induction is still not a complete proof system with respect to the higher order initial model.

The signature used for the verification of the programmable convolver will be the same as that used for the verification of the non-programmable convolver Σ^{Ver}. However, the equational theory E^{Ver} must be redefined to take into account the equations for the programmable convolver.

5.5 Definition. Let E^{Ver} be a second order equational theory over Σ^{Ver} defined by

$$E^{Ver} = E^{Conv^n} \cup E^{PC^n}.$$

□

Again, regarding the soundness of the rule of free variable induction, observe that if $E^{Ver} \vdash_{ind} e$ then $I(\Sigma^{Ver}, E^{Ver}) \models e$ for any equation e over Σ^{Ver}.

The verification proof for the programmable convolver is similar to that for the non-programmable convolver. The only difference is that E^{Ver} no longer contains equation 4.4.(12a) which was required for step (9) in the proof of Theorem 5.3. Instead we assume that for the initial values $\mathbf{x} \in \mathbf{Z}^{3n}$ we have $x_{3,n} = w_1, \ldots, x_{3,1} = w_n$, where $w_1, \ldots, w_n \in \mathbf{Z}$ are the weights to be used in convolution, and use the following result.

5.6 Lemma. *Let* $t \in X_T$, $s \in X_{(T \to ring)}$ *and* $\mathbf{x} = x_{1,1}, \ldots, x_{3,n} \in X_{ring}^{3n}$, *then*

$$E^{Ver} \vdash_{ind} V^{3,i}(t, s, \mathbf{x}) = x_{3,i}.$$

Proof. We use the rule of free variable induction. By equation 4.4.(11), reflexivity and substitution

$$E^{Ver} \vdash V^{3,i}(0, s, \mathbf{x}) = x_{3,i}.$$

Let hyp_i be the induction hypothesis $V^{3,i}(c,s,\mathbf{x}) = x_{3,i}$, where c is a new constant symbol of sort T. Then by equation 4.8.(12b)

$$E^{Ver} \vdash V^{3,i}(succ(c), s, \mathbf{x}) = V^{3,i}(c, s, \mathbf{x})$$

and so

$$E^{Ver} \cup \{hyp_i\} \vdash V^{3,i}(succ(c), s, \mathbf{x}) = x_{3,i}$$

using hyp_i and transitivity. Thus

$$E^{Ver} \vdash_{ind} V^{3,i}(t, s, \mathbf{x}) = x_{3,i}$$

using the rule of free variable induction. □

Let us consider the concept of correctness for the programmable convolver. In this case the correct operation of the algorithm depends upon loading the weights $w_1, \ldots, w_n$ into the modules $M_{3,n}, \ldots, M_{3,1}$ respectively during initialisation. Under such circumstances we claim that the stream transformer computed by the algorithm, according to the specified input and output schedules and network function, agrees with the user specification of convolution. This concept of correctness can be expressed by a second order equation

$$OS(NET(\mathbf{x}, IS(s))) = conv^n(x_{3,n}, \ldots, x_{3,1}, s). \tag{II}$$

We can formally derive the correctness equation (II) from the verification theory E^{Ver} using the rules of higher order equational logic and the rule of free variable induction.

5.7 Correctness Theorem. (Programmable convolver).

$$E^{Ver} \vdash_{ind} OS(NET(\mathbf{x}, IS(s))) = conv^n(x_{3,n}, \ldots, x_{3,1}, s),$$

for $\mathbf{x} = x_{1,1}, \ldots, x_{3,n} \in X^{3n}_{ring}$ *and* $s \in X_{(T \to ring)}$.

Proof. The proof is similar to the proof of Theorem 5.3. Thus, following steps (1), ..., (8) of the proof and using the transitivity rule we have

$$\begin{aligned} E^{Ver} \vdash OS(NET(\mathbf{x}, IS(s)))(t) &= (V^{3,1}(succ^{2n-1}(twice(t)), IS(s), \mathbf{x}) \times \\ &\quad V^{1,1}(succ^{2n-1}(twice(t)), IS(s), \mathbf{x})) + \cdots + \\ (V^{3,n}(succ^{2n-n}(twice(t)), IS(s), \mathbf{x}) &\times V^{1,n}(succ^{2n-n}(twice(t)), IS(s), \mathbf{x})). \end{aligned} \tag{9}$$

Then by Lemma 5.6 above, Lemma 5.2.(ii) (which still holds for our new E^{Ver}) and substitution

$$\begin{aligned} E^{Ver} \vdash_{ind} (V^{3,1}(succ^{2n-1}(twice(t)), IS(s), \mathbf{x}) &\times V^{1,1}(succ^{2n-1}(twice(t)), \\ IS(s), \mathbf{x})) + \cdots + (V^{3,n}(succ^{2n-n}(twice(t)), IS(s), \mathbf{x}) &\times \\ V^{1,n}(succ^{2n-n}(twice(t)), IS(s), \mathbf{x})) &= (x_{3,1} \times IS(s)(succ^{2n-2}(twice(t)))) \\ + \cdots + (x_{3,n} &\times IS(s)(succ^{2n-2n}(twice(t)))), \end{aligned} \tag{10}$$

and by reflexivity, Lemma 5.2.(i) (which also still holds for E^{Ver}), symmetry and substitution we have

$$E^{Ver} \vdash (x_{3,1} \times IS(s)(succ^{2n-2}(twice(t)))) + \cdots + (x_{3,n} \times IS(s)(succ^{2n-2n}(twice(t)))) = (x_{3,1} \times IS(s)(twice(succ^{n-1}(t)))) + \cdots + (x_{3,n} \times IS(s)(twice(t))). \quad (11)$$

Also

$$E^{Ver} \vdash (x_{3,1} \times IS(s)(twice(succ^{n-1}(t)))) + \cdots + (x_{3,n} \times IS(s)(twice(t))) = (x_{3,1} \times s(succ^{n-1}(t))) + \cdots + (x_{3,n} \times s(t)), \quad (12)$$

by reflexivity, equation 4.4.(21) and substitution, and

$$E^{Ver} \vdash (x_{3,1} \times s(succ^{n-1}(t))) + \cdots + (x_{3,n} \times s(t)) = conv^n(x_{3,n}, \ldots, x_{3,1}, s)(t), \quad (13)$$

by equation 3.8.(11), reflexivity, symmetry and substitution. So by transitivity and (9),(10),(11),(12),(13)

$$E^{Ver} \vdash_{ind} OS(NET(\mathbf{x}, IS(s)))(t) = conv^n(x_{3,n}, \ldots, x_{3,1}, s)(t). \quad (14)$$

Thus, using the evaluation rule we derive the correctness equation (II), that is

$$E^{Ver} \vdash_{ind} OS(NET(\mathbf{x}, IS(s))) = conv^n(x_{3,n}, \ldots, x_{3,1}, s). \quad (15)$$

□

Note that the rule of free variable induction is used only once in the correctness proof of the programmable convolver, in the proof of Lemma 5.6.

We conclude with a metamathematical result on the role of the free variable induction rule in Correctness Theorem 5.7. It is natural to consider whether the correctness equation (II) can be derived from the theory E^{Ver} using the rules of higher order equational logic alone, i.e. without the use of induction. We will prove an independence result which shows that this is not possible. That is to say, the additional proof theoretic strength of free variable induction is essential in the formal proof of correctness of the programmable convolver. We prove this fact by constructing a non-standard model

$$A \in Alg_{Ext}(\Sigma^{Ver}, E^{Ver}),$$

of the verification theory (based on a non-standard model of time) in which the correctness formula is invalid and then applying the Completeness Theorem 2.6.

We begin by defining the non-standard model A.

5.8 Definition. Define A to be the second order S-typed Σ^{Ver} algebra with

$$A_T = \{0, 1, 2, \ldots, \omega, \omega + 1, \omega + 2, \ldots\},$$

$A_{ring} = \mathbf{Z}$ and $A_{(T\to ring)} = [A_T \to A_{ring}]$. Let $0_A, 1_A, -_A, +_A$ and $\times_A$ be the corresponding constants and operations of the ring $\mathbf{Z}$ of integers and $0_A = 0 \in A_T$, for $0 \in \Sigma^{Ver}_{\lambda,T}$. For each $a : \mathbf{N} \to \mathbf{Z}$ let $\hat{a}_A = a'$ where $a' : A_T \to A_{ring}$ is an extension of a to the non-standard time elements $\omega, \omega+1, \ldots$ (note that the extension chosen is irrelevant). Define

$$succ_A : A_T \to A_T, \quad succ_A(x) = \begin{cases} 0+m+1, & \text{if } x = 0+m, \\ \omega+m+1, & \text{if } x = \omega+m; \end{cases}$$

$$twice_A : A_T \to A_T, \quad twice_A(x) = \begin{cases} 0+m+m, & \text{if } x = 0+m, \\ \omega+m+m, & \text{if } x = \omega+m, \end{cases}$$

for any $m \in \mathbf{N}$. Let $eval_A^{(T\to ring)}$ be evaluation on streams and define $conv_A^n$ by

$$conv_A^n(w_1, \ldots, w_n, a)(t) = (w_n \times a(t+(n-1))) + \cdots + (w_1 \times a(t)).$$

Let $t \in A_T$, $a : A_T \to A_{ring}$ and $\mathbf{x} = x_{1,1}, \ldots, x_{3,n} \in A^{3n}_{ring}$ and define the value functions $V_A^{i,j} : A_T \times A_{(T\to ring)} \times A^{3n}_{ring} \to A_{ring}$ for $i = 1,2,3$ and $j = 1, \ldots, n$ by

$$\begin{aligned} V_A^{i,j}(0, a, \mathbf{x}) &= x_{i,j}, \\ V_A^{1,j}(\omega, a, \mathbf{x}) &= x_{i,j}, \\ V_A^{2,j}(\omega, a, \mathbf{x}) &= x_{i,j}, \\ V_A^{3,j}(\omega, a, \mathbf{x}) &= x_{i,j} + 1, \\ V_A^{3,j}(t+1, a, \mathbf{x}) &= V_A^{3,j}(t, a, \mathbf{x}). \end{aligned}$$

For $j = 2, \ldots, n$ define

$$\begin{aligned} V_A^{1,1}(t+1, a, \mathbf{x}) &= a(t), \\ V_A^{1,j}(t+1, a, \mathbf{x}) &= V_A^{1,j-1}(t, a, \mathbf{x}). \end{aligned}$$

For $j = 1, \ldots, n-1$ define

$$V_A^{2,j}(t+1, a, \mathbf{x}) = (V_A^{3,j}(t, a, \mathbf{x}) \times V_A^{1,j}(t, a, \mathbf{x})) + V_A^{2,j+1}(t, a, \mathbf{x}),$$

and

$$V_A^{2,n}(t+1, a, \mathbf{x}) = V_A^{3,n}(t, a, \mathbf{x}) \times V_A^{1,n}(t, a, \mathbf{x}).$$

Finally, define

$$NET_A : A^{3n}_{ring} \times A_{(T\to ring)} \to A_{(T\to ring)}, \quad NET_A(\mathbf{x}, a)(t) = V_A^{2,1}(t, a, \mathbf{x}),$$

$$\begin{aligned} &IS_A : A_{(T\to ring)} \to A_{(T\to ring)}, \\ &IS_A(a)(twice_A(t)+1) = 0, \\ &IS_A(a)(twice_A(t)) = a(t), \end{aligned}$$

$$OS_A : A_{(T\to ring)} \to A_{(T\to ring)}, \quad OS_A(a)(t) = a(twice_A(t+n)).$$

□

5.9 Proposition.
(i) $A \models Ext$;
(ii) $A \models E^{Ver}$.

Proof.
(i) Trivial since $A_{(T\to ring)} = [A_T \to A_{ring}]$ and $eval_A$ is evaluation on streams.
(ii) Routine exercise. □

To prove our independence result we require the following simple technical fact.

5.10 Lemma. *Let* $t \in A_T$, $a \in A_{(T\to ring)}$ *and* $\mathbf{x} \in A^{3n}_{ring}$, *then for* $j = 1, \ldots, n$

$$V_A^{3,j}(t, s, \mathbf{x}) = \begin{cases} x_{3,j}, & \text{if } t = 0 + m, \\ x_{3,j} + 1, & \text{if } t = \omega + m. \end{cases}$$

Proof. Exercise. □

We can now show that the correctness equation (II) is not valid in the non-standard model A.

5.11 Proposition. *For* $\mathbf{x} \in X_T$ *and* $s \in X_{(T\to ring)}$

$$A \not\models OS(NET(\mathbf{x}, IS(s))) = conv^n(x_{3,n}, \ldots, x_{3,1}, s).$$

Proof. Let $\alpha : X \to A$ be any assignment such that $\alpha_T(t) = \omega$ and

$$\alpha_{(T\to ring)}(s)(\omega), \ldots, \alpha_{(T\to ring)}(s)(\omega + n - 1) > 0.$$

We show that

$$\overline{\alpha}(OS(NET(\mathbf{x}, IS(s)))(t)) \neq \overline{\alpha}(conv^n(x_{3,n}, \ldots, x_{3,1}, s)(t)).$$

Now

$$\overline{\alpha}(OS(NET(\mathbf{x}, IS(s)))(t)) = OS_A(NET_A(\alpha(\mathbf{x}), IS_A(\alpha(s)))(\omega),$$

by definition of term evaluation and α,

$$= NET_A(\alpha(\mathbf{x}), IS_A(\alpha(s)))(twice_A(\omega + n))$$

by definition of OS_A,

$$= NET_A(\alpha(\mathbf{x}), IS_A(\alpha(s)))(succ_A^{2n}(\omega))$$

by definition of $twice_A$,

$$= V_A^{2,1}(succ_A^{2n}(\omega), IS_A(\alpha(s)), \alpha(\mathbf{x}))$$

by definition of NET_A,

$$= (V_A^{3,1}(succ_A^{2n-1}(\omega), IS_A(\alpha(s)), \alpha(\mathbf{x})) \times V_A^{1,1}(succ_A^{2n-1}(\omega), IS_A(\alpha(s)), \alpha(\mathbf{x})))$$
$$+ \cdots + (V_A^{3,n-1}(succ_A^{2n-(n-1)}(\omega), IS_A(\alpha(s)), \alpha(\mathbf{x})) \times V_A^{1,n-1}(succ_A^{2n-(n-1)}(\omega),$$
$$IS_A(\alpha(s)), \alpha(\mathbf{x}))) + V_A^{2,n}(succ_A^{2n-(n-1)}(\omega), IS_A(\alpha(s)), \alpha(\mathbf{x}))$$

by Lemma 5.2.(iii) (which holds for the new E^{Ver}) and Proposition 5.9,

$$= (V_A^{3,1}(succ_A^{2n-1}(\omega), IS_A(\alpha(s)), \alpha(\mathbf{x})) \times V_A^{1,1}(succ_A^{2n-1}(\omega), IS_A(\alpha(s)), \alpha(\mathbf{x})))$$
$$+ \cdots + (V_A^{3,n}(succ_A^{2n-n}(\omega), IS_A(\alpha(s)), \alpha(\mathbf{x}))$$
$$\times V_A^{1,n}(succ_A^{2n-n}(\omega), IS_A(\alpha(s)), \alpha(\mathbf{x})))$$

by definition of $V_A^{2,n}$,

$$= ((\alpha(x_{3,1}) + 1) \times IS_A(\alpha(s))(succ_A^{2n-2}(\omega)) + \cdots +$$
$$((\alpha(x_{3,n}) + 1) \times IS_A(\alpha(s))(\omega))$$

by definition of $V_A^{3,1}, \ldots, V_A^{3,n}$ and $V_A^{1,1}, \ldots, V_A^{1,n}$,

$$= ((\alpha(x_{3,1}) + 1) \times \alpha(s)(\omega + n - 1)) + \cdots + ((\alpha(x_{3,n}) + 1) \times \alpha(s)(\omega))$$

by definition of IS_A,

$$> (\alpha(x_{3,1}) \times \alpha(s)(\omega + n - 1)) + \cdots + (\alpha(x_{3,n}) \times \alpha(s)(\omega))$$

by choice of the assignment α,

$$= \overline{\alpha}(conv^n(x_{3,n}, \ldots, x_{3,1}, s)(t))$$

by definition of $conv_A^n$. Therefore

$$A \not\models OS(NET(\mathbf{x}, IS(s)))(t) = conv^n(x_{3,n}, \ldots, x_{3,1}, s)(t),$$

and so

$$A \not\models OS(NET(\mathbf{x}, IS(s))) = conv^n(x_{3,n}, \ldots, x_{3,1}, s).$$

□

We conclude with the following independence result.

5.12 Independence Theorem.

$$E^{Ver} \not\vdash OS(NET(\mathbf{x}, IS(s))) = conv^n(x_{3,n}, \ldots, x_{3,1}, s),$$

for $\mathbf{x} = x_{1,1}, \ldots, x_{3,n} \in X_{ring}^{3n}$ *and* $s \in X_{(T \to ring)}$.

Proof. Immediate from Propositions 5.9 , 5.11 and Completeness Theorem 2.6 for higher order equational logic. □

6 Conclusions.

We have presented a case study of higher order algebraic methods applied to the specification and verification of two systolic algorithms for convolution. This case study has emphasised the relationship between proof theoretic and model theoretic methods since we see *both* these methods as indispensable for a sound understanding of the specification and verification process. Much further work is necessary in the area of case studies using higher order algebra. For example, the use of an uncountable signature to specify an uncountable algebra of streams in Definition 3.6 can be avoided by the use of topological methods described in Meinke [1993]. This approach will be further developed in Steggles [1994]. A wide variety of type constructions can be modelled algebraically (see Meinke [1992c]), but the role of these various type constructions in algebraic design methodology has yet to be studied. Furthermore, in making higher order algebra into a practical formal method the problems of specification in the large and parameterisation methods need to be considered. Other approaches to types based on algebra and equations are Mosses [1989] and Manca et al [1990].

Our work also contributes towards the specification and verification theory of SCAs and clocked digital hardware. Several other approaches to hardware design and verification exist based on higher order logic, e.g. Gordon [1986], Cohn and Gordon [1990] and Hanna and Daeche [1986], and equational logic, e.g. Stavridou [1993]. Our own approach, based on using the weakest possible fragment of higher order logic, namely higher order equational logic, is new. Not only does this fragment possess a simple syntax and proof theory (which will be described in Meinke [1994b]), as we have already observed in Section 2, it also possesses an elegant model theory.

As we have shown in this paper, pure higher order equational logic is not always adequate for verification proofs. Further research into induction principles needed for verification is important and techniques from non-standard models of arithmetic (see for example Mc Aloon [1980] and Kaye [1991]) seem likely to be useful. Work on automating verification proofs in higher order equational logic by means of *higher order term rewriting* is already under way and will be reported elsewhere.

Further results on the general theory of SCAs may be found in Thompson [1987] and Thompson and Tucker [1991]. Applications of this theory to clocked digital hardware may be found in McConnell and Tucker[1992a] and Harman and Tucker[1988a, 1988b] and to systolic algorithms in Hobley et al [1988], Eker and Tucker[1988,1989], Martin and Tucker [1987] and McConnell and Tucker[1992b].

We are grateful for the helpful comments and advice of B. M. Hearn, J. R. Hindley, B. McConnell and J. V. Tucker during the preparation of this paper. We also acknowledge the financial support of the British Council, the Deutsche Akademischer Austauschdienst and the Science and Engineering Research Council.

7 References.

Bergstra and Tucker [1987], J.A. Bergstra and J.V. Tucker. Algebraic specifications of computable and semicomputable data types, *Theoretical Computer Science*, 50:137–181, 1987.

Broy [1987], M. Broy. Equational specification of partial higher–order algebras. In: M. Broy (ed) *Logic of programming and calculi of discrete design*. Springer-Verlag, Berlin, 1987.

Cohn [1965], P.M. Cohn. *Universal Algebra*. Harper and Row, New York, 1965.

Cohn [1982], P.M. Cohn. *Algebra* , Volume 1. John Wiley, Chichester, 1982, second edition.

Cohn and Gordon [1990], A. Cohn and M. Gordon. A mechanised proof of correctness of a simple counter. In: K. McEvoy and J. V. Tucker (eds), *Theoretical Foundations of VLSI Design*, pages 65–96. Cambridge Tracts in Theoretical Computer Science 10, Cambridge University Press, 1990.

Ehrich [1992], H.D. Ehrich. Algebraic data types. To appear in: S. Abramsky, D. Gabbay and T.S.E. Maibaum (eds), *Handbook of Logic in Computer Science*, Volume V. Oxford University Press, Oxford, 1992.

Ehrig and Mahr [1985], H. Ehrig and B. Mahr. *Fundamentals of Algebraic Specification 1 – Equations and Initial Semantics*. EATCS Monographs on Theoretical Computer Science 6. Springer-Verlag, Berlin, 1985.

Eker and Tucker [1988], S. M. Eker and J. V. Tucker. Specification, derivation and verification of concurrent line drawing algorithms and architectures. In: R. A. Earnshaw (ed), *Theoretical foundations of computer graphics and CAD*, pages 449–516. Springer-Verlag, Berlin, 1988.

Eker and Tucker [1989], S. M. Eker and J. V. Tucker. Specification and verification of synchronous concurrent algorithms : a case study of the pixel planes architecture. In: P. M. Drew, R. A. Earnshaw and T. R. Heywood (eds), *Parallel processing for computer vision and display*, pages 16–49. Addison Wesley, 1989.

Goguen and Meseguer [1982], J.A. Goguen and J. Meseguer. Completeness of many-sorted equational logic. *Association for Computing Machinery SIGPLAN Notices*, 17:9–17, 1982.

Goguen and Meseguer [1985], J.A. Goguen and J. Meseguer. Initiality, induction and computability. In: M. Nivat and J.C. Reynolds (eds), *Algebraic Methods in Semantics* , pages 459–541. Cambridge University Press, 1985.

Gordon [1986], M. Gordon. Why higher order logic is a good formalism for specifying and verifying hardware. In: G. J. Milne and P. A. Subrahmanyan (eds), *Formal Aspects of VLSI*, pages 1–177. North-Holland, Amsterdam, 1986.

Hanna and Daeche [1986], F. K. Hanna and N. Daeche. Specification and Verification using Higher-Order Logic : A Case Study. In: G. J. Milne and P. A. Subrahmanyan (eds), *Formal Aspects of VLSI*, pages 179–213. North-Holland, Amsterdam, 1986.

Harman and Tucker [1988a], N. A. Harman and J. V. Tucker. Formal specifications and the design of verifiable computers. In: *Proceedings of 1988 UK IT Conference, held under the auspices of the Information Engineering Directorate of the Department of Trade and Industry, Institute of Electrical Engineers*, pages 500–503, 1988.

Harman and Tucker [1988b], N. A. Harman and J. V. Tucker. Clocks, retimings and the formal specification of a UART. In G. Milne (ed), *The fusion of hardware design and verification*, pages 375–396. Proceedings of IFIP Working Group 10.2 Working Conference. North-Holland, Amsterdam, 1988.

Hobley et al [1988], K. M. Hobley, B. C. Thompson and J. V. Tucker, Specification and verification of synchronous concurrent algorithms : a case study of a convolution algorithm. In G. Milne (ed), *The fusion of hardware design and verification*, pages 347–374. Proceedings of IFIP Working Group 10.2 Working Conference. North-Holland, Amsterdam, 1988.

Kaye [1991], R. Kaye. *Models of Peano Arithmetic.* Oxford Logic Guides 15, Oxford University Press, 1991.

Kung [1982], H. T. Kung. Why systolic architectures?. *Computer*, pages 37–46, January 1982.

Maibaum and Lucena [1980], T. S. E. Maibaum and C. J. Lucena. Higher-order data types. *International Journal of Computer and Information Sciences*, 9:31–53, 1980.

Manca et al [1990], V. Manca, A. Salibra and G. Scollo. Equational Type Logic, *Theoretical Computer Science*, 77:131–159, 1990.

Martin and Tucker [1987], A. R. Martin and J. V. Tucker. The concurrent assignment representation of synchronous systems. In: J. W. de Bakker, A. J. Nijman and P. C. Treleaven (eds), *PARLE : Parallel Architectures and Languages Europe, Vol II Parallel languages*, Lecture Notes in Computer Science 259, pages 369–386. Springer-Verlag, Berlin, 1987.
A revised and expanded edition appears in *Parallel Computing* 9 (1988/89), 227-256.

Mc Aloon [1980], K. Mc Aloon (ed). *Modèles de l'Arithmétique*, Astérisque 73, Société Mathémat-ique de France, 1980.

McConnell and Tucker [1992a], B. McConnell and J. V. Tucker. Infinite synchronous concurrent algorithms : The algebraic specification and verification of a hardware stack. In H. Schwichtenberg (ed), *Logic and algebra of specification.* Springer-Verlag, 1992.

McConnell and Tucker [1992b], B. McConnell and J. V. Tucker. *Direct limits of algebras and the parameterisation of synchronous concurrent algorithms.* Report CSR, Department of Computer Science, University College of Swansea, 1992. In preparation.

Meinke [1992a], K. Meinke. Universal algebra in higher types. *Theoretical Computer Science*, 100:385–417, 1992.

Meinke [1992b], K. Meinke. Subdirect representation of higher type algebras. In: K. Meinke and J.V. Tucker (eds), *Many-sorted Logic and its Applications*, pages 135–146. John Wiley, 1992.

Meinke [1992c], K. Meinke. Equational Specification of Abstract Types and Combinators. In: E. Börger et al (eds), *Proc. Computer Science Logic'91*, Lecture Notes in Computer Science 626, pages 257–271. Springer-Verlag, Berlin, 1992.

Meinke [1993] K. Meinke. Topological Methods for Algebraic Specification. Report TUM-I9331, TUM Institut für Informatik, 1993.

Meinke [1994a], K. Meinke. A recursive second order initial algebra specification of primitive recursion. *Acta Informatica*, 1994 (in press).

Meinke [1994b] K. Meinke. *Proof theory of higher order equational logic: normal forms, continuity and term rewriting.* Technical Report, Department of Computer Science, University College of Swansea, 1994. (to appear)

Meinke and Tucker [1992], K. Meinke and J.V. Tucker. Universal algebra. In: S. Abramsky, D. Gabbay and T.S.E. Maibaum, (eds) *Handbook of Logic in Computer Science*, Volume I, pages 189–412. Oxford University Press, Oxford, 1992.

Möller [1987a], B. Möller. Algebraic specifications with higher–order operators. In: L.G.L.T. Meertens (ed), *Program specification and transformation.* North Holland, Amsterdam, 1987.

Möller [1987b], B. Möller. Higher–order algebraic specifications. Fakultät für

Mathematik und Informatik, Technische Universität München, Habilitationsschrift, 1987.

Möller [1994], B. Möller. Ordered and Continuous Models of Higher–Order Algebraic Specifications. This conference proceedings, 1994.

Möller et al [1988], B. Möller, A. Tarlecki and M. Wirsing. Algebraic specifications of reachable higher–order algebras. In: D. Sannella and A. Tarlecki (eds), *Recent Trends in Data Type Specification*, Lecture Notes in Computer Science 332, pages 154–169, Springer-Verlag, Berlin, 1988.

Mosses [1989], P. D. Mosses. Unified algebras and institutions. In: *Proceedings of 4th IEEE Symp. on Logic in Computer Science*, Pacific Grove, CA, 1989.

Poigné [1986], A. Poigné. On specifications, theories and models with higher types. *Information and Control*, 68:1–46, 1986.

Stavridou [1993], V. Stavridou. *Formal methods in digital design.* Cambridge University Press, 1993.

Steggles [1994] L. J. Steggles. Higher order algebraic specification: fundamental theory and case studies. Ph. D. Thesis, Department of Computer Science, University College of Swansea, 1994. (in preparation)

Thompson [1987], B. C. Thompson. *Mathematical theory of synchronous concurrent algorithms.* Ph. D. Thesis, School of Computer Studies, University of Leeds, 1987.

Thompson and Tucker [1991], B. C. Thompson and J. V. Tucker. *Equational specification of synchronous concurrent algorithms and architectures.* Report CSR 9–91, Department of Computer Science, University College of Swansea, 1991.

Wechler [1992], W. Wechler. *Universal Algebra for Computer Scientists.* EATCS Monographs on Theoretical Computer Science 25, Springer-Verlag, Berlin, 1992.

Wirsing [1991], M. Wirsing. Algebraic specification. In J. van Leeuwen (ed) *Handbook of Theoretical Computer Science.* North Holland, Amsterdam, 1991.

Ordered and Continuous Models of Higher-Order Specifications

Bernhard Möller

Institut für Mathematik, Universität Augsburg, D-86135 Augsburg, Germany,
e-mail: moeller@uni-augsburg.de

Abstract. We investigate the existence of continuous and fixpoint models of higher-order specifications. Particular attention is paid to the question of extensionality. We use ordered specifications, a particular case of Horn specifications. The main tool for obtaining continuous models is the ideal completion. Unfortunately, it may destroy extensionality. This problem is inherent: we show that there is no completion method which is guaranteed to preserve extensionality. To restore it, generally a quotient has to be taken. It is shown that under certain conditions this preserves the existence of least fixpoints. Examples of the specification method include the essential concepts of Backus's FP and Hoare's CSP.

1 Introduction

During the last years a number of papers have dealt with the extension of first-order algebraic specifications to higher-order ones (cf. [21, 36, 13, 38, 39, 8, 29, 30, 31, 32, 45, 24, 25]). The basic questions about existence of models are answered by now. In this paper we want to treat one particular approach to the question about existence of models in which recursion equations are solvable. This is especially of interest in connection with the algebraic specification of infinite objects (cf. [28]), for which, in turn, non-strict operations are a crucial concept. Based on the ideas of [40], it has turned out that ordered and continuous algebras (cf. [34, 10, 15]) are a convenient framework for modelling these notions. Therefore we consider higher-order specifications that admit continuous models. This idea is also pursued in [13], but not in [38, 39]. The present treatment is based on [29, 30] where (extensional) higher-order algebras with partially ordered carrier sets are considered. The paper is a reworking of some chapters of [30] based on the first-order reduction in [32] and the model theory presented in [25]. For reasons of space we only include proofs of the most essential theorems; the remaining proofs can be found in [28, 30] and the other references.

2 First-Order Horn Specifications and Their Models

2.1 First-Order Languages, Structures and Terms

In the presentation we follow closely the papers [32] and [25] and use the reduction of higher-order notions to the first-order case. This allows a direct re-use of the results of [28] on ordered and continuous algebras. To accommodate the order more conveniently we use the framework of Horn specifications (cf. [22, 20, 35]).

A **(first-order) language** $L = (S, F, P, \phi, \psi)$ is a quintuple consisting of a set S of **sorts**, a set F of **operators**, a set P of **predicate symbols** and **arity functions** $\phi : F \to S^+$ and $\psi : P \to S^*$ such that for every $s \in S$ there is an equality predicate symbol $=_s \in P$ with $\psi(=_s) = ss$. An operator $f \in F$ with arity $\phi(f) = s \in S$ is called a **constant**. L is called **algebraic** if P contains no other symbols than the $=_s$ $(s \in S)$. An algebraic language corresponds to a signature.

In the sequel we shall frequently use families of sets, functions, relations etc. Rather than saying that E is a family $(E_i)_{i \in I}$ of entities E_i, we call E an **I-indexed** entity. For instance, and I-indexed set X is a family $(X_i)_{i \in I}$. All set-theoretic notions are extended componentwise to I-indexed entities; for instance, requiring X to be nonempty means requiring $X_i \neq \emptyset$ for all $i \in I$.

An **L-structure** (see e.g. [14]) A consists of the following items:

1. an S-indexed non-empty set, also denoted by A, called the **carrier** of A;
2. for each operator $f \in F$ of arity $\phi(f) = ws$ with $w \in S^*$ and $s \in S$ an **operation** $f^A : A_w \to A_s$, where for $w = s_1 \cdots s_n$ we set $A_w \stackrel{\text{def}}{=} A_{s_1} \times \cdots \times A_{s_n}$;
3. for each predicate symbol $p \in P$ a relation $p^A \subseteq A_{\psi(p)}$.

We call A **normal** if $=^A$ is the S-indexed identity relation on A. The interpretation of a constant f is a function $f^A : \{()\} \to A_s$ from the set consisting only of the empty tuple $()$; as usual it will be identified with its result value on $()$. The class of all normal L-structures is denoted by STR(L).

For every language L one can construct a **trivial** L-structure $\mathbb{1}_L$ by choosing as the carrier an S-indexed singleton set and assigning to all predicate symbols except $=$ the singleton relation.

Given a language $L = (S, F, P, \phi, \psi)$ and an S-indexed set X of **variables** disjoint from L, the S-indexed set $WL(X)$ of L-terms is the smallest S-indexed set satisfying the following inductive properties:

1. $X \subseteq WL(X)$.
2. For $f \in F$ with $\phi(f) = s_1 \cdots s_n s$ and $t_i \in WL(X)_{s_i}$ we have $f(t_1, \ldots, t_n) \in WL_s(X)$.

$WL(X)$ is made into a normal L-structure by defining the operations in the standard way, interpreting the equality predicate symbols by the respective identity relations and setting $p^{WL(X)} \stackrel{\text{def}}{=} \emptyset$ for all other $p \in P$.

For $X = \emptyset$ we call $WL \stackrel{\text{def}}{=} WL(\emptyset)$ the **Herbrand structure** over L. A sort s is **populated** in L if $WL_s \neq \emptyset$. For an algebraic language the Herbrand structure corresponds to the ground term algebra.

A **valuation** of X in an L-structure A is an S-indexed function $v : X \to A$ that assigns to every variable an element of the respective carrier of A. We denote the set of all valuations of X in A by A^X. A valuation of X in WL is called a **ground term valuation.**

Every valuation $v \in A^X$ is extended into an S-indexed function $\overline{v} : WL(X) \rightarrow A$ by

1. $\overline{v}_s(x) \stackrel{\text{def}}{=} v_s(x)$ for $x \in X_s$.
2. $\overline{v}_s(f(t_1, \ldots, t_n)) \stackrel{\text{def}}{=} f^A(\overline{v}_{s_1}(t_1), \ldots, \overline{v}_{s_n}(t_n))$ for $\phi(f) = s_1 \cdots s_n s$ and $t_i \in WL(X)_{s_i}$.

For a ground term t the value $\overline{v}(t)$ is independent of $v \in A^X$; we denote it by t^A.

2.2 Homomorphisms and Term-Generated Structures

An L-**homomorphism** $h : A \rightarrow B$ between two L-structures A and B consists of an S-indexed function $h : A \rightarrow B$ that satisfies $h_s(f^A(x_1, \ldots, x_n)) = f^B(h_{s_1}(x_1), \ldots, h_{s_n}(x_n))$ for all $f \in F$ with $\phi(f) = s_1 \cdots s_n s$ and $x_i \in A_{s_i}$ and $h(p^A) \subseteq p^B$ for all predicate symbols $p \in P$. An L-**epimorphism** is a surjective L-homomorphism, whereas an L-**isomorphism** is a bijective L-homomorphism. In working with L-homomorphisms we omit L when it is clear from the context.

We write $A \preceq B$ if there is a homomorphism from A to B. $\preceq$ is a pre-order on $\mathrm{STR}(L)$. Likewise, we write $A \cong B$ if there is an isomorphism between A and B.

The interaction between homomorphisms and valuations is given by the freeness of $WL(X)$ over X (see e.g. [14]), viz. by

Lemma 1. *Consider $A, B \in \mathrm{STR}(L)$ and a valuation $v \in A^X$.*

1. *$\overline{v} : WL(X) \rightarrow A$ is an L-homomorphism.*
2. *If $h : A \rightarrow B$ is an L-homomorphism then $h \circ \overline{v} = \overline{h \circ v}$.*

Let $K \subseteq \mathrm{STR}(L)$ be a class of L-structures. A structure $I \in K$ is **initial** in K if for all $A \in K$ there is exactly one homomorphism from I to A. $Z \in K$ is **terminal** in K if for all $A \in K$ there is exactly one homomorphism from A to Z. For every $A \in \mathrm{STR}(L)$ the functions $\iota^A_s : u \mapsto u^A$ define a unique L-homomorphism $\iota^A : WL \rightarrow A$, so that WL is initial in $\mathrm{STR}(L)$.

If ι^A is surjective then A is called **term-generated** (cf. e.g. [3, 4]), **reachable** (cf. e.g. [32]) or **minimal** (cf. e.g. [25]). This means that every element of the carrier of A can be denoted by an L-ground-term or, in other words, that it can be obtained by applying finitely many operations of the structure. By $\mathrm{GEN}(L)$ we denote the class of all term-generated L-structures. Since WL itself is term-generated, it is also initial in $\mathrm{GEN}(L)$.

2.3 Horn Specifications and Their Models

Consider a language $L = (S, F, P, \phi, \psi)$ and an S-indexed set X of variables. First-order formulas over L and X and the validity of a formula Φ in an L-structure A, denoted by $A \models \Phi$, are defined as usual. For binary relation symbols infix notation will be used.

A **(positive) literal** over L is a formula of the form $p(t_1, \ldots, t_m)$ where $p \in P$ with $\psi(p) = s_1 \cdots s_m$ and $t_i \in WL(X)_{s_i}$ $(1 \le i \le n)$. If $t_i \in WL_{s_i}$ for all i then $p(t_1, \ldots, t_m)$ is called a **ground** literal. A **Horn formula** has the form

$$\bigwedge_{i \in I} Q_i \Rightarrow Q ,$$

where I is a finite or infinite index set and the Q_i and Q are literals over L.

The **equivalence axioms** over L are the Horn formulas

$$x_s =_s x_s ,$$
$$x_s =_s y_s \wedge y_s =_s z_s \Rightarrow x_s =_s z_s ,$$
$$x_s =_s y_s \Rightarrow y_s =_s x_s$$

for variables $x_s, y_s, z_s \in X_s$. The **substitutivity axioms** over L are the Horn formulas

$$\bigwedge_{i=1}^{n} x_i =_{s_i} y_i \Rightarrow f(x_1, \ldots, x_m) =_s f(y_1, \ldots, y_m)$$

for all $f \in F$ with $\phi(f) = s_1 \cdots s_m s$ and variables $x_{s_i}, y_{s_i} \in X_{s_i}$, and

$$\bigwedge_{i=1}^{n} x_i =_{s_i} y_i \wedge p(x_1, \ldots, x_n) \Rightarrow p\langle y_1, \ldots, y_n\rangle$$

for all $p \in P$ with $\psi(p) = s_1 \cdots s_n$ and variables $x_{s_i}, y_{s_i} \in X_{s_i}$. The **congruence axioms** over L are the equivalence and substitutivity axioms over L.

A **Horn specification** $K = (L, E)$ consists of a language L and a set E of Horn formulas over L, called the **axioms** of K, which includes the congruence axioms for L. **Equational specifications** are Horn specifications over algebraic languages.

An L-structure A is called a **model** of K if all axioms in E are valid in A and A is normal. We denote by $\mathrm{MOD}(L)$ the class of all models of L, and by $\mathrm{GEN}(L)$ the class of term-generated models of L.

2.4 Quotients

Models for equational specifications can be constructed as quotients of the term algebra by suitable congruences. We generalise this notion to arbitrary structures.

Consider an L-structure A. A family $R = (p^R)_{p \in P}$ with $p^A \subseteq p^R \subseteq A_{\psi(p)}$ is called a **pre-congruence** on A. For a positive literal $p(t_1, \ldots, t_n)$, valuation $v \in A^X$ and pre-congruence R on A we define

$$v, R \models p(t_1, \ldots, t_n) \overset{\text{def}}{\Leftrightarrow} (\overline{v}(t_1), \ldots, \overline{v}(t_n)) \in p^R .$$

This relation is extended inductively to general first-order formulas as usual. Then R **satisfies** a first-order formula Φ iff $v, R \models \Phi$ for all valuations $v \in A^X$.

By $\mathbb{U}^A$ we denote the universal pre-congruence on an L-structure A with $p^{\mathbb{U}_A} \stackrel{\text{def}}{=} A_{\psi(p)}$ for all $p \in P$. We say that a set of pre-congruences on A is **lattice-forming** if it is closed under intersection and contains $\mathbb{U}^A$. Such a set forms a complete lattice under relation inclusion where the greatest lower bound of a set G of pre-conguences on A is their intersection $\cap G$. Fundamental for the theory of Horn specifications is

Theorem 2. *Consider an L-structure A and a set E of Horn formulas over L. The set of pre-congruences on A that satisfy E is lattice-forming.*

Taking $E = \emptyset$ we obtain that the set of pre-congruences on A is lattice-forming. Its least element is R^A with $p^{R^A} = p^A$ for all $p \in P$.

A pre-congruence R is called a **congruence** if it satisfies the congruence axioms. The set of pre-congruences on A is denoted by $\mathrm{CG}(A)$. By Theorem 2 it is lattice-forming. The **quotient** A/R of A by $R \in \mathrm{CG}(A)$ is defined by setting

1. $(A/R)_s \stackrel{\text{def}}{=} \{[x] : x \in A_s\}$ where $[x]$ is the equivalence class of x w.r.t. $=^R_s$;
2. $f^{A/R}([a_1], \ldots, [a_n]) \stackrel{\text{def}}{=} [f^A(a_1, \ldots, a_n)]$;
3. $([a_1], \ldots, [a_n]) \in p^{A/R} \stackrel{\text{def}}{\Leftrightarrow} (a_1, \ldots, a_n) \in p^R$.

The congruence axioms ensure well-definedness of this construction. It is clear that $\mathbb{U}^A$ is a congruence and that $A/\mathbb{U}^A \cong \mathbb{1}_H$. Moreover, we have

Lemma 3. *A Horn formula Q is valid in the quotient A/R iff R satisfies Q.*

For L-structures A and B and homomorphism $h : A \to B$ we define the **kernel** R_h on A by $(a_1, \ldots, a_n) \in p^{R_h} \stackrel{\text{def}}{\Leftrightarrow} (h(a_1), \ldots, h(a_n)) \in p^B$. Then R_h is a congruence on A.

Theorem 4. *Consider $A, B \in \mathrm{STR}(L)$.*

1. *If $h : A \to B$ is an epimorphism then $A/R_h \cong B$.*
2. *For $R_1, R_2 \in \mathrm{CG}(A)$ the inclusion $R_1 \subseteq R_2$ implies $A/R_1 \preceq A/R_2$.*
3. *$A \in \mathrm{GEN}(L)$ iff $A \cong WL/R$ for some $R \in \mathrm{CG}(WL)$. Moreover, then $R = R_{\iota A}$.*
4. *If $A, B \in \mathrm{GEN}(L)$ we have $A \preceq B$ iff $R_{\iota A} \subseteq R_{\iota B}$. Thus, $\mathrm{GEN}(L)/\preceq$ is order-isomorphic to $(\mathrm{CG}(WL), \subseteq)$ and hence a complete lattice.*
5. *There is at most one homomorphism from one term-generated structure to another. If there is one, it is an epimorphism.*
6. *If $A, B \in \mathrm{GEN}(L)$ then $A \cong B$ iff $A \preceq B$ and $B \preceq A$.*
7. *$\mathrm{GEN}(L)$ is closed under quotients.*

We can use quotients to construct models:

Theorem 5. *Consider a Horn specification $K = (L, E)$.*

1. *K has an initial model. It is the quotient of WL by the least congruence on WL that satisfies E.*

2. The set of isomorphism classes of term-generated models of L forms a complete lattice w.r.t. $\preceq$.

A special case is the well-known result that the isomorphism classes of term-generated models of an equational specification form a complete lattice under the homomorphism ordering.

2.5 A Deductive Calculus

Consider now a Horn specification $K = (L, E)$ with $L = (S, F, P, \phi, \psi)$. For a term valuation $v \in WL(X)^X$ we extend $\overline{v}$ to Horn formulas by setting for literals $\overline{v}(p(t_1, \ldots, t_n)) \stackrel{\text{def}}{=} p(\overline{v}(t_1), \ldots, \overline{v}(t_n))$ and then $\overline{v}(\bigwedge_{i \in I} Q_i \Rightarrow Q) \stackrel{\text{def}}{=} \bigwedge_{i \in I} \overline{v}(Q_i) \Rightarrow \overline{v}(Q)$. Then the following inference rules are sound in MOD(L):

$$\text{(Axiom)} \quad \frac{}{\Phi} \qquad \text{for } \Phi \in E$$

$$\text{(Instantiation)} \quad \frac{\Phi}{\overline{v}(\Phi)} \qquad \text{for } \Phi \in E \text{ and } v \in WL(X)^X$$

$$\text{(ModusPonens)} \quad \frac{\bigwedge_{j \in J} Q_j \Rightarrow Q \qquad Q_i \ (i \in I)}{\bigwedge_{j \in J \setminus I} Q_j \Rightarrow Q}$$

These rules form a complete calculus:

Theorem 6. *Let* $K = (L, E)$ *be a Horn specification,* I *an initial structure in* GEN(K) *and* Q *a literal over* L. *Then the following equivalences hold:*

1. $L \vdash Q$ *iff* MOD(L) $\models Q$.
2. *If* Q *is a ground literal then* $I \models Q$ *iff* $L \vdash Q$ *iff* GEN(K) $\models Q$.

This theorem shows also the distinction between term-generated models and arbitrary ones: In GEN(K) usually more formulas are valid than in MOD(K), since the principle of term-induction (see below) is available. Thus not all formulas valid in GEN(K) need be provable using only the above calculus. However, for ground terms provability and validity in GEN(K) coincide; moreover, they are equivalent to the validity in initial models.

To conclude this section, we now state the above-mentioned principle of **term induction**.

Theorem 7 (Term Induction). *Let* L *be a language and* Φ *be a first-order formula over* L *containing exactly one free variable* $x_s \in X_s$ *for some* $s \in S$. *Then* Φ *is valid in a class* $K \subseteq$ GEN(L) *of term-generated structures for* L *iff the following property holds: For all* $f \in F$ *with* $\phi(f) = s_1 \cdots s_n s$ *the assumptions* $K \models \Phi[x_j/x_s]$ $(j \in J)$ *imply* $K \models \Phi[u(x_1, \ldots, x_n)/x_s]$, *where* $x_i \in X_{s_i}$ $(1 \leq i \leq n)$ *are variables not occurring in* Φ *and* $J \stackrel{\text{def}}{=} \{i : s_i = s\}$.

3 The Ordered Case

We turn now to the special case where the only predicate symbols besides the equality symbols are order relation symbols.

3.1 Pre-Orders

A **pre-order** is a reflexive and transitive binary relation $\leq$ on some set M; an **order** is an antisymmetric pre-order. E.g. in [5] one finds

Lemma 8. *For a pre-ordered set $(M, \leq)$ the relation $\sim_{\leq}$ defined by*

$$x \sim_{\leq} y \overset{\text{def}}{\Leftrightarrow} x \leq y \wedge y \leq x$$

is an equivalence relation on M.

We denote by $[x]_{\leq}$ the equivalence class of x w.r.t. $\sim_{\leq}$ and by $[M]_{\leq}$ the set $\{[x]_{\leq} : x \in M\}$. If no ambiguity arises, the index $_{\leq}$ will be dropped. For a pre-ordered set $(M, \leq)$ the set $[M]$ together with the relation

$$[x] \leq [y] \overset{\text{def}}{\Leftrightarrow} x \leq y$$

is an ordered set called the **quotient** of $(M, \leq)$ by $\sim_{\leq}$; we denote it by $M/\leq$.

3.2 Ordered Specifications and Structures

A language $L = (S, F, P, \phi, \psi)$ is called **ordered** if $P = \{=_s, \leq_s : s \in S\}$ with $\psi(=_s) = \psi(\leq_s) = ss$. If L is ordered then all L-homomorphisms between L-structures are monotonic.

A Horn specification $K = (L, E)$ is an **ordered specification** if L is ordered and E contains the **order axioms**

$$x_s \leq_s x_s \ ,$$
$$x_s \leq_s y_s \wedge y_s \leq_s z_s \Rightarrow x_s \leq_s z_s \ ,$$
$$x_s \leq_s y_s \wedge y_s \leq_s x_s \Rightarrow x_s =_s y_s$$

for all $s \in S$ and variables $x_s, y_s, z_s \in X_s$. The congruence and order axioms are called the **standard axioms** of K, whereas the other axioms are **non-standard axioms.** In the sequel we shall omit the indices of the equality and order predicate symbols.

Note that by the substitutivity axioms the antisymmetry axiom can be strengthened to

$$x \leq y \wedge y \leq x \Leftrightarrow x = y \ .$$

So in any L-structure A satisfying the order axioms $\leq^A$ is a pre-order and $=^A$ coincides with $\sim_{\leq^A}$ as defined in Lemma 8. If L is normal then $\leq^A$ is an order. Therefore we define: an **ordered L-structure** is a normal L-structure satisfying the order axioms. Hence, again by Lemma 8 any quotient by a congruence that satisfies the order axioms is an ordered L-structure.

As a simple example for an ordered specification consider the one having as non-standard axioms only the formulas

$$x \leq y \Rightarrow y \leq x$$

for all sorts. The models of this specification are exactly the trivially ordered structures. Thus we retrieve here in a second way the above-mentioned result on isomorphism classes of models of equational specifications.

3.3 Groundedness and Strictness

In connection with programming language semantics frequently "undefined situations" are modelled by special $\bot$-elements in the respective domains; these elements are considered as carrying no information, i.e., as least elements in the information contents ordering (cf. [40]).

We call an ordered set **grounded** if it contains a least element. A language $L = (S, F, P, \phi, \psi)$ is **grounded** if for all $s \in S$ there are distinguished constants $\bot_s$ with $\phi(\bot_s) \stackrel{\text{def}}{=} s$. If L is grounded then all L-homomorphisms are **strict**: Since each $\bot_s$ is a special constant, we have for $A, B \in \text{STR}(L)$ and $h : A \to B$ that $h_s(\bot_s^A) = \bot_s^B$.

We call a specification $K = (L, E)$ **grounded** if L is grounded and E contains the **groundedness axioms**

$$\bot_s \leq_s x_s$$

for all $s \in S$ and variables $x_s \in X_s$. To abbreviate our specification texts in examples, we use the keyword **grounded**; the operators $\bot_s$ as well as the groundedness axioms are then omitted from the specification text.

For many operations the concept of strictness in a certain argument is very important: Operationally speaking, this argument is crucial for computing the operation; as long as no information about this argument is available, the operation will be undefined. Given a grounded language L and an operator $f \in F$ with $\phi(f) = s_1 \cdots s_n s$ we can specify strictness in the kth argument by the axiom

$$f(x_{s_1}, \ldots, x_{s_{k-1}}, \bot_{s_k}, x_{s_{k+1}}, \ldots, x_{s_n}) \leq \bot_s.$$

3.4 Monotonicity

A second class of specifications is concerned with monotonic structures.

We call a specification $K = (L, E)$ **monotonic** if $L = (S, F, P, \phi, \psi)$ is ordered and E includes the **monotonicity axioms**

$$\bigwedge_{i=1}^{n} x_i \leq y_i \Rightarrow f(x_1, \ldots, x_n) \leq f(y_1, \ldots, y_n).$$

for all $f \in F$ with with $\phi(f) = s_1 \cdots s_m s$ and variables $x_{s_i}, y_{s_i} \in X_{s_i}$. An L-structure is **monotonic** if it satisfies the monotonicity axioms. The **standard**

axioms for a monotonic specification are the congruence, order and monotonicity axioms. We use the keyword monotonic to mean that the respective specification implicitly contains the monotonicity axioms.

A congruence R on an L-structure A is **monotonic** if it satisfies the monotonicity axioms. From Lemma 3 we obtain

Corollary 9. *A/R is monotonic iff R is monotonic.*

The specialisation of Theorem 5 for monotonic specifications was obtained in [26, 28].

A special case of ordered specifications are **inequational** specifications, i.e., monotonic specifications in which the non-standard axioms are only of the form $u \leq w$ with $u, w \in L(X)$. For these we have the stronger result (see also [6])

Theorem 10. *Let $K = (L, E)$ be an inequational specification.*

1. *$\mathrm{MOD}(L)$ is closed under epimorphisms and hence under quotients.*
2. *$\mathrm{GEN}(K)/\preceq$ is a complete sublattice of $\mathrm{GEN}(L)/\preceq$.*

The specification with grounded language L requiring only the groundedness and monotonicity axioms has as its initial model the Herbrand structure WL ordered by the syntactic approximation relation $\sqsubseteq$ first described in [34]: For two ground terms u_1, u_2 we have $u_1 \sqsubseteq u_2$ iff u_2 results from u_1 by replacing zero or more occurrences of $\perp$ by other ground terms. This is due to the fact that the order in the initial model is the least order comprising the relation generated by the groundedness and monotonicity axioms.

As another example we give a specification of an ordering that will be used below:

Example 1. Consider

```
spec S
    grounded monotonic
    sort nat
    0, 1 : nat
    plustwo : nat → nat
    axioms plustwo(⊥) = ⊥
           0 ≤ plustwo(0)
           1 ≤ plustwo(1)
endspec
```

The initial model I of S has the carrier

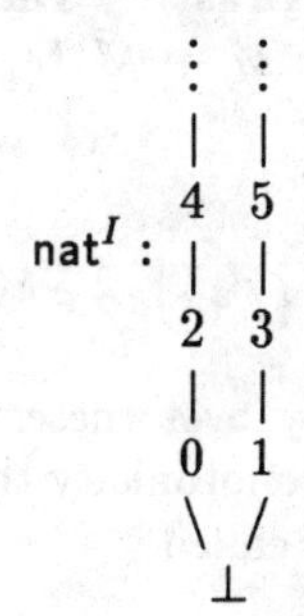

□

3.5 Completeness and Continuity

We head now for continuous structures, i.e., structures in which fixpoint equations are solvable; they allow thus a mathematical semantics for recursion equations (see e.g. [15]).

Let us give the necessary order-theoretic notions. Consider two subsets $S, T \subseteq M$ of a pre-ordered set $(M, \leq)$. We write $S \leq T$ if $S \times T \subseteq \leq$, i.e., if $\forall\, x \in S : \forall\, y \in T : x \leq y$. For singletons we write $x \leq T$ and $S \leq y$ rather than $\{x\} \leq T$ and $S \leq \{y\}$.

The sets of **upper** and **lower** bounds of a subset $N \subseteq M$ are defined by

$$\mathsf{upb}_{\leq} N \stackrel{\text{def}}{=} \{y \in M : N \leq y\}\ , \qquad \mathsf{lwb}_{\leq} N \stackrel{\text{def}}{=} \{x \in M : x \leq N\}\ .$$

An upper (lower) bound of N that lies in N is called a **greatest (least)** element of N, and we set

$$\mathsf{gst}_{\leq} N \stackrel{\text{def}}{=} N \cap \mathsf{upb}_{\leq} N\ , \qquad \mathsf{lst}_{\leq} N \stackrel{\text{def}}{=} N \cap \mathsf{lwb}_{\leq} N\ .$$

Note that all greatest (least) elements of a subset are identified in $M/\leq$.

If the set of all upper (lower) bounds of N possesses least (greatest) elements, these elements are called **least upper bounds (greatest lower bounds)** of N. We set

$$\mathsf{lub}_{\leq} N \stackrel{\text{def}}{=} \mathsf{lst}_{\leq}\ \mathsf{upb}_{\leq} N\ , \qquad \mathsf{glb}_{\leq} N \stackrel{\text{def}}{=} \mathsf{gst}_{\leq}\ \mathsf{lwb}_{\leq} N\ .$$

In an ordered set, least upper bounds and greatest lower bounds are unique if they exist. In the sequel we shall omit the indices $_\leq$ if $\leq$ is clear from the context.

A subset $D \subseteq M$ of a pre-ordered set $(M, \leq)$ is **directed** if $D \neq \emptyset$ and $D \cap \mathsf{upb}\,\{x, y\} \neq \emptyset$ for all $x, y \in D$. A pre-ordered set $(M, \leq)$ is **Δ-complete** if M is grounded and $\mathsf{lub}\ D \neq \emptyset$ for every directed subset $D \subseteq M$.

Let $(M_1, \leq_1)$, $(M_2, \leq_2)$ be pre-ordered sets. A function $f : M_1 \rightarrow M_2$ is **continuous** if for every directed $D \subseteq M_1$ we have $f(\mathsf{lub}_{\leq_1} D) \subseteq \mathsf{lub}_{\leq_2} f(D)$. In the case of an order we may replace the inclusion by an equality. A continuous function is monotonic.

The following well-known fixpoint theorem is the basis for the solution of recursion equations:

Theorem 11 (Knaster, Tarski, Kleene). *Let $(M, \sqsubseteq)$ be a Δ-complete ordered set and $f : M \rightarrow M$ be continuous. Then f has a least fixpoint, viz.*

$$\mathit{fix}(f) \stackrel{\text{def}}{=} \mathsf{lub}\ f^*(\bot)\ ,$$

where $f^*(x) \stackrel{\text{def}}{=} \{f^i(x) : i \in \mathbb{N}\}$.

Now we carry over these notions to structures. Contrary to groundedness, strictness and monotonicity they cannot be characterised by Horn formulas. An L-structure A is called

1. **pre-complete** if all its operations are continuous in all arguments;

2. **continuous** if it is pre-complete and its carrier is Δ-complete.

Note that a pre-complete structure is monotonic.

Lemma 12. *For a given language L the Herbrand structure WL is pre-complete.*

We consider now particular congruences that fit these specialised notions of structures. Consider an ordered language L and an ordered L-structure A. We call a congruence $R \in \mathrm{CG}(A)$

1. **pre-complete** if for every operation $f^A : A_w \to A_s$ and every $\leq^R$-directed subset $D \subseteq A_w$ we have $f^A(\mathsf{lub}_{\leq^R} D) \subseteq \mathsf{lub}_{\leq^R} f^A(D)$;
2. **lub-compatible** if for every $\leq^A$-directed subset $D \subseteq A_s$ we have $\mathsf{lub}_{\leq^A} D \subseteq \mathsf{lub}_{\leq^R} D$. This is equivalent to $\mathsf{lub}_{\leq^A} D \leq^R \mathsf{upb}_{\leq^R} D$.

Theorem 13. *Let R be a congruence on $A \in \mathrm{STR}(L)$.*

1. *A/R is pre-complete iff R is pre-continuous.*
2. *Let $h : A \to B$ be a homomorphism.*
   (a) *If h is continuous then R_h is lub-compatible.*
   (b) *If h is an epimorphism and R_h is lub-compatible then h is continuous. In particular, for lub-compatible R the canonical homomorphism $k : A \to A/R$ given by $k(a) \stackrel{\text{def}}{=} [a]_{=^R}$ is continuous.*

For the proof see [11]. We denote by Mon^A and Pre^A the sets of monotonic and pre-continuous congruences on A.

Lemma 14. Mon^A *and* Pre^A *are lattice-forming.*

3.6 Completions

Our method of obtaining continuous structures is via completing term-generated monotonic structures. The requirement of monotonicity can be relaxed (see [33]), but for simplicity we shall here work assuming it.

Call a Δ-complete set M a **completion** of an ordered set L if

1. there is an order-embedding $\iota : L \to M$;
2. for every continuous function $h : L \to N$ into a Δ-complete set N there is a unique continuous function $\hat{h} : M \to N$ with $\hat{h} \circ \iota = h$, i.e., which extends h to M.

The completion is called **lub-preserving** if the embedding ι is continuous. Then it preserves the least upper bounds already existing in L. Thus, as pointed out in [11], there can only be a lub-preserving completion of A if A is pre-complete, i.e., if its operations are already continuous at the existing least upper bounds.

An important, however in general not lub-preserving, completion is the ideal completion (see e.g. [5, 12]). To define it, we need some auxiliary notions.

Consider a pre-ordered set $(M, \leq)$. The **downward closure** of a subset $N \subseteq M$ is

$$N^{\leq} \stackrel{\text{def}}{=} \{x \in M : \exists\, y \in N : x \leq y\}\,.$$

By a **cone** of M we mean a downward closed subset, i.e., a set $N \subseteq M$ such that $N = N^{\leq}$. An **ideal** of an ordered set $(M, \leq)$ is a directed cone in M. For a directed subset $D \subseteq M$ the cone $D^{\leq}$ is the **ideal generated by** D. Note that $\mathsf{lub}\, D = \mathsf{lub}\, D^{\leq}$. For $x \in M$ we write $x^{\leq}$ instead of $\{x\}^{\leq}$.

An element x of M is **finite (compact)** if for every directed set $D \subseteq M$ with $x \leq \mathsf{lub}\, D$ we have also $x \leq z$ for some $z \in D$. Equivalently, x is finite iff for every ideal $I \subseteq M$ with $x \leq \mathsf{lub}\, I$ we have $x \in I$. $(M, \leq)$ is **inductive (algebraic)** if every element of M is the least upper bound of a directed set of finite elements. A non-finite element of an inductive set is called a **limit point** or an **infinite element**.

Theorem 15. *Let $(M, \leq)$ be an ordered set and $I(M)$ be the set of ideals of M.*

1. *The set $(I(M), \subseteq)$ ordered by set inclusion is an inductive completion of M, the finite elements being the ideals $x^{\leq}$ for $x \in M$. M is embedded into $I(M)$ via the function $i : x \mapsto x^{\leq}$.*
2. *For every monotonic (not necessarily continuous) function $h : M \to N$ into a Δ-complete set $(N, \leq)$ there is a unique continuous function $\hat{h} : I(M) \to N$ extending h, i.e., with $\hat{h}(x^{\leq}) = h(x)$. $\hat{h}$ is given by $\hat{h}(I) = \mathsf{lub}\, h(I)$ for $I \in I(M)$; hence $\hat{h}(D^{\leq}) = \mathsf{lub}\, h(D)$ for directed D.*

For the proof see e.g. [11]. We call $(I(M), \subseteq)$ the **ideal completion** of M.

Let us now apply these notions to L-structures. A continuous L-structure B is a **completion** of a monotonic L-structure A if there is an order-embedding homomorphism $\iota : A \to B$ such that every continuous homomorphism $h : A \to C$ from A into a continuous L-structure C extends uniquely to a continuous homomorphism $\hat{h} : B \to C$ in the sense that $\hat{h} \circ \iota = h$. Again, the completion is called **lub-preserving** if ι is continuous.

Now every grounded monotonic L-structure A may be embedded into an inductive continuous L-structure A^{∞} using the ideal completion. Define

$$A^{\infty}_{\mathsf{t}} \stackrel{\text{def}}{=} I(A_{\mathsf{t}}) \qquad\qquad \text{ordered by inclusion,}$$
$$f^{A^{\infty}}(I_1, \ldots, I_n) \stackrel{\text{def}}{=} f^{A}(I_1, \ldots, I_n)^{\leq^{A}} \ (f \in F)\,.$$

Theorem 16. *1. A^{∞} is an inductive completion of A.*

2. *Every* monotonic *(not necessarily continuous) homomorphism $h : A \to B$ into a continuous L-structure B extends uniquely into a continuous homomorphism $\hat{h} : A^{\infty} \to B$ such that $\hat{h} \circ \iota = h$.*

We call A^{∞} the **ideal completion** of A. Its construction may be applied to monotonic structures that are not pre-complete. It resolves non-continuities by introducing additional limit points into the carriers and by suitably extending the functions to these limit points. The price for this is that in general it is not lub-preserving.

A different completion method which is based on techniques in [23] and [11] and which is lub-preserving is discussed in [30]. We do not describe it here, since the construction is fairly complicated and still does not give a satisfactory construction of continuous models, as we shall see below.

3.7 Fixpoint Structures

We now study structures in which fixpoint equations are solvable.

Let L be a grounded language and X be an S-indexed set of variables disjoint from L. A **system of recursion equations** over L and X is a valuation e of X in $WL(X)$. It may be viewed as the S-indexed set of equations

$$(\{x = e_s(x) : x \in X_s\})_{s \in S}.$$

Over any L-structure A such a system can be conceived of as a set of conditions about a family of elements of A, viz. about a valuation v of X in A. Thus we say that a valuation $v \in A^X$ is a **solution** of the system e if for all x we have $v_s(x) = \overline{v}(e_s(x))$.

Given an L-structure A, we now associate a **valuation transformer** $e_A : A^X \to A^X$ with a system e by setting for $v \in A^X$

$$e_A(v) \stackrel{\text{def}}{=} \overline{v} \circ e.$$

Thus the transformed valuation associates with each variable the value of the corresponding right hand side in e under the given valuation v. Then $v \in A^X$ is a solution of e iff v is a fixpoint of e_A.

To this end we need to order valuations. We use the standard pointwise order on functions. More generally, for a family $(M_i, \leq_i)_{i \in I}$ of pre-ordered sets, the product $\prod_{i \in I} M_i$ carries the **pointwise** pre-order given by $(x_i)_{i \in I} \leq (y_i)_{i \in I}$ iff $\forall\, i \in I : x_i \leq_i y_i$. For elements f, g of a function space $M \to N = \prod_{x \in M} N$ with pre-ordered set $(N, \leq)$ we therefore have $f \leq g \Leftrightarrow \forall\, x \in M : f(x) \leq g(x)$.

Lemma 17. *Consider a continuous L-structure A.*

1. *The set A^X is Δ-complete under the pointwise order on valuations.*
2. *For any system e of recursion equations over L and X the valuation transformer e_A is continuous.*

Hence by Theorem 11 $\mathit{fix}(e_A)$ is the least solution of e in A. If we set

$$\Omega^A(x) \stackrel{\text{def}}{=} \bot^A \text{ for all } x \in X,$$

we obtain

$$\mathit{fix}(e_A) = \mathsf{lub}\; e_A^*(\Omega^A)\;.$$

Thus, over continuous structures fixpoint equations are solvable. However, for the mere process of solving equations a continuous structure generally contains much too many "superfluous" least upper bounds. More liberally, we call a grounded and monotonic structure A a **fixpoint structure** if for every system e of recursion equations $\mathsf{lub}\, e_A^*(\Omega^A) \neq \emptyset$ and consists of a fixpoint of e_A. Note that, by monotonicity of A, it then is the least fixpoint of e_A.

Theorem 18. *1. Every continuous structure is a fixpoint structure.*

2. Let A be a fixpoint structure and $h : A \to B$ be a strict continuous homomorphism. Then B is again a fixpoint structure. Moreover, for a system e of recursion equations we have $h \circ \mathit{fix}(e_A) = \mathit{fix}(e_B)$.

3. The class of fixpoint structures is closed under quotients modulo lub-compatible congruences.

Proof. 1. is trivial.

2. For a system e of recursion equations and a valuation $v \in A^X$ we have, by Lemma 1, $h \circ e_A(v) = h \circ \overline{v} \circ e = \overline{h \circ v} \circ e = e_B(h \circ v)$. By induction it now follows that also $h \circ e_A^*(v) = e_B^*(h \circ v)$.
Moreover, by strictness of h, $h \circ \Omega^A = \Omega^B$. Hence, by continuity of h,

$$h \circ \mathit{fix}(e_A) = h \circ \mathsf{lub}\, e_A^*(\Omega^A) \subseteq \mathsf{lub}\, h \circ e_A^*(\Omega^A) = \mathsf{lub}\, e_B^*(\Omega^B) = \mathit{fix}(e_B)\ .$$

3. is immediate from 2. and Theorem 13.3.

□

4 Adding Higher Order

4.1 Higher-Order Languages and Structures

We shall now, following [32, 25], introduce higher-order languages and structures as particular instances of the corresponding first-order notions. To keep the presentation simple, we do not introduce product types. The reason for this is that an n-ary product behaves exactly like a function type over a domain of cardinality n. So product types would lead to a duplication in all definitions without adding extra power. Moreover, in function arguments they can always be eliminated by currying. A formal treatment can be found in [25].

We first define (higher-order) types. Assume that a non-empty set B of **basic types** is given. The set $T(B)$ of **types** over B is the smallest set such that $B \subseteq T(B)$ and for $\mathsf{s}, \mathsf{t} \in T(B)$ also $\mathsf{s} \to \mathsf{t} \in T(S)$. As customary we assume that $\to$ associates to the right and hence abbreviate a type $\mathsf{s}_1 \to (\mathsf{s}_2 \to \cdots (\mathsf{s}_n \to \mathsf{s}_{n+1}) \cdots)$ by $\mathsf{s}_1 \to \mathsf{s}_2 \to \cdots \mathsf{s}_n \to \mathsf{s}_{n+1}$.

A **higher-order language (hol)** is an ordered language $H = (S, F, P, \phi, \psi)$ satisfying the following conditions:

1. $S \subseteq T(B)$ for some set B of basic types. This set can be uniquely reconstructed from S and is denoted by $B(H)$.

2. For every populated type $\mathsf{s} \to \mathsf{t} \in S$ there is an operator $\alpha^{\mathsf{s}\to\mathsf{t}} \in F$ with $\phi(\alpha^{\mathsf{s}\to\mathsf{t}}) = (\mathsf{s} \to \mathsf{t})\,\mathsf{s}\,\mathsf{t}$, called **application operator**.
3. All other operators in F are constants.
4. S is exactly the set of types populated in H.
5. All basic types are populated in H.
6. Whenever a type $\mathsf{s} \to \mathsf{t}$ is populated in H, so is s (and hence also t). Then for every term denoting a functional object there is also at least one term denoting an argument to that functional object. This can always be achieved by adding suitable constants of all types.

The latter three conditions serve to exclude empty sets in the carriers of H-structures. Note that a grounded language automatically satisfies these requirements.

In the sequel we shall drop the type superscripts as they will always be clear from the context. We will also use a simplified notation for terms over a hol: we abbreviate $\alpha(t, u)$ by $t\,u$ and assume, as customary, that juxtaposition associates to the left.

A hol is called **essentially of first order** if its constants are only of types of the form s or $\mathsf{s}_1 \to \cdots \mathsf{s}_n \to \mathsf{s}_{n+1}$, with $\mathsf{s}, \mathsf{s}_i \in B$, and its only non-constant operators are the respective application operators α. Such hols correspond to the first-order languages defined in Section 2.1: a constant $f : \mathsf{s}_1 \to \cdots \mathsf{s}_n \to \mathsf{s}_{n+1}$ represents an operator f with $\phi(f) = \mathsf{s}_1\mathsf{s}_2\cdots\mathsf{s}_n\mathsf{s}_{n+1}$.

Consider now a hol H. In every H-structure A each element $f \in A_{\mathsf{s}\to\mathsf{t}}$ of a carrier of functional type $\mathsf{s} \to \mathsf{t}$ induces a function $[\![f]\!]^A : A_\mathsf{s} \to A_\mathsf{t}$ given by

$$[\![f]\!]^A(x) \stackrel{\text{def}}{=} \alpha^A(f, x)$$

for $x \in A_\mathsf{s}$; it is the curried form of α^A. Note that this function cannot be denoted by an operator in the language, since the corresponding "higher-order arity" does not exist. We shall frequently use this function instead of α^A; if A is clear from the context we shall omit the superscript ^{A}.

The **function order axioms** over a hol H are the formulas

$$f \leq g \Rightarrow f\,x \leq g\,x$$

with $f, g \in X_{\mathsf{s}\to\mathsf{t}}$ and $x \in X_\mathsf{s}$ for all populated functional types $\mathsf{s} \to \mathsf{t}$ in H. An H-structure A is called a **higher-order H-structure (H-hos)** if it is ordered and satisfies the function order axioms. Equivalently, A is an H-hos if all functions

$$[\![\_]\!]^A : A_{\mathsf{s}\to\mathsf{t}} \to (A_\mathsf{s} \to A_\mathsf{t})$$

are monotonic. This condition ensures that the ordering on carriers of functional types conforms to the pointwise function space ordering. The carriers A_s with $\mathsf{s} \in B(H)$ are called the **basic carriers** of A. The class of all H-hos is denoted by $\mathrm{HOS}(H)$. A similar notion occurs already in [37].

From our general setting we also obtain a notion of term-generated hoss. Note that a homomorphism between hoss also is a homomorphism w.r.t. $[\![\_]\!]$.

Of particular interest are monotonic (pre-complete) hoss: there all functions $[\![f]\!]$ are monotonic (continuous). In a continuous hos in addition the functions $[\![\_]\!]$ are continuous.

Finally, we study quotients of hoss. Consider a hol $H = (S, F, P, \phi, \psi)$ and an H-hos A. A congruence $R \in \mathrm{CG}(A)$ is called a **higher-order congruence (hop)** if it satisfies the order and the function order axioms so that all functions $[\![\_]\!]^A$ are monotonic w.r.t. $\leq^R$. The set of all hops on A is denoted by $\mathrm{HCG}(A)$. Since the function order axioms are Horn formulas, we obtain from Lemma 3 and Theorem 2

Corollary 19. *Consider $A \in \mathrm{STR}(H)$ and $R \in \mathrm{CG}(A)$.*

1. *$A/R \in \mathrm{HOS}(A)$ iff $R \in \mathrm{HCG}(A)$.*
2. *$\mathrm{HCG}(A)$ is lattice-forming.*

4.2 Extensionality

Consider a hol $H = (S, F, P, \phi, \psi)$. The **extensionality axioms** over H are the formulas

$$(\forall\, \mathsf{s}\ x : f\,x \leq g\,x) \Rightarrow f \leq g$$

with $f, g \in X_{\mathsf{s} \to \mathsf{t}}$ and $x \in X_{\mathsf{s}}$ for all populated functional types $\mathsf{s} \to \mathsf{t}$ in H. They are the reverse implications of the function order axioms. An H-hos A is called **extensional** if it satisfies the extensionality axioms for H. Equivalently, A is extensional if all functions $[\![\_]\!]^A$ are order-embeddings. In particular we have that

$$[\![f]\!]^A \leq^A [\![g]\!]^A \Rightarrow f \leq^A g\ .$$

Then the carriers of functional types are isomorphic to subsets of the respective function spaces. Since we allow subsets rather than always full function spaces, we are working in the framework of "general" models (cf. [18]) rather than "standard" or "full" models of higher type logic. This is the reason why we can still obtain a complete deductive calculus. Moreover, it allows a notion of term-generatedness for extensional hos, which in general would not be possible for full models, since for a countable language the Herbrand structure and therefore all term-generated models have countable carriers, whereas full funcion spaces over infinite domains are uncountable.

A first example of an extensional structure is given by

Lemma 20. *For a given hol H the Herbrand structure WH is extensional.*

Although the extensionality axioms are not Horn formulas, for a particular H-hos A we can replace them by equivalent infinite Horn formulas. To this end we extend the language H by an S-indexed set C of new constants, where $C_{\mathsf{t}} \stackrel{\mathrm{def}}{=} \{c_x : x \in A_{\mathsf{t}}\}$.Denote the extended language by $H(C)$. Then A is made into an $H(C)$-structure by defining $c_x^A \stackrel{\mathrm{def}}{=} x$. This means that each element of the carrier of A is now denoted by a constant. Then the extensionality axiom for type $\mathsf{s} \to \mathsf{t} \in S$ can be replaced by the (in general infinite) Horn formula

$$\bigwedge_{c \in C_{\mathsf{s}}} f\,c \leq g\,c \Rightarrow f \leq g\ .$$

This observation is important in view of the results of [20, 42, 43]; it is the deeper reason for the nice structural properties of extensional specifications. If A is even term-generated, we can avoid the language extension and instead replace the extensionality axiom by the (again in general infinite) Horn formula

$$\bigwedge_{u \in WH_{\mathsf{s}}} f\,u \leq g\,u \Rightarrow f \leq g\ .$$

We call a congruence $R \in \mathrm{HCG}(A)$ **extensional** if it satisfies the extensionality axioms, i.e., if for every function type $\mathsf{s} \to \mathsf{t} \in S$ and all $f, g \in A_{\mathsf{s}\to\mathsf{t}}$ we have

$$[\![f]\!] \leq^R [\![g]\!] \Rightarrow f \leq^R g\ .$$

The set of extensional congruences on A is denoted by Ext^A .

From Lemma 3 and the above remark on Horn formulations of extensionality we have

Corollary 21. *Consider $A \in \mathrm{HOS}(H)$.*

1. *If $R \in \mathrm{HCG}(A)$ then A/R is extensional iff R is extensional.*
2. *Ext^A is lattice-forming.*

For $A \in \mathrm{HOS}(H)$ therefore $\cap\mathrm{Ext}^A$ is the least extensional congruence on A; it forms the "closest" extensional quotient of A. We now want to study the function $\mathrm{Ext} : A \mapsto A/\cap\mathrm{Ext}^A$ more thoroughly. First we note that Ext generally not only influences the carriers of types of higher order, but also the basic carriers.

Example 2. Consider a hol H with the basic types m and n and the constants

$$\begin{aligned} a\ &: \mathsf{m} \\ d, e &: \mathsf{n} \\ f, g &: \mathsf{m} \to \mathsf{m} \\ p\ &: (\mathsf{m} \to \mathsf{m}) \to \mathsf{n} \end{aligned}$$

and $A \in \mathrm{HOS}(H)$ with

$$\begin{aligned} & f(a)^A = g(a)^A = a^A \\ & d^A \neq e^A \\ & p(f)^A = d^A \\ & p(g)^A = e^A \end{aligned}$$

Then in $B \stackrel{\mathrm{def}}{=} \mathrm{Ext}(A)$ we have $f^B = g^B$ and hence, by the congruence property of $\cap\mathrm{Ext}^A$, also $d^B = e^B$. □

In general, this is an undesired effect. Thus we call $A \in \mathrm{HOS}(H)$ **extensionally well-behaved** if $\leq_{\mathsf{t}}^{\cap \mathrm{Ext}^A} = \leq_{\mathsf{t}}^{A}$ for all basic types t of H. This means that the extensional view of A does not induce additional order relations or equalities on the basic elements of A.

Lemma 22. *Assume $A \in \mathrm{HOS}(H)$ for $H = (S, F, P, \phi, \psi)$.*

1. *If A is extensionally well-behaved then $\cap\mathrm{Ext}^A$ is inductively given by*
   (a) $\leq_{\mathsf{t}}^{\cap \mathrm{Ext}^A} = \leq_{\mathsf{t}}^{A}$ *for* $\mathsf{t} \in B(H)$.
   (b) *For $f, g \in A_{\mathsf{s} \to \mathsf{t}}$ we have $f \leq_{\mathsf{s} \to \mathsf{t}}^{\cap \mathrm{Ext}^A} g$ iff $f =^A g$ or for all $a, b \in A_{\mathsf{s}}$ with $a =_{\mathsf{s}}^{\cap \mathrm{Ext}^A} b$ also $[\![f]\!](a) \leq_{\mathsf{t}}^{\cap \mathrm{Ext}^A} [\![g]\!](b)$.*
2. *Let A be continuous and R be an extensional congruence on A that is lub-compatible on the basic carriers. Then R is lub-compatible. In particular, if A is extensionally well-behaved then $\cap\mathrm{Ext}^A$ is lub-compatible.*

Proof. 1. Define a system R of relations on A inductively by
(a) $\leq_{\mathsf{t}}^{R} \overset{\mathrm{def}}{=} \leq_{\mathsf{t}}^{A}$ for $\mathsf{t} \in B(H)$,
(b) $f \leq_{\mathsf{s} \to \mathsf{t}}^{R} g$ iff $f =^A g$ or for all $a, b \in A_{\mathsf{s}}$ with $a =_{\mathsf{s}}^{R} b$ also $[\![f]\!](a) \leq_{\mathsf{t}}^{R} [\![g]\!](b)$,

and $a =_{\mathsf{t}}^{R} b \overset{\mathrm{def}}{\Leftrightarrow} a \leq_{\mathsf{t}}^{R} b \wedge b \leq_{\mathsf{t}}^{R} a$. To prove 1. we show now that $R = \cap\mathrm{Ext}^A$. To this end we first prove that R is a congruence on R.
By construction $=^R$ and $\leq^R$ are reflexive and $=^R$ is symmetric and substitutive w.r.t. $\leq^R$. Next, we show that $=^R$ is also substitutive w.r.t. the application operations α. Assume $f =_{\mathsf{s} \to \mathsf{t}}^{R} g$ and $a =_{\mathsf{s}}^{R} b$. By definition of $=^R$ then $f \leq_{\mathsf{s} \to \mathsf{t}}^{R} g$ and $g \leq_{\mathsf{s} \to \mathsf{t}}^{R} f$. By definition of $\leq_{\mathsf{s} \to \mathsf{t}}^{R}$ and symmetry of $=_{\mathsf{s}}^{R}$ we therefore have $[\![f]\!](a) \leq_{\mathsf{t}}^{R} [\![g]\!](b)$ and $[\![f]\!](b) \leq_{\mathsf{t}}^{R} [\![g]\!](a)$, i.e., $[\![f]\!](a) =_{\mathsf{t}}^{R} [\![g]\!](b)$.
We now show transitivity of $\leq_{\mathsf{t}}^{R}$ by induction on t. For $\mathsf{t} \in B(H)$ this is clear from the assumption. Assume $f \leq_{\mathsf{s} \to \mathsf{t}}^{R} g \wedge g \leq_{\mathsf{s} \to \mathsf{t}}^{R} h$ for $f, g, h \in A_{\mathsf{s} \to \mathsf{t}}$. If $f =_{\mathsf{s} \to \mathsf{t}}^{A} g$ or $g =_{\mathsf{s} \to \mathsf{t}}^{A} h$ then $f \leq_{\mathsf{s} \to \mathsf{t}}^{R} h$ is immediate. Otherwise consider arbitrary $a, b \in A_{\mathsf{s}}$ with $a =_{\mathsf{s}}^{R} b$. By definition of $\leq_{\mathsf{s} \to \mathsf{t}}^{R}$ and reflexivity of $=_{\mathsf{s}}^{R}$ we get $[\![f]\!](a) \leq_{\mathsf{t}}^{R} [\![g]\!](a)$ and $[\![g]\!](a) \leq_{\mathsf{t}}^{R} [\![h]\!](b)$. Since by the induction hypothesis $\leq_{\mathsf{t}}^{R}$ is transitive this implies $[\![f]\!](a) \leq_{\mathsf{t}}^{R} [\![h]\!](b)$. Since a, b were arbitrary we conclude from the definition of $\leq_{\mathsf{s} \to \mathsf{t}}^{R}$ also $f \leq_{\mathsf{s} \to \mathsf{t}}^{R} h$.
Hence $\leq^R$ is a preorder and thus, by Lemma 8, $=^R$ is an equivalence.
By construction, R satisfies the function order axioms. We next show that $\leq_{\mathsf{t}}^{A} \subseteq \leq_{\mathsf{t}}^{R}$ by induction on t. For the basic types this is trivial. Assume now $f \leq_{\mathsf{s} \to \mathsf{t}}^{A} g$ and $a =_{\mathsf{s}}^{R} b$ for $a, b \in A_{\mathsf{s}}$. By substitutivity and reflexivity of $=^R$ we get $[\![f]\!](a) \leq_{\mathsf{t}}^{R} [\![f]\!](b)$. Since A satisfies the function order axioms we have $[\![f]\!](b) \leq_{\mathsf{t}}^{A} [\![g]\!](b)$. By the induction hypothesis this implies $[\![f]\!](b) \leq_{\mathsf{t}}^{R} [\![g]\!](b)$, and transitivity of $\leq_{\mathsf{t}}^{R}$ shows $[\![f]\!](a) \leq_{\mathsf{t}}^{R} [\![g]\!](b)$. Since a, b were arbitrary we conclude $f \leq_{\mathsf{s} \to \mathsf{t}}^{R} g$.
Since substitutivity was shown above, R thus is a congruence on A.

Next we show that R is extensional. Suppose $[\![f]\!](c) \leq^R_{\mathsf{t}} [\![g]\!](c)$ for all $c \in A_{\mathsf{s}}$. By substitutivity and reflexivity of $=^R$ we have for arbitrary $a, b \in A_{\mathsf{s}}$ with $a =^R_{\mathsf{s}} b$ that $[\![f]\!](a) =^R_{\mathsf{t}} [\![f]\!](b) \leq^R_{\mathsf{t}} [\![g]\!](b) =^R_{\mathsf{t}} [\![g]\!](b)$. Since a, b were arbitrary we conclude $f \leq^R_{\mathsf{s} \to \mathsf{t}} g$.

Let now R' be another extensional congruence on A. We show by induction on t that $\leq^R_{\mathsf{t}} \subseteq \leq^{R'}_{\mathsf{t}}$ for all $\mathsf{t} \in S$.

(a) For $\mathsf{t} \in B(H)$ we have $\leq^R_{\mathsf{t}} = \leq^A_{\mathsf{t}} \subseteq \leq^{R'}_{\mathsf{t}}$.

(b) Let $f \leq^R_{\mathsf{s} \to \mathsf{t}} g$. We have for all $a \in A_{\mathsf{s}}$ that $[\![f]\!](a) \leq^R_{\mathsf{t}} [\![g]\!](a)$, and hence, by the induction hypothesis, also $[\![f]\!](a) \leq^{R'}_{\mathsf{t}} [\![g]\!](a)$ showing $f \leq^{R'}_{\mathsf{s} \to \mathsf{t}} g$ by extensionality of R'.

2. The proof again uses induction on the types involved. On the basic types the lub-compatibility of R is assumed. Let now $D \subseteq A_{\mathsf{s} \to \mathsf{t}}$ be $\leq^A_{\mathsf{s} \to \mathsf{t}}$-directed. For all $a \in A_{\mathsf{s}}$ we have, by $\leq^A$-continuity of $[\![\_]\!]$, that $[\![\mathsf{lub}_{\leq^A_{\mathsf{s} \to \mathsf{t}}} D]\!](a) = \mathsf{lub}_{\leq^A_{\mathsf{t}}} [\![D]\!](a)$. By the induction hypothesis $\mathsf{lub}_{\leq^A_{\mathsf{t}}} [\![D]\!](a) \leq^R_{\mathsf{t}} \mathsf{upb}_{\leq^R_{\mathsf{t}}} [\![D]\!](a)$. Since R satisfies the function order axioms, $[\![\_]\!]$ is $\leq^R$-monotonic and so also $[\![\mathsf{lub}_{\leq^A_{\mathsf{s} \to \mathsf{t}}} D]\!](a) \leq^R_{\mathsf{t}} [\![\mathsf{upb}_{\leq^R_{\mathsf{s} \to \mathsf{t}}} D]\!](a)$. Now extensionality of R shows $\mathsf{lub}_{\leq^A_{\mathsf{s} \to \mathsf{t}}} D \leq^R_{\mathsf{s} \to \mathsf{t}} \mathsf{upb}_{\leq^R_{\mathsf{s} \to \mathsf{t}}} D$.

□

Property 2. will be instrumental for the existence of extensional structures that allow solving fixpoint equations. However, the structure $\mathrm{Ext}(A)$ need not be continuous even if A is:

Example 3. Let A_{t_1}, A_{t_2} and $A_{\mathsf{t}_1 \to \mathsf{t}_2}$ be given by

$A_{\mathsf{t}_1 \to \mathsf{t}_2}$: $a_0 \; a_1 \; a_2 \; \dots$ (discrete points)

A_{t_1} : a

A_{t_2} : $b_0 < b_1 < b_2 < \dots$ (chain b_0 — b_1 — b_2 — ⋮)

and define $[\![a_i]\!]^A(a) \stackrel{\mathrm{def}}{=} b_i$. Then $A_{\mathsf{t}_1 \to \mathsf{t}_2}$ is Δ-complete, but $\mathrm{Ext}(A)_{\mathsf{t}_1 \to \mathsf{t}_2}$ is order-isomorphic to A_{t_2} which is not Δ-complete. □

4.3 Higher-Order and Extensional Specifications

A **higher-order specification** is an ordered specification $K = (H, E)$ with a higher-order signature $H = (S, F, P, \phi, \psi)$ such that the set E of axioms includes the function order axioms. K is **extensional** if E also includes the extensionality axioms. It is clear that all models of an extensional higher-order specification are extensional. We use the keyword extensional to express that the respective specification implicitly contains the extensionality axioms.

Consider now specifications that are **almost Horn**, i.e., higher-order specifications $K = (H, E)$ such that E consists of Horn axioms except possibly for extensionality axioms. Note that an almost Horn specification may even be Horn. As an immediate consequence of Theorem 2 and Horn expressibility of extensionality on term-generated algebras we obtain

Theorem 23. *For a higher-order specification K that is*
- *almost Horn,*
- *almost Horn and grounded,*
- *almost Horn and monotonic,*
- *almost Horn, monotonic, and grounded,*

the set $\mathrm{GEN}(K)/\preceq$ *of isomorphism classes forms a complete lattice. In particular, K has an initial term-generated model.*

For a specification with an extensionality axiom for a type $\mathsf{s} \to \mathsf{t} \in S$ the following additional inference rule is sound:

$$\text{(Extensionality)}\ \frac{u\,x \leq w\,x}{u \leq w} \qquad \text{for } u, w \in WH(X)_{\mathsf{s}\to\mathsf{t}} \text{ and } x \in X_{\mathsf{s}} \text{ not occurring in } u, w$$

As shown in [30] and generalising the respective result in [25] we have the following completeness result:

Theorem 24. *Let* $K = (H, E)$ *be an almost Horn specification and* $u, w \in WH(X)$. *Then*

$$K \vdash u \leq w \textit{ iff } \mathrm{MOD}(K) \models u \leq w.$$

Note that the analogue of Theorem 10 fails for almost Horn specifications.

4.4 Particular Higher-order Specifications

So far we have not used λ notation. This decision was taken to avoid the usual problems with bound variables. This does not imply a loss in power: A term $\lambda x.t$ can always be simulated by a new constant f with the axiom $f\,x = t$. Another way to realise λ notation is to include the classical combinators S, K, I as additional constants with the usual axioms and to replace λ terms by their combinator equivalents.

In the case of a higher-order specification $K = (H, E)$, we call K **grounded** if E comprises the groundedness axioms and all axioms of the form $\bot(x) = \bot$. Strictness is expressed as in the first-order case: for a hol $H = (S, F, P, \phi, \psi)$ and $f \in F$ with $\phi(f) = \mathsf{t}_1 \to \cdots \to \mathsf{t}_{k-1} \to \mathsf{t}_k \to \mathsf{t}_{k+1} \to \cdots \to \mathsf{t}_n \to \mathsf{t}$, strictness in the k-th argument is required by the axiom

$$f\,x_{\mathsf{t}_1} \cdots x_{\mathsf{t}_{k-1}} \bot_{\mathsf{t}_k} x_{\mathsf{t}_{k+1}} \cdots x_{\mathsf{t}_n} \leq \bot_{\mathsf{t}}.$$

4.5 Continuous Higher-Order Structures

Since hoss are just a special case of structures, we can apply completion techniques to monotonic hoss as well. However, the ideal completion may destroy extensionality as is shown by

Example 4. Let $\mathbb{N}_\perp$ be the flat domain of natural numbers extended by $\perp$ and define the functions $f_i : \mathbb{N}_\perp \to \mathbb{N}_\perp$ $(i \in \mathbb{N} \cup \{\infty\})$ by

$$f_i(x) \stackrel{\text{def}}{=} \begin{cases} x & \text{if } 0 \le x < i \\ \perp & \text{otherwise} \end{cases}$$
$$f_\infty(x) \stackrel{\text{def}}{=} x \,.$$

The set $M \stackrel{\text{def}}{=} \{f_i : i \in \mathbb{N} \cup \{\infty\}\}$ is extensional. The ideals of M are the sets $I_i \stackrel{\text{def}}{=} \{f_j : j \le i\}$ $(i \in \mathbb{N} \cup \{\infty\})$ and $J \stackrel{\text{def}}{=} \{f_j : j < \infty\}$. We have $I_i \subseteq I_k$ iff $i \le k$, $I_i \subseteq J$ iff $i < \infty$, and $J \subseteq I_\infty$. In order to extend function application in a continuous way to $(I(M), \subseteq)$ we need to set for $I \in I(M)$

$$I(x) \stackrel{\text{def}}{=} \mathsf{lub}\,\{f(x) : f \in I\} \,.$$

We have for all $x \in \mathbb{N}_\perp$ that $I_\infty(x) = J(x)$, but $I_\infty \neq J$. Hence $I(M)$ is not extensional. □

This example shows also that the embedding of M into $I(M)$, in general, is not continuous; hence the ideal completion is not a lub-preserving completion.

It may seem that the loss of extensionality is due to the additional limit point, i.e., to the fact that the ideal completion is not lub-preserving. However, even under a lub-preserving completion extensionality may be lost. In fact, it turns out that there is *no* completion technique which is guaranteed to preserve extensionality. To prove this formally, we first study a particular ordered set:

Lemma 25. *Consider the following ordered set M and its completion M^∞:*

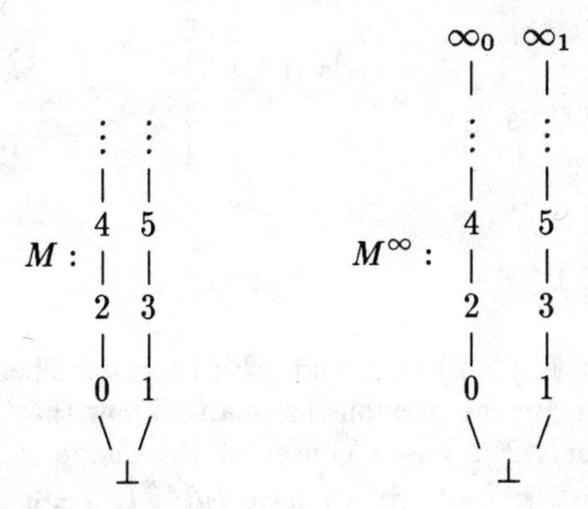

Then in every completion Q of M we have

$$\mathsf{lub}_Q\,\{\iota(n) : n \text{ even}\} \neq \mathsf{lub}_Q\,\{\iota(n) : n \text{ odd}\}$$

where $\iota : M \to Q$ is the embedding of M into Q.

Proof. Consider the embedding $h : M \to M^\infty$. Since Q is a completion of M and M^∞ is Δ-complete, h extends uniquely into a continuous function $\hat{h} : Q \to M^\infty$. We calculate, for $k \in \{0,1\}$

$$\begin{aligned}
& \hat{h}(\mathsf{lub}_Q\,\{\iota(2i+k) : i \in \mathbb{N}\}) \\
= \quad & \{\!\!\{ \text{ continuity } \}\!\!\} \\
& \mathsf{lub}_{M^\infty}\,\{\hat{h}(\iota(2i+k)) : i \in \mathbb{N}\} \\
= \quad & \{\!\!\{ \ Q \text{ completion } \}\!\!\} \\
& \mathsf{lub}_{M^\infty}\,\{h(2i+k) : i \in \mathbb{N}\} \\
= \quad & \{\!\!\{ \text{ ideal completion } \}\!\!\} \\
& \infty_k \,.
\end{aligned}$$

Since $\infty_0 \neq \infty_1$ and $\hat{h}$ is a function the claim follows. □

Now we can show

Theorem 26. *There is no completion technique which is guaranteed to preserve extensionality.*

Proof. Consider a hos A and types s,t with $A_\mathsf{s} = \{\perp\}$ and A_t and $A_{\mathsf{s}\to\mathsf{t}}$ given by the following diagrams.

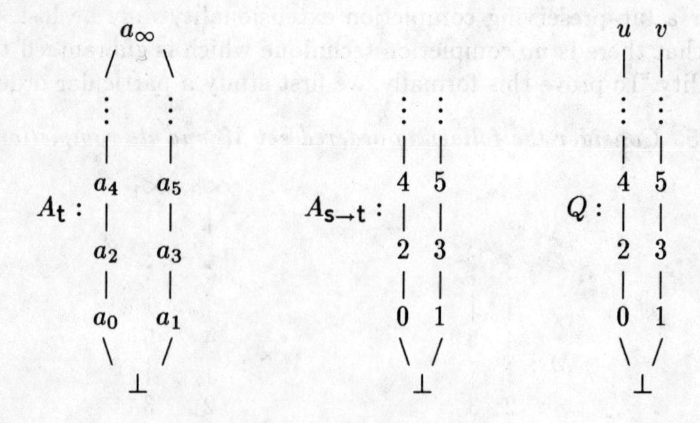

Assume further that $[\![\perp]\!]^A(\perp) = \perp$ and $[\![i]\!]^A(\perp) = a_i$. Then $A_{\mathsf{s}\to\mathsf{t}}$ satisfies the extensionality axiom. By the previous lemma we know that in every completion A^∞ of A the carrier $A^\infty_{\mathsf{s}\to\mathsf{t}}$ has a subset of the shape of Q. For the unique continuous extension $[\![\_]\!]^{A^\infty}$ of $[\![\_]\!]^A$ we have $[\![u]\!]^{A^\infty}(\perp) = a_\infty = [\![u]\!]^{A^\infty}$ although $u \neq v$. Hence A^∞ is not extensional. □

A remedy for this unpleasant situation will be discussed in the following section.

4.6 Higher-Order Fixpoint Structures

We now want to apply the results of Section 3.7 to the case of higher-order structures. Given a hol H and a grounded structure $A \in \text{GEN}(H)$, we can carry out the following construction:

1. Pass to the monotonic structure $B \stackrel{\text{def}}{=} \text{Mon}(A)$.
2. Form the ideal completion $C \stackrel{\text{def}}{=} B^\infty$.
3. If $\cap\text{Ext}^C$ is lub-compatible on the basic types, pass to the extensional structure $D \stackrel{\text{def}}{=} \text{Ext}(C)$.

By Lemma 22 and Theorem 18.3 then D is an extensional fixpoint structure.

To conclude this section, we want to discuss the aspect of generatedness for this construction. Consider a term-generated structure $A \in \text{GEN}(H)$. In the ideal completion A^∞ every carrier element is the least upper bound of a directed set of (injections of) elements of A, i.e., of term-denotable elements, and the behaviour of A^∞ is, through continuity, determined by this term-denotable set. Hence in the ideal completion A^∞ the infinite entities may just be viewed as a "way of speaking" about certain sets of finitely denotable elements. Note that this behaviour of A^∞ carries over to quotients by lub-compatible congruences. Thus, unlike in the approaches of [44, 1], no transfinite terms are involved here. This seems intuitively much more appealing, in particular, since we want to interpret the finitely denotable elements as approximations through which we handle infinite elements, as is the case on actual computers.

4.7 Continuous and Fixpoint Models of Higher-Order Specifications

In this section we want to construct **fixpoint models** of specifications, i.e., models that are fixpoint structures. In keeping with the idea of approximability, we want to use the construction of the previous section. To ensure that, starting from a model, this construction leads again to a model, the axioms of the specifications should remain valid in the completed structure.

From [28] we know

Lemma 27. *Consider an ordered and grounded language L and a formula Φ of the shape $\Phi = (u_1 \leq u_2)$ or $\Phi = (u_1 = u_2)$. Then for any grounded and monotonic ordered structure $A \in \text{STR}(L)$ we have $A \models \Phi$ iff $A^\infty \models \Phi$.*

Together with the principle of term-induction this gives a very powerful means of proving properties of completions of term-generated models of an inequational specification. Further classes of axioms that are preserved under the ideal completion are discussed in [28]. As we shall, however, see below, more general axioms are not useful in the higher-order case.

Example 5. Consider the specification S from Example 1. The ideal completion I^∞ of the initial model I of S has a carrier which is order-isomorphic to the set M^∞ in Lemma 25. □

For the inequational specification with grounded language L and groundedness and monotonicity axioms only (cf. Section 3.4) we get as a continuous model the structure $WL^\infty \stackrel{\text{def}}{=} (WL/\sqsubseteq)^\infty$. It can be thought of as the completion of the term structure WL by "infinite terms". For a grounded hol H that is essentially of first order, it is isomorphic to the free Δ-complete F-magma defined in [10]. WL^∞ can thus be interpreted as a generalised term structure comprising finite and infinite L-terms. In the case of a general hol we may also have infinite terms of functional type.

Finally, we need to consider the case where a quotient has to be taken after completion. To this end, the class of models of the respective specification should be closed under quotients. This is the case for inequational specifications, as we know from Theorem 10.1. Hence we have

Corollary 28. *Let K be a grounded and monotonic inequational specification and assume $A \in \mathrm{GEN}(\mathrm{Mon}(K))$. If $\cap \mathrm{Ext}^{A^\infty}$ is lub-compatible on the basic types then* $\mathrm{Ext}(A^\infty)$ *is an extensional fixpoint model of K.*

5 Examples

In this section we give a number of examples illustrating various uses of higher-order specifications. For modularisation, we use the phrase include SPC′ within a specification SPC to indicate that SPC′ is a subspecification of SPC.

On some occasions we use overloading of operator identifiers. Also, we freely use mixfix notation; the positions of actual parameters are marked by underscores.

We assume grounded and monotonic specifications BOOL and NAT that define the basic type bool with the truth values *true* and *false* and the standard operations on them, and the basic type nat with 0, successor *succ*, predecessor *pred*, zero test *iszero*, addition *add*, equality test *eq* and the doubling function *double*. All these operators are supposed to be strict.

Finally, for every populated type s we assume a constant

$$\text{if \_then \_else \_ fi} : \text{bool} \to \text{s} \to \text{s} \to \text{s}$$

with the axioms

$$\begin{aligned}\text{if } \bot \text{ then } a \text{ else } b \text{ fi} &= \bot,\\ \text{if } \mathit{true} \text{ then } a \text{ else } b \text{ fi} &= a,\\ \text{if } \mathit{false} \text{ then } a \text{ else } b \text{ fi} &= b.\end{aligned}$$

5.1 A Small Functional Programming Language

In this example we specify a small language showing some of the essential features of Backus's FP (cf. [2]), viz. constant function $\bar{\_}$, lifting $\hat{\_}$ of basic functions and predicates, function composition $\_ \circ \_$, conditional $\_ \longrightarrow \_;\_$ and function application $\_ : \_$ We restrict ourselves to unary functions on natural numbers and assume a superspecification OPS of NAT that provides the primitive functions and predicates from which to build the functional programs.

spec FUNCT
grounded monotonic
include OBS
basic type funct, bfunct
$\bar{\_}$: nat $\rightarrow$ funct
$\hat{\_}$: (bool $\rightarrow$ nat) $\rightarrow$ bfunct
$\hat{\_}$: (nat $\rightarrow$ nat) $\rightarrow$ funct
$\_ \circ \_$: funct $\rightarrow$ funct $\rightarrow$ funct
$\_ \rightarrow \_ ; \_$: bfunct $\rightarrow$ funct $\rightarrow$ funct $\rightarrow$ funct
$\_ : \_$: bfunct $\rightarrow$ nat $\rightarrow$ bool
$\_ : \_$: funct $\rightarrow$ nat $\rightarrow$ nat
axioms $\bot : x = \bot$
$\overline{y} : x = y$
$\hat{p} : x = p\,x$
$\hat{f} : x = f\,x$
$(f \circ g) : x = f : (g : x)$
$(p \rightarrow f \,;\, g) : x =$ if $p : x$ then $f : x$ else $g : x$ fi
endspec

The only higher-order objects are the combining forms $\bar{\_}$, $\hat{\_}$, $\_ \circ \_$ and $\_ \rightarrow \_ ; \_$ whereas all functional programs are basic objects. Therefore questions of extensionality do not arise.

Unlike the first-order approaches in [9, 27], the present framework allows a proper treatment of the lifting operators $\hat{\_}$ that assign to every operation of type nat $\rightarrow$ bool and nat $\rightarrow$ nat a functional program denoting that operation. In the approaches mentioned, this had to be done *outside* the respective specifications by adding for each operator of type $s_1 \rightarrow s_2$ a new constant of sort funct s_1 s_2.

The specification is inequational and hence our completion techniques apply. A typical recursion equation over FUNCT is

$$\text{funct } zero = \widehat{iszero} \rightarrow \overline{0} \,;\, (zero \circ \widehat{pred})$$

defining the recursive constant zero function.

5.2 Function Composition

As a building block for further specifications we define a generic composition operator for functions. Note that this is quite different from the purely syntactic composition operator in the functional language of the previous example. This specification by itself does not have a hol in the sense defined in Section 4.1. However, we only want to use it as a building block for larger specifications which then will have proper hols.

spec COMP
$\_ \circ \_$: $(t_1 \rightarrow t_2) \rightarrow (t_2 \rightarrow t_3) \rightarrow (t_1 \rightarrow t_3)$
axioms $(f \circ g)\, x = f\,(g\,x)$
endspec

We have

$$\text{COMP} \vdash (f \circ (g \circ h))\, x = ((f \circ g) \circ h)\, x$$

with variables f, g, h, and x. In extensional models A of specifications that include COMP we have therefore also

$$A \models f \circ (g \circ h) = (f \circ g) \circ h,$$

i.e., composition is associative in A.

5.3 Fixpoints and Function Iteration

We now define a fixpoint operation as used in the semantics of recursively routines or objects; contrary to Section 3.7, we do this *within* rather than *over* a specification. We compare the fixpoint with the elements obtained by functional iteration starting from $\bot$.

spec FIX
  grounded
  include NAT
  include COMP
  $\mathit{fix} : (\mathsf{t} \to \mathsf{t}) \to \mathsf{t}$
  $\mathit{id} : \mathsf{t} \to \mathsf{t}$
  $\_^{\_} : (\mathsf{t} \to \mathsf{t}) \to \mathsf{nat} \to (\mathsf{t} \to \mathsf{t})$
  axioms $\mathit{fix}\, f = f\,(\mathit{fix}\, f)$
    $x = f\,x \Rightarrow \mathit{fix}\, f \leq x$
    $\mathit{id}\, x = x$
    $f^0 = \mathit{id}$
    $f^{\mathit{succ}\, n} = f \circ f^n$
endspec

Note that the specification of *fix* is not complete relative to the other functions. Therefore, in an initial model of FIX, the interpretation of $\mathit{fix}(f)$ will not coincide with that of any other term; it provides a proper extension of the carrier of type $\mathsf{t} \to \mathsf{t}$ (cf. [16]).

Using the proof rules from Section 2.5 we can deduce $f^i \bot \leq \mathit{fix}\, f$. Hence $\mathit{fix}\, f$ is an upper bound of the $f^i \bot$. However, since we have no assertion of continuity (in fact not even a notion of continuity at this level), we cannot be sure that it is the least upper bound of the $f^i \bot$. Note that this property cannot even be specified by a Horn axiom.

This unsatisfactory behaviour is the reason why we consider recursion equations *over* specifications.

5.4 Communicating Sequential Processes

In [19] already an algebraic approach to the specification of communicating sequential processes is taken. We want to show with this Example, how that description fits into the present framework, and how higher-order concepts can be used to smoothen the description.

The ordering on processes used in the specification is the refinement ordering: It is intended that $p \leq q$ should hold iff q is a refinement of p, i.e., iff the behaviour of q is at least as determinate and defined as that of p. In the terminology of [19], $p \leq q$ states that q is at least as reliable as p. Actually, we define the ordering conversely to the one used in [19] to make it fit in with the ideal completion.

We assume a primitive specification DATA that defines the basic type data of data values to be used in the communications. Based on DATA, we want to build our description of CSP in several levels to bring out the structure of that language and its semantics more clearly. First we define the totally unreliable and underspecified process $\top$ that is refined by any process.

Moreover, we define the output construct $!d \to p$ that describes a process that outputs the element d of type data and then behaves like the process p.

The sort of processes is denoted by pcs. The corresponding specification reads

spec TRACES
  monotonic
  include DATA
  basic type pcs
  $\top$: pcs
  $!\_ \to \_$: data $\to$ pcs $\to$ pcs
  axioms $\top \leq p$
endspec

The only terms of type pcs in TRACES are of the form

$$!d_1 \to (!d_2 \to \cdots (!d_k \to \top) \cdots)$$

with primitive terms d_i of type data. Since the only axioms besides leastness of $\top$ are monotonicity axioms, in an initial model I of TRACES we have

$$!d_1 \to (!d_2 \to \cdots (!d_m \to \top) \cdots)^I \leq^I$$
$$!e_1 \to (!e_2 \to \cdots (!e_n \to \top) \cdots)^I$$

iff $m \leq n$ and for all $i \leq m$ then $d_i^I = e_i^I$. Thus the carrier pcs^I is order-isomorphic to the set of all finite sequences of data elements under the prefix ordering. Hence, for the completion I^∞, pcs^{I^∞} is order-isomorphic to the set of all finite and infinite sequences of data elements, again under the prefix ordering. In the terminology of CSP, these sequences are also called traces, whence the name of our specification. More frequently, these sequences are also called streams.

In the next step, we add the operator $\_ \vee \_$ of disjunction of processes. It is used to form processes that offer alternative behaviour. The operator $\_ \vee \_$ is associative, commutative, and idempotent; moreover, a disjunction is at most as reliable as either of its alternatives.

spec TRACESETS
monotonic
include TRACES
$\_ \vee \_ : \mathsf{pcs} \to \mathsf{pcs} \to \mathsf{pcs}$
axioms $(p \vee q) \vee r = p \vee (q \vee r)$
$p \vee q = q \vee p$
$p \vee p = p$
$p \vee q \leq p$
$(!d \to p) \vee (!d \to q) \leq\, !d \to (p \vee q)$
endspec

The last axiom, together with the derivable property

$$!d \to (p \vee q) \leq (!d \to p) \vee (!d \to q)$$

implies

$$!d \to (p \vee q) = (!d \to p) \vee (!d \to q),$$

i.e., that output distributes over disjunction, as is stipulated in [19]. This property allows us to write every process as a disjunction of traces.

Another derivable property is

$$p \leq q \Leftrightarrow p \vee q = p\,. \tag{1}$$

In [19] this property is taken as the definition of the refinement ordering, since there the ordering is a derived concept, whereas in the present framework it is, of course, a basic notion. This property allows omitting extensions of traces in a disjunction from that disjunction without changing the result. Together with the associativity, commutativity, and idempotence of choice this implies that the carrier pcs^I in an initial model of TRACESETS is isomorphic to the set of finite sets of traces that are not prefixes of each other. Hence in the completion I^∞ the carrier pcs^{I^∞} is isomorphic to the Smyth power domain (cf. [41]) over the set of finite and infinite traces.

From (1) one obtains immediately

$$\top \vee q = \top\,, \tag{2}$$

which states that disjunction is "demonic", i.e., that a process offering a totally unreliable alternative is totally unreliable itself. Note that (2) depends crucially on the axiom $p \vee q \leq p$. By using the dual axiom $p \leq p \vee q$ we would obtain an "angelic" disjunction with $\top \vee p = p$.

One can also show that the disjunction of two processes is the greatest lower bound of these processes w.r.t. the refinement ordering; hence in all models of TRACESET the carrier of sort pcs is a lower semilattice.

In the next step we introduce an input command. In CSP this is coupled with a binding mechanism: The input value is bound to an identifier under which it is available throughout the remaining process. However, to keep the description simple, we want to avoid the introduction of identifiers and the problems of binding. Rather, we use a combinator variant of input and represent a process depending on a free identifier for data elements as a function from data to pcs. The input command may only be applied to such a parameterised process *pp*; after input of a data value *d* the process then behaves like *pp*(*d*). We extend disjunction pointwise to parameterised processes. The operator ©. lifts a process to a constant parameterised process.

spec INPUT
  monotonic
  include TRACESETS
  $? \rightarrow \_ : (\mathsf{data} \rightarrow \mathsf{pcs}) \rightarrow \mathsf{pcs}$
  $\_ \vee \_ : (\mathsf{data} \rightarrow \mathsf{pcs}) \rightarrow (\mathsf{data} \rightarrow \mathsf{pcs}) \rightarrow (\mathsf{data} \rightarrow \mathsf{pcs})$
  $©\_ : \mathsf{pcs} \rightarrow (\mathsf{data} \rightarrow \mathsf{pcs})$
  axioms $? \rightarrow (pp \vee qq) = (? \rightarrow pp) \vee (? \rightarrow qq)$
    $(pp \vee qq)(d) = pp(d) \vee qq(d)$
    $(©p)(d) = p$
endspec

Let us now specify a process representing an unbounded buffer. In this, we follow [7]. Since a buffer process internally maintains the sequence of buffered values, we first specify sequences of data values.
[] denotes the empty sequence, [_] is the singleton sequence former, and _+_ denotes concatenation.

spec SEQU =
  include DATA
  basic type sequ
  $[] : \mathsf{sequ}$
  $[\_] : \mathsf{data} \rightarrow \mathsf{sequ}$
  $\_ + \_ : \mathsf{sequ} \rightarrow \mathsf{sequ} \rightarrow \mathsf{sequ}$
  axioms $[] + s = s$
    $s + [] = s$
    $r + (s + t) = (r + s) + t$
endspec

spec BUFFER
  monotonic
  include SEQU
  $buf\_ : \mathsf{sequ} \rightarrow \mathsf{pcs}$
  $enterbuf\_ : \mathsf{sequ} \rightarrow (\mathsf{data} \rightarrow \mathsf{pcs})$
  axioms $buf_{[]} = ? \rightarrow enterbuf_{[]},$

$$buf_{s+[x]} = (? \rightarrow enterbuf_{s+[x]}) \vee (!x \rightarrow buf_s)$$
$$enterbuf_t\, y = buf_{[y]+t}$$

endspec

These axioms may be interpreted as follows: A buffer with empty internal sequence can only input and then behave like the enterbuf process with empty internal sequence. A buffer with non-empty internal sequence either may input another value and pass it to the enterbuf process or output the last buffer value and behave like a buffer with a shortened internal sequence. The enterbuf process merely attaches the input value to the front of its internal sequence and then behaves like a buffer. This specification shows how, using higher-order concepts, we can give the use of indices in [7] a precise foundation. Further CSP constructs such as chaining and interleaving can be introduced similarly following [19].

Since the whole specification is inequational, grounded (with $\top$ instead of $\bot$), and monotonic, the constructions of Sections 4.6 and 4.7 yield fixpoint models of PCSCOMB in which solutions to these recursion equations exist. As an example for such a recursion equation consider

$$\text{pcs } dinf = !d \rightarrow dinf$$

for some fixed $d \in$ **data**. Its least solution is the process that emits an infinite number of ds. However, this equation is of first order and so already the techniques of [26, 28] would have sufficed to give its semantics. A higher-order recursion is

$$\text{pcs } finf = ? \rightarrow \lambda x.\, !f\, x \rightarrow finf$$

for some fixed $f \in$ **data** $\rightarrow$ **data**. This is an infinitary version of the apply-to-all operation from functional languages: an infinite process fed into *finf* by chaining (see [19]) gets all its output transformed by f.

Finally it should be remarked that the description given leads to problems with fairness when infinite processes are considered. However, this is due to the model chosen in [19] and not to the underlying framework of higher-order specification. The present example was meant to show how the semi-formal treatment of [7, 19] can be made fully formal within the proposed specification style; it is not an attempt at giving a new semantics for parallelism.

6 Conclusion and Outlook

The approach to higher-order specifications using a generation principle also for higher types has proved to lead to a framework which preserves the simplicity and clarity of the first-order case. Higher-order specifications can be formulated in a straightforward way; for Horn axioms the existence of extensional models is guaranteed, so that in this case one can use the higher-order framework quite naively and actually conceive of the higher-order objects as functions. We deem this of high importance for the practical acceptance of the idea of formal specification in general and of higher-order specifications in particular. Only if the framework is simple, programmers will actually want to use it.

Although the construction of fixpoint models turns out to be more awkward than in the first-order case, the way how specifications need to be written is still fairly straightforward.

It will be interesting to see whether from Kleene's approximation sequence for the solution of recursion equations over specifications one can derive busy and lazy operational semantics as was done in [27] for the first-order case. Such an investigation should go hand in hand with a search for deduction-oriented sufficient criteria for hierarchy-completeness extending the ones in [17] to the higher-order case. Similarly, deduction-oriented sufficient criteria for relative consistency w.r.t. subspecifications should be developed.

Of course, a large field of further research is the investigation of wider classes of axioms that admit extensional and fixpoint models.

Acknowledgement. This work has been initiated by discussions with my colleagues from the project CIP and with members of IFIP WG 2.1. I am most grateful to F.L. Bauer, M. Broy, W. Dosch, H. Ehler, K. Meinke, F. Nickl, H. Partsch, O. Paukner, A. Poigné, A. Tarlecki, and M. Wirsing for their valuable comments.

References

1. J. Adámek, A.H. Mekler, E. Nelson, J. Reiterman: On the logic of continuous algebras. Notre Dame J. Formal Logic **29**, 365–380 (1988)
2. J. Backus: Can programming be liberated from the von Neumann style? A functional style and its algebra of programs. Commun. ACM **21**, 613-641 (1978)
3. F.L. Bauer, H. Wössner: Algorithmic language and program development. Berlin: Springer 1982
4. F.L. Bauer, R. Berghammer, M. Broy, W. Dosch, F. Geiselbrechtinger, R. Gnatz, E. Hangel, W. Hesse, B. Krieg-Brückner, A. Laut, T. Matzner, B. Möller, F. Nickl, H. Partsch, P. Pepper, K. Samelson, M. Wirsing, H. Wössner: The Munich project CIP. Volume I: The wide spectrum language CIP-L. Lecture Notes in Computer Science **183**. Berlin: Springer 1985
5. G. Birkhoff: Lattice theory, 3rd edition. American Mathematical Society Colloquium Publications, Vol. XXV. Providence, R.I.: AMS 1967
6. S. Bloom: Varieties of ordered algebras. J. Computer Syst. Sci. **13**, 200–212
7. S.D. Brookes, C.A.R. Hoare, A.W. Roscoe: A theory of communicating sequential processes. J. ACM **31**, 560-599 (1984)
8. M. Broy: Partial interpretations of higher order algebraic types. In: M. Broy (ed.): Logic of Programming and Calculi of Discrete Design. NATO ASI Series. Series F: Computer and Systems Sciences, Vol. 36. Berlin: Springer 1987, 185–241
9. M. Broy, M. Wirsing: Initial versus terminal algebra semantics for partially defined abstract types. Institut für Informatik der TU München, TUM-I8018, 1980
10. B. Courcelle, M. Nivat: Algebraic families of interpretations. 17th Annual IEEE Symposium on Foundations of Computer Science, 1976, 137-146
11. B. Courcelle, J.-C. Raoult: Completions of ordered magmas. Fundamenta Informaticae **3**, 105-111 (1980)

12. B.A. Davey, H.A. Priestley: Introduction to lattices and order. Cambridge: Cambridge University Press 1990
13. P. Dybjer: Category-theoretic logics and algebras of programs. Chalmers University of Technology at Göteborg, Dept. of Computer Science, Ph.D. Thesis, 1983
14. H.B. Enderton: A mathematical introduction to logic. New York: Academic Press 1972
15. J.A. Goguen, J. Thatcher, E.G. Wagner, J.B. Wright: Initial algebra semantics and continuous algebras. J. ACM **24**, 68-95 (1977)
16. T. Grünler: Spezifikationen höherer Ordnung. Fakultät für Mathematik und Informatik der Universität Passau, Dissertation 1990
17. J.V. Guttag: The specification and application to programming of abstract data types. PH. D. Thesis, University of Toronto, Department of Computer Science, Rep. CSRG-59, 1975
18. L. Henkin: Completeness in the theory of types. J. Symbolic Logic **15**, 81-91 (9150)
19. C.A.R. Hoare: Algebraic specifications and proofs for communicating sequential processes. In: M. Broy (ed.): Logic of programming and calculi of discrete design. NATO ASI Series. Series F: Computer and Systems Sciences, Vol. 36. Berlin: Springer 1987, 277–300
20. B. Mahr, J.A. Makowsky: Characterizing specification languages which admit initial semantics. Theoretical Computer Science **31**, 49-59 (1984)
21. T.S.E. Maibaum, C.J.Lucena: Higher order data types. International Journal of Computer and Information Sciences **9**, 31-53 (1980)
22. A.I. Mal'cev: Ob umnoženii klassov algebraičeskih sistem. Sibir. Mat. Ž. **8**, 346–365 (1967). English translation: A few remarks on quasi varieties of algebraic systems. In A.I. Mal'cev: The metamathematics of algebraic systems. Collected papers 1936–1967. Studies in Logic and the Foundations of Mathematics, Vol. 66. Amsterdam: North-Holland 1971, 416–421
23. G. Markowsky: Chain-complete posets and directed sets with applications. Algebra Universalis **6**, 53-68 (1976)
24. K. Meinke: A recursive second order initial algebra specification of primitive recursion. Report CSR 8-91, Dept. of Computer Science, University College of Swansea, 1991. Also: Acta Informatica (to appear)
25. K. Meinke: Universal algebra in higher types. Theoretical Computer Science **100**, 385–417 (1992)
26. B. Möller: Unendliche Objekte und Geflechte. Fakultät für Mathematik und Informatik der TU München, Dissertation, Report TUM-I8213, 1982
27. B. Möller: An algebraic semantics for busy (data-driven) and lazy (demand-driven) evaluation and its application to a functional language. In: J. Diaz (ed.): Automata, languages and programming. Lecture Notes in Computer Science **154**. Berlin: Springer 1983, 513-526
28. B. Möller: On the algebraic specifications of infinite objects - Ordered and continuous models of algebraic types. Acta Informatica **22**, 537-578 (1985)
29. B. Möller: Algebraic specifications with higher-order operators. In L.G.L.T. Meertens (ed.): Program Specification and Transformation. Amsterdam: North-Holland 1987, 367–398
30. B. Möller: Higher-order algebraic specifications. Fakultät für Mathematik und Informatik der TU München, Habilitationsschrift 1987
31. B. Möller, A. Tarlecki, M. Wirsing: Algebraic specifications with built-in domain constructions. In M. Dauchet, M. Nivat (eds.): Proc 13th CAAP, Nancy, March

21–24, 1988. Lecture Notes in Computer Science **299**. Berlin: Springer 1988, 132–148

32. B. Möller, A. Tarlecki, M. Wirsing: Algebraic specifications of reachable higher-order algebras. In: D. Sannella, A. Tarlecki (eds.): Recent trends in data type specification. Lecture Notes in Computer Science **332**. Berlin: Springer 1988, 154–169
33. F. Nickl: Algebraic specifications for domain theory. In: B. Monien, R. Cori (eds.): Theoretical aspects of computer science. Lecture Notes in Computer Science **349**. Berlin: Springer 1989, 360–375. Extended version: Algebraic specifications of domain constructions. Fakultät für Mathematik und Informatik der Universität Passau, Report MIP-8815, 1988
34. M. Nivat: On the interpretation of recursive polyadic program schemes. Istituto Nazionale di Alta Matematica, Symposia Matematica XV. London: Academic Press 1975, 255-281
35. P. Padawitz: computing in Horn clause theories. EATCS Monographs on Theoretical Computer Science **16**. Berlin: Springer 1988
36. K. Parsaye-Ghomi: Higher-order abstract data types. Dept. of Computer Science, University of California at Los Angeles, Ph.D. Thesis, 1981
37. G.D. Plotkin: LCF considered as a programming language. Theoretical Computer Science **5**, 223-255 (1977)
38. A. Poigné: Higher-order data structures - cartesian closure versus λ-calculus. In: M. Fontet, K. Mehlhorn (eds.): Theoretical aspects of computer science. Lecture Notes in Computer Science **166**. Berlin: Springer 1984, 174-185
39. A. Poigné: On specifications, theories, and models with higher types. Information and Control **68**, 1-46 (1986)
40. D.S. Scott: Outline of a mathematical theory of computation. Proc. 4th Annual princeton Conference on Information Sciences and Systems, 1970, 169-176, and Technical Monograph PRG-2, Oxford University, Computing Laboratory, Programming Research Group, 1970
41. M.B. Smyth: Power domains. J. Computer Syst. Sci. **16**, 23-36 (1978)
42. A. Tarlecki: On the existence of free models in abstract algebraic institutions. Theoretical Computer Science **37**, 269–304 (1985)
43. A. Tarlecki: Quasi-varieties in abstract algebraic institutions. Journal of Computer and System Sciences **33**, 333–360 (1986)
44. A. Tarlecki, M. Wirsing: Continuous abstract data types. Fundamenta Informaticae **9**, 95-125 (1986)
45. Q. Zhenyu: An algebraic semantics of higher-order types with subtypes. Acta Informatica **30**, 569–607 (1993)

Rewriting Properties of Combinators for Rudimentary Linear Logic

Monica Nesi[1], Valeria de Paiva[1] and Eike Ritter[2]

[1] Computer Laboratory, University of Cambridge
[2] Computing Laboratory, Oxford University

1 Introduction

In this paper we investigate the possibility of developing a (semi-)automatic rewriting tool for manipulating and reasoning about combinators for intuitionistic linear logic. In particular, we develop a term rewriting system for a theory of categorical combinators for rudimentary linear logic. We make use of the Knuth-Bendix completion algorithm [8] to transform the equational theory for the combinators into an equivalent canonical rewrite system. This means that a set of categorical combinators for linear logic has first to be derived, and then the resulting system of combinators can be checked for rewriting properties using rewriting techniques.

Intuitionistic linear logic was introduced by Girard in [11]. The basic assumption of linear logic is that one should be able to have a logical control of the resources available for a derivation. This resource-sensitiveness of linear logic is its main claim to applicability, and applications have sprung up in all areas of computer science. An introduction to linear logic and some of its applications can be found in [22]. Since (multiplicative-exponential) intuitionistic linear logic is a refinement of intuitionistic logic, it makes sense to develop a linear λ-calculus [2] and a system of linear combinators [17, 21]. Our main concern here is with a calculus of (linear) categorical combinators that models intuitionistic linear logic faithfully and efficiently. The relationship of our categorical combinators to category theory and intuitionistic linear logic is explained in [21]; here we mention only the key ideas to make the paper self-contained.

Curien's categorical combinators for the λ-calculus, based on cartesian closed categories, were used to derive the categorical abstract machine CAM [5, 6]. Lafont [17] used symmetric monoidal categories to derive combinators for linear logic. Both Curien's and Lafont's combinators model substitutions and terms as morphisms in the appropriate categories. These combinators do not make explicit the conceptual distinction between terms and substitutions. Such a distinction is instead present in the $\lambda\sigma$-calculus, a λ-calculus with explicit substitutions [1], and in the categorical combinators for the Calculus of Constructions [19]. Because substitutions model what is usually called an environment (and abstract machines are by their nature environment machines), the categorical combinators that do not identify terms and substitutions make categorical abstract machines easier to design and to show correct [19]. Our combinators based on syntactic multicategories do not make the identification, and context morphisms model substitutions precisely. Therefore, we obtain combinators which should yield better behaved categorical abstract machines.

The process of deriving categorical combinators has been a relatively long one, with a fruitful interaction between the more abstract side of the work and its mechanized version. We started with a (theoretically correct) formulation of the combinators for each one of the several fragments of the logic under consideration. As usual, the use of an automatic rewriting tool to derive and check properties such as local confluence, termination and canonicity of a system, has led us to "improve" the first axiomatic characterization of the categorical combinators, until an equivalent presentation was derived with the nice feature that rewriting techniques can be applied to it successfully. This process of transformation of the data provided by the theoretical considerations follows a definite pattern that is described in Section 3.

The paper is organized as follows. First, we mention some basic facts about intuitionistic linear logic, categorical combinators and term rewriting systems. We then describe how our system of categorical combinators has been derived. The next sections discuss some of the rewriting properties of such a system, namely termination and normal forms. Finally, we draw some conclusions and mention some further work.

2 Background

In this section we recall briefly intuitionistic linear logic, and some basic facts about categorical combinators and term rewriting systems.

2.1 Intuitionistic Linear Logic

Intuitionistic linear logic [11] is a subsystem of classical linear logic, which consists of multiplicative, additive and exponential fragments. The heart of intuitionistic linear logic is its multiplicative-exponential fragment, sometimes referred to as *MELL*. In Fig. 1 we recall the (structural and logical) rules for multiplicative-exponential intuitionistic linear logic using a sequent calculus system.

The system of (multiplicative-exponential) intuitionistic linear logic is a refinement of intuitionistic logic, where besides the linear implication '$\multimap$' and (a special kind of) conjunction, namely the tensor product '$\otimes$', we have the modal operator '!', which controls the use of hypotheses in the derivations. The fragment of *MELL* without the rules for the modality '!', i.e. the tensor-linear implication fragment, is called *rudimentary* linear logic.

2.2 Categorical Combinators

The Curry-Howard correspondence has been used by Curien [5, 6] to develop categorical combinators for the simply typed λ-calculus (and, by omitting the type information, also for the untyped λ-calculus). These combinators are an

$$\frac{}{A \vdash A}\,(Identity) \qquad \frac{\Gamma, A, B, \Delta \vdash C}{\Gamma, B, A, \Delta \vdash C}\,(Exchange)$$

$$\frac{\Gamma \vdash B \qquad B, \Delta \vdash C}{\Gamma, \Delta \vdash C}\,(Cut)$$

$$\frac{\Gamma, A, B \vdash C}{\Gamma, A \otimes B \vdash C}\,(\otimes_{\mathcal{L}}) \qquad \frac{\Gamma \vdash A \qquad \Delta \vdash B}{\Gamma, \Delta \vdash A \otimes B}\,(\otimes_{\mathcal{R}})$$

$$\frac{\Gamma \vdash A \qquad \Delta, B \vdash C}{\Gamma, \Delta, A \multimap B \vdash C}\,(\multimap_{\mathcal{L}}) \qquad \frac{\Gamma, A \vdash B}{\Gamma \vdash A \multimap B}\,(\multimap_{\mathcal{R}})$$

$$\frac{\Gamma \vdash B}{\Gamma, !A \vdash B}\,(weakening) \qquad \frac{\Gamma, !A, !A \vdash B}{\Gamma, !A \vdash B}\,(contraction)$$

$$\frac{\Gamma, A \vdash B}{\Gamma, !A \vdash B}\,(dereliction) \qquad \frac{!\Gamma \vdash A}{!\Gamma \vdash !A}\,(promotion)$$

Fig. 1. Multiplicative-Exponential Intuitionistic Linear Logic

equational presentation of the cartesian closed categorical structure and can be used to derive categorical abstract machines, e.g. the CAM, as follows: first, turn the equations into reduction (or rewrite) rules, then define a suitable reduction strategy and finally construct an environment machine implementing the chosen strategy [5, 19].

But the use of cartesian closed categories to construct environment machines is somewhat problematic because several concepts, that are important for the design of the abstract machines, are not properly modelled in cartesian closed categories:

- environments and terms are both represented by morphisms although they are conceptually different;
- composition has two roles, namely substitution in a term with respect to environments and application of a function to an argument;
- product types and contexts are both modelled by products in the cartesian closed category, so again two separate issues are merged into one construction.

As a consequence, the correctness proof of the CAM becomes rather complicated [19]. These problems have been addressed by Ehrhard [9], who proposed to model the simply typed λ-calculus by means of (full constant split) D-categories. The derived categorical abstract machine is then more efficient, as argued in [19].

A similar kind of situation arises with intuitionistic linear logic and the linear λ-calculus, which can be derived from the logic via the Curry-Howard correspondence [2]. The usual categorical model of intuitionistic linear logic is given in terms of symmetric monoidal closed categories, and for this model Lafont derived combinators in [17]. Lafont's combinators model contexts as tensor products in the symmetric monoidal categories, and substitutions and terms as morphisms in these categories. Hence this approach suffers from the same problems as the one based on cartesian closed categories. Our combinators, in the spirit of Ehrhard's D-categories, are instead based on *syntactic multicategories* [21], an improvement of Lambek's multicategories. These combinators do make the distinction between terms and substitutions and context morphisms model substitutions precisely. We therefore obtain combinators that should yield better behaved categorical abstract machines. To construct the machine, we must first turn the categorical equations into rewrite rules. The transformation of the categorical equations into rewrite rules is described in Section 3.3.

2.3 Term Rewriting Systems

We assume that the reader is familiar with the basic notions of term rewriting systems. Below we summarize the most relevant definitions, and we refer to [8] for more details.

Let $F = \bigcup_n F_n$ be a set of function symbols, where F_n is the set of symbols of arity n. Let $\mathcal{T}(F,X)$ be the set of (finite, first order) terms with function symbols F and variables X. An *equation* is an unordered pair (s,t) of terms, written $s = t$. The terms s and t may contain variables, which are understood as being universally quantified. Given a (finite or infinite) set of equations E over $\mathcal{T}(F,X)$, the *equational theory* of E, $Th(E)$, is the set of equations that can be obtained by taking reflexivity, symmetry, transitivity and context application as inference rules and all instances of equations in E as axioms.

A *term rewriting system* (TRS) or *rewrite system* R is any (finite or infinite) set $\{(l_i,r_i) \mid l_i, r_i \in \mathcal{T}(F,X), Var(r_i) \subseteq Var(l_i)\}$. The pairs (l_i,r_i) are called *rewrite rules* and written $l_i \to r_i$. The *rewriting relation* $\to_R$ over $\mathcal{T}(F,X)$ is defined as the smallest relation containing R that is closed under context application and substitution. A term t *rewrites* to a term s, written $t \to_R s$, if there exists $l \to r$ in R, a substitution σ and a subterm $t|_u$ at the position u, called *redex*, such that $t|_u = l\sigma$ and $s = t[r\sigma]_u$. A term t is said to *overlap* a term t' if t unifies with a non-variable subterm of t' (after renaming the variables in t so as not to conflict with those in t'). If $l_i \to_R r_i$ and $l_j \to_R r_j$ are two rewrite rules (with distinct variables), u is the position of a non-variable subterm of l_i, and σ is a most general unifier of $l_i|_u$ and l_j, then the equation $(l_i\sigma)[r_j\sigma]_u = r_i\sigma$ is a *critical pair* formed from those rules.

Let $\overset{+}{\to}_R$ and $\overset{*}{\to}_R$ denote the transitive and reflexive-transitive closure of $\to_R$ respectively. A TRS R is *terminating* if there is no infinite sequence $t_1 \to_R t_2 \to_R \ldots$ in R. A TRS R is *confluent* if whenever $s \;{}_R\!\overset{*}{\leftarrow} t \overset{*}{\to}_R q$, there exists a term t' such that $s \overset{*}{\to}_R t' \;{}_R\!\overset{*}{\leftarrow} q$, and R is *locally confluent* if whenever $s \;{}_R\!\leftarrow t \to_R q$, there exists a term t' such that $s \overset{*}{\to}_R t' \;{}_R\!\overset{*}{\leftarrow} q$. A term t is *in R-normal form* if there is no term s such that $t \rightarrow_R s$. A term s is an *R-normal form* of t if $t \overset{*}{\rightarrow}_R s$ and s is in R-normal form. A TRS R is said to be *canonical* (or *complete*) if it is terminating and confluent.

Finite rewrite systems have a finite number of critical pairs, and local confluence can be decided by the *Critical Pair Lemma* using the *superposition test* by checking that there exists a term t' such that $s \overset{*}{\rightarrow}_R t' \;{}_R\overset{*}{\leftarrow} q$ for any peak situation $s \;{}_R{\leftarrow} t \rightarrow_R q$. A terminating relation is confluent if and only if it is locally confluent (*Diamond* or *Newman Lemma*). Thus, the confluence of a finite rewrite system is decidable, provided it is terminating. A *completion* procedure, such as the Knuth-Bendix completion algorithm, is a program which, given a finite set E of equations and a term ordering $\succ$, generates a canonical rewrite system R (if it exists) equivalent to the equational theory for E. A completion procedure *diverges* if it generates a rewrite system with infinitely many rules [15, 16].

3 A Rewrite System for the Categorical Combinators

Before we describe the development of the rewrite system in detail, we give some explanation about the categorical structure on which our combinators are based, and indicate how the combinators can be derived from this structure.

3.1 Categorical Models

The categorical combinators for rudimentary linear logic are based on syntactic multicategories, called S-multicategories. The key idea behind the definition of an S-multicategory is that it provides separate notions for environments, i.e. lists of terms, and terms themselves. An S-multicategory consists of the following data:

- types, which contain ground types, tensor product and linear implication;
- contexts, which are lists of types;
- for every context Γ, there is a collection of multimaps t of type A, written $\Gamma \xrightarrow{t} A$;
- context morphisms, which correspond to lists of multimaps. Contexts and context morphisms form a category.

We also have the following operations:

- a substitution operation '$*$', assigning a multimap $f * t$ to every context morphism $f = (t_n, \ldots, t_0)$ and every multimap t;
- a list-forming operation '$(\_, \_)$', taking a context morphism f and a multimap t and producing a context morphism (f, t).

We refer to Appendix A for the requirements that these operations must satisfy.

The relation to the term assignment system for linear logic is as follows:

Calculus	Multicategory
Type A	Type A
Context $(x_n\colon A_n, \ldots, x_0\colon A_0)$	Context $(A_n, \ldots, A_0)$
Term $\Gamma \vdash t\colon A$	Multimap $\Gamma \xrightarrow{t} A$

As already mentioned, one important aspect of S-multicategories is that they model explicit substitutions. The term assignment system admits substitution as an implicit operation, corresponding to a multiple Cut in the logic:

$$\frac{\Gamma \vdash t_i\colon A_i \ (0 \leq i \leq n) \qquad (x_n\colon A_n, \ldots, x_0\colon A_0) \vdash t\colon A}{\Gamma \vdash t[t_i/x_i]\colon A} \ (Cut^*)$$

The term $t[t_i/x_i]$ corresponds to the multimap $(((\mathsf{Id}, t_n), \ldots), t_0) * t$, where Id represents the identity substitution. The substitution $[t_i/x_i]$ is modelled by a different part of the categorical structure than a term t, namely by a context morphism $(((\mathsf{Id}, t_n), \ldots), t_0)$ rather than a multimap.

Tensor product and linear implication are modelled by the following natural bijections:

$$\mathsf{Multi}((\Gamma, A, B); C) \cong \mathsf{Multi}((\Gamma, A \otimes B); C)$$
$$\mathsf{Multi}((\Gamma, A); B) \cong \mathsf{Multi}(\Gamma; A \multimap B)$$

where $\mathsf{Multi}(\Gamma; A)$ denotes the set of all multimaps $\Gamma \xrightarrow{t} A$. The naturality of these bijections means intuitively that they respect substitution.

3.2 Developing the Categorical Combinators

Based on the categorical structure described above, we have four kinds of combinators corresponding to the four kinds of constructions of an S-multicategory:

- types A, which consist of ground types G, tensor product and linear implication;
- contexts Γ, which are lists of types;
- multimaps t, which correspond "essentially" to linear terms;
- lists of multimaps f, which are the morphisms between contexts.

The original syntax for the combinators was the following:

Definition 1 (Raw linear combinators)

$$\begin{aligned} A &::= G \mid A \otimes A \mid A \multimap A \\ \Gamma &::= () \mid (\Gamma, A) \\ t &::= n \mid \otimes \mid \mathsf{L} \mid \mathsf{R} \mid \mathsf{Cur}(t) \mid \mathsf{App} \mid f * t \\ f &::= \mathsf{Id} \mid (f, t) \mid f; f \end{aligned}$$

The intuition behind the combinators is as follows. The combinators $()$, (Γ, A), (f, t) and $f * t$ model the corresponding structure of a multicategory, and the combinators Id and $f; g$ correspond to the categorical identity and composition.

The modelling of the part of the categorical structure that represents variables is closely related to de Bruijn-numbers. The multimap n corresponds to the de Bruijn-number n, which indicates the position of a variable in a context. These numbers are associated to the right, i.e. the number 0 corresponds to the rightmost variable x_0 in the context $(x_n: A_n, \ldots, x_0: A_0)$. A permutation of a context of length n, which is a special context morphism corresponding to the exchange rule, can be simply modelled by its effect on the natural numbers 0 to $n-1$.

The combinators $\otimes$, L and R model the tensor product. The combinator $\otimes$ corresponds to the identity map $(A\otimes B) \to A\otimes B$ under the bijection defining the tensor product, and L and R decompose a tensor product in its components. The combinators Cur and App model the bijection defining the linear implication.

The well-formed linear combinators are defined in two steps. First, we define well-formed combinators (not necessarily linear)

$$\Gamma \vdash t: A$$
$$\Gamma \vdash f: \Delta$$

meaning that in context Γ the multimap t has type A and the context morphism f has type Δ (a list of types). All these judgements are listed in Fig. 2.

- All types and contexts are well-formed.
- Multimaps:

$$\frac{m \geq n}{(A_m, \ldots, A_0) \vdash n: A_n} \qquad \frac{}{(\Gamma, A, B) \vdash \otimes: A\otimes B}$$

$$\frac{}{(\Gamma, A\otimes B) \vdash \mathsf{L}: A} \qquad \frac{}{(\Gamma, A\otimes B) \vdash \mathsf{R}: B}$$

$$\frac{(\Gamma, A) \vdash t: B}{\Gamma \vdash \mathsf{Cur}(t): A \multimap B} \qquad \frac{}{(\Gamma, A \multimap B, A) \vdash \mathsf{App}: B}$$

$$\frac{\Gamma \vdash f: \Delta \qquad \Delta \vdash t: A}{\Gamma \vdash f * t: A}$$

- Lists of multimaps:

$$\frac{}{\Gamma \vdash \mathsf{Id}: \Gamma} \qquad \frac{\Gamma \vdash f: \Delta \qquad \Gamma \vdash t: A}{\Gamma \vdash (f, t): (\Delta, A)} \qquad \frac{\Gamma \vdash f: \Gamma' \qquad \Gamma' \vdash g: \Gamma''}{\Gamma \vdash f; g: \Gamma''}$$

Fig. 2. Rules for well-formed combinators

In a second step we annotate these judgements with *occurrence items* that list all parts of the context that a multimap or a context morphism uses. For example, the judgement

$$(A_m, \ldots, A_0) \mid n \vdash n: A_n \quad (m \geq n)$$

is annotated by a natural number n indicating that the multimap n uses the type A_n and nothing else. The well-formed linear combinators are defined as those well-formed combinators that use every part of the context exactly once. This is made precise by the use of an occurrence check. Note that L and R are not linear, but the occurrence check will ensure that every component of a tensor product is used exactly once. Because the occurrence items do not interact with the rewrite system, they are not explained here; we refer to [18] for more details.

3.3 Developing the Term Rewriting System

The above combinators can be characterized through a set of equations suggested by category theory. We developed these equations and studied their rewriting properties in a modular way by first considering the structural rules, then adding the logical rules for the tensor product, and finally the ones for the linear implication. The aim was to obtain a canonical rewrite system in which normal forms are the translations of those of the term calculus. We also wanted to avoid non-left-linear rewrite rules as much as possible, as the implementation of non-left-linear rules requires a costly equality check. The LP (Larch Prover) system [10] was used for executing the Knuth-Bendix completion procedure.

The Structural Rules These rules describe the structure of an S-multicategory. Such a structure is captured by the equations for the identity Id in the base category CM and for associativity of composition

$$\mathsf{Id}; f = f \tag{1}$$

$$f; \mathsf{Id} = f \tag{2}$$

$$f; (g; h) = (f; g); h \tag{3}$$

and the equations for the functor '$*$'

$$\mathsf{Id} * t = t \tag{4}$$

$$f * (g * t) = (f; g) * t \tag{5}$$

The combinators for the bijection defining context extensions (clause 1 in Definition 9 in Appendix A) are derived in a way similar to the manner combinators for adjunctions are obtained [4]. There is a combinator $(\_, \_)$, representing the categorical operation $(\_, \_)$. The other direction of the bijection is represented by its effect on the identity morphism. This is sufficient because the naturality conditions can be used to define this direction of the bijection for all morphisms. The equations between the combinators can in principle be derived according to the pattern for adjunctions, too. However, the naturality condition 2 of Definition 9 causes a problem when we try to derive equations for it. The equations in the definition involve typing information, hence they are *not* suitable as equations between raw combinators. We solve this problem by using combinators for non-linear logics and specifying the linear combinators as a subset of them. More precisely, we introduce a combinator n for every natural number n to indicate which part of a context is accessed. Then the morphisms f^c is modelled by the combinator $(f * n, \ldots, f * 1)$ and the multimap f^t by $f * 0$.

The terminology for the equations is adapted to the case of the λ-abstraction, which can also be characterized by an adjunction. First, we have the β-equations, saying in this case that the application of $(\_,\_)$ first and then the other direction of the bijection yields the combinator we started with. In this case, these equations model variable access:

$$(f,t) * (n+1) = f * n$$
$$(f,t) * 0 = t$$

The adjunction also yields an equation expressing that the composition of the two maps in the other order is the identity. This equation specifies an extensionality condition and asserts in this case that two context morphisms are equal if and only if they consist of the same list of terms. It is the following one:

$$(\mathsf{Id}, n, \ldots, 0) = \mathsf{Id}$$

where both sides are context morphisms from Γ to Γ. This equation is not necessary for the correspondence between the type theory and the combinators, but can be useful for optimizations in the categorical abstract machine. It is omitted here because it requires again typing information to be included in the combinators. Finally, the naturality condition is captured by the following equation, which justifies the introduction of the non-linear combinators:

$$f;(g,t) = (f;g, f * t) \tag{6}$$

The above equations (except the second but last one) oriented from left to right are a locally confluent rewrite system.

The Tensor Fragment The combinators for the tensor product are derived from the categorical bijection

$$\mathsf{Multi}((\Gamma, A, B); C) \cong \mathsf{Multi}((\Gamma, A{\otimes}B); C)$$

as follows. The direction from right to left can be easily captured by its effect on the identity multimap $(A{\otimes}B) \xrightarrow{\mathsf{Id}} (A{\otimes}B)$ and appropriate naturality conditions. This yields a combinator $\otimes$. The other direction of the bijection cannot be captured by its effect on the identity morphism because there is no multimap from (A, B) to itself. Again, a detour via non-linear logic solves the problem. We introduce two (non-linear) multimaps $(A{\otimes}B) \xrightarrow{\mathsf{L}} A$ and $(A{\otimes}B) \xrightarrow{\mathsf{R}} B$. The context morphism (L, R) (which is non-linear, because the context $A{\otimes}B$ is used twice) is used as the replacement for the non-existing multimap.

The difference between the derivation of the combinators for context extensions and those for the tensor product is the modelling of the operation that describes the new construction to be introduced. In the former case we introduced a new operation whereas in the latter we describe only its effect on identities. The advantage of this version is a purely technical one: if we had introduced a new operation, the naturality conditions would have required typing information again. The equivalence between these two versions can be easily shown by using a standard categorical argument about adjunctions.

The naturality rules, as derived from category theory, are as follows:

$$((f,t),s) * \otimes = ((\mathsf{Id},t),s) * \otimes$$
$$(f,t) * \mathsf{L} = (\mathsf{Id},t) * \mathsf{L}$$
$$(f,t) * \mathsf{R} = (\mathsf{Id},t) * \mathsf{R}$$

As soon as these equations are turned into rewrite rules from left to right, they obviously lead to a non-terminating system whenever $f \equiv \mathsf{Id}$. We solved this problem of non-termination by introducing three new combinators, $\mathsf{tens}(t,s)$, $\mathsf{L}'(t)$ and $\mathsf{R}'(t)$ as abbreviations for $((\mathsf{Id},t),s) * \otimes$, $(\mathsf{Id},t) * \mathsf{L}$ and $(\mathsf{Id},t) * \mathsf{R}$ respectively.

The new version of the (linear) categorical combinators for rudimentary linear logic and the judgements for the new combinators are the following:

Definition 2 (Raw linear combinators, second version)

$$A ::= G \mid A \otimes A \mid A \multimap A$$
$$\Gamma ::= () \mid (\Gamma, A)$$
$$t ::= n \mid \otimes \mid \mathsf{tens}(t,t) \mid \mathsf{L} \mid \mathsf{R} \mid \mathsf{L}'(t) \mid \mathsf{R}'(t) \mid \mathsf{Cur}(t) \mid \mathsf{App} \mid f * t$$
$$f ::= \mathsf{Id} \mid (f,t) \mid f;f$$

$$\frac{\Gamma \vdash t : A \qquad \Gamma \vdash t' : B}{\Gamma \vdash \mathsf{tens}(t,t') : A \otimes B} \qquad \frac{\Gamma \vdash t : A \otimes B}{\Gamma \vdash \mathsf{L}'(t) : A} \qquad \frac{\Gamma \vdash t : A \otimes B}{\Gamma \vdash \mathsf{R}'(t) : B}$$

Fig. 3. Rules for the combinators tens, L' and R'

The equational presentation for the tensor fragment consisted of the above equations (1) ÷ (6) plus the equations for variable access, and the following equations for the tensor product which can be split into three classes.
The naturality rules

$$(f,t) * \mathsf{L} = \mathsf{L}'(t)$$
$$(f,t) * \mathsf{R} = \mathsf{R}'(t)$$
$$f * \otimes = \mathsf{tens}(f * 1, f * 0)$$
$$f * \mathsf{tens}(t,t') = \mathsf{tens}(f * t, f * t')$$

the β-rules

$$(f, \otimes) * \mathsf{L} = 1$$
$$(f, \otimes) * \mathsf{R} = 0$$
$$(f, \mathsf{tens}(t, t')) * \mathsf{L} = t$$
$$(f, \mathsf{tens}(t, t')) * \mathsf{R} = t'$$

and the η-rule

$$\mathsf{tens}(\mathsf{L}, \mathsf{R}) = 0$$

These equations follow the same pattern as the corresponding ones for the structural rules. The equational theory for the tensor fragment can be completed into a locally confluent rewrite system. The completion process generates new rules which rewrite the combinators L and R in terms of L' and R' respectively.

The Linear Implication The combinators for the linear implication follow the pattern described for the structural rules. The categorical bijection is

$$\mathsf{Multi}((\Gamma, A); B) \cong \mathsf{Multi}(\Gamma; A \multimap B)$$

and we obtain two combinators Cur and App. The first represents the direction from left to right, and the other the effect of the other direction on the identity morphism. Again, a naturality rule for application

$$((f, t), s) * \mathsf{App} \rightarrow ((\mathsf{Id}, t), s) * \mathsf{App}$$

gives rise to a non-terminating relation whenever $f \equiv \mathsf{Id}$. Similarly to the naturality rules for $\otimes$, L and R, a new combinator $\mathsf{App}'(t, s)$ was introduced as an abbreviation for $((\mathsf{Id}, t), s) * \mathsf{App}$, thus solving the non-termination problem.

Another problem was the incompatibility between the naturality rule for Cur and the representation of variables in the term calculus. It is not possible to formulate the naturality rule for Cur in this version as a rewrite rule because in the rule

$$f * \mathsf{Cur}(t) \rightarrow \mathsf{Cur}((((\mathsf{Id}, n), \ldots, 1); f, 0) * t)$$

the number n can only be derived if we consider typing information. Hence, we have to use one of the categorical combinators for the simply typed λ-calculus, namely Fst, which is equal to $((\mathsf{Id}, n), \ldots, 1)$ for some n inferred from the typing rules. In this way we obtain the rule

$$f * \mathsf{Cur}(t) \rightarrow \mathsf{Cur}((\mathsf{Fst}; f, 0) * t)$$

If we apply the Knuth-Bendix algorithm to the system with the natural numbers and Fst, the completion diverges into an infinite rewrite system, whose infinitely many new rules are characterized by the following divergence pattern [15]:

$$(\underbrace{((((f; \mathsf{Fst}); \mathsf{Fst}); \ldots); \mathsf{Fst}}_{(n+1)\ times}, f * n) \rightarrow \underbrace{((f; \mathsf{Fst}); \ldots); \mathsf{Fst}}_{n\ times}$$

The solution was to add another combinator from the simply typed λ-calculus, namely Snd, and remove the natural numbers n in favour of $\mathsf{Fst}^n * \mathsf{Snd}$. In this way the divergence disappears because the left and right sides of the above rule pattern become equal. This way of representing variables has the additional advantage that the η-rule for the bijection describing context extensions can now be stated without any typing information, hence it is included in the presentation.

The (linear) categorical combinators for rudimentary linear logic are finally defined as follows:

Definition 3 (Raw linear combinators, final version)

$$\begin{aligned} A &::= G \mid A \otimes A \mid A \multimap A \\ \Gamma &::= () \mid (\Gamma, A) \\ t &::= \mathsf{Snd} \mid \otimes \mid \mathsf{tens}(t,t) \mid \mathsf{L} \mid \mathsf{R} \mid \mathsf{L}'(t) \mid \mathsf{R}'(t) \\ &\quad \mid \mathsf{Cur}(t) \mid \mathsf{App} \mid \mathsf{App}'(t,t) \mid f * t \\ f &::= \mathsf{Fst} \mid \mathsf{Id} \mid (f,t) \mid f;f \end{aligned}$$

and the judgements for the new combinators are the following:

– Multimaps:

$$\frac{\Gamma \vdash t\colon A \multimap B \qquad \Gamma \vdash s\colon A}{\Gamma \vdash \mathsf{App}'(t,s)\colon B} \qquad\qquad \frac{}{(\Gamma, A) \vdash \mathsf{Snd}\colon A}$$

– Lists of multimaps:

$$\frac{}{(\Gamma, A) \vdash \mathsf{Fst}\colon \Gamma}$$

Fig. 4. Rules for the combinators App′, Snd and Fst

From the linear logic point of view, the combinators Fst and Snd are very surprising, as one does not have projections from a (categorical) tensor product. They also seem to be unnatural at first because they are non-linear, but the typing restrictions ensure that they can be used only as subexpressions in linear expressions, e.g. in the expression (Fst, Snd).

Thus, the equations for accessing parts of the context become the equations for the multimap Snd and the context morphism Fst. The equations for the constants Fst and Snd are

$$(f,t); \mathsf{Fst} = f \tag{7}$$

$$(f,t) * \mathsf{Snd} = t \tag{8}$$

$$(f; \mathsf{Fst}, f * \mathsf{Snd}) = f \tag{9}$$

$$(\mathsf{Fst}, \mathsf{Snd}) = \mathsf{Id} \tag{10}$$

Since the primitive combinators n have been replaced by $\mathsf{Fst}^n * \mathsf{Snd}$, the equations for the tensor product have to be reformulated. They become the following equations.
The naturality rules

$$\otimes = \mathsf{tens}(\mathsf{Fst} * \mathsf{Snd}, \mathsf{Snd}) \tag{11}$$
$$f * \mathsf{tens}(t, s) = \mathsf{tens}(f * t, f * s) \tag{12}$$
$$\mathsf{L} = \mathsf{L}'(\mathsf{Snd}) \tag{13}$$
$$\mathsf{R} = \mathsf{R}'(\mathsf{Snd}) \tag{14}$$
$$f * \mathsf{L}'(t) = \mathsf{L}'(f * t) \tag{15}$$
$$f * \mathsf{R}'(t) = \mathsf{R}'(f * t) \tag{16}$$

the β-rules

$$\mathsf{L}'(\mathsf{tens}(t, s)) = t \tag{17}$$
$$\mathsf{R}'(\mathsf{tens}(t, s)) = s \tag{18}$$

and the η-rule

$$\mathsf{tens}(\mathsf{L}'(t), \mathsf{R}'(t)) = t \tag{19}$$

The equations corresponding to the logical rules for the linear implication are as follows.
The naturality rules for application

$$\mathsf{App} = \mathsf{App}'(\mathsf{Fst} * \mathsf{Snd}, \mathsf{Snd}) \tag{20}$$
$$f * \mathsf{App}'(t, s) = \mathsf{App}'(f * t, f * s) \tag{21}$$

the β-rule

$$\mathsf{App}'(\mathsf{Cur}(t), s) = (\mathsf{Id}, s) * t \tag{22}$$

and the naturality rule for currying

$$f * \mathsf{Cur}(t) = \mathsf{Cur}((\mathsf{Fst}; f, \mathsf{Snd}) * t) \tag{23}$$

As usual, when dealing with categorical combinators, we do not consider an η-rule for implication. In fact, if we add the η-rule for the linear implication

$$\mathsf{Cur}(\mathsf{App}'(\mathsf{Fst} * \mathsf{Snd}, \mathsf{Snd})) = \mathsf{Snd}$$

to the above equations, the completion procedure generates infinitely many new rules according to different divergence patterns. These new rules derive from critical pairs involving the η-rule for the linear implication and the naturality rules for application and currying.

We are now able to obtain the following result:

Theorem 4 *The equational theory of categorical combinators for rudimentary linear logic (equations (1)÷(23)) admits a locally confluent rewrite system.*

Proof. By running the Knuth-Bendix completion on the equations $(1) \div (23)$, a locally confluent rewrite system is derived, the rules of which are the equations oriented from left to right. The whole rewrite system is shown in Appendix B.

We conclude this section with a comparison of our combinators to those defined by Curien [6]. These combinators do not make a distinction between terms and context morphisms, which is a key aspect of the system of combinators presented here. Curien's combinators can be obtained by taking the subsystem that consists of Fst, Snd, Id, Cur and App and merging terms and context morphisms. This merger removes the need for the combinator App′ because the combinator $((\mathsf{Id}, t), s); \mathsf{App}$ is replaced by $(t, s); \mathsf{App}$. Hardin [13] uses the Knuth-Bendix procedure to show the local confluence of the untyped system of combinators for cartesian closed categories. This result requires no modification of the rewrite system. The main aspect of [13] is the confluence of a subsystem of these combinators. This is a much harder problem because the untyped system is strong enough to code well-known counterexamples for confluence of the untyped λ-calculus with surjective pairing.

4 Termination of the Rewrite System

We now examine the termination of the rewrite system for the categorical combinators for rudimentary linear logic. From now on, we will refer to the subsystem which consists of the rules $(1) \div (19)$ (see Appendix B) as T, i.e. the rewrite system for the tensor fragment.

An application of rewriting techniques shows the termination of T. Zantema [24] has given two proofs for this result, one being a nice application of distribution elimination as proposed in [23], the other one being a simpler proof based on polynomial interpretations. We present the latter below.

Theorem 5 *The rewrite system T is terminating.*

Proof. Define a weight $w(e)$ for every combinator e as follows:

$$
\begin{aligned}
w(\mathsf{Id}) = w(\mathsf{Fst}) = w(\mathsf{Snd}) &= 2 \\
w(\otimes) &= 8 \\
w(\mathsf{L}) = w(\mathsf{R}) &= 4 \\
w(\mathsf{L}'(t)) = w(\mathsf{R}'(t)) &= w(t) + 1 \\
w(\mathsf{tens}(t, s)) &= w(t) + w(s) + 1 \\
w((f, t)) &= w(f) + w(t) + 1 \\
w(f; g) &= w(f) \times w(g) \\
w(f * t) &= w(f) \times w(t)
\end{aligned}
$$

With this interpretation the weight of the left-hand side of all rules except (4) and (6) is greater than that of the right-hand side, and for rules (4) and (6) we obtain equality. If we define another interpretation by $w'(f; g) = w'(f) + 2 \times w'(g)$ and $w'(f * t) = w'(f) + 2 \times w'(t)$ and $w'(e) = w(e)$ otherwise, the left-hand side of the rules (4) and (6) has a greater weight than the right-hand side. Hence, the function $e \mapsto (w(e), w'(e))$ maps any expression e into a pair of natural numbers such that for each rule its application to the left-hand side is greater in the lexicographic ordering than its application to the right-hand side. The rewrite system T is thus terminating.

Newman Lemma says that a locally confluent and terminating rewrite system is also confluent. Hence, we obtain from Theorems 4 and 5:

Corollary 6 *The rewrite system T is canonical.*

Note that the proof of Theorem 5 does not involve any use of the typing information. Therefore, even the rewrite system for the *raw* combinators for the tensor fragment terminates. This is no longer true if we consider the whole rewrite system for the combinators, i.e. the system with the linear implication. The reason is that we can model the untyped λ-calculus in this system of raw combinators. Hence, we obtain the following reduction of a raw combinator to itself, which corresponds to the reduction of $(\lambda x.xx)(\lambda x.xx)$ to itself in the untyped λ-calculus:

$$\begin{array}{ll} & \mathsf{App}'(\mathsf{Cur}(\mathsf{App}'(\mathsf{Snd},\mathsf{Snd})),\mathsf{Cur}(\mathsf{App}'(\mathsf{Snd},\mathsf{Snd}))) \\ \stackrel{(22)}{\rightarrow} & (\mathsf{Id},\mathsf{Cur}(\mathsf{App}'(\mathsf{Snd},\mathsf{Snd}))) * \mathsf{App}'(\mathsf{Snd},\mathsf{Snd}) \\ \stackrel{(21)}{\rightarrow} & \mathsf{App}'((\mathsf{Id},\mathsf{Cur}(\mathsf{App}'(\mathsf{Snd},\mathsf{Snd}))) * \mathsf{Snd},(\mathsf{Id},\mathsf{Cur}(\mathsf{App}'(\mathsf{Snd},\mathsf{Snd}))) * \mathsf{Snd}) \\ \stackrel{(8),(8)}{\rightarrow} & \mathsf{App}'(\mathsf{Cur}(\mathsf{App}'(\mathsf{Snd},\mathsf{Snd})),\mathsf{Cur}(\mathsf{App}'(\mathsf{Snd},\mathsf{Snd}))) \end{array}$$

So Zantema's proof of Theorem 5, which is an application of rewriting techniques for raw combinators, cannot be applied in this case. Moreover, this counterexample implies that any proof of the termination of the whole system of combinators has to use the typing information in an essential way. Such a proof can only work if a reduction of any well-formed expression yields a well-formed expression of the same type. This property is known under the (rather obscure) name of *Subject Reduction* and holds for this calculus.

The obvious approach to show termination is to try to apply the *reducibility method* [12], which has been used for various typed λ-calculi and to show the strong normalization of the linear term calculus [3]. However, this method cannot cope adequately with the extra reductions that arise from the explicit substitution which is captured by the operation '$*$'. A similar problem arises if one tries to prove strong normalization of the typed $\lambda\sigma$-calculus. In fact, the termination of the typed $\lambda\sigma$-calculus and of the rewrite system for the well-formed combinators for rudimentary linear logic is still open. It is only known that the substitution fragment of the $\lambda\sigma$-calculus, i.e. all the reduction rules modelling substitution, but not the β-reduction, is terminating [14, 7, 23]. However, it should be possible to carry over the approach presented in [20] for the $\lambda\sigma$-calculus to the combinators presented here, and show that all strategies used in actual implementations terminate.

As far as the confluence of the rewrite system of combinators for rudimentary linear logic is concerned, it is possible to prove it by following the approach described in [19]. Note that the termination of the subsystem T constitutes an important part of this proof, because we can identify terms that are equal modulo the equations of T.

5 Normal Forms

The normal forms of the rewrite system T as well as the ones of the whole rewrite system for rudimentary linear logic can easily be characterized. Because the system T is canonical, every term can be reduced to a unique normal form.

Theorem 7

- *For the rewrite system T, every combinator reduces to exactly one combinator from the set $\mathcal{N}$ of normal forms, which is inductively defined as follows:*

Lists of multimaps

$$\frac{}{\mathsf{Fst}^k \in \mathcal{N}} \qquad \frac{}{\mathsf{Id} \in \mathcal{N}}$$

$$\frac{f \in \mathcal{N} \qquad t \in \mathcal{N} \qquad \not\exists k : f \equiv \mathsf{Fst}^{k+1}, t \equiv \mathsf{Fst}^k * \mathsf{Snd}}{(f,t) \in \mathcal{N}}$$

Multimaps

$$\frac{}{\mathsf{Fst}^k * \mathsf{Snd} \in \mathcal{N}} \qquad \frac{t, t' \in \mathcal{N} \qquad \not\exists s : t \equiv \mathsf{L'}(s), t' \equiv \mathsf{R'}(s)}{\mathsf{tens}(t,t') \in \mathcal{N}}$$

$$\frac{t \in \mathcal{N} \qquad \not\exists s\ s' : t \equiv \mathsf{tens}(s,s')}{\mathsf{L'}(t) \in \mathcal{N}} \qquad \frac{t \in \mathcal{N} \qquad \not\exists s\ s' : t \equiv \mathsf{tens}(s,s')}{\mathsf{R'}(t) \in \mathcal{N}}$$

- *The set $\mathcal{N}$ of normal forms of the rewrite system of combinators for rudimentary linear logic is defined by the above clauses plus the two rules*

$$\frac{t \in \mathcal{N}}{\mathsf{Cur}(t) \in \mathcal{N}} \qquad \frac{t, t' \in \mathcal{N} \qquad \not\exists s : t \equiv \mathsf{Cur}(s)}{\mathsf{App'}(t,t') \in \mathcal{N}}$$

Proof. Show by an induction over the structure of the raw combinators that a combinator is irreducible if and only if it is in the set $\mathcal{N}$. Corollary 6 then implies the claim. We consider one case for illustration, namely the case of a context morphism (f,t). If f or t are not in normal form, then (f,t) is not either. If both are normal forms, then (f,t) can be reduced only by rule (9). By induction hypothesis we know that f and t are in normal form exactly if and only if they are in the set $\mathcal{N}$. Hence, the only possible redex with respect to rule (9) occurs if $f \equiv \mathsf{Fst}^{k+1}$ and $t \equiv \mathsf{Fst}^k * \mathsf{Snd}$. But this is ruled out in the clause for (f,t).

6 Conclusions and Future Work

In this paper we have presented a canonical rewrite system associated to an equational theory of categorical combinators for the tensor fragment of intuitionistic linear logic, and a locally confluent system of combinators for rudimentary linear logic. The development of the system of categorical combinators profited from the application of rewriting techniques, and led to a better understanding of the theory behind it. Not only can such an equational theory be completed successfully in a locally confluent rewrite system, but the termination proofs for some of its subsystems are nice applications of very recent techniques for proving termination of rewrite systems.

The results obtained so far show that it is promising to pursue this project, and we aim at extending this approach to more complex logical systems. First we need to prove the termination and the confluence for the whole rewrite system of categorical combinators for rudimentary linear logic. It should be possible to do it following the approach described in [19]. Then we want to extend our approach to the whole of intuitionistic linear logic, by dealing with the modal operator '!' and with the additive connectives. Once this is done, we plan to design and implement an abstract categorical machine for intuitionistic linear logic.

Acknowledgements

We thank Martin Hyland for his guidance with this work, Hans Zantema for providing us with the termination proofs, and Cosimo Laneve and the anonymous referees for useful comments on the presentation of the paper. We should also like to acknowledge financial support from the CLICS-II project.

References

1. M. Abadi, L. Cardelli, P.-L. Curien, J.-J. Lévy, 'Explicit substitutions', in *Journal of Functional Programming*, 1991, Vol. 1, No. 4, pp. 375–416.
2. N. Benton, G. Bierman, V. de Paiva, M. Hyland, 'Term assignment for intuitionistic linear logic', Technical Report No. 262, Computer Laboratory, University of Cambridge, 1992.
3. G. Bierman, 'On Intuitionistic Linear Logic', Draft of Ph.D. Thesis, Computer Laboratory, University of Cambridge, 1993.
4. T. Coquand, Th. Ehrhard, 'An equational presentation of higher order logic', in Proceedings of Category Theory and Computer Science, Lecture Notes in Computer Science 283, Springer-Verlag, 1987, pp. 40–56.
5. G. Cousineau, P.-L. Curien, M. Mauny, 'The categorical abstract machine', in *Science of Computer Programming*, 1987, Vol. 8, pp. 173–202.
6. P.-L. Curien, '*Categorical Combinators, Sequential Algorithms and Functional Programming*', Birkhäuser, 1993.
7. P.-L. Curien, T. Hardin, A. Ríos, 'Strong normalisation of substitutions', in Proceedings of the 17th International Symposium on Mathematical Foundations of Computer Science 1992, Lecture Notes in Computer Science 629, Springer-Verlag, 1992, pp. 209–217.

8. N. Dershowitz, J.-P. Jouannaud, 'Rewrite Systems', in *Handbook of Theoretical Computer Science, Vol. B: Formal Models and Semantics*, J. van Leeuwen (ed.), North-Holland, 1990, pp. 243–320.
9. Th. Ehrhard, 'A categorical semantics of constructions', in Proceedings of the 3rd Annual Symposium on Logic in Computer Science, IEEE, 1988, pp. 264–273.
10. S. J. Garland, J. V. Guttag, 'A Guide to LP, The Larch Prover', Release 2.2 LP Documentation, MIT, November 1991.
11. J.-Y. Girard, 'Linear logic', in *Theoretical Computer Science*, 1987, Vol. 50, pp. 1–102.
12. J.-Y. Girard, P. Taylor, Y. Lafont, '*Proofs and Types*', Cambridge University Press, 1989.
13. T. Hardin, 'Confluence results for the pure strong categorical logic CCL. λ-calculi as subsystems of CCL', in *Theoretical Computer Science*, 1989, Vol. 65, pp. 291–342.
14. T. Hardin, A. Laville, 'Proof of termination of the rewriting system Subst on CCL', in *Theoretical Computer Science*, 1986, Vol. 46, pp. 305–312.
15. M. Hermann, 'Vademecum of Divergent Term Rewriting Systems', Technical Report CRIN 88-R-082, Centre de Recherche en Informatique de Nancy, 1988.
16. M. Hermann, 'Chain Properties of Rule Closures', in *Formal Aspects of Computing*, 1990, Vol. 2, pp. 207–225.
17. Y. Lafont, 'The linear abstract machine', in *Theoretical Computer Science*, 1988, Vol. 59, pp. 157–180.
18. V. de Paiva, E. Ritter, 'Categorical combinators for symmetric multicategories', unpublished manuscript, 1994.
19. E. Ritter, 'Categorical Abstract Machines for Higher-Order Lambda Calculi', to appear in *Theoretical Computer Science.* Also available as Technical Report No. 297, Computer Laboratory, University of Cambridge, 1993.
20. E. Ritter, 'Normalization for Typed Lambda Calculi with Explicit Substitution', to appear in Proceedings of the 1993 Annual Conference of the European Association for Computer Science Logic CSL 93, Lecture Notes in Computer Science.
21. E. Ritter, V. de Paiva, 'Syntactic Multicategories and Categorical Combinators for Linear Logic', presented at the Fifth Biennial Meeting on Category Theory and Computer Science CTCS-5, Amsterdam, September 1993.
22. A. Scedrov, 'A Brief Guide to Linear Logic', in Bulletin of EATCS, No. 41, June 1990, pp. 154–165.
23. H. Zantema, 'Termination of term rewriting by interpretation', in Proceedings of the 3rd International Workshop on Conditional Term Rewriting Systems CTRS-92, Lecture Notes in Computer Science 656, Springer-Verlag, 1993, pp. 155–167.
24. H. Zantema, Private Communication, June 1993.

A S-multicategories

This appendix includes the definition of an S-multicategory. For the motivation behind the definition, see Section 3.1. Such a structure is given in two parts.

Definition 8 *A* syntactic pre-multicategory *is a functor* $E: CM \rightarrow \mathbf{Set}$, *where* CM *is any category and* **Set** *is the category of sets and functions.*

Such a functor is given by

- a collection of basic objects (or types) T usually denoted by $A_1, A_2, \ldots$
- a collection of contexts Γ, Δ which are finite (or empty) strings of types $(A_n, \ldots, A_1)$.
- a collection of multimaps (or multiarrows) M of the form $\Gamma \xrightarrow{t} A$ where Γ is a context and A a type. We identify the type A with the context (A).
- a collection of context morphisms CM of the form $\Gamma \xrightarrow{f} \Delta$ where both Γ and Δ are contexts.
- an operation '$*$' which takes a context morphism $\Gamma \xrightarrow{f} \Delta$ in CM and a multimap $\Delta \xrightarrow{t} A$ in M and produces a multimap $\Gamma \xrightarrow{f*t} A$ in M.
- an operation ';' of composition of context morphisms.

Furthermore, the following conditions hold:

1. Contexts and contexts morphisms form a category CM with the identity morphism of Δ denoted by Id_Δ.
2. Given context maps $f\colon \Delta' \to \Gamma$, $g\colon \Gamma \to \Delta$ and a multimap $t\colon \Delta \to A$, we have:

$$\mathsf{Id}_\Delta * t = t$$
$$(f;g) * t = f * (g * t)$$

Note that we use a standard convention in the area of indexed category theory and denote the function $E(f)(\_)$ by $f * \_$.

The additional conditions for a syntactic multicategory are as follows:

Definition 9 *A syntactic multicategory $\mathcal{M}$ (or S-multicategory) is a syntactic pre-multicategory such that*

1. *There exists a bijection between a context morphism of the form*

$$\Gamma \xrightarrow{\quad f \quad} (\Delta, A)$$

and a pair

$$(f^c\colon \Gamma_1 \to \Delta,\ f^t\colon \Gamma_2 \to A)$$

where Γ_1 and Γ_2 form a partition of Γ, f^c is a context morphism (we write 'c' to remind ourselves of contexts) and f^t is a multimap. For any type A, let Id_A denote the multimap corresponding to the identity morphism on A. Furthermore, we write $(\_,\_)$ for the inverse map of the bijection, i.e. (f,t) denotes the context morphism g such that $g^c = f$ and $g^t = t$.

2. *This bijection is natural in Γ, i.e. given context maps $g_1\colon \Delta_1 \to \Gamma_1$, $g_2\colon \Delta_2 \to \Gamma_2$ and $f\colon \Gamma \to (\Delta, A)$ with a corresponding pair $(f^c, f^t)\colon (\Gamma_1, \Gamma_2) \to (\Delta, A)$, we have:*

$$((g_1, \mathsf{Id}_{\Gamma_2}); f)^c = g_1; f^c$$
$$((\mathsf{Id}_{\Gamma_1}, g_2); f)^t = g_2 * f^t$$

The context morphism $(g_1, \mathsf{Id}_{\Gamma_2})$ *is defined by induction over the length of* Γ_2 *as*

$$(g_1, \mathsf{Id}_{()}) \stackrel{\text{def}}{=} g_1$$
$$(g_1, \mathsf{Id}_{(\Gamma,A)}) \stackrel{\text{def}}{=} ((g_1, \mathsf{Id}_\Gamma), \mathsf{Id}_A)$$

If $|\Gamma_2| = k$*, then the bijection in clause 1 yields multimaps* $t_k, \ldots, t_0$ *such that*

$$g = (\mathsf{Id}_{()}, t_k, \ldots, t_0)$$

and then (f, g) *is defined as* $((\cdots((f, t_k), t_{k-1}), \ldots), t_0)$.

B The Rewrite System for the Categorical Combinators for Rudimentary Linear Logic

The locally confluent rewrite system for the equational theory of categorical combinators for rudimentary linear logic is recalled below.

$$\mathsf{Id}; f \to f \tag{1}$$
$$f; \mathsf{Id} \to f \tag{2}$$
$$f; (g; h) \to (f; g); h \tag{3}$$
$$\mathsf{Id} * t \to t \tag{4}$$
$$f * (g * t) \to (f; g) * t \tag{5}$$
$$f; (g, t) \to (f; g, f * t) \tag{6}$$
$$(f, t); \mathsf{Fst} \to f \tag{7}$$
$$(f, t) * \mathsf{Snd} \to t \tag{8}$$
$$(f; \mathsf{Fst}, f * \mathsf{Snd}) \to f \tag{9}$$
$$(\mathsf{Fst}, \mathsf{Snd}) \to \mathsf{Id} \tag{10}$$
$$\otimes \to \mathsf{tens}(\mathsf{Fst} * \mathsf{Snd}, \mathsf{Snd}) \tag{11}$$
$$f * \mathsf{tens}(t, s) \to \mathsf{tens}(f * t, f * s) \tag{12}$$
$$\mathsf{L} \to \mathsf{L'}(\mathsf{Snd}) \tag{13}$$
$$\mathsf{R} \to \mathsf{R'}(\mathsf{Snd}) \tag{14}$$
$$f * \mathsf{L'}(t) \to \mathsf{L'}(f * t) \tag{15}$$
$$f * \mathsf{R'}(t) \to \mathsf{R'}(f * t) \tag{16}$$
$$\mathsf{L'}(\mathsf{tens}(t, s)) \to t \tag{17}$$
$$\mathsf{R'}(\mathsf{tens}(t, s)) \to s \tag{18}$$
$$\mathsf{tens}(\mathsf{L'}(t), \mathsf{R'}(t)) \to t \tag{19}$$
$$\mathsf{App} \to \mathsf{App'}(\mathsf{Fst} * \mathsf{Snd}, \mathsf{Snd}) \tag{20}$$
$$f * \mathsf{App'}(t, s) \to \mathsf{App'}(f * t, f * s) \tag{21}$$
$$\mathsf{App'}(\mathsf{Cur}(t), s) \to (\mathsf{Id}, s) * t \tag{22}$$
$$f * \mathsf{Cur}(t) \to \mathsf{Cur}((\mathsf{Fst}; f, \mathsf{Snd}) * t) \tag{23}$$

Comparing Combinatory Reduction Systems and Higher-order Rewrite Systems

Vincent van Oostrom[1] and Femke van Raamsdonk[2]

[1] Department of Mathematics and Computer Science, Vrije Universiteit, De Boelelaan 1081a, 1081 HV Amsterdam, The Netherlands, email: oostrom@theory.ntt.jp ***

[2] CWI, P.O. Box 94079, 1009 GB Amsterdam, The Netherlands, email: femke@cwi.nl

Abstract. In this paper two formats of higher-order rewriting are compared: Combinatory Reduction Systems introduced by Klop and Higher-order Rewrite Systems defined by Nipkow. Although it always has been obvious that both formats are closely related to each other, up to now the exact relationship between them has not been clear. This was an unsatisfying situation since it meant that proofs for much related frameworks were given twice. We present two translations, one from Combinatory Reduction Systems into Higher-Order Rewrite Systems and one vice versa, based on a detailed comparison of both formats. Since the translations are very 'neat' in the sense that the rewrite relation is preserved and (almost) reflected, we can conclude that as far as rewrite theory is concerned, Combinatory Reduction Systems and Higher-Order Rewrite Systems are equivalent, the only difference being that Combinatory Reduction Systems employ a more 'lazy' evaluation strategy. Moreover, due to this result it is the case that some syntactic properties derived for the one class also hold for the other.

The research of the second author is supported by NWO/SION project 612-316-606.

1 Introduction

This paper is concerned with a comparison of two formats of higher-order rewriting: Combinatory Reduction Systems (CRSs) as introduced by Klop [Klo80] and Higher-order Rewrite Systems (HRSs) as introduced by Nipkow [Nipa].

Inspired by Aczel [Acz78], Klop defined CRSs in [Klo80] as first-order term rewriting systems possibly with bound variables, so as to include both first-order rewrite systems such as Curry's Combinatory Logic and rewrite systems with bound variables such as Church's λ-calculus. The point was that a large amount of syntactic rewrite theory could be developed for this framework.

*** Current address: NTT Basic Research Laboratories 3-1, Morinosato Wakamiya, Atsugi-Shi, Kanagawa Pref., 243-01, Japan

In [Nipa], Nipkow introduces HRSs as a generalisation of first-order rewrite systems to terms with higher-order functions and bound variables. Furthermore, HRSs were designed to have the same logical basis as systems like Isabelle [Pau90] and λProlog [NM88]. That is, a typed λ-calculus is used as a meta-language.

These different objectives have led to surprisingly large differences in the presentation of these systems. For CRSs the meta-language, i.e. the language in which the notions of term, substitution and rewrite step are expressed, is left implicit in the presentation. For HRSs the meta-language is Church's $\lambda^{\rightarrow}$-calculus of simply typed λ-terms with β as rewrite rule. In the case of CRSs, the introduction of a special purpose meta-language makes the definition quite involved. However, a closer inspection shows that in fact the meta-language of CRSs is (a polyadic version of) λ-calculus with developments (or `let`-expressions), denoted by $\underline{\lambda}$. See [Klo80, Sec. I.3.5] or [Bar84, §11.1.3] for details.

Once we have made the meta-language of CRSs explicit, we can compare both formats by comparing their respective meta-languages. Comparing is done by giving encodings of one system into the other and vice versa. The encoding of CRSs into HRSs is straightforward because $\underline{\lambda}$-calculus can be encoded into $\lambda^{\rightarrow}$-calculus. The encoding of HRSs into CRSs is somewhat more involved; $\lambda^{\rightarrow}$-calculus cannot be encoded directly into $\underline{\lambda}$-calculus. For example, the latter does enjoy the ***disjointness property*** (rewriting preserves disjointness, cf. [Klo80, pg. 38]), while the former doesn't. In general, in $\lambda^{\rightarrow}$-calculus rewrite sequences can be longer than in $\underline{\lambda}$-calculus. Our solution is to add an explicit β-rule (and a symbol for application) to the encoding of an HRS. A rewrite step in the HRS is then simulated by a rewrite step in the CRS possibly followed by an explicit β-reduction to normal form. More precisely, let $\mathcal{C}$ be a CRS and $\mathcal{H}$ be a HRS. We write $\rightarrow_{\mathcal{C}}$ and $\rightarrow_{\mathcal{H}}$ for their rewrite relations. Translating is denoted by $\langle \_ \rangle$, and reduction to normal form with respect to the explicit β-rule is written as $\rightarrow^{!}_{\beta}$. Then we have

$$\langle \rightarrow_{\mathcal{C}} \rangle = \rightarrow_{\langle \mathcal{C} \rangle}$$
$$\langle \rightarrow_{\mathcal{H}} \rangle = \rightarrow_{\langle \mathcal{H} \rangle} \cdot \xrightarrow[\beta]{!}$$

if the relations are restricted to the set of translated terms.

The naturality of an an encoding can be measured by the properties it preserves and reflects. Our encoding of CRSs into HRSs both preserves and reflects the main property of rewrite systems, i.e. whether one term rewrites (in one step) to another. This allows for a confluence proof for orthogonal CRSs via a proof of confluence for orthogonal HRSs. As noted above the translation the other way around is not that nice. The HRS is simulated by a more refined CRS; 'giant' HRS-steps are simulated by many 'small' CRS-steps. This is analogous to the way in which λ-calculus is simulated by the $\lambda\sigma$-calculus defined in [ACCL90]. Of course, not every step in the refined system is reflected in the original HRS, but still we can say something: every rewrite sequence between encodings of HRS-terms is reflected in the original HRS. Again, this allows for a confluence proof for orthogonal HRSs via a proof of confluence for orthogonal CRSs. For the moment being, we have only considered use of our translation for confluence results.

Our comparison only considers CRSs versus HRSs. There are some more alternatives for higher-order rewriting, such as Khasidashvili's Expression Reduction Systems [Kha90] and Takahashi's Conditional Lambda Calculi [Tak]. We claim that the main differences between these and CRSs (or HRSs) are of a syntactic nature.

The paper is organised as follows. In section 2 we will discuss in detail the difference between CRSs and HRSs by first considering only terms, and next also the rewrite relation on terms. In section 3 we define a translation from CRSs into HRSs and, using this translation, we give a confluence proof for orthogonal CRSs. The translation from HRSs into CRSs is presented in section 4, again the translation is used to give a confluence proof, now for orthogonal HRSs. Section 5 concludes the paper with some discussion on higher-order rewriting. The reader is assumed to be familiar with term rewriting and (simply typed) λ-calculus. For the induction proofs we refer the reader to CWI Report CS-R9361 or VU Report IR-333 [OR93], both with the same title as this paper.

NOTATION. We adhere mostly to the notations introduced by Klop for CRSs, and Nipkow for HRSs. Since their introduction both formats have been subject to some change and we will use their most recent presentations, viz. [KOR93] for CRSs and [Nipb] for HRSs. The most notable change is the use of the functional format for CRSs instead of the applicative one of [Klo80]. The reason for choosing the functional format is that it is closer to the usual notation for term rewriting systems. Moreover, in applicative CRSs the object-language application symbol is left implicit in the notation, while for HRSs the meta-language application symbol is left implicit, which would possibly give rise to confusion in comparing these formats.

2 Comparing the Syntax

We first restrict attention to term formation, since already on that level some important differences between CRSs and HRSs are manifest. Next, we consider rule formation and finally the generation of the rewrite relation.

2.1 Term Formation

CRS Terms. A CRS $\mathcal{C}$ is a pair $(\mathcal{A}, \mathcal{R})$, where $\mathcal{A}$ is its alphabet and $\mathcal{R}$ its set of *rewrite* or *reduction* rules. (Because of the termination connotation of the word 'reduction' we will use it only in the case of normalising rewrites.) In a CRS a distinction is made between metaterms and terms. The left- and right-hand side of a rule are metaterms, but the rewrite relation is a relation on terms.

The alphabet $\mathcal{A}$ of a CRS $(\mathcal{A}, \mathcal{R})$ consists of

- symbols for variables $x\ y\ z \ldots$,
- a symbol $[\_]\_$ for the abstraction operator,
- symbols for operators with a fixed arity $F\ G\ H\ \ldots$,
- symbols for metavariables with a fixed arity $Z\ Z_0\ Z_1\ \ldots$.

The set MTerms of metaterms is the least set such that
(1) $x \in$ MTerms for every variable x,
(2) $[x]t \in$ MTerms for a variable x and $t \in$ MTerms,
(3) $F(t_1, \ldots, t_n) \in$ MTerms if $t_1, \ldots, t_n \in$ MTerms and F is an n-ary operator,
(4) $Z(t_1, \ldots, t_n) \in$ MTerms if $t_1, \ldots, t_n \in$ MTerms and Z is an n-ary metavariable.

The set Terms of terms consists of all metaterms without metavariables. In a term or metaterm of the form $[x]t$, we call t the *scope* of $[x]$. A variable x occurs *free* in a term or metaterm if it is not in the scope of an occurrence of $[x]$. A variable x occurs *bound* otherwise. A term is called *closed* if all variables occur bound. Only variables (and no metavariables) can be bound by the abstraction operator. We will sometimes write $[x_1 \ldots x_n]t$ for $[x_1] \ldots [x_n]t$.

Let $\Box$ be a fresh symbol. A *context* is a term with one or more occurrences of $\Box$. A context with exactly one occurrence of $\Box$ is written as $C[\,]$, and one with n occurrences of $\Box$ as $C[, \ldots,]$. If $C[, \ldots,]$ is a context with n occurrences of $\Box$ and $t_1, \ldots, t_n$ are terms, then $C[t_1, \ldots, t_n]$ denotes the result of replacing from left to right the occurrences of $\Box$ by $t_1, \ldots, t_n$.

Example 1. The alphabet of λ-calculus in CRS format contains two symbols for operators: a unary operator λ for λ-abstraction and a binary operator @ for application. Examples of some λ-terms written in CRS notation:

$\lambda([x]x)$ for $\lambda x.x$,
$@(x, y)$ for xy,
$\lambda([x]@(y, x))$ for $\lambda x.yx$, and
$@(\lambda([x]x), y)$ for $(\lambda x.x)y$.

Because of the very liberal term formation in the CRS framework, many terms can be formed from the alphabet consisting of λ and @ that do not correspond to any λ-term. These kind of terms are called 'junk'. In general, it is often necessary to consider a CRS with a restricted set of terms, that has to be closed under rewriting. If one wants to stress the point that only a subset of the set of terms is considered, one speaks about *sub*-CRSs.

HRS Terms. A HRS $\mathcal{H}$ is a pair $(\mathcal{A}, \mathcal{R})$ whith $\mathcal{A}$ an alphabet and $\mathcal{R}$ a set of rules. Term formation is specified using $\lambda^{\rightarrow}$-calculus, Church's simply typed λ-calculus. (Simple) types are formed from base types and the function type constructor $\rightarrow$. Types are denoted by $\sigma, \tau, \ldots$.

The alphabet $\mathcal{A}$ of a HRS $\mathcal{H} = (\mathcal{A}, \mathcal{R})$ consists of
- symbols for typed variables $x\ y\ z\ \ldots$,

- a distinguished symbol λ for abstraction,
- symbols for typed operators $F\ G\ H \ldots$.

Typed terms are formed from abstraction and application, which is written by juxtaposition, according to the following rules:

$$\text{(var)}\ \frac{}{x:\tau} \qquad \text{(const)}\ \frac{}{F:\tau} \qquad \text{(abstr)}\ \frac{\begin{array}{c}[x:\sigma]\\ \vdots \\ t:\tau\end{array}}{\lambda x.t:\sigma\to\tau} \qquad \text{(appli)}\ \frac{t:\sigma\to\tau \quad t':\sigma}{tt':\tau}$$

Although environments are not made explicit, we take it for granted that variables and constants cannot have more than one type. So every typable term has a unique type. Like in λ-calculus, a variable x occurs *bound* in a term if it occurs in the scope of a λx, and it occurs *free* otherwise.

Let $\Box$ be a fresh symbol of some base type. A *context* is a term with one or more occurrences of $\Box$. Like in the CRS case, a context with exactly one occurrence of $\Box$ is written as $C[\,]$, and one with n occurrences of $\Box$ as $C[,\ldots,]$. If $C[,\ldots,]$ is a context $C[,\ldots,]$ with n occurrences of $\Box$ and $t_1,\ldots,t_n$ are terms of appropriate base type, then $C[t_1,\ldots,t_n]$ denotes the result of replacing from left to right the occurrences of $\Box$ by $t_1,\ldots,t_n$.

Only terms (and contexts) in long η-normal form will be considered.

Definition 1. The long η-normal form of a $\lambda^{\to}$-term s is obtained as follows. Replace repeatedly subterms t of s by $\lambda x.tx$ provided x doesn't occur free in t, t is not of the form $\lambda y.t_0$ and t doesn't occur in a subterm of the form tu. This has the effect that all subterms are provided with the right number of arguments.

Example 2. In the representation of untyped λ-calculus as a HRS, we have only one base type 0. The alphabet contains two symbols for operators, namely $\mathsf{app} : 0 \to (0 \to 0)$ for application and $\mathsf{abs} : (0 \to 0) \to 0$ for λ-abstraction. Some examples of λ-terms in this notation are:

$\mathsf{abs}\,(\lambda x.x)$ for $\lambda x.x$,
$\mathsf{app}\,xy$ for xy,
$\mathsf{abs}\,(\lambda x.\mathsf{app}yx)$ for $\lambda x.yx$,
$\mathsf{app}\,(\mathsf{abs}\,(\lambda x.x))y$ for $(\lambda x.x)y$.

Like in the CRS case, the HRS-representation of λ-calculus contains junk. For instance, the term $\lambda x.x$ doesn't correspond to any λ-term. Note that all λ-terms and the variables occurring in them have type 0 in this notation. This example illustrates that a notion of sub-HRS, analogous to the notion of sub-CRS, is called for. Furthermore, the example shows that properties, such as strong normalisation, of the meta-language ($\lambda^{\to}$-calculus) have no bearing on properties of the object language (λ-calculus).

Comparing Term Formation. We discuss the two most important differences between both formats.

In CRSs metaterm formation is given by a direct inductive definition. Function symbols and metavariables come equipped with an arity and metaterms are formed by supplying these symbols with the right number of arguments. Terms are metaterms not containing metavariables.

In HRSs a direct inductive definition of terms is circumvented by making use of $\lambda^{\rightarrow}$-calculus term formation. Function symbols come as constants equipped with a type and are combined using the formation rules of $\lambda^{\rightarrow}$-calculus. Attention is then restricted to terms in long η-normal form. Most of the time, except at intermediate stages of a computation, attention is further restricted to terms in long $\beta\eta$-normal form.

Note that the typing does not mean that only typed systems can be written as HRSs; the typing takes place on metalevel. If an untyped system is represented as a HRS, then only one base type 0 is used and all terms of the HRS corresponding to a term in the untyped system we are considering, are of type 0. The base type 0 can be thought of as the set of all well-formed terms. The statement $t : 0$ can be read as 't is a well-formed term'.

Typing in this way, such that well-formed terms are of base type, actually establishes two things. For discussing them, first the *arity* and the *order* of a type are defined.

Definition 2. The arity $\mathsf{Ar}(\sigma)$ of a type σ is inductively defined as follows:

$$\mathsf{Ar}(\sigma) = 0 \text{ (if } \sigma \text{ is a base type)}$$
$$\mathsf{Ar}(\sigma \rightarrow \tau) = 1 + \mathsf{Ar}(\tau)$$

The order $\mathsf{Ord}(\sigma)$ of a type σ is defined as follows:

$$\mathsf{Ord}(\sigma) = 0 \text{ (if } \sigma \text{ is a base type)}$$
$$\mathsf{Ord}(\sigma \rightarrow \tau) = \mathsf{max}(1 + \mathsf{Ord}(\sigma), \mathsf{Ord}(\tau))$$

The arity (order) of a term is defined to be the arity (order) of its type.

First, in a term every operator has exactly as many arguments as prescribed by the arity of its type. This is because terms must be in long η-normal form. For instance, an operator $F : 0 \rightarrow (0 \rightarrow 0)$ can form a term only if it is provided with two arguments t_1 and t_2 of type 0. So the type of an operator, like the arity of an operator in CRSs, determines how many arguments it should have. Second, in a term all the arguments have the right *order*, indicating how active they are, or, whether they can be applied to other terms. For example an operator $G : (0 \rightarrow 0) \rightarrow 0$ should have one argument of order 1. The order of an operator cannot be directly expressed in the CRS framework. The arity of an CRS operator only prescribes how many arguments this operator should get, but nothing is specified about the orders these arguments should have.

The second difference is that in CRSs a distinction is made between metavariables and variables and metaterms and terms. Metavariables occur only in metaterms, which in turn occur only as the left- or right-hand side of rewrite rules. The objects which are rewritten are terms. This distinction is made in order to stress the point that a rewrite rule acts as a scheme, so its left- and right-hand side are not ordinary terms. Taking this point of view, x in $F(x)$-as-a-term is a variable, and x in $F(x)$-as-a-left-or-right-hand-side is a metavariable. In CRS notation, the former is written as $F(x)$ and the latter as $F(Z)$. In HRSs no distinction is made between metavariables and variables, and no distinction is made between ordinary terms on the one hand and left- and right-hand sides of rules on the other hand; they both can be rewritten. The metavariables in CRS-rules correspond to free variables in HRS-rules.

2.2 Rule Formation

In this section we will compare the rule formation of CRSs with the one of HRSs. We show that rewrite rules in both formats satisfy equivalent requirements.

CRS Rules. In a CRS, a rewrite rule $l \to r$ must satisfy the following:
(1) l and r are metaterms,
(2) the head-symbol of l is an operator symbol,
(3) all metavariables in r occur in l as well, l and r are closed,
(4) a metavariable Z in l occurs only in the form $Z(x_1, \ldots, x_n)$ with $x_1, \ldots, x_n$ distinct bound variables.
We call the last condition the *pattern*-condition.

Example 3. The β-rule of λ-calculus, $(\lambda x.M)N \to M[x := N]$ is written in CRS format as

$$@(\lambda([x]Z(x)), Z') \to Z(Z')$$

The head-symbol of the left-hand side is @, and the metavariables Z and Z' occur in both sides.

HRS Rules. A rewrite rule $l \to r$ in a HRS must meet the following requirements:
(1) l and r are both long $\beta\eta$-normal forms of the same base type,
(2) l is not η-equivalent to a free variable,
(3) all free variables in r occur free in l as well,
(4) a free variable z in l occurs only in the form $zt_1 \ldots t_n$ with $t_1, \ldots, t_n$ η-equivalent to n distinct bound variables.
Like for CRSs, the last condition is called the *pattern*-condition.

Example 4. The β-rewrite rule in HRS notation is

$$\mathsf{app}\,(\mathsf{abs}\,(\lambda x.yx))z \to yz$$

with $x, z : 0$ and $y : 0 \to 0$.

Comparing Rule Formation. Remembering that metavariables in rewrite rules of CRSs correspond to free variables in rewrite rules, it is not difficult to see that the requirements (1)–(4) of CRS rules correspond to the same ones of HRS rules.

The first condition specifies that rules are built from metaterms for the CRS case. The second one states that left-hand sides must have some structure and the third one that rewriting cannot introduce arbitrary terms. These conditions are familiar from first-order rewriting. The last condition is the pattern-condition. By that condition only names (simple objects), not values (compound objects) can occur as arguments of free variables. Both in the case of CRSs and of HRSs it establishes decidability of unification of patterns, and computability of the rewrite relation, a result of [Mil]. Intuitively this is the case since an instance of a pattern has the same 'global structure' as the pattern itself.

2.3 Rewrite Step Generation

Once we know what requirements the rewrite rules should satisfy, we have to define for both formats how rewrite rules are instantiated in order to obtain an actual rewrite step. In both cases, we have to plug in some term in the 'holes' of the rule. In CRSs, the holes in the rule are the metavariables, and in HRSs the free variables. The ways in which metavariables and free variables are assigned a value, are related, but nevertheless essentially different.

For defining substitution for CRSs, a polyadic version of $\underline{\lambda}$-calculus is used. The substitution is performed by replacing a metavariable by a (special form of a) λ-term, and by reducing, in the term obtained by this replacement, all residuals of β-redexes that are present in the initial term, i.e. by performing a development (or expanding `let`-constructs). The well-known result in λ-calculus that all developments are finite, guarantees that the substitution is well-defined.

For defining substitution for HRSs, like for defining term and rule formation, $\lambda^{\rightarrow}$-calculus is used as a metalanguage. The substitution is performed by replacing a free variable by a term of the same type, and reducing the result of the replacement to β-normal form. In this case, substitution is well-defined since in $\lambda^{\rightarrow}$-calculus all β-rewrite sequences eventually terminate.

CRS Rewrite Steps. In order to define assignments for CRSs we first introduce a new concept: the so-called *substitutes* (cf. [Kah92]). An n-ary substitute is an expression of the form $\underline{\lambda}(x_1,\ldots,x_n).s$, where s is a term, $\underline{\lambda}$ a 'metalambda' and $(x_1,\ldots,x_n)$ a tuple of n distinct variables, which are considered to be bound by $\underline{\lambda}$ and may be renamed in the usual way. A substitute $\underline{\lambda}(x_1,\ldots,x_n).s$ can be applied to an n-tuple of terms $(t_1,\ldots,t_n)$, yielding s with $x_1,\ldots,x_n$ simultaneously replaced by $t_1,\ldots,t_n$ respectively:

$$(\underline{\lambda}(x_1,\ldots,x_n).s)(t_1,\ldots,t_n) = s[x_1 := t_1 \ldots x_n := t_n]$$

An *assignment* σ is a mapping from n-ary metavariables to n-ary substitutes:

$$\sigma(Z) = \underline{\lambda}(x_1,\ldots,x_n).s \quad (Z \text{ an } n\text{-ary metavariable})$$

It is extended to a mapping from metaterms to terms in the following way:

$$x^\sigma = x$$
$$([x]t)^\sigma = [x]t^\sigma$$
$$(F(t_1,\ldots,t_n))^\sigma = F(t_1^\sigma,\ldots,t_n^\sigma)$$
$$(Z(t_1,\ldots.t_n))^\sigma = \sigma(Z)(t_1^\sigma,\ldots,t_n^\sigma)$$

Note that the result of applying $\sigma(Z)$ to $(t_1^\sigma,\ldots,t_n^\sigma)$ in the last clause is indeed a term.

A variable in an instance of a metavariable should be bound only if it is bound in the occurrence of the metavariable. Unintended bindings occur for instance in $(F[x]Z)^\sigma$ if $\sigma(Z) = x$, and in $(Z(Z'))^\sigma$ if $\sigma(Z) = \underline{\lambda}(x).[y]x$ and $\sigma(Z') = y$. These problems can be avoided by renaming bound variables. In the following we will assume that this is done whenever necessary.

NOTATION. In this paper we stick to the definition of [KOR93] of substitution as a one-stage process. If we would use $\underline{\lambda}$-calculus as a meta-language, we would obtain substitution as a two-stage process: first replacing the metavariables by the terms assigned to them, and then explicitly developing the $\underline{\beta}$-redexes. This would yield a presentation closer to the one of HRSs.

Rewrite rules generate a rewrite relation $\rightarrow$ on terms in the following way. If $l \rightarrow r$ is a rewrite rule and σ an assignment, then $C[l^\sigma] \rightarrow C[r^\sigma]$ is a *rewrite* or *reduction step*, where $C[\,]$ is some context. A *contraction* is defined as $l^\sigma \rightarrow r^\sigma$. The reflexive-transitive closure of $\rightarrow$ is called *rewriting* and is denoted as $\twoheadrightarrow$. If $s \twoheadrightarrow t$ then we say that s *rewrites* to t. If we want to make explicit that a rewrite rule R is applied in a rewrite step we write $\rightarrow_R$ instead of $\rightarrow$.

HRS Rewrite Steps. In a HRS, an assignment is a finite mapping from variables to terms in long $\beta\eta$-normal form of the same type. Using the variable convention of λ-calculus, an assignment σ is extended to a mapping from terms to terms, in the following way:

$$F^\sigma = F \quad \text{(for a constant } F\text{)}$$
$$x^\sigma = \sigma(x) \text{ (for a variable } x\text{)}$$
$$(\lambda x.t)^\sigma = \lambda x.t^\sigma$$
$$(tt')^\sigma = t^\sigma t'^\sigma$$

We assume bound variables to be renamed whenever necessary. A rewrite relation $\rightarrow$ on terms in long $\beta\eta$ normal form is generated in the following way. If $l \rightarrow r$ is a rewrite rule and σ an assignment, then $C[l^\sigma\downarrow_\beta] \rightarrow C[r^\sigma\downarrow_\beta]$ is a rewrite step. Here $\downarrow_\beta$ denotes β-reduction to normal form. Such a normal form indeed exists since simply typed λ-calculus is considered. A contraction is defined as $l^\sigma\downarrow_\beta \rightarrow r^\sigma\downarrow_\beta$. The terminology of rewriting is the standard one like in the CRS case.

Comparing Rewrite Step Generation. In both formats it is the case that the first step in performing a substitution is to replace a 'hole' in the rewrite rule by a kind of 'λ-term'. Then we compute the result of this replacement. And here the difference lies: since in the case of CRSs we perform only a development of the λ-terms, there is no reduction of created redexes. On the other hand, to compute the result for HRSs full fledged $\lambda^{\rightarrow}$-calculus is used, that is, redexes that are created during rewriting are also contracted.

To get an idea of what kind of difference in the rewrite relations we have, due to these distinct evaluation mechanisms, consider the following example. The HRS rule $F(\lambda y.z(\lambda x.yx)) \to z(\lambda x.x)$ with assignment $\sigma : z \mapsto \lambda u.uK$. We have the rewrite step:

$$\begin{aligned} F(\lambda y.yK) &= F(\lambda y.(\lambda x.yx)K){\downarrow_\beta} \\ &= F(\lambda y.(\lambda u.uK)(\lambda x.yx)){\downarrow_\beta} \\ &= (F(\lambda y.z(\lambda x.yx)))^\sigma{\downarrow_\beta} \\ &\to (z(\lambda x.x))^\sigma{\downarrow_\beta} \\ &= (\lambda u.uK)\lambda x.x{\downarrow_\beta} \\ &= (\lambda x.x)K{\downarrow_\beta} \\ &= K \end{aligned}$$

Observe how the complete development of the λ-redexes of the assignment creates a new redex which is also contracted (in the last line). This redex is 'created downwards', so for this process to end, we cannot rely on termination of developments or even superdevelopments (cf. [Raa93]), but really need strong normalisation of simply typed λ-calculus. On the other hand, the corresponding CRS rule $F([y]Z(y)) \to Z([x]x)$ and assignment $\sigma : Z \mapsto \underline{\lambda}(u).@(u, K)$, act more lazily:

$$\begin{aligned} F([y]@(y, K)) &= (F([y]Z(y)))^\sigma \\ &\to (Z([x]x))^\sigma \\ &= @([x]x, K) \end{aligned}$$

The substitution is evaluated by a complete development of the $\underline{\lambda}$-redex. We have to add an explicit β-reduction step, namely

$$@([x]x, K) \to K$$

in order to simulate the HRS rewrite step completely.

An other way of looking at it is to view $[\_]\_$ really as an abbreviation of $\Lambda(\underline{\lambda}\_.\_)$, for some fresh symbol Λ. Now, although it seems that $\underline{\beta}$-redexes can be created in the substitution process above due to the presence of $\underline{\lambda}$'s in terms, this is not the case because they are always 'blocked' by the Λ. In the example, we end up with the term $@(\Lambda(\underline{\lambda}x.x), K)$. A 'rule' like $@(\Lambda(Z), Z') \to ZZ'$ is needed to 'unblock' the metalanguage redex $(\underline{\lambda}x.x)K$. This is the only thing used in the translation of CRSs into HRSs.

The same 'blocking' idea of this translation can also be used to show that developments of terms in λ-calculus must terminate: put fresh variables 'in front of' abstractions and applications not taking part in a β-redex. This gives a trivial typable $\lambda^{\rightarrow}$-term which exactly simulates developments. Creating new redexes is prevented by the presence of the fresh variables.

3 Translating a CRS into a HRS

In this section we will show that a CRS can be translated into a HRS such that there is a one-to-one correspondence between rewritings in the CRS and in its translation. We use $\langle\_\rangle$ as notation for the translation. The mapping $\langle\_\rangle$ is chosen to be injective.

Definition 3. The HRS alphabet $\langle\mathcal{A}\rangle$ associated with a CRS alphabet $\mathcal{A}$ consists of

- the symbol $\Lambda : (0 \to 0) \to 0$ (meant to 'collapse' a functional type),
- a variable $\langle x\rangle = x : 0$, for each variable x in $\mathcal{A}$,
- a constant $\langle F\rangle = F : 0 \to \ldots \to 0 \to 0$ ($\mathsf{Ar}(F)$ times an $\to$), for each operator F in $\mathcal{A}$,
- a variable $\langle Z\rangle = z : 0 \to \ldots \to 0 \to 0$ ($\mathsf{Ar}(Z)$ times an $\to$), for each metavariable Z in $\mathcal{A}$,

and the ordinary symbols in a HRS alphabet.

Note that only one base type, namely 0, is used. The translation between CRS metaterms and contexts is defined by extending the translation of symbols as follows.

Definition 4. (1) $\langle [x]t\rangle = \Lambda(\lambda x.\langle t\rangle)$

Abstractions are translated as projected λ-abstractions.

(2) $\langle F\rangle = F$ for F 0-ary, and $\langle F(t_1,\ldots,t_n)\rangle = F\langle t_1\rangle \ldots \langle t_n\rangle$ for F n-ary with $n \geq 1$

$\langle Z\rangle = z$ for Z 0-ary, and $\langle Z(t_1,\ldots,t_n)\rangle = z\langle t_1\rangle \ldots \langle t_n\rangle$ for Z n-ary with $n \geq 1$

Functional terms are translated by currying.

(3) $\langle \Box\rangle = \Box : 0$

Holes are of base type.

The translation of a context $\langle C[\,]\rangle$ is denoted by $\langle C\rangle[\,]$. The translation of CRS rule $R = l \to r$ is defined as $\langle R\rangle = \langle l\rangle \to \langle r\rangle$.

For a CRS $\mathcal{C}$, the HRS $\langle\mathcal{C}\rangle$ is obtained by translating the alphabet and the set of rules of $\mathcal{C}$. We often restrict attention to the sub-HRS of $\langle\mathcal{C}\rangle$ where only terms that are translations of terms in $\mathcal{C}$ are considered. We first give the translation of the main ingredient needed in a rewrite step: assignment.

Definition 5. The translation $\langle\sigma\rangle$ of an assignment σ is defined as follows: if $\sigma(Z) = \underline{\lambda}(x_1,\ldots,x_n).s$, then $\langle\sigma\rangle(z) = \lambda x_1 \ldots x_n.\langle s\rangle$.

We now show that these translations are correct in the sense that a CRS concept yields the corresponding HRS concept.

Proposition 6. *Let s' be the translation $\langle s\rangle$ of a CRS metaterm s. Then*

a $s' : 0$, *moreover there is a bijective correspondence between subterms of s and subterms of type* 0 *of* s',

b s' *is in long $\beta\eta$-normal form,*

c *if s satisfies the pattern-condition, then s' satisfies the pattern-condition,*

d $\langle \mathsf{Fvar}(s)\rangle = \mathsf{Fvar}(s')$, *where* Fvar *denotes the set of free variables and of metavariables in a CRS (or HRS) metaterm.*

PROOF. The four properties are proved simultaneously, by structural induction. □

The bijective correspondence in **a** can be made more precise using the notion of position.

Proposition 7. *The translations of CRS rules, contexts, and assignments yield the corresponding concepts in the associated HRS.*

Next we state some propositions expressing the interaction between forming contexts and applying assignments on the one hand and translating on the other hand.

Proposition 8.

a $\langle C[t]\rangle = \langle C\rangle[\langle t\rangle]$

b $\langle C[t_1,\ldots,t_n]\rangle = \langle C\rangle[\langle t_1\rangle,\ldots,\langle t_n\rangle]$

c $\langle s[x_1 := t_1 \ldots x_n := t_n]\rangle = \langle s\rangle[x_1 := \langle t_1\rangle \ldots x_n := \langle t_n\rangle]$

d $\langle t^\sigma\rangle = \langle t\rangle^{\langle\sigma\rangle}\downarrow_\beta$.

PROOF.

a The proof proceeds by induction on the structure of $C[\,]$.

b By repeatedly applying **a**.

c Choose a context $C[,\ldots,]$ such that $s = C[x_{i_1},\ldots,x_{i_j}]$ and precisely the occurrences of the variables $x_1,\ldots,x_n$ are being displayed. Then

$$\begin{aligned}\langle s[x_1 := t_1 \ldots x_n := t_n]\rangle &= \langle C[t_{i_1},\ldots,t_{i_j}]\rangle \\ &= \langle C\rangle[\langle t_{i_1}\rangle,\ldots,\langle t_{i_j}\rangle] && \text{(by b)} \\ &= \langle s\rangle[x_1 := \langle t_1\rangle \ldots x_n := \langle t_n\rangle] && \text{(use Proposition 6 a)}\end{aligned}$$

d The statement is proved by induction on the structure of the metaterm t. Note that all λ's in $\langle t\rangle$ that are introduced by translating the abstraction operator do not yield a β-redex since they are 'blocked' by their big brother Λ. □

Now we show that rewrite steps are naturally preserved by the translation.

Theorem 9. *If $s \to_R t$ in a CRS $\mathcal{C}$ with $R = l \to r$ a rewrite rule, then we have $\langle s\rangle \to_{\langle R\rangle} \langle t\rangle$ in the corresponding HRS $\mathcal{H}$.*

PROOF. Let $s \to_{\langle R \rangle} t$ in $\mathcal{C}$, where $s = C[l^\sigma]$ and $t = C[r^\sigma]$, for some context $C[\,]$ and some assignment σ. Then,

$$\begin{aligned}
\langle s \rangle &= \langle C[l^\sigma] \rangle \\
&= \langle C \rangle [\langle l^\sigma \rangle] && \text{(by Proposition 8 a)} \\
&= \langle C \rangle [\langle l \rangle^{\langle \sigma \rangle} \downarrow_\beta] && \text{(by Proposition 8 d)} \\
&\to \langle C \rangle [\langle r \rangle^{\langle \sigma \rangle} \downarrow_\beta] \\
&= \langle C \rangle [\langle r^\sigma \rangle] && \text{(by Proposition 8 d)} \\
&= \langle C[r^\sigma] \rangle && \text{(by Proposition 8 a)} \\
&= \langle t \rangle
\end{aligned}$$

□

This proves that for every rewrite step in a CRS $\mathcal{C}$ a rewrite step in the associated HRS $\mathcal{H}$ can be performed. Now we will show that a rewrite step in the translation of a term must originate from a rewrite step in $\mathcal{C}$ itself. For this, we will use that both contexts and assignments in $\mathcal{H}$ can be translated back into the corresponding concepts in $\mathcal{C}$, under the proper restrictions.

Proposition 10.

a *If $C'[t'] = \langle s \rangle$, then there exist $C[\,]$ and t such that $\langle C[\,] \rangle = C'[\,]$ and $\langle t \rangle = t'$.*

b *If $\langle l \rangle^{\sigma'} \downarrow_\beta = \langle s \rangle$ and l satisfies the conditions for a left-hand side of a CRS rule, then there exists a σ such that $\langle \sigma \rangle = \sigma'$.*

PROOF.

a From Proposition 6 **a** we have $C'[\,] : 0$ and by the definition of a HRS context $t' : 0$. Then the bijective correspondence of Proposition 6 **a** provides us with suitable $C[\,]$ and t.

b By Proposition 6 **d**, we know that a free variable z in $\langle l \rangle$ stems from a metavariable Z in l. By the pattern-condition, each free variable occurs only in subterms of the form $zx_1 \ldots x_n$ in $\langle l \rangle$, which have type 0 by Proposition 6 **a**, where n is the arity of Z. Note that all the x_i have type 0, so this subterm is η-expanded. If $\sigma'(z) = \lambda y_1 \ldots y_m.t'$, then $m = n$, because $\sigma'(z)$ is by definition in long $\beta\eta$-normal form. Hence, $(zx_1 \ldots x_n)^{\sigma'} \downarrow_\beta = t'[y_1 := x_1 \ldots y_n := x_n] \downarrow_\beta = t'[y_1 := x_1 \ldots y_n := x_n]$, because renaming doesn't create redexes. It is easy to show that $t'[y_1 := x_1 \ldots y_n := x_n]$ must in fact be a subterm (of type 0!) of $\langle s \rangle$ and therefore a translation of some term $\hat{t}$. Now we can define $\sigma(Z) = \underline{\lambda}(y_1, \ldots, y_n).\hat{t}[x_1 := y_1 \ldots x_n := y_n]$, which meets the requirements. Note that by the pattern-condition all the x_i are distinct.

□

Theorem 11. *Let $\mathcal{C}$ be a CRS and $\mathcal{H}$ its associated HRS. If $\langle s \rangle \to_{\langle R \rangle} t'$ in $\mathcal{H}$ by rewrite rule $\langle R \rangle = \langle l \rangle \to \langle r \rangle$, then $s \to_R t$ for some CRS term t such that $\langle t \rangle = t'$.*

PROOF. Let $\langle s \rangle \to_{\langle R \rangle} t'$ in $\mathcal{H}$ with $\langle R \rangle = \langle l \rangle \to \langle r \rangle$. Then $\langle s \rangle = C'[\langle l \rangle^{\sigma'} \downarrow_\beta]$ and $t' = C'[\langle r \rangle^{\sigma'} \downarrow_\beta]$ for some context $C'[\,]$ and some assignment σ'. Then

$$\begin{aligned}
\langle s \rangle &= C'[\langle l \rangle^{\sigma'} \downarrow_\beta] \\
&= \langle C \rangle [\langle l \rangle^{\sigma'} \downarrow_\beta] \quad \text{(by Proposition 10 } \mathbf{a}\text{)} \\
&= \langle C \rangle [\langle l \rangle^{\langle \sigma \rangle} \downarrow_\beta] \quad \text{(by Proposition 10 } \mathbf{b}\text{)} \\
&= \langle C \rangle [\langle l^\sigma \rangle] \quad \text{(by Proposition 8 } \mathbf{d}\text{)} \\
&= \langle C[l^\sigma] \rangle \quad \text{(by Proposition 8 } \mathbf{a}\text{)}
\end{aligned}$$

By injectivity we have $s = C[l^\sigma]$. If we take $t = C[r^\sigma]$, then $s \to t$ and

$$\begin{aligned}
\langle t \rangle &= \langle C[r^\sigma] \rangle \\
&= \langle C \rangle [\langle r^\sigma \rangle] \quad \text{(by Proposition 8 } \mathbf{a}\text{)} \\
&= \langle C \rangle [\langle r \rangle^{\langle \sigma \rangle} \downarrow_\beta] \quad \text{(by Proposition 8 } \mathbf{d}\text{)} \\
&= \langle C \rangle [\langle r \rangle^{\sigma'} \downarrow_\beta] \\
&= t'
\end{aligned}$$

□

The naturality of a translation is can be measured by the properties which it preserves and reflects. Theorems 9 and 11 state that the main property of CRSs and HRSs, i.e. whether one term rewrites (in one step) to another, is both preserved and reflected. Combining this with the fact that orthogonality is preserved, we obtain a confluence proof for orthogonal CRSs via confluence of their associated HRS. A system is *orthogonal* if it is left-linear and non-ambiguous. For a precise definition of orthogonality we refer the reader to [Klo80] and [Nipb].

Corollary 12. *Orthogonal CRSs are confluent.*

PROOF. Let $s \twoheadrightarrow_{\mathcal{C}} t_1$ and $s \twoheadrightarrow_{\mathcal{C}} t_2$ be rewrites in an orthogonal CRS $\mathcal{C}$. By Theorem 9, we can lift these to rewrites $\langle s \rangle \twoheadrightarrow_{\mathcal{H}} \langle t_1 \rangle$ and $\langle s \rangle \twoheadrightarrow_{\mathcal{H}} \langle t_2 \rangle$ in the HRS $\mathcal{H}$ associated to $\mathcal{C}$. Because $\mathcal{H}$ is easily seen to be orthogonal, we conclude from [Nipb, Cor. 4.9] that it is confluent, hence there exist rewrites $\langle t_1 \rangle \twoheadrightarrow_{\mathcal{H}} r'$ and $\langle t_2 \rangle \twoheadrightarrow_{\mathcal{H}} r'$, for some r'. These sequences can be projected again to form $t_1 \twoheadrightarrow_{\mathcal{C}} r$ and $t_2 \twoheadrightarrow_{\mathcal{C}} r$ by Theorem 11, also showing that $\langle r \rangle = r'$. The proof is expressed by the following diagram.

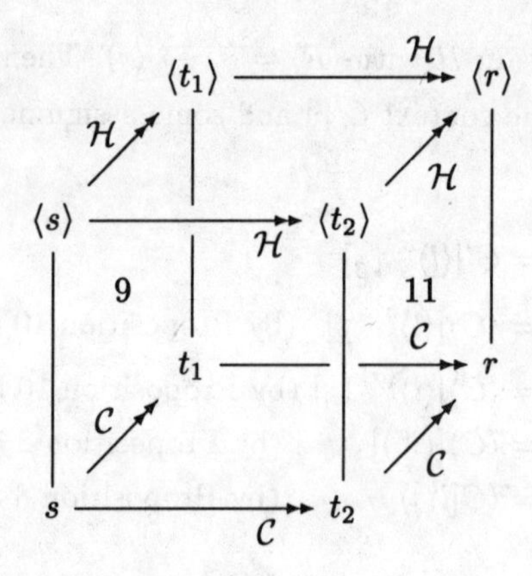

□

4 Translating a HRS into a CRS plus explicit β

In this section we define a translation from HRSs into CRSs. This translation is not as straightforward as the one the other way round, due to the fact that the metalanguage of HRSs, $\lambda^{\rightarrow}$, has more 'rewrite power' than the (hidden) metalanguage of CRSs, $\underline{\lambda}$. In order to be able to simulate every rewrite of a HRS in its associated CRS, a β-reduction rule and a binary symbol @ for application have to be added to the translation. It is given as

$$@([x]Z(x), Z') \rightarrow_{\beta} Z(Z')$$

We will denote β-reduction to normal form as $\rightarrow^{!}_{\beta}$. To simplify the notation a bit we sometimes use $@_n$ to abbreviate n applications, for instance $@_2(A, B, C)$ stands for $@(@(A, B), C)$. Formally, $@_n$ for $n \geq 1$ is defined as

$$@_1(t, t_1) = @(t, t_1)$$
$$@_{n+1}(t, t_1, \ldots, t_n, t_{n+1}) = @(@_n(t, t_1, \ldots, t_n), t_{n+1})$$

Again, the translation is denoted as $\langle \_ \rangle$ and is chosen to be injective. We do not obtain a $1-1$-correspondence between rewrite steps in a HRS and rewrite steps in its encoding, but the translation does satisfy a weaker property. Let $\mathcal{H}$ be a HRS and let $\mathcal{C}$ be its encoding, having as rewrite rules the translated rules of $\mathcal{H}$ and the β-rule. We have that if $s \rightarrow_{\mathcal{H}} t$ in a HRS $\mathcal{H}$, then $\langle s \rangle \rightarrow_{\mathcal{C}\beta} \langle t \rangle$, where $\rightarrow_{\mathcal{C}\beta}$ is defined to be a rewrite in $\mathcal{C}$ consisting of one step via a translated $\mathcal{H}$-rule followed by a β-reduction to β-normal form. Moreover, we obtain that a rewrite in the encoding of $\mathcal{H}$ starting with the encoding of some term of $\mathcal{H}$ can be extended to a rewrite corresponding to a rewrite in the original HRS.

First to an alphabet $\mathcal{A}$ of a HRS a CRS alphabet $\langle \mathcal{A} \rangle$ is associated.

Definition 13. The CRS alphabet $\langle \mathcal{A} \rangle$ associated with a HRS alphabet $\mathcal{A}$ consists of

- a symbol @ for application,
- for every symbol $F \in \mathcal{A}$ for an operator of type τ, a symbol F for an operator with arity $n = \mathsf{Ar}(\tau)$,
- the ordinary symbols of a CRS alphabet, i.e. symbols for variables $x\ y\ z \ldots$, symbols for metavariables with a fixed arity $Z\ Z_0\ Z_1 \ldots$ and a symbol for abstraction, $[\_]\_$.

Definition 14. The translation of terms in long $\beta\eta$-normal form is defined inductively as follows:

- $\langle \lambda x_1 \ldots x_m.xt_1 \ldots t_n \rangle = \begin{cases} [x_1]\ldots[x_n]x & \text{if } n = 0 \\ [x_1]\ldots[x_n]@_n(x, \langle t_1 \rangle, \ldots, \langle t_n \rangle) & \text{if } n \geq 1 \end{cases}$
- $\langle \lambda x_1 \ldots x_m.Ft_1 \ldots t_n \rangle = \begin{cases} [x_1]\ldots[x_n]F & \text{if } n = 0 \\ [x_1]\ldots[x_n]F(\langle t_1 \rangle, \ldots, \langle t_n \rangle) & \text{if } n \geq 1 \end{cases}$

It is extended to contexts by defining $\langle \Box \rangle = \Box$. We write $\langle C \rangle[\,]$ for the translation of $C[\,]$.

Free variables in terms of HRSs correspond to free variables in terms of CRSs. Free variables in rules of HRSs correspond to metavariables in rules of CRSs. Therefore a separate definition of the translation of a rule has to be given, in which free variables are translated in another way than in the translation of a term.

Definition 15. The translation $\langle l \to r \rangle$ of a HRS rule $l \to r$ is defined as $\langle l \rangle \to \langle r \rangle$, where $\langle l \rangle$ and $\langle r \rangle$ are defined inductively as follows.

a The left-hand side l of a HRS rewrite rule is of the form $l = Ft_1 \ldots t_k$. Here $t_1, \ldots, t_k$ are long $\beta\eta$-normal forms in which inputs of free variables are η-equivalent to distinct bound variables. The translation $\langle l \rangle$ of l is defined by induction on the structure of such a long $\beta\eta$-normal form.

- $\langle \lambda x_1 \ldots x_m.xt_1 \ldots t_n \rangle = \begin{cases} [x_1 \ldots x_m]x & \text{if } n = 0 \\ [x_1 \ldots x_m]@_n(x, \langle t_1 \rangle, \ldots, \langle t_n \rangle) & \text{if } n \geq 1 \end{cases}$
- $\langle \lambda x_1 \ldots x_m.zt_1 \ldots t_n \rangle = \begin{cases} [x_1 \ldots x_m]Z & \text{if } n = 0 \\ [x_1 \ldots x_m]Z(t_1{\downarrow_\eta}, \ldots, t_n{\downarrow_\eta}) & \text{if } n \geq 1 \end{cases}$

  if z is a variable which is free in l (note that $t_1, \ldots, t_n$ are η-equivalent to distinct bound variables by the pattern-condition),
- $\langle \lambda x_1 \ldots x_m.Ft_1 \ldots t_n \rangle = \begin{cases} [x_1 \ldots x_m]F & \text{if } n = 0 \\ [x_1 \ldots x_m]F(\langle t_1 \rangle, \ldots, \langle t_n \rangle) & \text{if } n \geq 1 \end{cases}$

b The right-hand side r of a HRS rewrite rule is of the form $r = st_1 \ldots t_k$ with s a symbol standing for a free variable or an operator and $t_1, \ldots, t_k$ in long $\beta\eta$-normal form. The translation $\langle r \rangle$ of r is by induction on the structure of a long $\beta\eta$-normal form.

- $\langle \lambda x_1 \ldots x_m.xt_1 \ldots t_n \rangle = \begin{cases} [x_1 \ldots x_m]x & \text{if } n = 0 \\ [x_1 \ldots x_m]@_n(x, \langle t_1 \rangle, \ldots, \langle t_n \rangle) & \text{if } n \geq 1 \end{cases}$
- $\langle \lambda x_1 \ldots x_m.zt_1 \ldots t_n \rangle = \begin{cases} [x_1 \ldots x_m]Z & \text{if } n = 0 \\ [x_1 \ldots x_m]Z(\langle t_1 \rangle, \ldots, \langle t_n \rangle) & \text{if } n \geq 1 \end{cases}$
- $\langle \lambda x_1 \ldots x_m.Ft_1 \ldots t_n \rangle = \begin{cases} [x_1 \ldots x_m]F & \text{if } n = 0 \\ [x_1 \ldots x_m]F(\langle t_1 \rangle, \ldots, \langle t_n \rangle) & \text{if } n \geq 1 \end{cases}$

As in the translation from CRSs to HRSs, we show that rewrite steps in a HRS can be simulated by essentially the same step in the associated CRS. To that end, the translation is extended to assignments.

Definition 16. An assignment of a HRS assigns to a variable a term in long $\beta\eta$-normal form of the same type. So an assignment assigns to a variable y of type τ a term of the form $\lambda x_1 \ldots x_n.t$ with $n = \mathsf{Ar}(\tau)$ and t not a λ-abstraction. The translation $\langle\sigma\rangle$ of an assignment σ is defined as follows: if $\sigma(y) = \lambda x_1 \ldots x_n.t$, then $\langle\sigma\rangle(Y) = \underline{\lambda}(x_1, \ldots, x_n), \langle t\rangle$.

First we show that the translation produces correct terms and that the translation of a rewrite rule is well-defined.

Proposition 17.
a *If t is a HRS term in long $\beta\eta$-normal form, then $\langle t\rangle$ is well-defined as a CRS term.*
b *The translation $\langle l \to r\rangle$ of a HRS rewrite rule $l \to r$ satisfies the definition of a CRS rewrite rule.*

Then we show that decomposing a term by a context, commutes with the translation.

Proposition 18. $\langle C[t]\rangle = \langle C\rangle[\langle t\rangle]$.

PROOF. The proof proceeds by induction on the definition of $C[\,]$. □

Finally we show that decomposing a term into a (meta)term and an assignment almost commutes with the translation. For a decomposition into a left-hand side, which is a pattern, the commutation is perfect, but for right-hand sides we need additional β-steps. This is proved in the following two propositions.

Proposition 19.
a *Let s be a term in long $\beta\eta$-normal form, and $u_1, \ldots, u_n$ terms that are η-equivalent to distinct variables. Then $\langle s\rangle[z_1 := u_1{\downarrow_\eta} \ldots z_n := u_n{\downarrow_\eta}] = \langle s[z_1 := u_1 \ldots z_n := u_n]{\downarrow_\beta}\rangle$.*
b $\langle l\rangle^{\langle\sigma\rangle} = \langle l^\sigma{\downarrow_\beta}\rangle$ *for l a term in long $\beta\eta$-normal form satisfying the pattern condition.*

PROOF.
a The proof proceeds by induction on the structure of s.
b The statement is proved by induction on the structure of a long $\beta\eta$-normal form, in which arguments of free variables are η-equivalent to distinct bound variables.

□

Combining the last two propositions, we observe that the 'matching power' (or complexity of matching, depending on one's point of view) of HRSs is equally present in CRSs, making a natural encoding of the former into the latter possible. This is due to the pattern-condition of HRSs. For HRSs not satisfying the pattern-condition (cf. [Wol93]) this is no longer the case, and an encoding doesn't seem to be straightforward anymore, even if we would lift some of the restrictions on left-hand sides of CRS-rules. The next proposition shows that although CRSs and HRSs have the same matching power, HRSs have more 'rewrite power', i.e. they can do more in one step.

Proposition 20.

a *Let s and $u_1, \ldots, u_n$ be terms in long $\beta\eta$-normal form. Then we have* $\langle s\rangle[z_1 := \langle u_1\rangle \ldots z_n := \langle u_n\rangle] \rightarrow^!_\beta$ $\langle s[z_1 := u_1 \ldots z_n := u_n]\downarrow_\beta\rangle$.

b $\langle r\rangle^{\langle\sigma\rangle} \rightarrow^!_\beta \langle r^\sigma\downarrow_\beta\rangle$.

PROOF.

a The proof proceeds by induction on the maximal length of the β-reduction of $s[z_1 := u_1 \ldots z_1 := u_n]$ to normal form.

b The proof proceeds by induction on the structure of r.

□

Now we can collect the results of this section to show that every rewrite step in a HRS $\mathcal{H}$ can be simulated in its corresponding CRS $\mathcal{C}$.

Theorem 21. *If $s \rightarrow_R t$ by rewrite rule $R = l \rightarrow r$ in $\mathcal{H}$, then we have $\langle s\rangle \rightarrow_{\langle R\rangle} \rightarrow^!_\beta \langle t\rangle$, or $\langle s\rangle \rightarrow_{\mathcal{C}\beta} \langle t\rangle$, in the corresponding CRS $\mathcal{C}$.*

PROOF. The term s is of the form $C[l^\sigma\downarrow_\beta]$, and we have $s = C[l^\sigma\downarrow_\beta] \rightarrow_R C[r^\sigma\downarrow_\beta] = t$. We have

$$\begin{array}{rll}
\langle s\rangle & = & \langle C[l^\sigma\downarrow_\beta]\rangle \\
 & = & \langle C\rangle[\langle l^\sigma\downarrow_\beta\rangle] \quad \text{(by Proposition 18)} \\
 & = & \langle C\rangle[\langle l\rangle^{\langle\sigma\rangle}] \quad \text{(by Proposition 19 b)} \\
 & \rightarrow_{\langle R\rangle} & \langle C\rangle[\langle r\rangle^{\langle\sigma\rangle}] \\
 & \rightarrow^!_\beta & \langle C\rangle[\langle r^\sigma\downarrow_\beta\rangle] \quad \text{(by Proposition 20 b)} \\
 & = & \langle C[r^\sigma\downarrow_\beta]\rangle \quad \text{(by Proposition 18)} \\
 & = & \langle t\rangle
\end{array}$$

□

The next thing to be done is to connect somehow a rewrite step in the translation of a HRS with a rewrite step in the original HRS itself. Since the translation of a HRS $\mathcal{H}$ acts as a refinement of $\mathcal{H}$, we cannot hope for a result as neat as in the previous section. But still something can be said. First we will show that if we have the rewrite $\langle s\rangle \rightarrow_{\mathcal{C}\beta} t'$ in the translation of a HRS, then we can project it to a rewrite step $s \rightarrow_{\mathcal{H}} t$, such that $\langle t\rangle = t'$.

The first observation we need is that there is a 1-1 correspondence between functional subterms in $\langle s\rangle$, i.e. subterms with a function symbol (also taking the $@_n$ for $n \geq 1$ into account) or a variable as head, and subterms of type 0 in s. Further, we need two propositions.

Proposition 22. *Suppose $C'[t'] = \langle s \rangle$ with t' a functional term. Then a context $C[\,]$ and a term t exist such that $\langle C \rangle[\,] = C'[\,]$ and $\langle t \rangle = t'$.*

PROOF. Via the correspondence we obtain an appropriate subterm of s, i.e. a context $C[\,]$ and a term t such that $s = C[t]$ and $\langle t \rangle = t'$. Using Proposition 18 we have that $\langle C \rangle[\,] = C'[\,]$. □

Proposition 23. *Let l be the left-hand side of a HRS rewrite rule. If $\langle l \rangle^{\sigma'} = \langle s \rangle$, then there is an assignment σ with $\langle \sigma \rangle = \sigma'$.*

PROOF. Metavariables Z_i occur in $\langle l \rangle$ in the form $Z(x_1, \ldots x_n)$. Suppose σ' is defined as $\sigma'(Z_i) = \underline{\lambda}(u_1, \ldots, u_n).t'$, hence $(Z(x_1, \ldots x_n))^{\sigma'} = t'[u_1 := x_1 \ldots u_n := x_n]$. We know that $t' \neq [x]t''$, because otherwise we cannot have $\langle l \rangle^{\sigma'} = \langle s \rangle$ due to the typing. Hence, $t'[\ldots]$ is a functional term, i.e. of one of the forms $F(\ldots)$ or $@_n(\ldots)$ with $n \geq 1$ and so in s there is a corresponding subterm t of type 0, such that $\langle t \rangle = t'[u_1 := x_1 \ldots u_n := x_n]$. Define σ as $\sigma z = \lambda x_1 \ldots x_n.t$. Then $\langle \sigma \rangle = \sigma'$. □

Theorem 24. *If $\langle s \rangle \rightarrow_{\langle R \rangle} \rightarrow^!_\beta t'$ in the CRS $\mathcal{C}$ by rewrite rule $\langle R \rangle = \langle l \rightarrow r \rangle$, then $s \rightarrow_R t$ in $\mathcal{H}$, for some t such that $\langle t \rangle = t'$.*

PROOF. We have $\langle s \rangle = C'[\langle l \rangle^{\sigma'}] \rightarrow_{\mathcal{C}} t_0 = C'[\langle r \rangle^{\sigma'}] \rightarrow^!_\beta t'$, for some context $C'[\,]$, assignment σ'. Now we have

$$\begin{aligned}
\langle s \rangle &= C'[\langle l \rangle^{\sigma'}] \\
&= \langle C \rangle[\langle l \rangle^{\sigma'}] && \text{(by Proposition 22)} \\
&= \langle C \rangle[\langle l \rangle^{\langle \sigma \rangle}] && \text{(by Proposition 23)} \\
&= \langle C \rangle[\langle l^\sigma \downarrow_\beta \rangle] && \text{(by Proposition 19 b)} \\
&= \langle C[l^\sigma \downarrow_\beta] \rangle && \text{(by Proposition 18)}
\end{aligned}$$

By injectivity of $\langle \_ \rangle$, we have $s = C[l^\sigma \downarrow_\beta]$. Take $t = C[r^\sigma \downarrow_\beta]$, then $s \rightarrow_R t$ and

$$\begin{aligned}
\langle t \rangle &= \langle C[r^\sigma \downarrow_\beta] \rangle \\
&= \langle C \rangle[\langle r^\sigma \downarrow_\beta \rangle] && \text{(by Proposition 18)} \\
&\leftarrow^!_\beta \langle C \rangle[\langle r \rangle^{\langle \sigma \rangle}] && \text{(by Proposition 20 b)} \\
&= \langle C \rangle[\langle r \rangle^{\sigma'}] \\
&= t_0
\end{aligned}$$

By confluence of β and injectivity of $\langle \_ \rangle$, we have $\langle t \rangle = t'$. □

If we want to prove the Church-Rosser property for orthogonal HRSs via the same property for CRSs, Theorem 24 is not quite enough. The β-rule, by construction, indeed is orthogonal to the other rules and coinitial rewrites can be lifted, but only $\mathcal{C}\beta$-steps can be projected, not arbitrary $\mathcal{C}$-rewrites. We now show that every rewrite in an *arbitrary* CRS C starting with a term which is the translation of some HRS-term, can be completed, by performing a β-reduction to β-normal form, to a rewrite which can be simulated by a 'standard' rewrite consisting of $C\beta$-steps.

The proof follows the strategy employed for proving $\Sigma \models \mathrm{WCR}^+$ in [Klo80, pp. 144–148]. However, some difficulties arise. First, because of the possible non-left-linearity of the rules. Second, because simply typed λ-calculus doesn't satisfy the disjointness property in contrast to underlined λ-calculus.

The main property to be proved is that β-reductions to normal form do not interfere with rewrite steps. To do this we first need to define some tracing mechanisms.

Definition 25.

a Let $R = l \to r$ be a rewrite rule. Its *conditional version* $R_c = l_c \to r$ is obtained by repeatedly replacing occurrences of a metavariable Z which occurs at least twice in l by a fresh metavariable Z' and adding the condition $Z \equiv Z'$ to the rule. Its *linearisation* is R_l obtained from R_c by omitting the conditions.

b Let r be a metaterm, and $\mathbf{Z} = Z_1, \ldots, Z_n$ be a list of metavariables containing the ones in r, then the *freezing* r_f of $\mathbf{Z}$ in r is defined by $r_f = @_n([\mathbf{z}]r', [\mathbf{x_1}]Z_1(\mathbf{x_1}), \ldots [\mathbf{x_n}]Z_n(\mathbf{x_n}))$, where $\mathbf{z} = z_1, \ldots, z_n$ is a list of fresh variables, and r' is obtained by replacing in r all occurrences of $Z(\mathbf{t})$ by $@(z, \mathbf{t})$. For a rule $R = l \to r$, the freezing $R_f = l \to r_f$ is defined by freezing the metavariables of l in r. We define $R_{f\beta}$ to be the CRS with rules R_f and β

c Let $R = l \to r$ be a rewrite rule. Its *underlining* $\underline{R}$ is obtained from R_{cf} by underlining the head-symbol of l_c. (So R is first made conditional, then frozen and finally its head-symbol is underlined.)

d An (R-)underlining of a term s is a term containing some underlined symbols, which are the head-symbols of $\underline{R}$-redexes, and which is equal to s after removing the underlining.

e A rewrite $s \twoheadrightarrow_C t$ is an (R-)*development* if there is some underlining $\underline{s}$ of s, such that $\underline{s} \twoheadrightarrow_{\underline{R}+\beta} \underline{t}$ and the underlined rewrite 'projects' onto the original one. Note that, due to non-left-linearity, the terms in the underlined rewrite need not be underlinings of terms.

NOTATION. This underlining of (head symbols of) redexes might be be considered confusing, because underlinings were also used in $\underline{\lambda}$-calculus. Yet, we think it is the right notation because the underlinings express the same idea of marking both times.

Example 5. Let R be a rule defined by $R = \nu([x]Z(x), [x]Z(x)) \to Z(\nu([x]Z(x)))$. Then we have

$$R_l = \nu([x]Z(x), [x]Z'(x)) \to Z(\nu[x]Z(x))$$
$$R_f = \nu([x]Z(x), [x]Z(x)) \to @([z]@(z, \nu[x]@(z, x)), [x]Z(x))$$
$$\underline{R} = \underline{\nu}([x]Z(x), [x]Z'(x)) \to @([z, z']@(z, \nu[x]@(z, x)), [x]Z(x), [x]Z'(x)) \text{ if } Z \equiv Z'$$

The idea of freezing is the one of [Lan93], postponing both duplication of the metavariables and substitution into the metavariables. It is more extensive than the one in [Klo80], where only substitution is postponed. Both postponed actions can be performed by β-reduction:

Proposition 26.

a $s \rightarrow_{C\beta} t$ *is a development.*

b *Let* $\underline{s}$ *be equal to* s *after removal of underlinings. If* $\underline{s} \rightarrow_{\underline{R}\beta} \underline{t}$, *then* $s \rightarrow_{C\beta} t$. *Here* $\rightarrow_{\underline{R}\beta} = \rightarrow_{\underline{R}} \rightarrow^{!}_{\beta}$.

PROOF.

a Idea. Underline the redex to obtain an underlining of s. Rewrite it with the underlined rule and then to β-nf.

b Idea. In the β-reduction to normal form, we can do the postponed duplication and substitution steps first and then the others. This rewrite can be projected to a $C\beta$-step.

□

It is not difficult to see that explicit β-reductions in translated terms can be made to correspond to β-reductions in $\lambda^{\rightarrow}$-calculus. (Define a suitable forgetful map, forgetting explicit @'s and replacing [_] by λ, giving typable terms). Hence β is terminating. In the following we only consider rewrites that start with the translation of some term in the HRS.

Proposition 27. *Every rewrite in* $\underline{R} + \beta$ *terminates.*

PROOF. Sketch. Let $\underline{R} = \underline{l} \rightarrow \underline{r}$. Let $\underline{s}'$ be obtained from $\underline{s}$ by replacing all $\underline{R}$ redexes $\underline{l}^{\sigma}$ by $@([x]x, \underline{r}^{\sigma})$. This can be done unambiguously because $\underline{R}$ doesn't have overlap with itself. The idea is that we have replaced left-hand sides by their right-hand sides in advance, but have put an extra identity in, to keep in mind that we have to do soto simulate a step. (This replacement works only because the rule is (left- and right-)linear). Now we have that every rewrite starting from $\underline{s}$ can be simulated by a rewrite of the same length starting from $\underline{s}'$. More precisely, each β-step is simulated by a β-step and an $\underline{R}$-step $C[\underline{l}^{\sigma}] \rightarrow_{\underline{R}} C[\underline{r}^{\sigma}]$ is simulated by the corresponding β-step $C'[@([x]x, \underline{r}^{\sigma'})] \rightarrow_{\beta} C'[\underline{r}^{\sigma'}]$. Since β-rewriting terminates, the $\underline{R} + \beta$-rewrite must be finite. □

Corollary 28. *Every development is finite.*

Proposition 29. β *commutes with* $\underline{R} + \beta$. *That is*

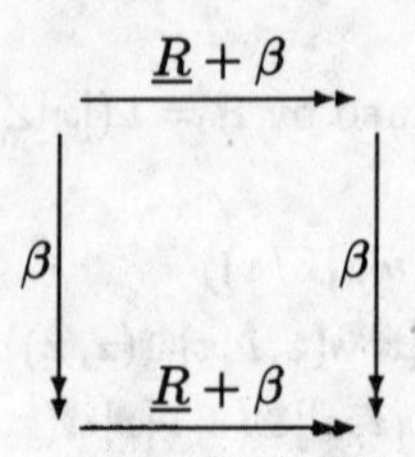

PROOF. By termination of $\underline{R}+\beta$ (Proposition 27) and Newman's Lemma (see e.g. [Oos94a]) it suffices to consider only local divergences in proving commutativity. The case of a local divergence of β-steps is covered by confluence of β. The other case follows by considering the relative positions of the β- and $\underline{R}$-redex. If the β-redex is inside the $\underline{R}$-redex, we have the following diagram.

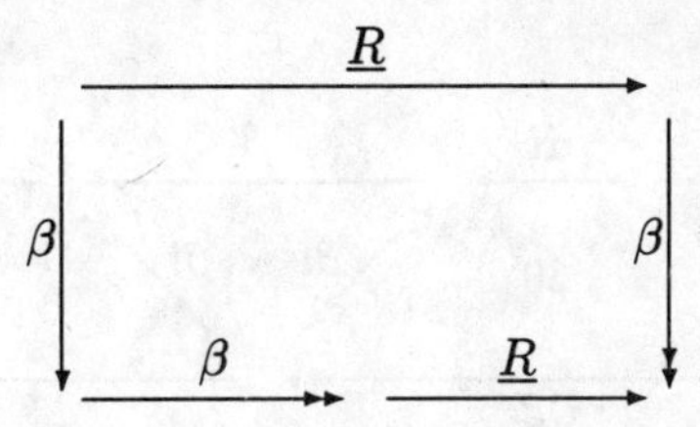

We may need some 'compensating' β-steps due to possible conditions on $\underline{R}$ (which originate from non-left-linearity of R). If the β-redex is outside the $\underline{R}$-redex, we have the diagram:

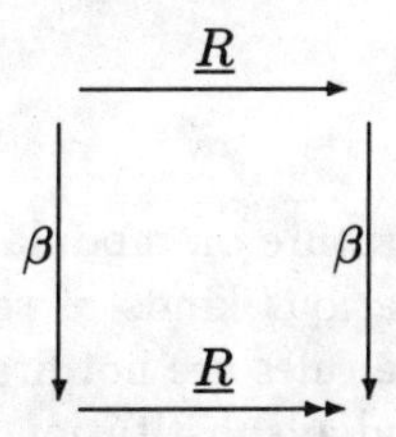

The β-step may duplicate and even nest $\underline{R}$-redexes, but this does not cause much trouble. A parallel inside-out reduction of $\underline{R}$-redexes works. (This is just as easy as for combinations of λ-calculus and first-order rewrite systems, as $\underline{R}$ is linear.) □

Proposition 30. *If $s \to_{C\beta} t$ and $s \to^{!}_{\beta} s'$, then $s' \twoheadrightarrow_{C\beta} t$.*

PROOF. By Proposition 26 we can construct the reduction $\underline{s} \to_{\underline{R}} \to^{!}_{\beta} t$, for some rule R and underlining $\underline{s}$ of s. We prove that if $\underline{s} \to_{\underline{R}} \twoheadrightarrow_{\underline{R}+\beta} t$, such that t is in β-normal form, and $\underline{s} \to^{!}_{\beta} \underline{s}'$, then $\underline{s}' \to_{\underline{R}\beta} t$. The proof is by induction on the maximal length of a $\underline{R}+\beta$ reduction sequence starting from $\underline{s}$ and expressed by the following diagram.

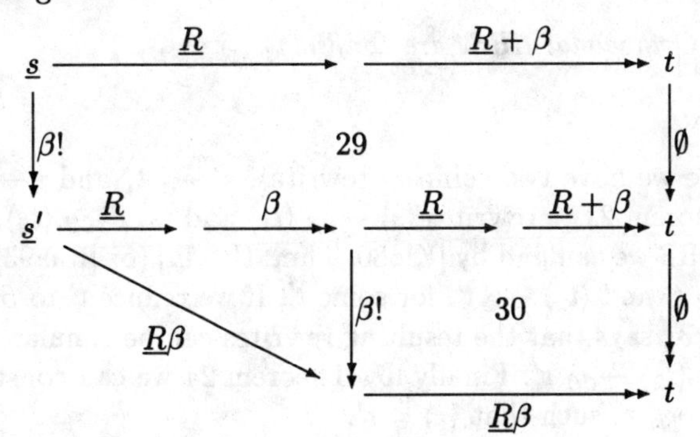

By applying Proposition 26 again (in the reverse direction), we are done. □

Lemma 31. *Suppose $s \twoheadrightarrow_C t$, and s and t are in β-normal form, then $s \twoheadrightarrow_{C\beta} t$.*

PROOF. The proof is expressed by the following diagram

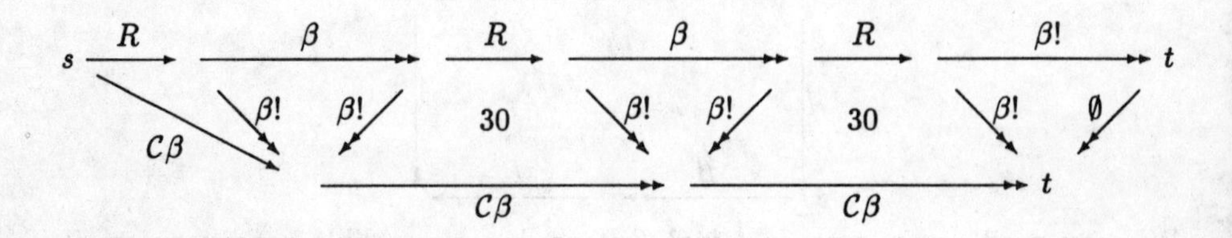

□

Remark. The results in the literature on modularity of confluence for combinations of typed λ-calculi with various kinds of rewriting do not seem to apply here. This is because the rewrite rules are not first-order. They can be frozen as above into a 'first-order part' and a 'substitution part', but the former may contain rules with β-redexes on their right-hand sides, which is not allowed for the systems studied in the literature. On the other hand, the method employed here seems to be quite flexible, since it makes use only of completeness of (typed) β. For example, the confluence result of [BTG89] should be an easy consequence.

In [Oos94b] a technique which is related to the one in this section is used to prove a Finite Developments Theorem for the large class of orthogonal Higher-Order Rewriting Systems.

Corollary 32. *Orthogonal HRSs are confluent.*

PROOF. Suppose we have two coinitial rewrites, $s \twoheadrightarrow_{\mathcal{H}} t_1$ and $s \twoheadrightarrow_{\mathcal{H}} t_2$, we can lift them by Theorem 21 to rewrites $\langle s\rangle \twoheadrightarrow_{C\beta} \langle t_1\rangle$ and $\langle s\rangle \twoheadrightarrow_{C\beta} \langle t_2\rangle$. Because C is an orthogonal CRS we can find by [Klo80, Thm. II.3.11] (or [Raa93]), convergent rewrites $\langle t_1\rangle \twoheadrightarrow_C t'$ and $\langle t_2\rangle \twoheadrightarrow_C t'$, for some t'. If we reduce t' to β-normal form r', then Lemma 31 says that the resulting rewrites can be simulated by rewrites $\langle t_1\rangle \twoheadrightarrow_{C\beta} r'$ and $\langle t_2\rangle \twoheadrightarrow_{C\beta} r'$. Finally, by Theorem 24 we can construct rewrites $t_1 \twoheadrightarrow_{\mathcal{H}} r$ and $t_2 \twoheadrightarrow_{\mathcal{H}} r$, such that $\langle r\rangle = r'$. □

The proof is expressed by the following diagram.

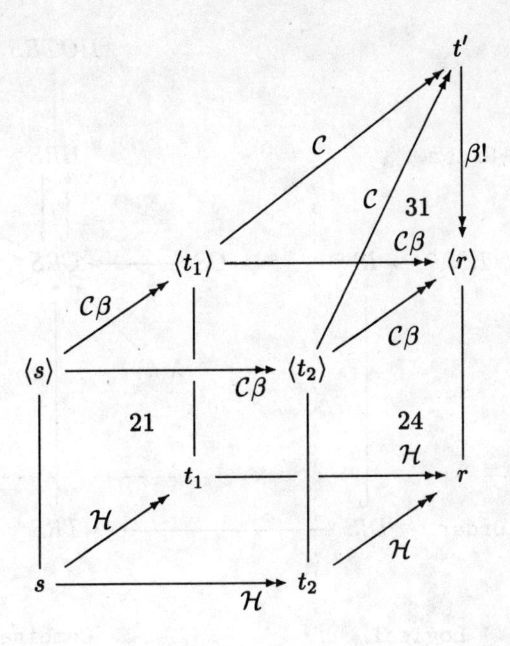

In fact, the reduction sequence $t' \to_{\beta}^{!} \langle r \rangle$, can be shown to be empty, giving a somewhat stronger result.

5 Discussion

In the picture below we show the relationships between some classes of rewriting systems occurring in the literature. We have classified them along two dimensions. Horizontally, we distinguish between *logical* and *combinatorial* systems. The logical systems are the ones for which the Curry-Howard Isomorphism still makes sense, i.e. the left-hand sides of rules satisfy a constructor-destructor discipline. If the left-hand sides consist of possibly complex combinations of symbols, then we call the system combinatorial. Vertically, we distinguish between *first-order* and *higher-order* in the usual way. One could add a third dimension: *functional* versus *communicational*. Apart from the IINs which generalise to INs, it is not clear to us which systems are obtained when lifting the functional systems below to communicational ones.

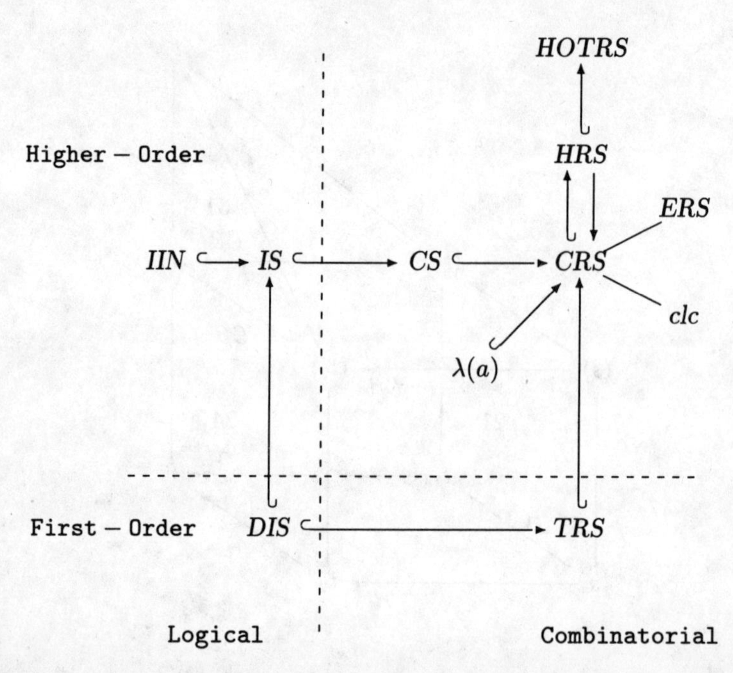

We will not discuss these systems here, but give references to the literature instead. For an overview of systems until 1980 see also [Klo80, pp. 132,133]. In (as far as we know) historical order we have:

- *TRS* = Term Rewriting System. We don't know who introduced this name, but they were known at the end of the seventies. Cf. also Rosen [Ros73].
- *CS* = Contraction Scheme. Introduced by Aczel [Acz78].
- λ(a)-reductions were introduced by Hindley [Hin78].
- *CRS* = Combinatory Reduction System. Introduced by Klop [Klo80].
- *HOTRS* = Higher-Order Term Rewriting System. Introduced by Wolfram in his PhD thesis, see [Wol93].
- *ERS* = Expression Reduction System. Introduced by Khasidashvili [Kha90].
- *(I)IN* = (Intuitionistic) Interaction Net. Introduced by Lafont [Laf90].
- *HRS* = Higher-order Rewrite System. They were introduced by Nipkow [Nipa].
- *(D)IS* = (Discrete) Interaction System. Introduced by Asperti and Laneve [AC92].
- *clc* = conditional λ-calculi. Introduced by Takahashi [Tak].

In general, if one system is encoded into another, it makes sense to check which syntactic properties are preserved by the translation. We will briefly review whether the (syntactic) properties (local) confluence and (weak) termination are preserved by the two encodings that are presented in this paper.

First we will consider the translation from CRSs into HRSs, say a CRS $\mathcal{C}$ is encoded into a HRS $\langle\mathcal{C}\rangle$. Since term formation in HRSs is quite liberal we will first restrict attention to the sub-HRS $\langle\mathcal{C}\rangle_{tr}$ with only the terms that are a translation of a term in $\mathcal{C}$. This is the HRS which we usually call 'the translation' of $\mathcal{C}$.

It is easily shown that if $\mathcal{C}$ is (locally) confluent, then its translation $\langle \mathcal{C} \rangle_{tr}$ is (locally) confluent. This is a consequence of Theorem 3.11 and 3.8. It is also easily shown that $\langle \mathcal{C} \rangle_{tr}$ is (weakly) terminating if $\mathcal{C}$ is (weakly) terminating. Note that we have as a corollary of Theorem 3.11 that the translation of a normal form is again a normal form.

If we, just out of curiousity, consider not the translation $\langle \mathcal{C} \rangle_{tr}$ of a CRS $\mathcal{C}$ but the full HRS $\langle \mathcal{C} \rangle$, then the situation is completely different. This is due to the fact that the rewrite relation for CRS is defined only on the set of terms (not containing metavariables), whereas in a HRS there is no syntactic difference between metavariables and variables.

Confluence nor local confluence are preserved if a CRS $\mathcal{C}$ is encoded into a HRS $\langle \mathcal{C} \rangle$. Consider the CRS $\mathcal{C}_1$:

$$
\begin{aligned}
F(F(Z)) &\to F(Z) \\
F([x]Z(x)) &\to F([x]Z(Z(x))) \\
F([x]Z(x)) &\to F(Z(A))
\end{aligned}
$$

It is confluent. This is a consequence of the fact that we have only one operator symbol F, for which we have the first rule. The corresponding HRS $\langle \mathcal{C}_1 \rangle$

$$
\begin{aligned}
F(Fz) &\to Fz \\
F(\Lambda\lambda x.zx) &\to F(\Lambda\lambda x.z(zx)) \\
F(\Lambda\lambda x.zx) &\to F(zA)
\end{aligned}
$$

is not confluent: the term $F(\lambda x.zx)$ can be rewritten to $F(zA)$ by applying the last rule, and to $F(z(zA))$ by applying the second and then the third rule. Note that $F(\lambda x.zx)$ is not the translation of a term in $\mathcal{C}_1$. Note that $\langle \mathcal{C}_1 \rangle$ is not locally confluent either.

The same holds for the properties weak termination and termination: they are not preserved by the translation from CRSs into HRSs. Consider the CRS $\mathcal{C}_2$:

$$
F([x][y]Z(x,y)) \to Z([x][y]Z(y,x),[x]x)
$$

It is terminating and thus weakly terminating. This can be understood by remarking that the alphabet contains only one operator symbol, which is unary. So an instance of $Z(x,y)$ will contain exactly one variable, and never both x and y. The corresponding HRS $\langle \mathcal{C}_2 \rangle$

$$
F(\Lambda\lambda y.zxy) \to z(\Lambda\lambda x\Lambda\lambda y.z(\Lambda\lambda x\Lambda\lambda y.zyx)(\Lambda\lambda x.x)
$$

is not terminating: the term $F(\Lambda\lambda x.\Lambda\lambda y.z(Fx)(Fy))$ permits an infinite rewrite sequence. Note moreover that it hasn't got a normal form, so it is not weakly terminating either. If we would consider the properties for rewriting on metaterms, then we conjecture that the properties are all preserved.

Now we will consider the translation from HRSs into CRSs, say $\mathcal{H}$ is translated into $\langle\mathcal{H}\rangle$. Again we consider first the sub-CRS $\langle\mathcal{H}\rangle_{tr}$ of $\langle\mathcal{H}\rangle$ where we only consider the terms that are a translation of a term in $\mathcal{H}$. This is the CRS which we usually call 'the translation'.

As a consequence of Lemma 4.21 and Theorem 4.10, we obtain that $\langle\mathcal{H}\rangle_{tr}$ is confluent if $\mathcal{H}$ is confluent. It is also easily shown that weak termination and termination are preserved.

Again, the situation is completely different if we consider the full CRS $\langle\mathcal{H}\rangle$. In this case, this is due to the fact that the untyped β-rule is always present in the CRS which is associated to a HRS.

The HRS $\mathcal{H}_1$

$$Dxx \to E$$

is confluent, but the associated CRS $\langle\mathcal{H}_1\rangle$

$$D(Z,Z) \to E$$
$$@([x]Z(x), Z') \to Z(Z')$$

is not confluent. This has been proven by Klop (see [Klo80]).

For an arbitrary HRS $\mathcal{H}_2$, its associated CRS $\langle\mathcal{H}_2\rangle$ is terminating nor weakly terminating. This is the case since $\langle\mathcal{H}_2\rangle$ will contain for instance the term $@([x]@(x,x),[x]@(x,x))$. So in particular the CRS associated to a HRS that is terminating is not terminating, and the same holds for weak termination.

In this paper we have shown two extensions of first-order rewriting to higher-order, CRSs and HRSs, to be almost equivalent. The difference lies in the meta-language used; they employ different flavours of the λ-calculus to generate their rewrite relations. For CRSs the underlined λ-calculus is used, while for HRSs the simply typed λ-calculus is used.

The translations from one system to the other are relatively simple because both are based on λ-calculus. The situation would be different for arbitrary meta-languages. But in fact it is hard to imagine a meta-language essentially different from λ-calculus. The basic steps of a rewrite (or redex-reaction) are: decomposing an object into a context and a redex, decomposing a redex into a pattern and a substitution, replacing the pattern with some other pattern, and then composing everything in the reverse order. The λ-calculus can be viewed as a 'calculus of (de)composing', so seems to be basic to any meta-language. If we look at other higher order rewrite formalisms, such as the Expression Reduction Systems of Khasidashvili [Kha90] and the Conditional Lambda Calculi of Takahashi [Tak], this claim seems to be supported. The precise interrelation is left to future work. We do note however that the similarities between these systems are obfuscated by the surprisingly large syntactical differences.

The work in this paper seems to suggest that only two basic properties are required for the flavour of λ-calculus one uses for the meta-language: confluence and termination. One can view CRSs and HRSs then as special cases of such a unifying theory of Higher Order Rewriting Systems (HORS). A large part of the syntactic rewrite theory should carry over to higher-order rewriting with more powerful meta-languages such as higher-order λ-calculi. Progress in this respect has been made in [Oos94b] and [OR94].

6 Acknowledgements

We would like to thank Fer-Jan de Vries for comments on an earlier version of this paper. We have benefitted from discussions with and between Jan Willem Klop, Tobias Nipkow and Stefan Kahrs on this subject.

References

[AC92] A. Asperti and Laneve C. Interaction systems I: the theory of optimal reductions. Technical Report 1748, INRIA-Rocquencourt, September 1992.

[ACCL90] M. Abadi, L. Cardelli, P.-L. Curien, and J.-J. Lévy. Explicit substitutions. In *Proceedings of the ACM Conference on Principles of Programming Languages*, San Francisco, 1990.

[Acz78] Peter Aczel. A general Church-Rosser theorem. Technical report, University of Manchester, July 1978.

[Bar84] H.P. Barendregt. *The Lambda Calculus, its Syntax and Semantics*, volume 103 of *Studies in Logic and the Foundations of Mathematics*. North-Holland Publishing Company, revised edition, 1984. (Second printing 1985).

[BG93] M. Bezem and J.F. Groote, editors. *Proceedings of the International Conference on Typed Lambda Calculi and Applications*, volume 664 of *Lecture Notes in Computer Science*, Utrecht, The Netherlands, March 1993. Springer-Verlag.

[BTG89] V. Breazu-Tannen and J. Gallier. Polymorphic rewriting conserves algebraic strong normalization and confluence. In *Proceedings of the 16th International Colloquium on Automata, Languages and Programming*, volume 372 of *Lecture Notes in Computer Science*, pages 137–150, 1989.

[Hin78] R. Hindley. Reductions of residuals are finite. *Transactions of the American Mathematical Society*, 240:345–361, June 1978.

[Kah92] S. Kahrs. Context rewriting. In M. Rusinowitch and J.L. Rémy, editors, *Proceedings of the Third International Workshop on Conditional and Typed Rewriting Systems*, pages 21–35, 1992.

[Kha90] Z.O. Khasidashvili. Expression reduction systems. In *Proceedings of I. Vekua Institute of Applied Mathematics*, volume 36, pages 200–220, Tbilisi, 1990.

[Klo80] J.W. Klop. *Combinatory Reduction Systems*. Mathematical Centre Tracts Nr. 127. CWI, Amsterdam, 1980. PhD Thesis.

[KOR93] J. W. Klop, V. van Oostrom, and F. van Raamsdonk. Combinatory reduction systems, introduction and survey. *Theoretical Computer Science*, 121(1–2):279–308, December 1993.

[Laf90] Yves Lafont. Interaction nets. In *Proceedings 17th ACM Symposium on Principles of Programming Languages*, pages 95–108, 1990.

[Lan93] C. Laneve. *Optimality and Concurrency in Interaction Systems*. PhD thesis, dipartimento di informatica università di pisa, March 1993.

[LIC91] Amsterdam, The Netherlands. *Proceedings of the sixth annual IEEE Symposium on Logic in Computer Science*, Los Alamitos, July 1991. IEEE Computer Society Press.

[Mil] D. Miller. A logic programming language with lambda-abstraction, function variables, and simple unification. In [Sie91].

[Nipa] T. Nipkow. Higher-order critical pairs. In [LIC91].

[Nipb] T. Nipkow. Orthogonal Higher-Order Rewrite Systems are Confluent. In [BG93].

[NM88] G. Nadathur and D. Miller. An overview of λProlog. In R.A. Kowalski and K.A. Bowen, editors, *Proc. 5th Int. Logic Programming Conference*, pages 810–827. MIT Press, 1988.

[Oos94a] V. van Oostrom. Confluence by decreasing diagrams. *Theoretical Computer Science*, 1994.

[Oos94b] V. van Oostrom. *Confluence for Abstract and Higher-Order Rewriting.* PhD thesis, Vrije Universiteit, March 1994.

[OR93] V. van Oostrom and F. van Raamsdonk. Comparing combinatory reduction systems and higher-order rewrite systems. Technical Report CS-R9361, CWI, September 1993. also available as VU technical Report IR-333.

[OR94] V. van Oostrom and F. van Raamsdonk. Weak orthogonality implies confluence: the higher-order case. In *Proceedings of the Symposium on Logical Foundations of Computer Science 1994*, 1994. to appear.

[Pau90] L.C. Paulson. Isabelle: The next 700 theorem provers. In P. Odifreddi, editor, *Logic and Computer Science*, pages 361–385. Academic Press, 1990.

[Raa93] F. van Raamsdonk. Confluence and superdevelopments. In C. Kirchner, editor, *Proceedings of the 5th International Conference on Rewrite Techniques and Applications*, 1993.

[Ros73] Barry K. Rosen. Tree-manipulating systems and Church-Rosser theorems. *Journal of the Association for Computing Machinery*, 20(1):160–187, January 1973.

[Sie91] J. Siekmann, editor. *Extensions of Logic Programming*, volume 475 of *Lecture Notes in Artificial Intelligence.* Tübingen, FRG, Springer-Verlag, December 1991.

[Tak] M. Takahashi. λ-calculi with conditional rules. In [BG93].

[Wol93] D.A. Wolfram. *The Clausal Theory of Types*, volume 21 of *Cambridge Tracts in Theoretical Computer Science.* Cambridge University Press, 1993.

Termination Proofs for Higher-order Rewrite Systems

Jaco van de Pol

Department of Philosophy, Utrecht University
Heidelberglaan 8, 3584 CS Utrecht, The Netherlands
email: jaco@phil.ruu.nl

Abstract. This paper deals with termination proofs for Higher-Order Rewrite Systems (HRSs), introduced in [12]. This formalism combines the computational aspects of term rewriting and simply typed lambda calculus. The result is a proof technique for the termination of a HRS, similar to the proof technique "Termination by interpretation in a well-founded monotone algebra", described in [8, 19]. The resulting technique is as follows: Choose a higher-order algebra with operations for each function symbol in the HRS, equipped with some well-founded partial ordering. The operations must be strictly monotonic in this ordering. This choice generates a model for the HRS. If the choice can be made in such a way that for each rule the interpretation of the left hand side is greater than the interpretation of the right hand side, then the HRS is terminating. At the end of the paper some applications of this technique are given, which show that this technique is natural and can easily be applied.

1 Introduction

In the field of automated proof verification one sees a development towards higher-order concepts. In the generic theorem prover Isabelle [15], typed lambda calculus is used as the syntax for the formulae. In other systems, as Coq [14], typed lambda calculus is even used for the logic, using the Curry-Howard isomorphism which links formulae to types and proofs to terms.

This development is mirrored in the research on Term Rewriting Systems (TRS). There are different formalisms dealing with the combination of term rewriting and an abstraction mechanism. In [11] the concept of Combinatory Reduction Systems (CRS) was introduced. These systems essentially are TRSs with bound variables. In [12, 13] the formalism of Higher-order Rewrite Systems (HRS) is described, which is very similar to CRSs in essence, but rather different in presentation. A precise comparison is given in [17]. A more general setting is given in [18]. Quite other approaches can be found in [4, 9].

Two important issues concerning rewrite systems are termination and confluence. For results about local confluence of HRSs and confluence of orthogonal HRSs the reader is referred to [12] and [13] respectively. In [11, ch. II.3] the confluence of regular CRSs is proved. However, the question of termination of the higher-order frameworks seems hardly to have been explored. As far as we know, only in [11, ch. II.6.2] a sufficient condition for termination of regular CRSs is given. With this condition, stated in terms of redexes and descendants, a termination proof for CRSs remains a syntactical matter. Other work on this line is done in [10]. We also refer to [9] where a recursion scheme for higher-order rules is given that guarantees termination.

Termination of first-order Term Rewriting is already an undecidable problem. But as the termination of TRSs is an interesting question, many semi-algorithms and characterisations of termination are proposed in the literature. A nice characterisation of termination is given in [19]. It builds on the "Increasing Interpretation Method" of [8, p. 367]. The function symbols of a TRS $\mathcal{R}$ have to be interpreted as strictly monotonic operations in some well-founded algebra. This interpretation is extended to closed terms as a usual algebraic homomorphism. Now the associated rewrite relation is terminating if every left hand side is greater (under the chosen interpretation) than the belonging right hand side, for each possible interpretation of the variables in that rule.

The strength of this characterisation is that one can concentrate on the "intuitive reason" for termination. This intuition can be translated in suitable operations on well-founded orderings, thus using semantical arguments. The real termination proof consists of testing a simple condition on the rules only instead of on all possible rewrite steps or all possible redexes. This semantical approach is more convenient than a syntactical technique.

The aim of this paper is to generalise this semantical characterisation of termination for TRSs to one for HRSs. The definition of Higher-order Rewriting we use, is close to [18], so it is an extension of [12, 13]. The main result is that such a generalisation is possible. The interpretation of terms can be extended to the interpretation of higher-order terms. The orderings and the notion of strictness can also be generalised. The techniques to achieve this are similar to those used in [5, 6]. Moreover, the result that termination proofs can be given with a well-founded monotone algebra in [19] carries over to HRSs with simple conditions on the well-founded ordering. With this technique some natural HRSs are proved to be terminating (see Section 6).

I like to thank Jan Friso Groote, who supplied some crucial ideas. I am also grateful to Jan Bergstra, Marc Bezem, Tonny Hurkens, Vincent van Oostrom, Alex Sellink and Jan Springintveld for reading previous versions of this document and for their suggestions. Finally, I thank H. Schwichtenberg and both referees for their close reading of the preliminary version, resulting in several improvements.

2 Term Rewriting in the Simply Typed Lambda Calculus

2.1 Types and Terms

In this section the sets of types and terms of simply typed lambda calculus are defined. The types are constructed from a set of base types. Let $\mathcal{B}$ be the set of *base types*. Then the set $\mathbb{T}(\mathcal{B})$ of *simple types* over these base types is defined as:

Definition 1. $\mathbf{T}(\mathcal{B})$ is the smallest set satisfying

- $\mathcal{B} \subseteq \mathbf{T}(\mathcal{B})$.
- If $\sigma, \tau \in \mathbf{T}(\mathcal{B})$ then also $\sigma \to \tau \in \mathbf{T}(\mathcal{B})$.

Terms are constructed from typed constants and typed variables. Let $\mathcal{C}_\tau$ be disjoint sets of *function symbols* (also called constants) of type τ. The union $\bigcup \mathcal{C}_\tau$ is written as $\mathcal{C}$. Similarly, $\mathcal{V}_\tau$ and $\mathcal{V}$ are sets of *typed variables.* Each $\mathcal{V}_\tau$ is supposed to be infinite. Furthermore, $\mathcal{C}$ and $\mathcal{V}$ are disjoint. To specialise particular sets of constants and variables, the notion of a *signature* is introduced:

Definition 2. A signature ($\mathcal{F}$) is a triple $(\mathcal{B}, \mathcal{C}, \mathcal{V})$, where $\mathcal{B}$ are the base types, $\mathcal{C}$ is a set of typed constants and $\mathcal{V}$ is a set of typed variables.

Given a signature $\mathcal{F}$ we can define the set of *simply typed lambda terms*:

Definition 3. Let $\mathcal{F}$ be the signature $(\mathcal{B}, \mathcal{C}, \mathcal{V})$. The sets $\Lambda^{\to}_\tau(\mathcal{F})$ (with $\tau \in \mathbf{T}(\mathcal{B})$) are defined inductively by:

- $\mathcal{V}_\tau \subseteq \Lambda^{\to}_\tau(\mathcal{F})$,
- $\mathcal{C}_\tau \subseteq \Lambda^{\to}_\tau(\mathcal{F})$,
- If $m \in \Lambda^{\to}_{\sigma\to\tau}(\mathcal{F})$ and $n \in \Lambda^{\to}_\sigma(\mathcal{F})$, then $mn \in \Lambda^{\to}_\tau(\mathcal{F})$ (application),
- If $x \in \mathcal{V}_\sigma$ and $m \in \Lambda^{\to}_\tau(\mathcal{F})$, then $\lambda x.m \in \Lambda^{\to}_{\sigma\to\tau}(\mathcal{F})$ (abstraction).

The set of simply typed lambda terms over the signature $\mathcal{F}$ is $\bigcup_{\tau\in\mathbf{T}(\mathcal{B})} \Lambda^{\to}_\tau(\mathcal{F})$ and is denoted by $\Lambda^{\to}(\mathcal{F})$.

The set of free variables of a well-typed term t is defined as usual and denoted by $FV(t)$. In the sequel we often abbreviate $\mathbf{T}(\mathcal{B})$ and $\Lambda^{\to}(\mathcal{F})$ to $\mathbf{T}$ and $\Lambda^{\to}$. Terms without free variables are called closed. A variable x is called *bound* in a term t if it occurs in a subterm of t of the form $\lambda x.s$. Terms that only differ in the renaming of bound variables (known as α-conversion) are identified. This permits us to stick to the convention that variables never occur free and bound as well in any mathematical context. See [2, p. 26] for details about the *variable convention.*

To express the complexity of a term, the notion of *type level* is defined inductively on types:

Definition 4. The type level of a term in $\Lambda^{\to}_\sigma$ is $TL(\sigma)$, where

- $TL(b) = 0$, for $b \in \mathcal{B}$,
- $TL(\sigma \to \tau) = max(TL(\sigma) + 1, TL(\tau))$.

A *substitution* is a mapping from variables to terms of the same type. More precisely:

Definition 5. A substitution θ is a finite function $\{x_1 \mapsto t_1, \ldots, x_n \mapsto t_n\}$, with $t_i \in \Lambda^{\to}_\tau$ if $x_i \in \mathcal{V}_\tau$. The set $\{x_1, \ldots, x_n\}$ is called the domain of this substitution, denoted by $DOM(\theta)$.

Substitutions are extended to homomorphisms on terms in the standard way. Due to the variable convention, substitution cannot create new bindings. We denote $\theta(t)$ by t^θ or, in case $\theta = \{x \mapsto a\}$, by $t[x := a]$.

We recall the β- and η-reduction schemes for the lambda calculus, denoted by $\rightarrow_\beta$ and $\rightarrow_\eta$ respectively:

Definition 6. The relation $\rightarrow_\beta$ is the smallest compatible relation satisfying $(\lambda x.m)n \rightarrow_\beta m[x := n]$. The relation $\rightarrow_\eta$ is the smallest compatible relation satisfying $\lambda x.(fx) \rightarrow_\eta f$. The reflexive, symmetric and transitive closure of $\rightarrow_\eta$ is denoted by $=_\eta$.

Standard theory tells us that every $\Lambda^{\rightarrow}$-term has a unique *normal form* with respect to β-conversion. This β-normal form is denoted by $t\downarrow_\beta$. Normal forms are always of the form $\lambda x_1.\cdots\lambda x_n.(at_1\cdots t_m)$, where $a \in \mathcal{V} \cup \mathcal{C}$.

2.2 Higher-order Rewrite Systems

There are various definitions of higher-order rewrite mechanisms in recent literature [11, 4, 12, 13, 18]. The definition in this subsection is not meant to add a new formalism to the existing ones. Most conditions on the rules are dropped, because they are not necessary in the proof. The rewrite relation is as liberal as possible. Of course, the main result applies to formalisms admitting fewer rules and fewer rewrite steps. The chosen formalism is much like the formalism in [18, ch. 4.1], but we made some choices.

A *Higher-order Rewrite System* is given by a signature and a set of rules:

Definition 7. A *Higher-order Rewrite System* $\mathcal{R}$ is a tuple $(\mathcal{F}, R)$, where $\mathcal{F}$ is a signature and R is a set of rules in this signature. A *rule* is a pair (l, r), with $l, r \in \Lambda^{\rightarrow}_{\tau}(\mathcal{F})$, both closed β-normal forms.

Rules (l, r) are denoted by $l \rightarrow r$. In this rule, l is called the left hand side and r the right hand side.

The definition of a rule is the same as in [18, p. 46], except that we don't use η-long forms. Free variables are not admitted in a rule, for reasons explained below. In the examples, we mostly leave out the λ-binders in front of the left- and right hand side of the rules for shortness, thus introducing free variables par abuse. The definition of HRSs in [13, p. 308] is a special case of Definition 7, because it has some extra conditions on the occurrences of the free variables (which should be abstracted from in the present definition) to guarantee decidability of a rewrite step. This restriction is not needed in the termination result.

Definition 8. A *context* is a term $\lambda\Box.C$, where C is in β-normal form and $\Box \in FV(C)$.

Definition 9. Let $\mathcal{R} = (\mathcal{F}, R)$ be an HRS. The rewrite relation is defined by: $s \rightarrow_\mathcal{R} t$ if and only if there exist a context C and a rule $l \rightarrow r \in R$, such that $(Cl)\downarrow_\beta =_\eta s$ and $(Cr)\downarrow_\beta =_\eta t$. Only s and t in β-normal form will be considered.

Definition 9 differs in three aspects from the definition of the higher-order rewrite relation in [18, p. 46]. Firstly, η-long forms are not required, but furthermore the notion of positions is circumvented. In this way the variable convention can be upheld[1]. To be able to rewrite subterms with bound variables the rules are required to be closed.

The proposed definition of rewriting is inspired by the Leibniz-equality in higher-order logic: $l =_D r$ abbreviates $\forall P : D \to Prop.P(r) \to P(l)$. An application of this formula reduces any goal $P(l)$ to a goal $P(r)$. To have a proper notion of reduction, the condition $\Box \in FV(C)$ is added. To ensure that C really depends on $\Box$, the condition that C is in β-normal form is added. Without these conditions we would never have termination:

Example 1. Let $l \to r$ be a rule of HRS $\mathcal{R}$. If $\lambda\Box.a$ were a context, we would have the rewrite step $a = ((\lambda\Box.a)l)\downarrow_\beta \to_\mathcal{R} ((\lambda\Box.a)r)\downarrow_\beta = a$. The same example can be given if $\lambda\Box.(\lambda y.a)\Box$ were a context.

The last difference with [18] is that more occurrences of $\Box$ in a context are admitted. This is another advantage of seeing contexts as ordinary λ-terms and is close to the pragmatics of Leibniz equality. The transitive closure of the rewrite relation is not changed if we allow this form of parallelism, so confluence and termination are maintained.

Here is an example of rewriting:

Example 2. Let $+$ be a binary function symbol. If there is a rule $\lambda X.X + X \to \lambda X.X$, then we have the following rewrite steps:

Term		Reduct	Context used
$P + P$	$\to$	P	$\lambda\Box.\Box P$
$(P + P) + (P + P)$	$\to$	$(P + P)$	$\lambda\Box.\Box(P + P)$ or $\lambda\Box.(\Box P) + (\Box P)$
$(P + P) + (P + P)$	$\to$	P	$\lambda\Box.\Box(\Box P)$
$\lambda qy.(qy) + (qy)$	$\to$	$\lambda qy.(qy)$	$\lambda\Box qy.\Box(qy)$

3 The Model of Hereditarily Monotonic Functionals

We try to apply the general idea of the proof technique "termination by interpretation" for TRSs in [8, 19] to HRSs. The outline of this technique is as follows: The function symbols are interpreted by operations of the same arity in an algebra, equipped with a well-founded partial order. This interpretation is extended to the terms of the TRS in an algebraic way. The interpretation is chosen in such a way that for all rules, the left hand side is interpreted by a greater value than the right hand side. If such an interpretation can be found, the TRS is terminating. To prove the correctness of this technique, first we have to show that the ordering on terms is closed under substitution. The other step is to show that the ordering is closed under placing terms into a context. This can be proved

[1] In the notion of positions and occurrences, a replacement can introduce new bindings. Therefore contexts cannot be seen modulo α-conversion. See [12, p. 343]

using the fact that the function symbols are interpreted by strictly monotonic functions, thus preserving the ordering. Now we have for any substitution θ and context C that $C[l^\theta] > C[r^\theta]$. This is exactly the form of a rewrite step, thus showing that a rewrite step can be translated to a decrease in the well-founded ordering. In this way termination of rewriting is guaranteed.

In our definition of higher-order rewriting, we don't need closure under substitution, as the rules consist of closed terms. We only need closure under placing terms into a context; but this doesn't help us, because substitutions can easily be coded as special contexts. Matters are more complicated than in the first-order case. The notions of interpretation, strictness and ordering have to be extended to higher-order concepts. After these definitions the same idea can be used. It will turn out that we have to use two different orderings to show termination. The condition on the rules of an HRS is stated in terms of one ordering, so we try to show that for every rule $(l \to r)$, $l >_1 r$. Now we use a second ordering to show that for any context C, $Cl >_2 Cr$. This second ordering will be well-founded, thus proving termination of the HRS.

In this chapter we will define an interpretation for the terms into the hereditarily monotonic functions. A similar idea occurs already in [6] and [5]. One difference is that in this paper two β-equivalent terms have the same interpretation. This difference stems from the fact that we don't count β-steps. They are only used in the meta-language.

3.1 Interpretation of types by monotonic function spaces

Let us start with a type interpretation $\mathcal{I}$ of the set of base types $\mathcal{B}$:

Definition 10. A type interpretation $\mathcal{I}$ is a set of non-empty strict partial orders $\{(\mathcal{M}_B, >_B) | B \in \mathcal{B}\}$.

Starting from a base type interpretation $\mathcal{I}$, we define the interpretation of all types as sets of hereditarily monotonic functionals. A similar idea occurs in [6, p. 457], which shows that terms of the typed λ-I calculus denote strictly monotonic functions. The sets of ***hereditarily monotonic functions*** (monotonic functions for short) are denoted by $\mathcal{M}_\rho$ and depend on an ordering, denoted by $_{\text{mon}}\geq$. This partial order is inherited from the partial order on the interpretation of the base types in the way defined below. This ordering itself depends on the notion of monotonicity, so we give a definition by simultaneous induction:

Definition 11.

- For $\rho \in \mathcal{B}$:
  - $\mathcal{M}_\rho$ is as given by $\mathcal{I}$,
  - $a \;_{\text{mon}}\geq_\rho b$ iff $a, b \in \mathcal{M}_\rho$ and $a >_\rho b$ or $a = b$.
- For $\rho = \sigma \to \tau$:
  - $\mathcal{M}_\rho = \{f \in \mathcal{M}_\sigma \to \mathcal{M}_\tau \mid \text{for all } x, y \in \mathcal{M}_\sigma, \text{if } x \;_{\text{mon}}\geq_\sigma y \text{ then } f(x) \;_{\text{mon}}\geq_\tau f(y)\}$,
  - $f \;_{\text{mon}}\geq_\rho g \iff f, g \in \mathcal{M}_\rho$ and for all $x \in \mathcal{M}_\sigma$, $f(x) \;_{\text{mon}}\geq_\tau g(x)$.

Instead of $a \in \mathcal{M}_\rho$ for some ρ we say: a is *monotonic*. Furthermore, a set $\mathcal{M}_\rho$ is called a *domain*. Now we can define the following strict partial order:

Definition 12. The relation ${}_{\text{mon}}{>}_\rho$ on domains $\mathcal{M}_\rho$ with $\rho \in \mathbf{T}$ is defined with induction on ρ:

- If $\rho \in \mathcal{B}$ and $a, b \in \mathcal{M}_\rho$ then $a \,{}_{\text{mon}}{>}_\rho\, b$ if and only if $a >_\rho b$.
- If $\rho = \sigma \to \tau$ and $f, g \in \mathcal{M}_\rho$ then $f \,{}_{\text{mon}}{>}_\rho\, g$ if and only if for all $x \in \mathcal{M}_\sigma$, $f(x) \,{}_{\text{mon}}{>}_\tau\, g(x)$.

The type subscripts in ${}_{\text{mon}}{>}_\rho$ and ${}_{\text{mon}}{\geq}_\rho$ are omitted when this is not confusing. Note that $f \,{}_{\text{mon}}{>}\, g$ means that f is pointwise greater than g; only monotonic points are in the domain of f and g. The reader is warned not to confuse $f \,{}_{\text{mon}}{\geq}\, g$ with $f \,{}_{\text{mon}}{>}\, g \vee f = g$. We have the following fact about ${}_{\text{mon}}{>}$ and ${}_{\text{mon}}{\geq}$ only in one direction:

Lemma 13. *For all $\rho \in \mathbf{T}$ and $f, g \in \mathcal{M}_\rho$, if $f \,{}_{\text{mon}}{>}\, g$ or $f = g$, then $f \,{}_{\text{mon}}{\geq}\, g$.*

Proof. (Induction on the type of f and g) Let $f \,{}_{\text{mon}}{>}\, g$, or $f = g$. If $\rho \in \mathcal{B}$ it is trivial. If $\rho = \sigma \to \tau$, take an $x \in \mathcal{M}_\sigma$. Then $f(x) \,{}_{\text{mon}}{>}\, g(x)$ (in case $f \,{}_{\text{mon}}{>}\, g$) or $f(x) = g(x)$. But then, by induction hypothesis, $f(x) \,{}_{\text{mon}}{\geq}\, g(x)$. Now by Definition 11, $f \,{}_{\text{mon}}{\geq}\, g$. □

Lemma 14. *Transitivity between ${}_{\text{mon}}{>}$ and ${}_{\text{mon}}{\geq}$:*

- *If $f \,{}_{\text{mon}}{\geq}\, g$ and $g \,{}_{\text{mon}}{\geq}\, h$ then $f \,{}_{\text{mon}}{\geq}\, h$.*
- *If $f \,{}_{\text{mon}}{\geq}\, g$ and $g \,{}_{\text{mon}}{>}\, h$ then $f \,{}_{\text{mon}}{>}\, h$.*
- *If $f \,{}_{\text{mon}}{>}\, g$ and $g \,{}_{\text{mon}}{\geq}\, h$ then $f \,{}_{\text{mon}}{>}\, h$.*
- *If $f \,{}_{\text{mon}}{>}\, g$ and $g \,{}_{\text{mon}}{>}\, h$ then $f \,{}_{\text{mon}}{>}\, h$.*

Proof. Simple induction on the type of f. □

3.2 Interpretation of terms in domains

Now we can take our next step: A term in $\Lambda_\tau^{\to}$ has to be interpreted as a value in the domain $\mathcal{M}_\tau$. To interpret the terms we have to specify what the interpretation of the free variables and the constants will be. The free variables are dealt with by ***valuations***: mappings from variables to values. The interpretation of constants is given by a ***constant interpretation***.

Definition 15. A valuation α is a family of mappings α_ρ, with $\alpha_\rho : \mathcal{V}_\rho \to \mathcal{M}_\rho$ for $\rho \in \mathbf{T}$.

With $\alpha[x := a]$ we denote the valuation that behaves like α on all variables except x, where it returns a. To compare two valuations we define:

Definition 16. $\alpha_1 \geq \alpha_2$ if and only if for each $x \in \mathcal{V}$, $\alpha_1(x) \,{}_{\text{mon}}{\geq}\, \alpha_2(x)$.

The constants have to be interpreted by functionals of the right domain. Therefore the following notion is introduced:

Definition 17. A constant interpretation $\mathcal{J}$ is a family of functions $\mathcal{J}_\sigma : \mathcal{C}_\sigma \to \mathcal{M}_\sigma$ for $\sigma \in \mathbf{T}$.

Now we are ready to define the interpretation of terms, depending on a particular choice for the constant interpretation $\mathcal{J}$.

Definition 18. The interpretation of a term under the valuation α is defined inductively on the structure of the term:

- $[\![x]\!]_\alpha = \alpha(x)$ for $x \in \mathcal{V}$.
- $[\![c]\!]_\alpha = \mathcal{J}(c)$ for $c \in \mathcal{C}$.
- $[\![mn]\!]_\alpha = [\![m]\!]_\alpha [\![n]\!]_\alpha$.
- $[\![\lambda x.m]\!]_\alpha = \boldsymbol{\lambda} a \in \mathcal{M}_\sigma . [\![m]\!]_{\alpha[x:=a]}$ for $x \in \mathcal{V}_\sigma$.

If s is a closed term, then $[\![s]\!]_\alpha$ does not depend on α, so we may suppress this subscript.[2] The following theorem states that the interpretation of each term is monotonic. The proof is by simultaneous induction of the following conjunct:

Proposition 19. *For each $s \in \Lambda^{\to}$ and valuations α, β:*

1. *$[\![s]\!]_\alpha$ is monotonic.*
2. *If $\alpha \geq \beta$ then $[\![s]\!]_\alpha \;{}_{\text{mon}}\!\geq [\![s]\!]_\beta$.*

Proof. (induction on the structure of s)
If $s = x \in \mathcal{V}$:

1. $[\![s]\!]_\alpha = \alpha(x)$ is monotonic by Definition 15.
2. $[\![s]\!]_\alpha = \alpha(x) \;{}_{\text{mon}}\!\geq \beta(x) = [\![s]\!]_\beta$. (From Definition 16).

If $s = c \in \mathcal{C}$:

1. $[\![s]\!]_\alpha = \mathcal{J}(c)$ is monotonic by Definition 17.
2. $[\![s]\!]_\alpha = \mathcal{J}(c) = [\![s]\!]_\beta$, so they are ${}_{\text{mon}}\!\geq$-related (Lemma 13).

If $s = mn$:

1. By induction hypothesis (1) both $[\![m]\!]_\alpha$ and $[\![n]\!]_\alpha$ are monotonic. Then by Definition 11, also $[\![s]\!]_\alpha = [\![m]\!]_\alpha [\![n]\!]_\alpha$ is monotonic.
2. By induction hypothesis (2) $[\![m]\!]_\alpha \;{}_{\text{mon}}\!\geq [\![m]\!]_\beta$. By induction hypothesis (1) $[\![n]\!]_\alpha$ is monotonic, so we have, with Definition 11, that $[\![m]\!]_\alpha [\![n]\!]_\alpha \;{}_{\text{mon}}\!\geq [\![m]\!]_\beta [\![n]\!]_\alpha$. We also get from the induction hypotheses (1,2) that $[\![n]\!]_\alpha \;{}_{\text{mon}}\!\geq [\![n]\!]_\beta$ and $[\![m]\!]_\beta$ is monotonic. Therefore, with Definition 11, $[\![m]\!]_\beta [\![n]\!]_\alpha \;{}_{\text{mon}}\!\geq [\![m]\!]_\beta [\![n]\!]_\beta$. Now, using transitivity of ${}_{\text{mon}}\!\geq$ we have $[\![s]\!]_\alpha \;{}_{\text{mon}}\!\geq [\![s]\!]_\beta$.

If $s = \lambda x.t$: (Say $x \in \mathcal{V}_\sigma$)

1. Firstly, choose $a \in \mathcal{M}_\sigma$, then $[\![\lambda x.t]\!]_\alpha(a) = [\![t]\!]_{\alpha[x:=a]}$. This is in $\mathcal{M}_\tau$ by induction hypothesis (1), so $[\![s]\!]_\alpha \in \mathcal{M}_\sigma \to \mathcal{M}_\tau$. Furthermore, if $a \;{}_{\text{mon}}\!\geq_\sigma b$, then $\alpha[x := a] \geq \alpha[x := b]$, so by induction hypothesis (2) $[\![t]\!]_{\alpha[x:=a]} \;{}_{\text{mon}}\!\geq [\![t]\!]_{\alpha[x:=b]}$. This is equivalent to $[\![s]\!]_\alpha(a) \;{}_{\text{mon}}\!\geq [\![s]\!]_\alpha(b)$, so $[\![s]\!]_\alpha$ is monotonic.

[2] The existence of at least one valuation is guaranteed because the base domains are non-empty

2. Let $a \in \mathcal{M}_\sigma$. We have $\alpha[x := a] \;_{\mathrm{mon}}\!\geq \beta[x := a]$. So using induction hypothesis (2) we can compute: $[\![\lambda x.t]\!]_\alpha(a) = [\![t]\!]_{\alpha[x:=a]} \;_{\mathrm{mon}}\!\geq [\![t]\!]_{\beta[x:=a]} = [\![\lambda x.t]\!]_\beta(a)$. So indeed, $[\![s]\!]_\alpha \;_{\mathrm{mon}}\!\geq [\![s]\!]_\beta$. □

Corollary 20. *Given a valuation α, $[\![\cdot]\!]_\alpha$ is a family of functions from $\Lambda^{\rightarrow}_\sigma$ to $\mathcal{M}_\sigma$ for each $\sigma \in \mathbf{T}$.*

We have an ordering on the domains and an interpretation mapping terms into domains. Using this interpretation we can define an ordering on terms. The free variables are dealt with by valuations:

Definition 21. For terms s, t in $\Lambda^{\rightarrow}$, $s \;_{\mathrm{mon}}\!> t$ if and only if for each valuation α, $[\![s]\!]_\alpha \;_{\mathrm{mon}}\!> [\![t]\!]_\alpha$.

The next lemma shows that free and bound variables are treated similarly with respect to the ordering $_{\mathrm{mon}}\!>$:

Lemma 22. *If for two $\Lambda^{\rightarrow}_\tau$-terms $s \;_{\mathrm{mon}}\!> t$ then $\lambda x.s \;_{\mathrm{mon}}\!> \lambda x.t$.*

Proof. Let $s \;_{\mathrm{mon}}\!> t$, say $x \in \mathcal{V}_\sigma$. Choose $k \in \mathcal{M}_\sigma$ arbitrarily. Let α be any valuation. Then

$$[\![\lambda x.s]\!]_\alpha(k) = [\![s]\!]_{\alpha[x:=k]} \;_{\mathrm{mon}}\!>_\tau [\![t]\!]_{\alpha[x:=k]} = [\![\lambda x.t]\!]_\alpha(k) \; .$$

The equalities hold by Definition 18, the inequality by Definition 21. The calculation holds for any k, so $[\![\lambda x.s]\!]_\alpha \;_{\mathrm{mon}}\!>_{\sigma\rightarrow\tau} [\![\lambda x.t]\!]_\alpha$. This holds for any α, so $\lambda x.s \;_{\mathrm{mon}}\!> \lambda x.t$ □

3.3 Hereditarily monotonic functionals serve as a model of $\lambda^{\rightarrow}_{\beta,\eta}$

We show that the monotonic functionals are a model of $\lambda^{\rightarrow}_{\beta,\eta}$. Lemma 24 establishes a link between substitutions and valuations. As a corollary we get that the $_{\mathrm{mon}}\!>$-ordering is closed under substitution. Furthermore, the interpretation in the model of two β,η-convertible terms is the same. Together with Lemma 19 this proves that $\mathcal{M}$ forms a $\lambda^{\rightarrow}$-model. No new technique is used here, the proofs resemble those in [6, 5, 19].

Definition 23. Let θ be a substitution and α a valuation. Then the composition $\alpha \circ \theta$ is a new valuation, mapping $x \mapsto [\![x^\theta]\!]_\alpha$.

The mapping $\alpha \circ \theta$ is indeed a valuation, because $[\![x^\theta]\!]_\alpha$ is monotonic by Proposition 19. We have the following connection between valuations and substitutions:

Lemma 24 (Substitution Lemma). *Let θ be a substitution and α a valuation. Then for each term $s \in \Lambda^{\rightarrow}$ we have $[\![s^\theta]\!]_\alpha = [\![s]\!]_{\alpha\circ\theta}$.*

Proof. (induction on the structure of s)

- $s = x \in \mathcal{V}$: by definition of $\alpha \circ \theta$.

– $s = c \in \mathcal{C}$: $[\![c^\theta]\!]_\alpha = [\![c]\!]_\alpha = \mathcal{J}(c) = [\![c]\!]_{\alpha\circ\theta}$.
– $s = mn$: Using the induction hypothesis for m and n, we have

$$[\![s^\theta]\!]_\alpha = [\![m^\theta]\!]_\alpha [\![n^\theta]\!]_\alpha = [\![m]\!]_{\alpha\circ\theta}[\![n]\!]_{\alpha\circ\theta} = [\![s]\!]_{\alpha\circ\theta} \ .$$

– $s = \lambda x.m$: Say $x \in \mathcal{V}_\sigma$. Let $a \in \mathcal{M}_\sigma$.

Claim. *In this situation we have* $(\alpha[x := a]) \circ \theta = (\alpha \circ \theta)[x := a]$.

Proof. By the variable convention, $x \notin DOM(\theta)$, so $x^\theta = x$, and for all $y \in DOM(\theta)$, $x \notin FV(y^\theta)$. Under this convention, $((\alpha[x := a]) \circ \theta)(x) = [\![x^\theta]\!]_{\alpha[x:=a]} = a = ((\alpha \circ \theta)[x := a])(x)$. And for $y \neq x$: $((\alpha[x := a]) \circ \theta)(y) = [\![y^\theta]\!]_{\alpha[x:=a]} = [\![y^\theta]\!]_\alpha = (\alpha \circ \theta)(y) = ((\alpha \circ \theta)[x := a])(y)$. □

Now this case can be proved by the following calculation:

$$\begin{aligned}[\![s^\theta]\!]_\alpha &= \boldsymbol{\lambda} a \in \mathcal{M}_\sigma.([\![m^\theta]\!]_{\alpha[x:=a]})\\ &= \boldsymbol{\lambda} a \in \mathcal{M}_\sigma.([\![m]\!]_{(\alpha[x:=a])\circ\theta}) \text{ (by i.h.)}\\ &= \boldsymbol{\lambda} a \in \mathcal{M}_\sigma.([\![m]\!]_{(\alpha\circ\theta)[x:=a]}) \text{ (by claim)}\\ &= [\![s]\!]_{\alpha\circ\theta} \ .\end{aligned}$$

□

From the Substitution Lemma we have two corollaries, one with respect to the ordering $_{\text{mon}}>$ and one (Proposition 26) with respect to β-equality.

Corollary 25. *If for terms s, t in $\Lambda^\to$, $s \ _{\text{mon}}> t$ then for each substitution θ, $s^\theta \ _{\text{mon}}> t^\theta$.*

Proof. Let $s \ _{\text{mon}}> t$. Then by Lemma 24, $[\![s^\theta]\!]_\alpha = [\![s]\!]_{\alpha\circ\theta} \ _{\text{mon}}> [\![t]\!]_{\alpha\circ\theta} = [\![t^\theta]\!]_\alpha$ for any valuation α and substitution θ, so indeed $s^\theta \ _{\text{mon}}> t^\theta$. □

Proposition 26. *If $s \to_\beta t$ for two terms in $\Lambda^\to$, then $[\![s]\!]_\alpha = [\![t]\!]_\alpha$ for each valuation α.*

Proof. (induction to the depth of the β-redex in s.)

Let $s \to_\beta t$. The only interesting case is that the β-step takes place on top-level: Let $s = (\lambda x.s_1)s_2$, then $t = s_1[x := s_2]$. We have the following calculation:

$$[\![(\lambda x.s_1)s_2]\!]_\alpha = [\![s_1]\!]_{\alpha[x:=[\![s_2]\!]_\alpha]} = [\![s_1[x := s_2]]\!]_\alpha \ .$$

The first equality holds by Definition 18, the second because of Lemma 24, applied on α and the substitution $\{x := s_2\}$. □

A similar result holds for η-reduction:

Proposition 27. *If $s \to_\eta t$ for two terms in $\Lambda^\to$, then $[\![s]\!]_\alpha = [\![t]\!]_\alpha$ for each valuation α.*

Proof. This is obvious, because the domains have extensional equality. □

4 An ordering which is closed under context

4.1 On strictness

In section 3 we saw that the ordering $_{\text{mon}}>$ on terms is closed under substitution. We would like that this ordering is also closed under placing a term into a context. The first objection to this is the interpretation of the constants. We have to ensure that this interpretation is order preserving. The proof in [19] also uses the condition that the constants have to be interpreted by *strictly monotonic* operations. Therefore we define the following notion:

Definition 28. The predicate "f is strictly $_{\text{mon}}>$-monotonic" with $f \in \mathcal{M}_\rho$ is defined with induction on ρ:

- For $\rho \in \mathcal{B}$: $f \in \mathcal{M}_\rho$ is always strictly $_{\text{mon}}>$-monotonic.
- For $\rho = \sigma \to \tau$: $f \in \mathcal{M}_\rho$ is strictly $_{\text{mon}}>$-monotonic if and only if for all $x \in \mathcal{M}_\sigma$, $f(x)$ is strictly $_{\text{mon}}>$-monotonic, and for all $x, y \in \mathcal{M}_\sigma$, if $x \;_{\text{mon}}>_\sigma y$ then $f(x) \;_{\text{mon}}>_\tau f(y)$.

Unfortunately this notion of strict monotonicity is not strong enough. This is shown in the following example:

Example 3. In this example there is one base type, o, which is interpreted by the natural numbers. Type $o \to o$ is written as 1, type $1 \to o$ is written as 2. Consider the following functional: $S := \lambda G \in \mathcal{M}_2.\lambda g \in \mathcal{M}_1.G(g) + g(0)$. This is strictly $_{\text{mon}}>$-monotonic: Let $G \in \mathcal{M}_2$ and take $f \;_{\text{mon}}>_1 g$, Then $S(G)(f) = G(f) + f(0) \;_{\text{mon}}> G(f) + g(0) \;_{\text{mon}}\geq G(g) + g(0) = S(G)(g)$. The last inequality holds, because G is monotonic. This shows that $S(G)$ is strictly $_{\text{mon}}>$-monotonic. Now, take $G \;_{\text{mon}}>_2 F$ and $g \in \mathcal{M}_1$, then $S(G)(g) = G(g) + g(0) \;_{\text{mon}}> F(g) + g(0) = S(F)(g)$. This shows that $S(G) \;_{\text{mon}}> S(F)$. So S is strictly $_{\text{mon}}>$-monotonic.

But in fact, this is undesired. Consider $G := \lambda f.fA$ and $g := \lambda x.5$. Then $S(G)(g) = 10$, so the A leaves no traces.

This example can be used to "prove" that the following non-terminating HRS is terminating:

Example 4. The definitions of Example 3 are used. Take the signature with one base type o and constants $\mathcal{C} = \{c, d : o; @ : 2 \to 1 \to o\}$. The set of variables $\mathcal{V}$ includes $\{x, y : o; f, g : 1\}$. In this signature we define the HRS with one rule:

$$@(\lambda f.fx)(\lambda y.c) \to @(\lambda g.g(@(\lambda f.fx)(\lambda y.c)))(\lambda y.d).$$

First of all, this HRS is not terminating, because the left hand side is a subterm of the right hand side. Now we choose as interpretation:

$$\mathcal{I}(o) = \mathbb{N} \qquad \mathcal{J}(c) = 2$$
$$\mathcal{J}(d) = 1 \qquad \mathcal{J}(@) = S\ .$$

Under this interpretation we can compute the left hand side and the right hand side. It will turn out that the left hand side equals 4 and the right hand side equals 2. This clearly shows that S cannot be called "strict".

Another objection against strict ${}_{\text{mon}}{>}$-monotonicity of contexts, is the interpretation of the variables:

Example 5. Let $l\ {}_{\text{mon}}{>}\ r$ for some terms l and r. Then it is not the case that $xl\ {}_{\text{mon}}{>}\ xr$: Take a valuation α with $\alpha(x) = \boldsymbol{\lambda} a.c$, for some constant c, then $[\![xl]\!]_\alpha = [\![xr]\!]_\alpha$.

This example shows that although $l\ {}_{\text{mon}}{>}\ r$, we cannot expect that $xl\ {}_{\text{mon}}{>}\ xr$ for any value for x. So this order is not closed under placing terms into a context. Fortunately, we can weaken our desires. We don't need this order for all values for x, because the free variables of the context (which is in β-normal form) cannot be instantiated during a rewrite step. Furthermore, if we have the rewrite step $xl \rightarrow_{\mathcal{R}} xr$, then it is not the case, that we have the same step for all substitutions for x: Some substitutions lead to a non legal context. So we have the freedom to restrict the condition and to look at some particular value of x. The idea, due to Jan Friso Groote, is to look at precisely those x that preserve the order, that is for the *strictly monotonic* x. This leads to a new ordering, ${}_{\text{str}}{>}$, which (intuitively) runs: $f\ {}_{\text{str}}{>}\ g$ if and only if for all strictly monotonic x, $f(x)\ {}_{\text{str}}{>}\ g(x)$. This new ordering is used to compare the terms of the rewrite sequence.

Below a simultaneous definition of a new notion of *strictness* and a new ordering ${}_{\text{str}}{>}$ is given, in such a way that strictness is stronger than strictly ${}_{\text{mon}}{>}$-monotonic and ${}_{\text{str}}{>}$ is weaker than ${}_{\text{mon}}{>}$. Here we diverge from [6, 5].

Definition 29. The relation ${}_{\text{str}}{>}_\rho$ and the set $\mathcal{S}_\rho$ of hereditarily strict functionals are defined with induction on $\rho \in \mathbf{T}$:

- For $\rho \in \mathcal{B}$:
  - $a\ {}_{\text{str}}{>}_\rho\ b$ if and only if $a, b \in \mathcal{M}_\rho$ and $a >_\rho b$ (according to a fixed type interpretation $\mathcal{I}$).
  - $\mathcal{S}_\rho = \mathcal{M}_\rho$.
- For $\rho = \sigma \rightarrow \tau$: Let $f, g \in \mathcal{M}_\rho$, then
  - $f\ {}_{\text{str}}{>}_\rho\ g$ if and only if $f\ {}_{\text{mon}}{\geq}_\rho\ g$[3] and for all $x \in \mathcal{S}_\sigma$, $f(x)\ {}_{\text{str}}{>}_\tau\ g(x)$.
  - $f \in \mathcal{S}_\rho$ if and only if
    - $f \in \mathcal{M}_\rho$ and
    - for all $x \in \mathcal{M}_\sigma$, $f(x) \in \mathcal{S}_\tau$ and
    - for all $x, y \in \mathcal{M}_\sigma$, if $x\ {}_{\text{str}}{>}_\sigma\ y$ then $f(x)\ {}_{\text{mon}}{>}_\tau\ f(y)$.

If $f \in \mathcal{S}_\rho$ for some $\rho \in \mathbf{T}$, we say that f is *strict.* We will omit the type subscripts for ${}_{\text{str}}{>}_\rho$. Now we can prove the following relation between the different partial orders:

Lemma 30. *Let $\rho \in \mathbf{T}$ and $f, g \in \mathcal{M}_\rho$. If $f\ {}_{\text{mon}}{>}_\rho\ g$ then $f\ {}_{\text{str}}{>}_\rho\ g$.*

Proof. (induction on ρ):

If $\rho \in \mathcal{B}$ the two orderings coincide. Let $f\ {}_{\text{mon}}{>}_\rho\ g$ for $\rho = \sigma \rightarrow \tau$. By Lemma 13, $f\ {}_{\text{mon}}{\geq}\ g$. Take an arbitrary strict $x \in \mathcal{S}_\sigma$. Then also $x \in \mathcal{M}_\sigma$,

---

[3] This addition is due to H. Schwichtenberg, and appears to be necessary in applications with arbitrary high order.

so $f(x)\ {}_{\text{mon}}\!>_\tau g(x)$. By induction hypothesis, $f(x)\ {}_{\text{str}}\!>_\tau g(x)$, giving $f\ {}_{\text{str}}\!> g$ (Definition 29). □

Corollary 31. *Let f be strict, then both f is strictly ${}_{\text{mon}}\!>$-monotonic and f is strictly ${}_{\text{str}}\!>$-monotonic.*

Proof. If $x\ {}_{\text{mon}}\!> y$ then $x\ {}_{\text{str}}\!> y$ (Lemma 30) and $f(x)\ {}_{\text{mon}}\!> f(y)$, by strictness of f. If $x\ {}_{\text{str}}\!> y$ then by strictness of f: $f(x)\ {}_{\text{mon}}\!> f(y)$ and by Lemma 30 $f(x)\ {}_{\text{str}}\!> f(y)$. □

Thus strict is indeed stronger than strictly ${}_{\text{mon}}\!>$-monotonic, and ${}_{\text{str}}\!>$ is indeed weaker than ${}_{\text{mon}}\!>$. The following example shows that the problem from Example 3 is solved:

Example 6. The functional $S = \boldsymbol{\lambda} G \in \mathcal{M}_2.\boldsymbol{\lambda} g \in \mathcal{M}_1.G(g)+g(0)$ from Example 3 is not strict: Take

$$G = \boldsymbol{\lambda} g.\begin{cases} 5 & \text{if } g(0) = g(1) = 0 \\ 10 & \text{else,} \end{cases}$$

and $F = \boldsymbol{\lambda} g.5$. Then F and G are both monotonic, clearly $G\ {}_{\text{str}}\!> F$, but the relation $S(G)\ {}_{\text{mon}}\!> S(F)$ doesn't hold, because $S(G)(\boldsymbol{\lambda} x.0) = S(F)(\boldsymbol{\lambda} x.0)$.

The type level of functional S is 3. In the next two lemmas this complexity is shown to be essential.

Lemma 32. *Let $f, g \in \mathcal{M}_\rho$. If $TL(\rho) \leq 1$ then $f\ {}_{\text{str}}\!>_\rho g$ implies $f\ {}_{\text{mon}}\!>_\rho g$.*

Proof. (induction on ρ).

- $\rho \in \mathcal{B}$: By Definition 29.
- $\rho = \sigma \to \tau$: $TL(\rho)$ must be 1, so $\sigma \in \mathcal{B}$ and $TL(\tau) \leq 1$. Let $f\ {}_{\text{str}}\!> g$ and $x \in \mathcal{M}_\sigma$, then x is strict (Definition 29 for base type), so $f(x)\ {}_{\text{str}}\!>_\tau g(x)$. By induction hypothesis $f(x)\ {}_{\text{mon}}\!> g(x)$, so indeed $f\ {}_{\text{mon}}\!> g$. □

Lemma 33. *Let $f \in \mathcal{M}_\rho$. If $TL(\rho) \leq 2$ then f strictly ${}_{\text{mon}}\!>$-monotonic implies f is strict.*

Proof. (induction on ρ).

- If $\rho \in \mathcal{B}$ then f is strict by Definition 29.
- If $\rho = \sigma \to \tau$ then $TL(\sigma) \leq 1$ and $TL(\tau) \leq 2$. Let f be strictly ${}_{\text{mon}}\!>$-monotonic, then for $x, y \in \mathcal{M}_\sigma$ we have $f(x)$ is strictly ${}_{\text{mon}}\!>$-monotonic and by induction hypothesis $f(x)$ is strict. Furthermore, if $x\ {}_{\text{str}}\!>_\sigma y$ then $x\ {}_{\text{mon}}\!>_\sigma y$ (Lemma 32), so $f(x)\ {}_{\text{mon}}\!> f(y)$. These two facts yield that f is strict (Definition 29). □

The notion of strictness is extended to the interpretation of constants and variables:

Definition 34. A constant interpretation $\mathcal{J}$ is strict, if $\mathcal{J}(c)$ is strict for all $c \in \mathcal{C}$.

Definition 35. A valuation α is *strict* if for each $x \in \mathcal{V}$, $\alpha(x)$ is strict.

We can lift ${}_{\text{str}}\!>$ from the domain level into the term level in the same way as we lifted ${}_{\text{mon}}\!>$. To treat bound and free variables similarly, we now use strict valuations only:

Definition 36. For terms s, t of $\Lambda^{\rightarrow}$, $s \;{}_{\text{str}}\!> t$ if for each *strict* valuation α, $[\![s]\!]_{\alpha} \;{}_{\text{str}}\!> [\![t]\!]_{\alpha}$.

Lemma 37. *For all $s, t \in \Lambda^{\rightarrow}$, if $s \;{}_{\text{mon}}\!> t$ then $s \;{}_{\text{str}}\!> t$.*

Proof. If $s \;{}_{\text{mon}}\!> t$ for $s, t \in \Lambda^{\rightarrow}_{\tau}$ then for all valuations α, $[\![s]\!]_{\alpha} \;{}_{\text{mon}}\!>_{\tau} [\![t]\!]_{\alpha}$. This implies (using Lemma 30) that $[\![s]\!]_{\alpha} \;{}_{\text{str}}\!>_{\tau} [\![t]\!]_{\alpha}$. This holds for all α, so certainly for strict α this relation holds. Therefore $s \;{}_{\text{str}}\!> t$. □

We end this section with an example that the ordering ${}_{\text{str}}\!>$ cannot be used for the rules, so the ordering ${}_{\text{mon}}\!>$ is still necessary.

Example 7. Let for constants l and r of base type, the only rule of an HRS be $\lambda x.xl \rightarrow \lambda x.xr$. An interpretation can be chosen, such that $l \;{}_{\text{mon}}\!> r$. Then $\lambda x.xl \;{}_{\text{str}}\!> \lambda x.xr$, but yet the HRS is non-terminating: Choose as context $C = \lambda\Box.\Box(\lambda y.c)$. Then $(C(xl))\!\downarrow_{\beta} \rightarrow (C(xr))\!\downarrow_{\beta}$, which is equivalent to $c \rightarrow c$. So the ordering ${}_{\text{str}}\!>$ is too weak for the rules.

4.2 The ordering ${}_{\text{str}}\!>$ is well-founded

The definition of strictness (Definition 29) is rather complex and it is not a priori clear that strict functionals exist. Some extra conditions are needed to guarantee that strict functionals exist in every domain. These conditions are collected in the following notion:

Definition 38. A type interpretation $\mathcal{I} = \{(\mathcal{M}_B, >_B) | B \in \mathcal{B}\}$ is called a *well-founded type interpretation* if the following conditions on the interpretation of base types are satisfied:

- For each $b \in \mathcal{B}$, $\mathcal{M}_b$ is non-empty.
- For each $b \in \mathcal{B}$, $>_b$ is well-founded.
- For each $b, c \in \mathcal{B}$, there exists a strict function in $\mathcal{M}_{b \rightarrow c \rightarrow c}$.

We suppose that for each combination (b, c) of base types one strict function is chosen and called $+_{b,c}$ (written infix when possible). This suggestive definition is justified by Proposition 40 which states that under a well-founded type interpretation, the domains are well-founded. But first, we need a lemma, saying that strict inhabitants exist for all domains, if we have a well-founded type interpretation:

Lemma 39. *If the underlying type interpretation is well-founded, then for every $\rho \in \mathbf{T}$ there is at least one inhabitant of $\mathcal{S}_{\rho}$, which we will call S_{ρ}.*

Proof. (with induction on the structure of ρ)

If ρ is of base type, we can choose any $S_\rho \in \mathcal{M}_\rho$, because the domains $\mathcal{M}_b$ are non-empty for base types b, and elements of $\mathcal{M}_b$ are always strict.

Now let $\rho = \sigma \to \tau$. Then σ and τ can be decomposed in $\sigma_1 \to \cdots \sigma_m \to c$ and $\tau_1 \to \cdots \tau_n \to d$, with $c, d \in \mathcal{B}$. By induction hypothesis there exist strict inhabitants S_{σ_j} and S_τ. Because the type interpretation is well-founded, we have the strict element $+_{c,d}$. Using this material we can define[4]:

$$S_\rho = \lambda y \in \mathcal{M}_\sigma . \lambda x_1 \in \mathcal{M}_{\tau_1} \cdots \lambda x_n \in \mathcal{M}_{\tau_n} . (y S_{\sigma_1} \cdots S_{\sigma_m}) +_{c,d} (S_\tau x_1 \cdots x_n) \ .$$

Clearly, this functional is in the appropriate domain and it is also strict in y and $x_1, \ldots, x_n$:

$$\begin{aligned} & y_1 \ {}_{\text{str}}{>}_\sigma \ y_2 \\ \Rightarrow\ & y_1 S_{\sigma_1} \cdots S_{\sigma_m} \ {}_{\text{str}}{>}_c \ y_2 S_{\sigma_1} \cdots S_{\sigma_m} \\ \Rightarrow\ & +_{c,d}(y_1 S_{\sigma_1} \cdots S_{\sigma_m}) \ {}_{\text{mon}}{>}_{d \to d} \ +_{c,d}(y_2 S_{\sigma_1} \cdots S_{\sigma_m}) \\ \Rightarrow\ & S_\rho y_1 \ {}_{\text{mon}}{>}_\tau \ S_\rho y_2 \ . \end{aligned}$$

A similar reasoning applies when some x_i is varied: Let $x_1, \ldots, x_n, x' \in \mathcal{M}$. Let $x_i \ {}_{\text{str}}{>} \ x'$. Because $S_\tau x_1 \cdots x_{i-1}$ is strict, $S_\tau x_1 \cdots x_n \ {}_{\text{mon}}{>} \ S_\tau x_1 \cdots x' \cdots x_n$, so for monotonic y, $S_\rho y x_1 \cdots x_i \ {}_{\text{mon}}{>} \ S_\rho y x_1 \cdots x_{i-1} x'$. □

Proposition 40. *If the underlying type interpretation is well-founded, then for every $\rho \in \mathbb{T}$, the partial order $(\mathcal{M}_\rho, {}_{\text{str}}{>}_\rho)$ is well-founded.*

Proof. (induction on the structure of ρ)

For $\rho \in \mathcal{B}$ well-foundedness is given by the well-foundedness of $>_\rho$. Let $\rho = \sigma \to \tau$. Assume that there exist $\{f_i | i \in \mathbb{N}\}$ with $f_i \ {}_{\text{str}}{>}_{\sigma \to \tau} \ f_{i+1}$. But now, using $S_\sigma \in \mathcal{S}_\sigma$ we can construct the descending chain $\{f_i(S_\sigma) | i \in \mathbb{N}\}$, with $f_i(S_\sigma) \ {}_{\text{str}}{>}_\tau \ f_{i+1}(S_\sigma)$, which contradicts the induction hypothesis. □

Proposition 41. *The relation ${}_{\text{str}}{>}$ is well-founded on terms of $\Lambda^\to$.*

Proof. Let $\{s_i | i \in \mathbb{N}\}$ of some type τ be given such that $s_i \ {}_{\text{str}}{>} \ s_{i+1}$. By Definition 36 this means that for every strict valuation α, $[\![s_i]\!]_\alpha \ {}_{\text{str}}{>}_\tau \ [\![s_{i+1}]\!]_\alpha$. Let α be the valuation $\{x \mapsto S_\sigma | x \in \mathcal{V}_\sigma, \sigma \in \mathbb{T}\}$ (see Lemma 39), then α is such a strict valuation, giving rise to a descending ${}_{\text{str}}{>}$ chain in $\mathcal{M}_\tau$, which contradicts Proposition 40. □

4.3 Contexts preserve some order

There is no difference between bound and free variables with respect to the ordering ${}_{\text{str}}{>}$. For ${}_{\text{mon}}{>}$ this is stated in Lemma 22.

Lemma 42. *If for two $\Lambda^\to$-terms $s \ {}_{\text{str}}{>} \ t$ then $\lambda x.s \ {}_{\text{str}}{>} \ \lambda x.t$.*

[4] See [6, p. 461] and [5, p. 83] for comparable functionals

Proof. Analogous to the proof of Lemma 22. Instead of arbitrary k and α, *strict* k and α have to be chosen. □

Now we are ready to prove that contexts preserve some order:

Proposition 43. *If $s \;_{\text{mon}}\!> t$ and $\mathcal{J}$ is a strict constant interpretation, then for all contexts C, $Cs \;_{\text{str}}\!> Ct$.*

Proof. Let $s \;_{\text{mon}}\!> t$ and let C be a context, then C is of the form $\lambda\Box.C'$, with $\Box \in FV(C')$. Because C' is in β-normal form, it is of the form $\lambda x_1 \cdots \lambda x_n.as_1 \cdots s_m$, with $a \in \mathcal{V} \cup \mathcal{C}$. (Note that a may be $\Box$.) So $Cs \to_\beta C'[\Box := s]$ and $Ct \to_\beta C'[\Box := t]$.

Now we prove with induction on the structure of the β-normal form of C': $\Box \in FV(C') \Rightarrow C'[\Box := s] \;_{\text{str}}\!> C'[\Box := t]$. Induction hypothesis (*) is: $\Box \in FV(s_i) \Rightarrow s_i[\Box := s] \;_{\text{str}}\!> s_i[\Box := t]$. Let α be an arbitrary strict valuation, then the following claim can be proved by induction on k:

Claim. *For $0 \le k \le m$, either $[\![(as_1 \cdots s_k)[\Box := s]]\!]_\alpha \;_{\text{mon}}\!> [\![(as_1 \cdots s_k)[\Box := t]]\!]_\alpha$, or $\Box \notin FV(as_1 \cdots s_k)$ and $[\![as_1 \cdots s_k]\!]_\alpha$ is strict.*

Before proving this claim we proceed with the main proof. We have that $\Box \in FV(C')$, so the claim yields $[\![(as_1 \cdots s_m)[\Box := s]]\!]_\alpha \;_{\text{mon}}\!> [\![(as_1 \cdots s_m)[\Box := t]]\!]_\alpha$. By Lemma 30 we may change $_{\text{mon}}\!>$ into $_{\text{str}}\!>$ and by Lemma 42 we may put the λ-binders in front, which yields $[\![C'[\Box := s]]\!]_\alpha \;_{\text{str}}\!> [\![C'[\Box := t]]\!]_\alpha$. After applying Proposition 26 on both sides we get $[\![Cs]\!]_\alpha \;_{\text{str}}\!> [\![Ct]\!]_\alpha$. The valuation α was an arbitrary strict one, so using Definition 29 we conclude that $Cs \;_{\text{str}}\!> Ct$. □

We still have to prove the claim:

Proof. (induction on k).

If $k = 0$:

- If $a = \Box$, then $[\![a[\Box := s]]\!]_\alpha = [\![s]\!]_\alpha \;_{\text{mon}}\!> [\![t]\!]_\alpha = [\![a[\Box := t]]\!]_\alpha$.
- If $a \in \mathcal{V} - \{\Box\}$, then $\Box \notin FV(a)$, and $[\![a]\!]_\alpha = \alpha(a)$ is strict, because α is a strict valuation.
- If $a \in \mathcal{C}$, then $\Box \notin FV(a)$, and $[\![a]\!]_\alpha = \mathcal{J}(a)$ is strict, because $\mathcal{J}$ is a strict constant interpretation.

If $k = j + 1$: We have the induction hypothesis, that either $[\![(as_1 \cdots s_j)[\Box := s]]\!]_\alpha \;_{\text{mon}}\!> [\![(as_1 \cdots s_j)[\Box := t]]\!]_\alpha$, or $\Box \notin FV(as_1 \cdots s_j)$ and $[\![as_1 \cdots s_j]\!]_\alpha$ is strict (**).

Firstly, $[\![s_{j+1}[\Box := s]]\!]_\alpha \;_{\text{mon}}\!\ge [\![s_{j+1}[\Box := t]]\!]_\alpha$ (***), as the following calculation shows:

$$\begin{array}{lll} & [\![s_{j+1}[\Box := s]]\!]_\alpha & \\ = & [\![(\lambda\Box.s_{j+1})s]\!]_\alpha & \text{by Proposition 26,} \\ = & [\![\lambda\Box.s_{j+1}]\!]_\alpha [\![s]\!]_\alpha & \\ {}_{\text{mon}}\!\ge & [\![\lambda\Box.s_{j+1}]\!]_\alpha [\![t]\!]_\alpha & \text{by Proposition 19,} \\ = & [\![(\lambda\Box.s_{j+1})t]\!]_\alpha & \\ = & [\![s_{j+1}[\Box := t]]\!]_\alpha & . \end{array}$$

From induction hypothesis (**) two cases can be distinguished:

If $(as_1 \cdots s_j)[\Box := s] \,_{\text{mon}}{>}\, (as_1 \cdots s_j)[\Box := t]$, then

$$\begin{array}{rl} & [\![(as_1 \cdots s_{j+1})[\Box := s]]\!]_\alpha \\ = & [\![(as_1 \cdots s_j)[\Box := s]]\!]_\alpha [\![s_{j+1}[\Box := s]]\!]_\alpha \\ _{\text{mon}}{>} & [\![(as_1 \cdots s_j)[\Box := t]]\!]_\alpha [\![s_{j+1}[\Box := s]]\!]_\alpha \\ _{\text{mon}}{\geq} & [\![(as_1 \cdots s_j)[\Box := t]]\!]_\alpha [\![s_{j+1}[\Box := t]]\!]_\alpha \text{ using (***)}, \\ = & [\![(as_1 \cdots s_{j+1})[\Box := t]]\!]_\alpha \ . \end{array}$$

In case $\Box \notin FV(as_1 \cdots s_j)$ and $[\![as_1 \cdots s_j]\!]_\alpha$ is strict, we distinguish two subcases:

$\Box \notin FV(s_{j+1})$: Then $\Box \notin FV(as_1 \cdots s_{j+1})$ and by Definition 29, we have strictness of $[\![as_1 \cdots s_j]\!]_\alpha [\![s_{j+1}]\!]_\alpha = [\![as_1 \cdots s_{j+1}]\!]_\alpha$.

$\Box \in FV(s_{j+1})$:

$$\begin{array}{rl} & [\![(as_1 \cdots s_{j+1})[\Box := s]]\!]_\alpha \\ = & [\![as_1 \cdots s_j]\!]_\alpha [\![s_{j+1}[\Box := s]]\!]_\alpha \\ _{\text{mon}}{>} & [\![as_1 \cdots s_j]\!]_\alpha [\![s_{j+1}[\Box := t]]\!]_\alpha \text{ using IH(*) and Definition 29}, \\ = & [\![(as_1 \cdots s_{j+1})[\Box := t]]\!]_\alpha \ . \end{array}$$

□

5 A Proof Technique for Termination

The notions of Definition 34 and 38 are collected and extended in such a way that we get the conditions for termination:

Definition 44. Let $\mathcal{R}$ be an HRS, with signature $\mathcal{F} = (\mathcal{B}, \mathcal{C}, \mathcal{V})$ and R a set of rules. We say that $\mathcal{R}$ has a *Termination Model* if there is a well-founded type interpretation $\mathcal{I}$ for $\mathcal{B}$ and a strict constant interpretation $\mathcal{J}$ for $\mathcal{C}$, such that for each rule $l \to r \in R$, $l \,_{\text{mon}}{>}\, r$.

Now we are able to state the relation between a rewrite step and the ordering of the domains:

Proposition 45. *Let an HRS $\mathcal{R}$ have a Termination Model. If $s \to_{\mathcal{R}} t$ then $s \,_{\text{str}}{>}\, t$.*

Proof. Let $s \to_{\mathcal{R}} t$. Then $s =_\eta (Cl)\downarrow_\beta$ and $t =_\eta (Cr)\downarrow_\beta$, for some context C and rule $(l \to r) \in R$. Because $\mathcal{R}$ has a Termination Model, $l \,_{\text{mon}}{>}\, r$, so by Proposition 43, $Cl \,_{\text{str}}{>}\, Cr$. Applying Proposition 26 and 27 yields: $s \,_{\text{str}}{>}\, t$. □

The main theorem of this paper uses the well-foundedness of the domains:

Theorem 46. *If an HRS $\mathcal{R}$ has a Termination Model, then $\mathcal{R}$ is terminating.*

Proof. If $\mathcal{R}$ is non-terminating, then there exists a sequence $(s_i)_{i \in \mathbb{N}}$, with $s_i \to_{\mathcal{R}} s_{i+1}$. By Proposition 45 we get an infinite descending chain $(s_i)_{i \in \mathbb{N}}$, with $s_i \,_{\text{str}}{>}\, s_{i+1}$. This is impossible because of the well-foundedness of $_{\text{str}}{>}$ (Proposition 41). Thus $\mathcal{R}$ must be terminating. □

The following is a recollection of the conditions that occur in the notion of a Termination Model. Theorem 46 suggests the following proof technique for the termination of an HRS:

- Find convenient partial orderings $\mathcal{I}(B)$ for each base type B, satisfying:
  - $\mathcal{I}(B)$ is non-empty.
  - $\mathcal{I}(B)$ is well-founded.
  - There exist strict functions with type $\mathcal{I}(a) \to \mathcal{I}(b) \to \mathcal{I}(b)$ for all combinations (a, b) of base types. (By Lemma 33 strict ${}_{\mathrm{mon}}>$-monotonicity suffices).
- Find a convenient strict interpretation $\mathcal{J}(c)$ for each constant symbol $c \in \mathcal{C}$.
- Prove that for each rule $(l \to r)$ in the system the interpretation of the left hand side is greater than the interpretation of the right hand side, in symbols $[\![l]\!] \;{}_{\mathrm{mon}}\!> [\![r]\!]$.

In the next section the applicability of this proof method is shown.

6 Applications

6.1 Process Algebra

The first application comes from Process Algebra, or better an extension of it: μCRL [7]. We only concentrate on the fragment of Process Algebra with choice (+), sequential composition $(\cdot)$ and deadlock (δ) and the data dependent choice (Σ) from μCRL. The Process Algebra part can be formulated in a first order Term Rewriting System (see for instance [1]). The rules for the Sum-operator require higher-order rewrite rules to deal with the bound variables. This reformulation of μCRL can be found in [16, p. 33].

There are two base types: {Proc, Data}. Furthermore, here is a list of function symbols with their types:

$$+ : \mathsf{Proc} \to \mathsf{Proc} \to \mathsf{Proc} \qquad \delta : \mathsf{Proc}$$
$$\cdot : \mathsf{Proc} \to \mathsf{Proc} \to \mathsf{Proc} \qquad \Sigma : (\mathsf{Data} \to \mathsf{Proc}) \to \mathsf{Proc}$$

$\{X, Y, Z, P, Q, D\}$ are used as free variables. Now we have the following set of rules, with the binary function symbols written infix:

$$\begin{array}{ll}
\text{A3:} & X + X \to X \\
\text{A4:} & (X + Y)\cdot Z \to (X\cdot Z) + (Y\cdot Z) \\
\text{A5:} & (X\cdot Y)\cdot Z \to X\cdot(Y\cdot Z) \\
\text{A6:} & X + \delta \to X \\
\text{A7:} & \delta\cdot X \to \delta \\
\text{Sum1:} & \Sigma(\lambda d : \mathsf{Data}.X) \to X \\
\text{Sum3:} & (\Sigma P) + (PD) \to (\Sigma P) \\
\text{Sum4:} & \Sigma(\lambda d : \mathsf{Data}.((Pd) + (Qd))) \to (\Sigma P) + (\Sigma Q) \\
\text{Sum5:} & (\Sigma P)\cdot X \to \Sigma(\lambda d : \mathsf{Data}.((Pd)\cdot X))
\end{array}$$

To prove termination of this system we interpret both base types Data and Proc by $\mathbb{N}_{\geq 1}$, with the usual ordering. This is a domain, because it is well-founded, non-empty and there exists a binary strictly monotonic function, ordinary "+" for instance. The function symbols are interpreted in the following way:

$$[\![+]\!] = \lambda a.\lambda b.a + b + 1 \qquad [\![\delta]\!] = 1$$
$$[\![\cdot]\!] = \lambda a.\lambda b.a \times b + a \qquad [\![\Sigma]\!] = \lambda f.3 \times f(1) + 1$$

This is an extension of the interpretation in [1] for the Process Algebra part of the system. The type level of all these function symbols is bounded by 2, so it is sufficient to check if the interpretations for them are strictly ${}_{\text{mon}}>$-monotonic (Lemma 33). The first three functions are clearly strictly monotonic. The last is also strictly ${}_{\text{mon}}>$-monotonic: Take $f \;{}_{\text{mon}}> g$, then $f(1) \;{}_{\text{mon}}> g(1)$. But then also $3 \times f(1) + 1 \;{}_{\text{mon}}> 3 \times g(1) + 1$. Now we compute the values of the left hand sides and right hand sides.

interpretation of the left hand side	interpretation of the right hand side
$2 \times [\![X]\!] + 1$	$[\![X]\!]$
$([\![X]\!] + [\![Y]\!] + 1) \times [\![Z]\!] + [\![X]\!] + [\![Y]\!] + 1$	$[\![X]\!] \times [\![Z]\!] + [\![X]\!] + [\![Y]\!] \times [\![Z]\!] + [\![Y]\!] + 1$
$[\![X]\!] \times ([\![Y]\!] \times [\![Z]\!] + [\![Z]\!] + [\![Y]\!] + 1)$	$[\![X]\!] \times ([\![Y]\!] \times [\![Z]\!] + [\![Y]\!] + 1)$
$[\![X]\!] + 2$	$[\![X]\!]$
$[\![X]\!] + 1$	1
$3 \times [\![X]\!] + 1$	$[\![X]\!]$
$3 \times [\![P]\!](1) + [\![PD]\!] + 2$	$3 \times [\![P]\!](1) + 1$
$3 \times ([\![P]\!](1) + [\![Q]\!](1)) + 4$	$3 \times ([\![P]\!](1) + [\![Q]\!](1)) + 3$
$3 \times [\![P]\!](1) \times [\![X]\!] + [\![X]\!] + 3 \times [\![P]\!](1) + 1$	$3 \times [\![P]\!](1) \times [\![X]\!] + 3 \times [\![P]\!](1) + 1$

The reader can verify that the interpretation of the left hand side is greater than the interpretation on the right hand side on each line in the table. So this system of Process Algebra- and Sum rules is terminating.

6.2 Quantifier reasoning

In [12] some HRSs concerning first order predicate logic are presented as an example. One of them is called *mini scoping*, pushing quantifiers inwards. The base types are {Term, Form}. The function symbols are $\{\vee, \wedge : \mathsf{Form} \to \mathsf{Form} \to \mathsf{Form};\ \forall : (\mathsf{Term} \to \mathsf{Form}) \to \mathsf{Form}\}$, the type level is at most 2.

The rules (with free variables $\{P, P', Q, Q'\}$) are:

$$\forall(\lambda x.P) \to P$$
$$\forall(\lambda x.((P'x) \wedge (Q'x))) \to (\forall P') \wedge (\forall Q')$$
$$\forall(\lambda x.((P'x) \vee Q)) \to (\forall P') \vee Q$$
$$\forall(\lambda x.(P \vee (Q'x))) \to P \vee (\forall Q')$$

Again we take as interpretation for both base types the positive natural numbers $\mathbb{N}_{\geq 1}$. The interpretation of the function symbols is as follows:

$$[\![\wedge]\!] = [\![\vee]\!] = \lambda a.\lambda b.a + b + 1$$
$$[\![\forall]\!] = \lambda f.2 \times f(1)$$

It is easily verified that under this interpretation the left hand side of each rule is greater than the interpretation of the right hand side.

6.3 Surjective Disjoint Union

We give one more example. The signature is given by the only base type Term, the function symbols:

$$\begin{aligned} \mathit{case} &: \mathsf{Term} \to (\mathsf{Term} \to \mathsf{Term}) \to (\mathsf{Term} \to \mathsf{Term}) \to \mathsf{Term} \\ \mathit{inl}, \mathit{inr} &: \mathsf{Term} \to \mathsf{Term} \end{aligned}$$

All function symbols have level ≤ 2. The free variables are $\{X, F, G, U\}$. The rules are:

$$\begin{aligned} case(inl(X), F, G) &\to F(X) \\ case(inr(X), F, G) &\to G(X) \\ case(U, \lambda x.F(inl(x)), \lambda x.F(inr(x))) &\to F(U) \end{aligned}$$

Note that this example does not fit in the framework of [12] (see page 347). Termination for this example is less trivial, because there is a real application in the interpretation of the function symbols. Furthermore it is not the case that the number of "case" occurrences decreases in every step: If X contains a "case" occurrence, then F can generate many copies of it in the right hand side of the first rule.

But the interpretation in a Termination Model is easy: Take $\mathcal{I}(\mathsf{Term}) = \mathbb{N}_{\geq 1}$. Furthermore, interpret:

$$\begin{aligned} [\![case]\!] &= \lambda a.\lambda f.\lambda g.f(a) + g(a) + a \\ [\![inl]\!] = [\![inr]\!] &= \lambda a.a + 1 \end{aligned}$$

These functions are strict: we only need to take into account monotonic functions for f and g. Take $a \,_{\mathrm{mon}}\!> b$. By monotonicity of f we have $f(a) \,_{\mathrm{mon}}\!\geq f(b)$ and the same holds for g. But then $f(a) + g(a) + a \,_{\mathrm{mon}}\!> f(b) + g(b) + b$. Furthermore, the interpretations of the left- and right hand sides can be computed:

Left hand side	Right hand side
$[\![F]\!]([\![X]\!] + 1) + [\![G]\!]([\![X]\!] + 1) + [\![X]\!] + 1$	$[\![F]\!]([\![X]\!])$
$[\![F]\!]([\![X]\!] + 1) + [\![G]\!]([\![X]\!] + 1) + [\![X]\!] + 1$	$[\![G]\!]([\![X]\!])$
$2 \times [\![F]\!]([\![U]\!] + 1) + [\![U]\!]$	$[\![F]\!]([\![U]\!])$

The left hand sides are all greater than the right hand sides, because we may restrict to monotonic functionals for F and G. So the system of "surjective disjoint union" is terminating.

References

1. G.J. Akkerman and J.C.M. Baeten. Term rewriting analysis in process algebra. Technical Report CS-R9130, Centre for Mathematics and Computer Science, June 1991.
2. H.P. Barendregt. *The Lambda Calculus. Its Syntax and Semantics.* North-Holland, Amsterdam, second, revised edition, 1984.
3. M. Bezem and J.F. Groote, editors. *Proceedings of the 1st International Conference on Typed Lambda Calculi and Applications, TLCA '93,* Utrecht, The Netherlands, volume 664 of *Lecture Notes in Computer Science.* Springer-Verlag, 1993.

4. V. Breazu-Tannen. Combining algebra and higher-order types. In *Proceedings* 3th *Annual Symposium on Logic in Computer Science,* Edinburgh, pages 82–90, July 1988.
5. R. de Vrijer. *Surjective Pairing and Strong Normalization: Two Themes in Lambda Calculus.* PhD thesis, University of Amsterdam, 1987.
6. R.O. Gandy. Proofs of strong normalization. In J.R. Hindley and J.P. Seldin, editors, *To H.B. Curry: Essays on Combinatory Logic, Lambda Calculus and Formalism*, pages 457–477. Academic Press, 1980.
7. J.F. Groote and A. Ponse. The syntax and semantics of μCRL. Report CS-R9076, CWI, Amsterdam, 1990.
8. G. Huet and D.C. Oppen. Equations and rewrite rules: A survey. In R. Book, editor, *Formal Language Theory: Perspectives and Open Problems*, pages 349–405. Academic Press, 1980.
9. J.P. Jouannaud and M. Okada. A computation model for executable higher-order algebraic specification languages. In *Proceedings* 6th *Annual Symposium on Logic in Computer Science,* Amsterdam, The Netherlands, pages 350–361, 1991.
10. Z. Khasidashvili. Perpetual reductions and strong normalization in orthogonal term rewriting systems. Technical Report CS-R9345, CWI, Amsterdam, July 1993.
11. J.W. Klop. *Combinatory Reduction Systems*, volume 127 of *Mathematical Centre Tracts.* Mathematisch Centrum, Amsterdam, 1980.
12. T. Nipkow. Higher-order critical pairs. In *Proceedings* 6th *Annual Symposium on Logic in Computer Science,* Amsterdam, The Netherlands, pages 342–349, 1991.
13. T. Nipkow. Orthogonal higher-order rewrite systems are confluent. In Bezem and Groote [3], pages 306–317.
14. C. Paulin-Mohring. Inductive definitions in the system Coq. Rules and properties. In Bezem and Groote [3], pages 328–345.
15. L. C. Paulson. Isabelle: The next 700 theorem provers. In P. Odifreddi, editor, *Logic and Computer Science*, pages 361–386. Academic Press Limited, London, 1990. In: The APIC-series 31.
16. M.P.A. Sellink. Verifying process algebra proofs in type theory. Technical Report 87, Logic Group Preprint Series, Utrecht University, March 1993.
17. V. van Oostrom and F. van Raamsdonk. Comparing combinatory reduction systems and higher-order rewrite systems. Technical Report IR-333, Vrije Universiteit Amsterdam, August 1993. Appears in this Volume.
18. D.A. Wolfram. *The Clausal Theory of Types,* volume 21 of *Cambridge tracts in theoretical computer science.* Cambridge University Press, Cambridge, 1993.
19. H. Zantema. Termination of term rewriting by interpretation. In M. Rusinowitch and J.L. Rémy, editors, *Conditional Term Rewriting Systems, proceedings third international workshop CTRS-92*, volume 656 of *Lecture Notes in Computer Science*, pages 155–167. Springer-Verlag, 1993. Full version appeared as report RUU-CS-92-14, Utrecht University.

Extensions of Initial Models and their Second-Order Proof Systems

Pierre-Yves Schobbens *

Fac. Univ. Notre-Dame de la Paix
Institut d'Informatique
Rue Grandgagnage 21
B-5000 Namur

Abstract. Besides explicit axioms, an algebraic specification language contains model-theoretic constraints such as initiality. For proving properties of specifications and refining them to programs, an axiomatization of these constraints is needed; unfortunately, no effective, sound and complete proof system can be constructed for initial models, and *a fortiori* for their extensions.

In this paper, we construct non-effective second-order axiomatizations for the initiality constraint, and its recently proposed extensions (minimal, quasi-free and surjective models) designed to deal with disjunction and existential quantification.

1 Introduction

Algebraic specification languages extend universal algebra as needed for their application to program specification:

- various typing disciplines have been proposed, the most known being many-sorted and order-sorted algebras;
- model-theoretic constraints (such as initiality) have been added in order to increase the expressive power of the language, and to eliminate trivial models.
- the specification logic, first restricted to equations, has been extended to equational Horn logic, equational clauses and first-order logic. Our results are thus designed to hold for these richer languages.

Most of these model-theoretic constraints increase the expressive power of the language beyond what is effectively axiomatizable, as shown for instance by [13] for the simple case of initial constraints on equational specifications; more cases are proved in [11].

Current research is trying to circumvent that impossibility:

---

* This research was mostly carried out at Centre National de la Recherche Scientifique, Centre de Recherche en Informatique de Nancy.

- Wirsing [18] uses rules with an infinite number of antecedents. His language, ASL, contains e.g. a reachability constraint which he axiomatizes by the rule:

$$\frac{\Phi[t/x] \quad \text{for all } T_F(X)}{\forall x : s.\Phi}$$

  where $T_F(X)$ is the (often infinite) set of generating terms. In practice, this infinite set of terms is often treated by an informal induction.
- Another solution uses non-logical antecedents in inference rules. For instance, to axiomatize initiality on equational Horn clauses, we might use the following rule in complement to some rule expressing term generation:

$$\frac{\text{for all } \sigma.\sigma t_1, \sigma t_2 \text{ ground} \qquad \nvdash \sigma t_1 = \sigma t_2}{\vdash t_1 \neq t_2}$$

  This rule uses non-provability of equations as an antecedent for provability of inequations. Thanks to the fact that the specification is Horn, this is not a circular definition. Yet, we need a decision procedure for equations to keep the system for disequations effective. In practice, we have only semi-decision procedures (such as narrowing) for some forms of specifications. When one of these complete procedures terminates without finding equal ground instances, the disequation follows. This method is thus suited to automation, but not very handy for manual proofs, since an exhaustive search is necessary to establish non-provability.
- In this article, in contrast, we are interested in second-order axiomatizations of such theories. This approach is adequate for a larger class of logics – including first-order logic – and constraints, and is suited to computer-assisted proofs but not to automated proofs.

2 Second-order logic

Second-order logic interprets predicate variables as ranging over all sets. It is not effectively axiomatizable, but the inference rules known for second-order logic are often sufficient in practice, and can sometimes be used to reduce a specification with constraints to a first-order specification.

To take a well-known example, the term-generation constraint can be axiomatized by a second-order induction rule. When no functions (except constants) are used, this rule is easily reduced to a first-order disjunction. The paper develops similar rules for more complex constraints that have been proposed recently.

The rules are directly based on semantic characterizations of the models; they can be divided into two types, the first one encoding the definition of homomorphisms, and the other one, the notion of congruence on a term algebra. The second type is often easier to use, but is only valid for universal specifications. These rules are obtained easily, as they just encode the set-theoretic definition of the constraints in second order logic. (The differences between set theory and second-order logic require just a few precautions, see 4.1.)

3 Relativization

As second-order logic is unsorted, while the logics used in most specification languages are sorted, we use through the paper the relativization technique [8]: we introduce a unary predicate for each sort, and replace quantification $\forall x : s.\phi$ by $\forall x.s(x) \Rightarrow R(\phi)$, where $R(\phi)$ is (recursively) the relativization of ϕ. The well-sortedness of operators has to be expressed by an axiom $OPS(O)$**3** ([1])

$$\bigwedge_{f:\mathbf{s}\to r\in O} \forall \mathbf{x}.\mathbf{s}(\mathbf{x}) \Rightarrow r(f(\mathbf{x}))$$

Our notations are the following:

- $\forall \mathbf{x}$ denotes the first-order quantification $\forall x_1, \ldots, x_n$.
- $\forall x_1, \ldots, x_n$ is itself an abbreviation $\forall x_1 \ldots . \forall x_n$, as usual.
- The scope of quantifiers and binders like $\bigwedge_{f\in O}$ extends as far as possible to the right. The second occurence of $\mathbf{x}$ is thus bound above.
- The precedence of operators is $=, \equiv, \wedge, \vee, \Rightarrow$. For instance $\forall x. \bigwedge_i x = k_i \wedge b_i(x) \Rightarrow c_i(x)$ means $\forall x.(\bigwedge_i(((x = k_i) \wedge b_i(x)) \Rightarrow c_i(x)))$.
- $f(\mathbf{x})$ denotes $f(x_1, \ldots, x_n)$. The number of elements $n = |\mathbf{x}|$ is determined by the arity of f (i.e. $|\mathbf{s}|$), as shown by the use of $f(\mathbf{x})$.
- $\mathbf{s}(\mathbf{x})$ denotes $(\bigwedge_{i\leq n} s_i(x_i))$.
- Here, s_i is the unary predicate symbol that translates the sort of the ith argument of f.

We define thus the translation $R(\mathcal{S})$ of a presentation $\mathcal{S} = (\Sigma, \phi)$, where $\Sigma = (S, O, P)$, as $OPS(O) \wedge R(\phi)$. Relativization also allows a simple treatment of subsorts.

4 Initiality and extensions

4.1 Initial model

To avoid trivial models, the semantics of a specification is often taken to be its initial model. The initial model is often intuitively explained as minimizing equality ("no confusion", or more precisely "minimum confusion"), on top of restricting to term-generated models ("no junk": every element can be constructed by some term). When predicates are present, a similar default is used to preserve uniqueness of the initial model [9]. Initial and free models are generalised in the last part in this section; we present them first for historical and paedagogical reasons.

[1] These bold numbers refer to the theorems of appendix B.

First-order logic The power of second-order logic allows a direct encoding of the definition of initial model, for axioms written in many-sorted first-order logic, a subset of which is used in Clear[3] and ACT-ONE[6]. A many-sorted signature will be encoded in second-order logic by a triple of second-order variables: (sort) predicates for S, functions for O, and predicates for P. The part of the interpretation corresponding to this triple will represent an arbitrary algebra. As a notational convention, we use the same name for the sorts of the multi-sorted logic, and for the sort predicates that represent them is second-order logic, and similarly for operators. Given two such triples, Σ and Σ', where the element corresponding to f is f' an homomorphism is a S-tuple of function $\mathbf{h}$ satisfying thus expressed as a S-tuple of functions $\mathbf{h} = (h_s)_{s \in S}$

- respecting sorts: $\bigwedge_{s \in S} \forall x.s(x) \Rightarrow s'(h_s(x))$
- compatible with functions: $\bigwedge_{f:\mathbf{s} \rightarrow r \in F} \forall \mathbf{x}.\mathbf{s}(\mathbf{x}) \Rightarrow h_r(f(\mathbf{x})) = f'(h_{\mathbf{s}}(\mathbf{x}))$
- preserving the truth of predicates: $\bigwedge_{p:\mathbf{s} \in P} \forall \mathbf{x}.\mathbf{s}(\mathbf{x}) \wedge p(\mathbf{x}) \Rightarrow p'(h_{\mathbf{s}}(\mathbf{x}))$

where $h_{\mathbf{s}}(\mathbf{x})$ abbreviates $h_{s_1}(x_1), \ldots, h_{s_n}(x_n)$.

These conditions together form the second-order formula $HOM(\mathbf{h}, \Sigma, \Sigma')$. Initiality seems thus directly translated as $*INI(\mathcal{S})$.

$$R(\mathcal{S}) \wedge \forall \Sigma', R(\mathcal{S})[\Sigma'/\Sigma] \Rightarrow \exists! \mathbf{h}.HOM(\mathbf{h}, \Sigma, \Sigma')$$

Example 1. Let us look more closely at the models of $*INI$ for:
spec $Nat =$
sort *nat*
opers $0 :\rightarrow nat$
$succ : nat \rightarrow nat$
end

The initial model of this presentation is known to be isomorphic to the natural numbers. When the domain contains a single element, $*INI$ is unexpectedly satisfied: there is a single function from the singleton *nat* to the singleton *nat'*. The problem is that Σ' ranges over subsets of the domain, while we would like it to range over sets of arbitrary cardinality. Knowing that a model of a first-order sentence is initial iff it is initial among models whose carriers are subsets of a domain that is at least denumerable[8], we just have to add an axiom DEN that ensures that the domain is at least denumerable.

$\exists 0, <, succ. \forall x, y, z :$
$0 < succ(x)$
$\wedge\, x < succ(x)$
$\wedge\, \neg(x < y \wedge y < x)$
$\wedge\, (x < y \wedge y < z \Rightarrow x < z)$

We conjoin it with our previous attempt to obtain an exact characterization.

Theorem 1. *$\mathcal{S} \models_{INI} \psi$ iff $INI(\mathcal{S}) \models R(\psi)$.*

When the specification is not Horn, there is often no initial model at all. This motivates the search for more sensible constraints, reviewed in section 4.3.

Furthermore, using the rule $INI(\mathcal{S})$ to make practical proofs is tedious: fortunately, a simpler rule can be used for universal sentences.

Universal sentences When the language is restricted to universal sentences, we know that the initial model is isomorphic to the quotient of T_Σ by the least congruence satisfying the axioms. This fact can be used to construct a more practical rule, where quantification over functions is eliminated. Assume that we can characterize algebras isomorphic to T_Σ by some axiom $TERM(\Sigma)$ (this will be done in section 4.2). Then we introduce a family of new binary predicates $\equiv_s$ to express the congruence on T_Σ, and we state the usual congruence axioms $EQ(\Sigma)$.

Of course, the presentation should use this congruence instead of equality, which is achieved by a simple substitution of the equality symbol by a congruence symbol of the adequate sort: $\phi[\equiv / =]$, where $\equiv$ is an abbreviation for a family $(\equiv_s)_{s \in S}$ of binary relations. The rule $LC(\mathcal{S})$, simply states that $\equiv$ is the least congruence:

$$\forall \equiv', EQ(\Sigma)[\equiv' / \equiv] \wedge R(\mathcal{S})[\equiv' / =] \Rightarrow (\bigwedge_s \forall x, y. x \equiv_s y \Rightarrow x \equiv'_s y)$$

The axiom we are looking for, $LCS(\mathcal{S})$, is thus[12]:

$$TERM(\Sigma) \wedge EQ(\Sigma) \wedge R(\mathcal{S})[\equiv / =] \wedge LC(\mathcal{S})$$

Of course, we also have to substitute $\equiv$ for$=$ in any goal. We will detail and generalise this axiom in the next section; what is important is to understand its construction and structure to follow easily the sequel.

The intersection of congruences gives an explicit definition of the desired congruence $EC_s(\mathcal{S})$:

$$x \equiv_s y \Leftrightarrow \forall \equiv'_s, EQ(\Sigma)[\equiv'_s / \equiv_s] \wedge R(\mathcal{S})[\equiv / =][\equiv'_s / \equiv_s] \Rightarrow x \equiv'_s y$$

Note that this intersection need not satisfy the axioms (except for Horn clauses); otherwise, the resulting axiom is unsatisfiable, as expected since there is no initial model. In this case, we can thus use $EC(\mathcal{S}) = TERM(\Sigma) \wedge \bigwedge_{s \in S} EC_s(\mathcal{S})$.

4.2 Free models

First-order logic Once again, the definition of initial model is not suitable for parameterized specifications, since it may put constraints on the parameter.

Example 2. We specify bags of some loosely specified sort of elements, plus a loose specification of containers (e.g. lists): The intended models of *Containers* are the free models wrt *Bag* on *Elem*; no freeness constraint should be put on *Elem*, because it would then be a singleton. Similarly, no constraint should be put on *Cont*, hence the importance of limiting the constraint to *Bag*.

An algebra A is said to be *free* over A_1 with respect to $\mathcal{S}_2$ on $\mathcal{S}_1$ ($\mathcal{S}_1 \subseteq \mathcal{S}_2$) iff there is an homomorphism $\eta : A_1 \rightarrow A|_{\Sigma_1}$ such that for all models A_2 of $\mathcal{S}_2$, and homomorphisms $h : A_1 \rightarrow A_2|_{\Sigma_1}$ there is a unique homomorphism $g : A|_{\Sigma_2} \rightarrow A_2$ satisfying $g|_{\Sigma_1} \circ \eta = h$.

spec *Elem* = **sort** *elem* **end**

spec *Bag* = **free with** *Elem*
sort *bag*
opns $add : elem \times bag \rightarrow bag$
$\quad empty : \rightarrow bag$
pred $. \in . : elem \times bag$
axioms $\forall e, f : elem, s : bag.$
$\quad e \in add(e, s)$
$\quad add(e, add(f, s)) = add(f, add(e, s))$
end

spec *Container* = **with** *Bag*
sort *cont*
opns $cons : elem \times cont \rightarrow cont$
$\quad bagify : cont \rightarrow bag$
pred $.in. : elem \times cont$
axioms $\forall e : elem, c : cont.$
$e \; in \; cons(e, s)$
$bagify(cons(e, s)) = add(e, bagify(s))$ **end**

Fig. 1. Bags and containers

This condition translates directly into the (complex!) second-order axiom $F^*(\mathcal{S}_2, \mathcal{S}_1)$:
$\exists \Sigma_1''.R(\mathcal{S}_1)[\Sigma_1''/\Sigma_1] \wedge \exists (\eta_s)_{s \in S_1}.HOM(\eta, \Sigma'', \Sigma_1) \wedge$
$\forall \Sigma_2'.R(\mathcal{S}_2)[\Sigma_2'/\Sigma_2] \Rightarrow \forall (h_s)_{s \in S_1}.HOM((h_s)_{s \in S_1}, \Sigma_1'', \Sigma_1') \Rightarrow$
$\exists ! (g_s)_{s \in S_2}.HOM((g_s)_{s \in S_2}, \Sigma_2, \Sigma_2') \wedge \bigwedge_{s \in S_1} \forall x.s''(x) \Rightarrow g_s(\eta_s(x)) = h_s(x)$

We have the same cardinality problem as for initiality, and here we have a slightly weaker result[8], since we have at least the cardinality of the "old" sorts (S_1). As each s_i is already included in the domain, it suffices to ensure a domain at least denumerable, for which our axiom DEN suffices. So our final axiom is $FREE(\mathcal{S}_2, \mathcal{S}_1) = F^*(\mathcal{S}_2, \mathcal{S}_1) \wedge DEN$[10].

Persistency This free model is usually considered *correct* [6] if persistency is insured, i.e. the Σ_1−reduct of the free extension wrt $\mathcal{S}_2$ of any model of $\mathcal{S}_1$ is isomorphic to that model.

Again, this proof obligation is directly expressible in second-order logic:
$\exists (i_s)_{s \in S_1}.HOM(i, \Sigma_1, \Sigma_1'')$
$\wedge \forall \bigwedge_{s \in S_1} \forall b.s''(b) \Rightarrow \exists a.s(a) \wedge i(a) = b$
$\wedge \forall a_1, a_2.s(a_1) \wedge s(a_2) \wedge a_1 \neq a_2 \Rightarrow i(a_1) \neq i(a_2)$

Universal sentences When the language is restricted to universal sentences, we can show that the Σ_2-reduct of a free model is isomorphic to the quotient of $T_{\Sigma_2}(I|_{\Sigma_1})$ by the least congruence satisfying the axioms of S_2. The free construction is again a term algebra, this time using the elements of S_1 as variables. We use a copy Σ_1' of Σ_1 disjoint from Σ_2, to represent the algebra we started from. This distinction is only temporary, for if we prove strong persistency (see below), the two algebras can be identified.

The following facts, $TERM(\Sigma, S_1')$ characterize term algebras[7]:

- the algebra is well-typed, $OPS(O)$.
- variables are terms, $INC(S, S_1')$: $\bigwedge_{s \in S} \forall x.s'(x) \Rightarrow s(x)$
- functions have disjoint images, $DIS(O)$.
  $\bigwedge_{f \neq g : \mathbf{s} \rightarrow r \in O} \forall \mathbf{x}, \mathbf{y}.\mathbf{s}(\mathbf{x}) \wedge \mathbf{s}'(\mathbf{y}) \Rightarrow \neg(f(\mathbf{x}) = g(\mathbf{y}))$

- each function is injective, $INJ(O)$.
  $\bigwedge_{f\in O} \forall \mathbf{x},\mathbf{y}.\mathbf{s}(\mathbf{x}) \wedge \mathbf{s}(\mathbf{y}) \wedge f(\mathbf{x}) = f(\mathbf{y}) \Rightarrow \bigwedge_{i\leq n} x_i = y_i$
- functions do not yield variables, $NOV(O, S_1')$.
  $\bigwedge_{f:\mathbf{s}\to r\in O} \forall \mathbf{x}, \mathbf{s}(\mathbf{x}) \Rightarrow \neg r'(f(\mathbf{x}))$
- the algebra is reachable $FG(\Sigma, S')$.
  $\forall (I_s)_{s\in S}.[\bigwedge_{f:\mathbf{s}\to r\in O} \forall \mathbf{x}, \mathbf{s}(\mathbf{x}) \wedge I_{\mathbf{s}}(\mathbf{x}) \Rightarrow I_r(f(\mathbf{x})) \wedge \bigwedge_{s\in S} \forall x.s'(x) \Rightarrow I_s(x)]$
  $\Rightarrow [\bigwedge_{s\in S} \forall x.s(x) \Rightarrow I_s(x)]$

The properties of congruences, $EQ(\Sigma)$, are:

- reflexivity: $\bigwedge_{s\in S} \forall x.s(x) \Rightarrow x \equiv_s x$
- symmetry: $\bigwedge_{s\in S} \forall x, y.s(x) \wedge s(y) \Rightarrow (x \equiv y \Rightarrow y \equiv x)$
- transitivity: $\bigwedge_{s\in S} \forall x, y, z.s(x) \wedge s(y) \wedge s(z) \Rightarrow (x \equiv y \wedge y \equiv z \Rightarrow x \equiv z)$
- compatibility:
  - for each operator $f : \mathbf{s} \to r \in O$:
    $\bigwedge_{s_i\in S} \forall \mathbf{x}, x_i'.\mathbf{s}(\mathbf{x}) \wedge s_i(x_i') \wedge x_i \equiv_{s_i} x_i' \Rightarrow f(\mathbf{x}) \equiv_r f(x_1, \ldots, x_i', \ldots, x_n)$
  - for each predicate $p : \mathbf{s} \in P$: $\bigwedge_{s_i\in S} \forall \mathbf{x}, x_i'.\mathbf{s}(\mathbf{x}) \wedge s_i(x_i') \wedge x_i \equiv_{s_i} x_i' \Rightarrow [p(\mathbf{x}) \Leftrightarrow p(x_1, \ldots, x_i', \ldots, x_n)]$

The congruence should respect the structure of Σ_1, as expressed by $RS(\Sigma_1')$.

- $\bigwedge_{f':\mathbf{s}'\to r'\in O_1'} \forall \mathbf{a}, b.\mathbf{s}'(\mathbf{a}) \wedge f'(\mathbf{a}) = b \Rightarrow f(\mathbf{a}) \equiv_s b$.
- $\bigwedge_{p':\mathbf{s}'\in P_1'} \forall \mathbf{a}.\mathbf{s}'(\mathbf{a}) \wedge p'(\mathbf{a}) \Rightarrow p(\mathbf{a})$.

The free model is characterized by the least congruence and by the least extension of predicates (i.e., $LC(\mathcal{S}_2)$)

$$LC(\mathcal{S}_2): \quad \forall \Delta'.EQ(\Sigma_2)[\equiv' / \equiv] \wedge R(\mathcal{S}_2)[\equiv / =][\Delta'/\Delta] \Rightarrow \Delta' \leq \Delta$$

where:

- Δ is a family of relations $(\equiv_s)_{s\in S_2}$ and a tuple of predicates $(p)_{p:\mathbf{s}\in P_2}$
- Δ' is a fresh copy of Δ, with $'$ giving the correspondence.
- $\Delta \leq \Delta'$ is $[\bigwedge_{s\in S_2} \forall x, y.s(x) \wedge s(y) \wedge x \equiv_s y \Rightarrow x \equiv_s' y] \wedge [\bigwedge_{p:\mathbf{s}\in P_2} \forall \mathbf{x}, \mathbf{s}(\mathbf{x}) \wedge p(\mathbf{x}) \Rightarrow p'(\mathbf{x})]$

So the adequate axiom $LCS(\mathcal{S}_2, \Sigma_1)$ is:

$$TERM(\Sigma_2, S_1') \wedge RS(\Sigma_1') \wedge EQ(\Sigma_2) \wedge R(\mathcal{S}_2)[\equiv / =] \wedge LC(\mathcal{S}_2)$$

Conjoining these conditions, we obtain LCS. This axiom allows us to derive exactly the same consequences than the original specification, provided we treat the goal similarly, i.e. we relativize it, and use $\equiv$ instead of $=$.[12]

Persistency This free model is considered *strongly correct* [6] if strong persistency is insured, i.e. the Σ_1−reduct of the free extension wrt $\mathcal{S}_2$ of any model of $\mathcal{S}_1$ is that model. Strong persistency can be ensured here has we construct the free model by an explicit technique.

With the assumptions of this section, strong persistency amounts to a simple check of sufficent completeness, $SUF(\Sigma_2, \Sigma_1)$ to be proved from LCS.

$$\bigwedge_{s \in S_1} \forall y.s(y) \Rightarrow \exists x.s''(x) \wedge x \equiv_s y$$

Knowing strong persistency, we may now use a simpler rule, where S_1 and S_1' are identified. FG can be simplified to $RCH(S_2 \setminus S_1, O_2)$,
$\forall (I_s)_{s \in S_2 \setminus S_1} \cdot [\bigwedge_{f:\mathbf{s} \to r \in O_2, r \notin S_1} \forall \mathbf{x}, \bigwedge s_i(x_i) \wedge \bigwedge_{s_i \notin S_1} I_{s_i}(x_i) \Rightarrow I_r(f(\mathbf{x}))]$
$\Rightarrow [\bigwedge_{s \in S_2 \setminus S_1} \forall x, s(x) \Rightarrow I_s(x)]$
where induction predicate are only provided for new sorts: $(I_s)_{s \in S_2 \setminus S_1}$, and f only ranges over functions yielding results in new sorts: $f : \mathbf{s} \to r \in O_2, r \notin S_1$.

The axioms INC and NOV can be simply suppressed; DIS and INJ are only applied for f with result outside S_1, and similarly for the EQ axioms.

Example 3. Our notations have now grown to a point where most readers have difficulties to unfold them mentally, so let's do the exercise for our *Bag* example (i.e., the left column of fig 1).

Here, the strong persistency proof obligation is trivial, since no new operator has an old result sort, and there are no equations on an old sort.

To simplify notation, we omit quantifications such as
$\forall e, f, s, t, x.elem(e) \wedge elem(f) \wedge bag(s) \wedge bag(t) \wedge bag(x) \Rightarrow ..$
around each line below.

OPS	$bag(empty)$
	$bag(add(e,s))$
$R(\mathcal{A}_1)$	$e \in add(e,s)$
$R(\mathcal{A}_2)$	$add(e, add(f,s)) \equiv add(f, add(e,s))$
DIS	$empty \neq add(e,s)$
INJ	$add(e,s) = add(f,t) \Rightarrow e = f \wedge s = t$
FG	$\forall I.[\forall e, s.I(s) \wedge bag(s) \wedge elem(e) \Rightarrow I(add(e,s))) \wedge I(empty)]$
	$\Rightarrow \forall s.bag(s) \Rightarrow I(s)$
EQ	$s \equiv s$
	$t \equiv s \Rightarrow s \equiv t$
	$x \equiv s \wedge s \equiv t \Rightarrow x \equiv t$
	$t \equiv s \Rightarrow add(e,s) \equiv add(e,t)$
	$t \equiv s \Rightarrow [e \in s \Leftrightarrow e \in t]$
LC	$\forall \equiv', \in' .R' \wedge EQ' \Rightarrow (s \equiv t \Rightarrow s \equiv' t) \wedge (e \in s \Rightarrow e \in' s)$

As an exercise, let use prove $\psi = \forall e : elem, s : bag.add(e,s) \neq empty$, which is not stated in the axioms but is non-monotonically derivable since the contrary is not stated. This goal translates to $R(\psi)[\equiv / =] = \forall e, s.elem(e) \wedge bag(s) \Rightarrow add(e,s) \not\equiv empty$. Let $x \equiv' y$ be $(x \equiv y) \wedge (x = empty \Leftrightarrow y = empty)$. It is easy to check that $R(\phi)$ and EQ are satisfied, so that we deduce by LC that

$\forall x, y.bag(x) \wedge bag(y) \wedge x \equiv y \Rightarrow [x = empty \Leftrightarrow y = empty]$. Take $x = empty, y = add(e, s)$. By DIS, $empty \neq add(e, s)$, hence the goal.

Now we may try to prove $\psi = \forall e : elem.\ e \notin empty$, which is also not stated in the axioms. The goal translates to $R(\psi)[\equiv / =] = \forall e.elem(e) \Rightarrow \neg(e \in empty)$. We instantiate $\equiv'$ by $\equiv$, and $\in'$ by $s \neq empty \wedge e \in s$ in LC. The antecedents can be eliminated:

- $EQ[\equiv' / \equiv]$ is derivable from EQ, except $t \equiv s \Rightarrow [(s \neq empty \wedge e \in s) \Leftrightarrow (t \neq empty \wedge e \in t)]$, which simplifies to $t \equiv s \Rightarrow (s = empty \Leftrightarrow t = empty)$, proved above.
- $R(\phi)[\equiv' / =][\in' / \in]$ is derivable using $TERM$ and $R(\phi)[\equiv / =]$.

We obtain thus $\forall e, s.elem(s) \wedge bag(s) \wedge e \in s \Rightarrow (s \neq empty \wedge e \in s)$, which simplifies to the goal.

We leave $add(e, add(e, s)) \neq add(e, s)$ and $f \in add(e, s) \Leftrightarrow (f = e \vee f \in s)$ as exercises to the reader.

4.3 Extensions of free models

The main problem of free models is that they are only guaranteed to exist for Horn specifications. In many cases there is no free model at all. Four approaches (minimal [1], quasi-free [10], circumscriptive [14] and surjective [17, 15]) have been designed to avoid this problem; all of them are expressible in second-order logic.

4.4 Minimal models

[1] propose to use instead "minimal" models, i.e. models that are source of a single morphism to any model that is the source of a morphism to them[2]. Their definition can be generalized by analogy with the generalization from initial to free models, and also by introducing predicates, giving: An algebra A is said to be *minimal* with respect to $\mathcal{S}_2$ on $\mathcal{S}_1$ ($\mathcal{S}_1 \subseteq \mathcal{S}_2$) iff for all models A_2 of $\mathcal{S}_2$ such that $A_2|_{\Sigma_1} = A|_{\Sigma_1}$, if there is an homomorphism $h : A_2 \rightarrow A|_{\Sigma_2}$ where $h|_{\Sigma_1} = id$, then there is a unique homomorphism $h_2 : A|_{\Sigma_2} \rightarrow A_2$ where $h_2|_{\Sigma_1} = id$.

Translating this definition in second-order logic is now an easy exercice, resulting in axiom MIN. This definition is axiomatized by $MIN(\mathcal{S}_2, \Sigma_1)$:
$DEN \wedge \forall \Sigma_2'.R(\mathcal{S}_2)[\Sigma_2'/\Sigma_2] \wedge \exists (h_s)_{s \in S_2 \setminus S_2}.HOM(\mathbf{h}, \Sigma_2', \Sigma_2)$
$\Rightarrow \exists! (h_s)_{s \in S_2 \setminus S_1}.HOM((h_s)_{s \in S_2}, \Sigma_2, \Sigma_2')$

Assuming strong persistency, we have that:

- Any free model is a minimal model[13].
- when there are free models, any minimal model is a free model[14].

[2] Note that they are not minimal in the usual sense.

Universal sentences When $\mathcal{S}_2$ is universal, then we note[16]:

- If $\mathcal{S}_2$ is satisfiable, it has a minimal model;
- each model of $\mathcal{S}_2$ is the target of a unique morphism from a minimal model;
- a model is minimal iff it is minimal among reachable models; specially, a minimal model is reachable;
- a minimal model is isomorphic to the quotient of the free term algebra by a minimal congruence.

Minimal models can thus be constructed from minimal (vs. minimum for free models) congruences.

This allows the use of the following axiom for minimal models, $MC(\mathcal{S}_2, \mathcal{S}_1)$:
$TERM(\Sigma_2, \Sigma_1) \wedge$
$\forall \Delta'.EQ(\Sigma_2, \Sigma_1)[\equiv' / \equiv] \wedge R(\phi_2)[\equiv' / =][p'/p] \wedge LE(\Delta', \Delta) \Rightarrow LE(\Delta, \Delta')$
where:

- Δ is $(\equiv_s)_{s \in S_2}, (p)_{p:\mathbf{s} \in P_2}$, and Δ' is a fresh copy of it.
- $LE(\Delta, \Delta')$ is $\bigwedge_{s \in S_2} \forall x, y.s(x) \wedge s(y) \wedge x \equiv_s y \Rightarrow x \equiv'_s y \wedge \bigwedge_{p:\mathbf{s} \in P_2} \forall \mathbf{x}, \mathbf{s}(\mathbf{x}) \wedge p(\mathbf{x}) \Rightarrow p'(\mathbf{x})$

Example 4. Minimal models, however, are not very satisfactory as soon as existential quantifiers are used. Let our specification be $\exists x, y : s.x \neq y$. Intuitively we would expect that an algebra with two elements would be an adequate model. However, as there are several morphisms from such a model to itself, this model is excluded: the specification has no minimal model. So the motto of minimal models could be: "no junk, minimal confusion for universal sentences, inconsistency elsewhere".

4.5 Quasi-free models

An older proposal, the *quasi-free* models of [10], has a better behaviour on existential sentences. Let $A \leq_{QM} B$ iff there is a unique homomorphism[3] from A to B. A model is *quasi-initial* if it is minimal for $\leq_{QM}$.

This definition is easily generalised to take into account predicates, and a fixed base set as for free models: An algebra A is said to be *quasi-free* with respect to $\mathcal{S}_2$ on $\mathcal{S}_1$ ($\mathcal{S}_1 \subseteq \mathcal{S}_2$) iff for all models A_2 of $\mathcal{S}_2$ such that $A_2|_{\Sigma_1} = A|_{\Sigma_1}$, if there is a unique homomorphism $h : A_2 \to A|_{\Sigma_2}$ where $h|_{\Sigma_1} = id$, then there is a unique homomorphism $h_2 : A|_{\Sigma_2} \to A_2$ satisfying $h_2|_{\Sigma_1} = id$. This time we have found no second-order axiom for quasi-free models of a first-order logic sentence, since we have no bound on the cardinality of quasi-free models.

For the universal sub-language, the same facts than for minimal models can be established[19], and these facts allow the use of exactly the same rule.

[3] Note that this relation is not an order, since it is not reflexive, nor antisymmetric, nor transitive.

Example 5. Retrying our previous example, we see now that *all* models of $\exists x, y : s.x \neq y$ are quasi-free models. Although this is better than leading to inconsistency, this is not the result that is intuitively expected. So the motto of quasi-free models could be: "no junk, minimal confusion for universal sentences, looseness elsewhere".

4.6 Equality circumscription

Comparing the two proposals above leads to a third definition [14], where models are minimal for the order "there is an homomorphism..".

Example 6. With our previous example, we see now that *all* models of $\exists x, y : s.x \neq y$ are preferred models, instead of those with just two elements.

4.7 Surjective models

This proposal is based on a different notion of model, that includes an internal naming scheme to distinguish the elements of the carriers.

A *model* A of a presentation $(\Sigma, \mathcal{A})$ is a pair (X_A, Alg_A) where X_A are the *generators* (a family of sets, one for each sort). The *internal signature* Σ_A is $(S, G_O \uplus X_A, P)$. Alg_A is an algebra of Σ and of Σ_A that satisfies the axioms $\mathcal{A}$ and is *surjective*, i.e. each element of a carrier s_A is the denotation of some ground term of $T^s_{\Sigma_A}$; equivalently, $e : T_{\Sigma_A} \to Alg_A$ is a surjective homomorphism, hence the name. Below, we will often note Alg_A simply by A to alleviate notation.

To increase flexibility, the user of our scheme can also declare a *policy* stating which sorts and predicates should be minimized. We partition the predicates of Σ into (M_P, Z_P, G_P): the predicates in M are minimized, the ones in Z are varied, and the ones in G are fixed. Sorts are similarly partitioned into (M_S, G_S). Operators are partitioned into (Z_O, G_O). (Minimized operators and varied sorts make no sense.) The variable Ω will range over policies. To obtain free models, all new sorts and predicates should be minimized, while old sorts and predicates are fixed. The policy where each symbol if fixed is noted $Fix(\Sigma)$.

We (pre)order models as follows: A is preferred to B, noted $A \leq_{SC(\Omega)} B$, if there is a family of relations $\sim_s$ between the carriers of A and B, such that:

- For any fixed sort $s \in G_S, X^s_A = X^s_B$;
- For any minimized sort $s \in M_S, X^s_A \subseteq X^s_B$;
- *compatibility with generators*: for any $x \in X^s_A, x_A \sim_s x_B$;
- *compatibility with fixed operators*: for any $f : s_1 \ldots s_n \to s \in G_O$: $\forall i, a_i \sim_{s_i} b_i \Rightarrow f_A(\mathbf{a}) \sim_s f_B(\mathbf{b})$;
- For any minimized predicate $p \in M_P$: if $p_A(\mathbf{a})$ then for all $\mathbf{b}$ such that $\forall i; a_i \sim b_i$, $p_B(\mathbf{b})$ holds.
- For any fixed predicate $g \in G_P$: whenever $\forall i; a_i \sim b_i$, $g_A(\mathbf{a})$ iff $g_B(\mathbf{b})$.

$\sim$ is then called an Ω-correspondence. It is total, and surjective when sorts are fixed.

The *surjective models* $SC(\phi; \Omega)$ are the minimal models of ϕ for the preorder above. $\phi \models_{SC(\Omega)} \psi$ and $SC(\phi;\Omega) \models \psi$ mean that all these minimal models satisfy ψ. (It is thus a preferential logic.)

To encode this semantic definition in second order logic, we use a tuple of symbols M that groups the minimized and varied symbols ($V_O \cup M_P \cup V_P \cup S \cup \{X_s | s \in M_S\}$). The fixed symbols are not included to ensure that their interpretation is identical in the models represented by M and M'.

We encode the preference order as $M \leq M'$, abbreviating
$\bigwedge_{p \in M_P} \forall \mathbf{x}.s(x) \wedge p(x) \Rightarrow p'(x)$
$\wedge \bigwedge_{s \in M_s} \forall x.X_s(x) \Rightarrow X'_s(x)$
As usual, $M < M'$ means $M \leq M' \wedge \neg M' \leq M$.

We can now build a second-order translation for surjective models [20]:
$\exists X.TERM(\Sigma, X) \wedge R(\phi) \wedge \neg \exists M'; TERM(\Sigma, X)' \wedge R(\phi)' \wedge (M' < M)$.

5 Conclusion

For most languages and constraints studied here, no effective proof system can exist [11]. In [16], we have shown how to finitely translate most known constraints to second-order logic. Here, we extend this work to deal with less known extensions of initial models, namely minimal, quasi-free and surjective models.

This translation allows one to prove the equivalence of two specifications, even written in different specification languages, a feature which is out of reach of infinite proof systems.

A Second-order axioms

In all axioms, we assume $\Sigma = (S, O, P)$.

$OPS(O)$	$\bigwedge_{f:\mathbf{s}\rightarrow s \in O} \forall \mathbf{x}.\mathbf{s}(\mathbf{x}) \Rightarrow s(f(\mathbf{x}))$
$FG((T,F),S')$	$\forall (I_s)_{s\in S}.[\bigwedge_{f:\mathbf{s}\rightarrow r \in F} \forall \mathbf{x}, \mathbf{s}(\mathbf{x}) \wedge I_{\mathbf{s}}(\mathbf{x}) \Rightarrow I_r(f(\mathbf{x})) \wedge \bigwedge_{s \in S} \forall x.s'(x) \Rightarrow I_s(x)] \Rightarrow [\bigwedge_{s\in T} \forall x.s(x) \Rightarrow I_s(x)]$
$HOM(\mathbf{h}, \Sigma, \Sigma')$	$HS((\mathbf{h}, S, S') \wedge HF(\mathbf{h}, \Sigma, \Sigma') \wedge HP(\mathbf{h}, \Sigma, \Sigma')$
$HS((\mathbf{h}, S, S')$	$\bigwedge_{s\in S} \forall x.s(x) \Rightarrow s'(h_s(x))$
$HF(\mathbf{h}, \Sigma, \Sigma')$	$\bigwedge_{f:\mathbf{r}\rightarrow s \in F} \forall \mathbf{x}.(\bigwedge_i s_i(x_i)) \Rightarrow h_r(f(\mathbf{x})) = f'(h_{\mathbf{s}}(\mathbf{x}))$
$HP(\mathbf{h}, \Sigma, \Sigma')$	$\bigwedge_{p:\mathbf{s}\in P} \forall \mathbf{x}.(\bigwedge_i s_i(x_i)) \wedge p(\mathbf{x}) \Rightarrow p'(h_{\mathbf{s}}(\mathbf{x}))$
$ISO(\Sigma, \Sigma')$	$\exists \mathbf{h}, \mathbf{g}.HOM(\mathbf{h}, \Sigma, \Sigma') \wedge HOM(\mathbf{g}, \Sigma', \Sigma) \wedge \bigwedge_{s\in S} \forall x.s(x) \Rightarrow g_s(h_s(x)) = x \wedge \forall x.s'(x) \Rightarrow h_s(g_s(x)) = x$
DEN	$\exists 0, <, succ. \forall x, y, z : 0 < succ(x) \wedge x < succ(x) \wedge \neg(x < y \wedge y < x) \wedge (x < y \wedge y < z \Rightarrow x < z)$
$FREE(S_2, S_1)$	$DEN \wedge F^*(S_2, S_1)$
$EQ(\Sigma)$	$RE(S) \wedge SY(S) \wedge TR(S) \wedge CF(\Sigma) \wedge CP(\Sigma)$
$RE(S)$	$\bigwedge_{s\in S} \forall x.s(x) \Rightarrow x \equiv_s x$
$SY(S)$	$\bigwedge_{s\in S} \forall x, y.s(x) \wedge s(y) \Rightarrow (x \equiv y \Rightarrow y \equiv x)$
$TR(S)$	$bwe_{s\in S} \forall x, y, z.s(x) \wedge s(y) \wedge s(z) \Rightarrow (x \equiv y \wedge y \equiv z \Rightarrow x \equiv z)$

$CF(\Sigma)$	$\bigwedge_{f:\mathbf{s}\to r\in O}\bigwedge_{s_i\in S}\forall \mathbf{x},x'_i.\mathbf{s}(\mathbf{x})\wedge s_i(x'_i)\wedge x_i\equiv_{s_i}x'_i\Rightarrow f(\mathbf{x})\equiv_r f(x_1,\ldots,x'_i,\ldots,x_n)$
$CP(\Sigma)$	$\bigwedge_{p:\mathbf{s}\in P}\bigwedge_{s_i\in S}\forall \mathbf{x},x'_i.\mathbf{s}(\mathbf{x})\wedge s_i(x'_i)\wedge x_i\equiv_{s_i}x'_i\Rightarrow [p(\mathbf{x})\Leftrightarrow p(x_1,\ldots,x'_i,\ldots,x_n)]$
$RS(\Sigma'_1)$	$\bigwedge_{f':\mathbf{s}'\to r'\in O'_1}\forall \mathbf{a},b.\mathbf{s}'(a)\wedge f'(a)=b\Rightarrow f(a)\equiv_s b\wedge \bigwedge_{p':\mathbf{s}'\in P'_1}\forall \mathbf{a},b.\mathbf{s}'(a)\wedge p'(a)\Rightarrow p(a)$
$LC(\Sigma_2)$	$\forall\Delta'.EQ(\Sigma_2)[\equiv'/\equiv]\wedge R(\mathcal{S}_2)[\equiv/=][\Delta'/\Delta]\Rightarrow\Delta'\leq\Delta$
Δ	$(\equiv_s)_{s\in S_2},(p)_{p:\mathbf{S}\in P_2}$
$\Delta\leq\Delta'$	$[\bigwedge_{s\in S_2}\forall x,y.s(x)\wedge s(y)\wedge x\equiv_s y\Rightarrow x\equiv'_s y]\wedge[\bigwedge_{p:\mathbf{S}\in P_2}\forall\mathbf{x},\mathbf{s}(\mathbf{x})\wedge p(\mathbf{x})\Rightarrow p'(\mathbf{x})]$
$TERM(\Sigma,S'_1)$	$OPS(O)\wedge INC(S,S'_1)\wedge INJ(O)\wedge NOV(O,S'_1)\wedge FG(\Sigma,S')$
$INC(S,S'_1)$	$\bigwedge_{s\in S}\forall x.s'(x)\Rightarrow s(x)$
$DIS(O)$	$\bigwedge_{f\neq g:\mathbf{s}\to r\in O}\forall\mathbf{x},\mathbf{y}.\mathbf{s}(\mathbf{x})\wedge\mathbf{s}'(\mathbf{y})\Rightarrow\neg(f(\mathbf{x})=g(\mathbf{y}))$
$INJ(O)$	$\bigwedge_{f\in O}\forall\mathbf{x},\mathbf{y}.\mathbf{s}(\mathbf{x})\wedge\mathbf{s}(\mathbf{y})\wedge f(\mathbf{x})=f(\mathbf{y})\Rightarrow\bigwedge_{i\leq n}x_i=y_i$
$NOV(O,S'_1)$	$\bigwedge_{f:\mathbf{s}\to r\in O}\forall\mathbf{x},\mathbf{s}(\mathbf{x})\Rightarrow\neg r'(f(\mathbf{x}))$
$LCS(\mathcal{S}_1,\mathcal{S}_2)$	$TERM(\Sigma_2,S'_1)\wedge RS(\Sigma'_1)\wedge EQ(\Sigma_2)\wedge R(\mathcal{S}_2)[\equiv/=]\wedge LC(\Sigma_2)$
$PER(\Sigma_2,\Sigma_1)$	$\exists(i_s)_{s\in S_1}.HOM(i,\Sigma_1,\Sigma'_1)\wedge\bigwedge_{s\in S_1}[\forall x,y.s(x)\wedge s(y)\Rightarrow[i_s(x)=i_s(y)\Leftrightarrow x\equiv_s y]\wedge\forall y.s(y)\Rightarrow\exists x.s(x)\wedge i_s(x)=y]$

B Theorems

Definition 2. $\mathcal{I}$, a second-order model of $OPS(\Sigma)$, is a *relativization* A, a Σ-algebra, (noted $R(A,\mathcal{I})$) iff:

- $s_A=s_{\mathcal{I}}$;
- the multi-sorted functions and predicates are a restriction of their counterparts: $f_A=f_{\mathcal{I}}|s_{1\mathcal{I}}\times\ldots\times s_{n\mathcal{I}}$;
- predicates also: $p_A(x)=p_{\mathcal{I}}(x)\wedge s_{1\mathcal{I}}(x)\wedge\ldots\wedge s_{n\mathcal{I}}(x)$.

This relation is total, because

- the domain can be defined as the union of carriers, plus a given element;
- the functions can be extended by setting their value to e.g. the given element.
- the predicates can be extended e.g. by setting them to false.

This definition gives a single A for any $\mathcal{I}$, called its *inverse relativization* $Ms(\mathcal{I})$,

Lemma 3. *[8] For any multi-sorted sentences ψ,ϕ: $\phi\models_{MS}\psi$ iff $OPS(\Sigma)\cup R(\phi)\models_2 R(\psi)$.*

Proof. By induction over formulae, we show that whenever $R(A,\mathcal{I})$, $A\models_V\phi$ iff $\mathcal{I}\models_V R(\phi)$ for any multi-sorted valuation V.

- $\phi=p(\mathbf{t})$: obvious, both sides are identical.
- $\phi=(t_1=t_2)$: obvious, both sides are identical.

- $\phi = \psi_1 \wedge \psi_2$: by inductive hypothesis.
- $\phi = \neg\psi$: by inductive hypothesis.
- $\phi = \forall x : s.\psi$: we need only to consider all second-order $V'_2 \approx_x V$ that satisfies $s_A(V(x))$; this is exactly the set of all many-sorted $V' \approx_x V$.

Definition 4. Let $\mathcal{S}$ be a specification (denoting a class of algebras), $R(\mathcal{S})$, a second-order sentence, is a *relativization* of $\mathcal{S}$ iff whenever $R(A,\mathcal{I})$, $A \models \mathcal{S}$ iff $\mathcal{I} \models_2 R(\mathcal{S})$, and $R(\mathcal{S}) \models_2 OPS(\Sigma)$.

Definition 5. In an algebra A, a sort s is said to be *reachable* by F and S', (where S' is a set of subsorts indexed by S; s' is the element of index s) if any element a of the carrier A_s, is the value of a term of $T_F(s'_A)_{s\in S}$. A is *reachable* on $T \subseteq S$ by F and S' if all sorts $s \in T$ are reachable.

Lemma 6. *Whenever $R(A,\mathcal{I})$, A is reachable on T with F and S' iff $\mathcal{I} \models_2 R(\mathcal{S}) \wedge FG((T,F),S')$.*

Proof. Given such $A,\mathcal{I}$, we prove that the set of reachable values is the least set satisfying the antecedent of FG, namely:

$$\forall (I_s)_{s\in S}.[\bigwedge_{f:\mathbf{s}\to r\in F} \forall \mathbf{x}, \mathbf{s}(\mathbf{x}) \wedge I_{\mathbf{s}}(\mathbf{x}) \Rightarrow I_r(f(\mathbf{x})) \wedge (\bigwedge_{s\in S} \forall x.s'(x) \Rightarrow I_s(x))]$$

So let $I_s(a) = (\exists V, t.V_{\mathcal{I}}(t) = a)$, where V ranges over multi-sorted valuations with values in s' for $x : s$, and t ranges over generating terms.

- This I_s satisfies the antecedent, namely:
  - $\bigwedge_{f:\mathbf{s}\to r\in F} \forall \mathbf{x}.(\mathbf{s}(\mathbf{x}) \wedge \bigwedge_i \exists V_i, t_i.V_{i\mathcal{I}}(t_i) = x_i) \Rightarrow I_s(\Rightarrow \exists V, t.I_r(f(\mathbf{x}))$. Indeed, since we have an infinite number of variables we can rename the variables of each t_i, giving t'_i, so that their variables are disjoint. Then V is constructed by taking any extension of $\bigcup_i V_i|_{vars(t'_i)}$, and $t = f(\mathbf{t}')$.
  - $s'(a) \Rightarrow \exists V, t.V_{\mathcal{I}}(a) = x$. Take $t = x : s$, $V(x) = x$.
- This I_s is the least such set. Let I'_s be a smaller set, so that there is a reachable $a : a \notin I'_s, s \in T$. Let t be the smallest term (for some total extension of the subterm ordering) such that $V(t) \notin I'_s$.
  - If $t = x \in X_s$: then $s'(x) \Rightarrow I_s(x)$ is violated.
  - If $t = f(\mathbf{t})$ then either OPS is violated, or $\mathbf{t} \notin \mathbf{s}$, contradicting the inductive hypothesis (t smallest).

Now if A is reachable on T, the set of reachable elements is the sort, so that we see that $FG((F,T),S')$ holds. Conversely, by taking I_s as above, we see that the algebra is reachable on T.

Lemma 7. *When $R(A,\mathcal{I})$, $\mathcal{I} \models_2 TERM(\Sigma, X)$ iff A is isomorphic to the algebra $T_\Sigma(\biguplus_{x\in X} x_{\mathcal{I}})$.*

Proof. $\Rightarrow$: We abbreviate $T_\Sigma(\biguplus_{x\in X} x_{\mathcal{I}})$ as T_Σ in this proof. We define the isomorphism $i : T_\Sigma \to A$ by recursion:

- $i_s(x) = x$;
- $i_r(f(\mathbf{t})) = f_{\mathcal{I}}(i_{\mathbf{s}}(\mathbf{t}))$, for each $f : \mathbf{s} \to r \in O$.

By OPS, $i_s(t) \in s_{\mathcal{I}} = s_A$. Let us prove that:

1. each i_s is injective, by induction on terms. If $i_s(t_1) = i_s(t_2)$, then, either:
   - $t_1 \in X_s, t_2 \in X_s$; then $t_1 = t_2$ by definition of i;
   - $t_1 \in X_s, t_2 = f(\mathbf{t})$; then $f(i_{\mathbf{s}}(\mathbf{t})) \in X_s$, which is impossible by NOV.
   - $t_2 \in X_s, t_1 = f(\mathbf{t})$; symmetrically.
   - $t_1 = f(\mathbf{t}), t_2 = g(\mathbf{t}')$; then either $f \neq g$, which contradicts DIS, or $f_{\mathcal{I}}(i_{\mathbf{s}}(\mathbf{t})) = f_{\mathcal{I}}(i_{\mathbf{s}}(\mathbf{t}'))$, so that $i_{\mathbf{s}}(\mathbf{t})) = i_{\mathbf{s}}(\mathbf{t}'))$ by SUR, which gives $\mathbf{t} = \mathbf{t}'$ by the inductive hypothesis, and thus $t_1 = f(\mathbf{t}) = f(\mathbf{t}') = t_2$.
2. each i_s is surjective. Let $I_s = \{i_s(t) | t \in T_{\Sigma}^s\}$. We easily check that the antecedent of FG is valid, and thus each $x \in s_{\mathcal{I}} = s_A = i(t)$ for some $t \in T_{\Sigma}^s\}$.
3. i is a homomorphism:
   - $T_\Sigma \to A$: by definition.
   - $A \to T_\Sigma$: Let $\mathbf{a} \in \mathbf{s}_A$. By bijectivity there is a unique $\mathbf{t} | i_{\mathbf{s}}(\mathbf{t}) = \mathbf{a}$, and a unique $t | i_r(t) = f_{\mathcal{I}}(\mathbf{a})$. Thus $f(\mathbf{t}) = t$.

   Note that I need *not* be isomorphic to the interpretation T_Σ.

$\Leftarrow$: – OPS: From $R(A, \mathcal{I})$.
- NOV: As i is injective, $f_{\mathcal{I}}(\mathbf{a}) = i_r(f(\mathbf{t})) = i_r(x) = a \in s'_A$ implies $f(\mathbf{t}) = x$, which contradicts the disjointness of O and X.
- SUR: Similarly.
- DIS: Similarly.
- FG: by lemma 6.

Lemma 8. *A model I is free with respect to $\mathcal{S}_2$ on Σ_1 iff it is free among the models of cardinality $max(\omega, card(I|_{\Sigma_1}))$, where $card(A) = max_{s \in S}(card(s_A))$.*

Proof. Let α be this cardinal.

$\Rightarrow$: Let I be free; so for all models A_2 of $\mathcal{S}_2$, and homomorphisms $h_1 : I|_{\Sigma_1} \to A_2|_{\Sigma_1}$ there is a unique homomorphism $h_2 : I|_{\Sigma_2} \to A_2$ satisfying $h_2|_{\Sigma_1} = h_1$. We skolemize the formula ϕ_2 to obtain a universal formula $Sk(\phi_2)$ for a larger signature where the function symbols F have been added. We retain only the part of I reachable on S_2 by $\Sigma \cup F$ and $I|_{\Sigma_1}$, that we call $Sk(I)$. It is a model of ϕ_2 of cardinality less than α. There is an homomorphism $i : Sk(I) \to I$, which is an inclusion, and $i|_{\Sigma_1}$ is an identity. Since I is free, there is a unique homomorphism $j : I|_{\Sigma_2} \to Sk(I)|_{\Sigma_2}$. The composition of $i|_{\Sigma_2} \circ j$ must be the unique identity from $I|_{\Sigma_2}$ to itself, so that $i|_{\Sigma_2}$ must be the identity, and $I|_{\Sigma_2} = Sk(I)|_{\Sigma_2}$ has a cardinality less than α.

$\Leftarrow$: Let I be free among models of cardinality lower than α, and M be any model of $\mathcal{S}_2$, h_1 be any homomorphism from $I|_{\Sigma_1}$ to $M|_{\Sigma_1}$. Thus, there is $j : I|_{\Sigma_2} \to Sk(M)$ extending h_1 and $i : Sk(M) \to M$, an inclusion obtained by the Skolemization. Thus $h = i \circ j : I \to M$, which extends h_1 because $i|_{\Sigma_1}$ is the identity. It remains to prove that it is unique. Let $h' \neq h$ extending h_1. The submodel of M reachable by $h(I|_{\Sigma_1}) = h'(I|_{\Sigma_1}) = h_1(I|_{\Sigma_1})$ and $\Sigma_2 \cup F$

is a model of ϕ_2 of cardinality less than α, and has two homomorphisms from $I|_{\Sigma_2}$ to it, contradicting our initial assumption.

Lemma 9. *If $\mathcal{I}$ is a denumerable relativization of A, then $A \models \mathcal{S}$* **free wrt** *$\mathcal{S}_2$* **on** *S_1 iff $\mathcal{I} \models_2 FREE(\mathcal{S}_2, \Sigma_1)$.*

Proof. If A is free, then we immediately check that $F*$ holds in any of its relativizations, and DEN holds by hypothesis. if A is not free, then there is a denumerable A' that shows it. We can then represent A' in the domain of $\mathcal{I}$, so that $F*$ is not satisfied.

Corollary 10. $\mathcal{S}$ **free wrt** $\mathcal{S}_2$ **on** $\mathcal{S}_1 \models \psi$ *iff* $FREE(\mathcal{S}_2, \mathcal{S}_1) \models R(\psi)$.

Lemma 11. *If $S_2 = (\Sigma_2, \phi_2)$, where ϕ_2 is a universal sentence, and $\mathcal{I}$ is a relativization of A, then $A \models \mathcal{S}$* **free wrt** *$\mathcal{S}_2$* **on** *S_1 iff $\mathcal{I} \models_2 R(\mathcal{S}) \wedge LCS(\mathcal{S}_2, \mathcal{S}_1)$*

Proof. We use the construction on the free algebra in [6], and lemma 7.

Corollary 12. *If ϕ_2 is a universal sentence, $\mathcal{S}$ free wrt $\mathcal{S}_2$ on $\mathcal{S}_1 \models \psi$ iff $R(\mathcal{S}) \wedge LCS(\mathcal{S}_2, \mathcal{S}_1) \models_2 R(\psi)[\equiv / =]$.*

Theorem 13. *Any strongly persistent free model is a minimal model.*

Proof. Let A be a free model, i.e., for all models A_2 of $\mathcal{S}_2$, and homomorphisms $h_1 : A|_{\Sigma_1} \rightarrow A_2|_{\Sigma_1}$ there is a unique homomorphism $h_2 : A|_{\Sigma_2} \rightarrow A_2$ satisfying $h_2|_{\Sigma_1} = h_1$. By taking h_1 to be the identity, we verify that it is a minimal model.

Lemma 14. *When there are free models, and persistency holds, any minimal model is a free model.*

Proof. Let M be a minimal model. Let F be the free extension of $M|_{\Sigma_1}$. By persistency, $F|_{\Sigma_1}$ is isomorphic to $M|_{\Sigma_1}$. Let F' be isomorphic to F, such that $F|_{\Sigma_1} = M|_{\Sigma_1}$. By freeness, $\exists! h : F' \rightarrow M$ extending i, so that $\exists! h_2 : M \rightarrow F'$ extending i. Therefore F', M are isomorphic, so that M is free.

Theorem 15. *For any specification $\mathcal{S}_2 = (\Sigma_2, \phi_2)$, where ϕ_2 is a first-order sentence, a model of $\mathcal{S}_2$ is minimal wrt $\mathcal{S}_1$ iff it is minimal among models of cardinality $\alpha = max(\omega, max_{s \in S_1}(card(s_A)))$.*

Proof. (similar to lemma 8)

$\Rightarrow$: Let A be minimal, let F be finitely many functions symbols introduced by skolemizing ϕ_2, Let G be the submodel of A reachable by $\Sigma_2 \cup F$ and $A|_{\Sigma_1}$. There is a single inclusion homomorphism from G to A, which is the identity on Σ_1. As A is minimal, there is a single $h : A \rightarrow G$. Thus A is isomorphic to G, which is of cardinality less than α.

$\Leftarrow$: Let A be minimal among such models, and M be a model of $\mathcal{S}_2$ such that $i : A|_{\Sigma_1} \approx M|_{\Sigma_1}$. Let G, the reachable submodel of M, is also a model, and $f : G \to M$ is an inclusion morphism. If there is a morphism $h : M \to A$ extending i, then $h \circ f$ is a morphism extending i, so that there is a unique $h_2 : A \to G$ extending i. So $f \circ h_2 : A \to M$. Let $h' : A \to M$ be a different morphism extending i. The submodel of M reachable by $h(I|_{\Sigma_1}) = h'(I|_{\Sigma_1})$ and $\Sigma_2 \cup F$ is a model of ϕ_2 of cardinality less than α, and thus should receive a single morphism.

Theorem 16. *Let $\mathcal{S}_2 = (\Sigma_2, \phi_2)$, where ϕ_2 is a universal sentence:*

1. *If $\mathcal{S}_2$ is satisfiable, it has a minimal model;*
2. *a model is minimal iff it is minimal among reachable (by O_2 and S_1 on S_2) models;*
3. *for any A model of $\mathcal{S}_2$, there is a minimal model M of $\mathcal{S}_2$ such that is a unique homomorphism from M to A;*
4. *A is a minimal model iff A is isomorphic to the quotient of a term algebra by a minimal congruence;*
5. $\mathcal{S}$, **min** $\mathcal{S}_2$ **wrt** $\mathcal{S}_1 \models_{MIN} \psi$ *iff* $R(\mathcal{S})[\equiv / =] \wedge \exists \Sigma_1'.MC(\mathcal{S}_2, \mathcal{S}_1) \models_2 R(\psi)[\equiv / =]$

Proof. 1. corollary of 3;

2. (similar to lemma 8)

   $\Rightarrow$: A minimal model is reachable: Assume A is minimal. Let G be its reachable submodel. There is a single inclusion homomorphism from G to A, which is the identity on Σ_1. As A is minimal, there is a single $h : A \to G$. Thus A is isomorphic to G.

   $\Leftarrow$: Let A be minimal among reachable models, and M be a model of $\mathcal{S}_2$ such that $i : A|_{\Sigma_1} \approx M|_{\Sigma_1}$. G, the reachable submodel of M, is also a model, and $f : G \to M$ is an inclusion morphism. If there is a morphism $h : M \to A$ extending i, then $h \circ f$ is a morphism, so that there is a unique $h_2 : A \to G$. So $f \circ h_2 : A \to M$. As A is reachable, this homomorphism is unique.

3. Among reachable models, we have $\exists h : A \to B$ iff $\exists! h : A \to B$. Thus this condition is a preorder. , and the associated equivalence is isomorphism. By Zorn's lemma, it suffices to prove that every strict chain of reachable models of $\mathcal{S}_2$ is minored. Let $C(A)$ be $\{p(\mathbf{t}) | p \in \Sigma_2, t \in T_{\Sigma_2}(A|_{\Sigma_1}), A \models_{id} p(t)\} \cup \{t_1 = t_2 | t_1, t_2 \in T_{\Sigma_2}(A|_{\Sigma_1}), A \models t_1 = t_2\}$ If A is reachable, given $i : A|_{\Sigma_1} \to A_2|_{\Sigma_1}$, there is a single extension $h : A|_{\Sigma_2} \to A_2$ defined inductively. This extension is a homomorphism iff $C(A) \subseteq C(A_2)$; we note this condition by $A \leq A_2$. Let $M_0 > M_1 > \ldots$ be a chain, where $A > B$ means that $C(A) \supset C(B)$. Let $H(A) = C(A) \cup \phi_2$. Consider the limit $H = \bigcup H(M_i)$. Let $F = \phi_1, \ldots, \phi_n$ be a finite subset of H. Each ϕ_i comes from some $H(M_{j(i)})$. Let j be the maximum of $\{j(i) | i \leq n\}$. By monotonicity of the chain, we know that $F \subseteq H(M_j)$, and thus that F is satisfiable. By compactness of first-order logic H is satisfiable, and has thus a model N. As H is universal,

every submodel of N satisfies H, and specially the reachable submodel N'. As $H(M_i) \subseteq H \subseteq H(N')$, there is an homomorphism from N' to each M_i. Let A be a model of ϕ_2; let M_0 be the reachable part of A, which is also a model of ϕ_2. We compose the homomorphism from $N' \to M_0$, with the inclusion homomorphism from M_0 to A. An homomorphism from a reachable algebra is always unique, so that N' is minimal.

4. (a) A reachable model is isomorphic to the quotient of a term algebra by the congruence defined by $t_1 \equiv_A t_2$ iff $e_A(t_1) = e_B(t_2)$.
   (b) There is an homomorphism $A \to B$, two reachable algebras, iff $\equiv_B \subseteq \equiv_A$ and $p_B \subseteq p_A$. This homomorphism is unique.
5. From 4 and lemma 7, similarly to 12.

Lemma 17. *The free models are quasi-free.*

Proof. By definition.

Lemma 18. *If a specification has free models, then the quasi-free models are the free models.*

Proof. Let F be a free model, and M be a minimal model. $\exists! h : F \to M$ extending id, so that $\exists! h_2 : M \to F$ extending id. Therefore F, M are isomorphic.

Theorem 19. *If ϕ_2 is a universal sentence, then quasi-free and minimal models are the same, and thus theorem 16 applies.*

Proof. Let A be a quasi-free model, and G be its reachable submodel. As $\exists! h_1 : G \to A$, $\exists! h_2 : A \to G$, so that A is isomorphic to G. So quasi-initial models are reachable here. Assume that $\exists h_3 : B \to G$. Let H be the reachable submodel of B. Then $\exists! h_4 : H \to G$, the restriction of h_3, and thus $\exists! h_5 : G \to H$. G and H are thus isomorphic.

Theorem 20. $SC(\phi; \Omega) \models \psi$ *iff*
$\exists X.TERM(\Sigma, X) \wedge R(\phi) \wedge \neg \exists M'.TERM(\Sigma, X)' \wedge R(\phi)' \wedge (M' <_\Omega M) \models R(\psi)$.

References

1. M. Bidoit and G. Bernot. Proving correctness of algebraically specified software: Modularity and observability issues. In M. Nivat, C. Rattray, T. Rus, and G. Scollo, editors, *AMAST'91*, pages 139–161. Springer-Verlag, 1992.
2. M. Broy and al. The requirement and design specification language spectrum: an introduction. Technical Report TUM-I9140, Technische Universität München, 1991.
3. R. Burstall and J. Goguen. Semantics of CLEAR, a Specification Language. In D. Bjorner, editor, *Abstract software specifications, Proc. 1979 Copenhagen Winter School*, volume 86, pages 292–332. Springer, 1980.
4. CIP Language Group. *The Munich Project CIP – Vol. I: The Language*, volume 183 of *Lecture Notes in Computer Science*. Springer, 1985.

5. N. Denyer. Pure second-order logic. *Notre-Dame Journal of Formal Logic*, 33(2):220, 1992.
6. H. Ehrig and B. Mahr. *Fundamentals of algebraic specification : Volume 1. Equations and initial semantics.* Springer Verlag, 1985.
7. H. Ehrig and B. Mahr. *Fundamentals of Algebraic Specification 2: Module Specifications and Constraints*, volume 21 of *EATCS Monographs on Theoretical Computer Science.* Springer-Verlag, 1990.
8. A. Schmidt, Die Zulässigkeit der Behandlung mehrsortiger Theorien mittels der üblichen einsortigen Prädikatenlogik, *Math. Ann.* 123, pages 187-200, 1951.
9. J. Goguen and R. Burstall. Institutions: Abstract model theory for specification and programming. *J. ACM*, 39(1):95–146, Jan. 1992.
10. S. Kaplan. Positive/negative conditional rewriting. In *Conditional Term Rewriting*, volume 308 of *Lecture Notes in Computer Science.* Springer, 1988.
11. D. MacQueen and D. Sannella. Completeness of proof systems for equational specifications. *IEEE TSE*, SE-11(5), May 1985.
12. P. Nivela and F. Orejas. Initial behaviour semantics for algebraic specifications. In *Recent Trends in Data Type Specification*, number 332 in Lecture Notes in Computer Science, pages 184–207. Springer-Verlag, 1987.
13. F. Nourani. On induction for programming logics: syntax, semantics, and inductive closure. *EATCS Bulletin*, 13:51–64, 1981.
14. P. Rathmann and M. Winslett. Circumscribing equality. In *Proc. of the 8th Nat. Conf. on Art. Int. (AAAI-89)*, pages 468–473, 1989.
15. P.-Y. Schobbens. Surjective Circumscription Proc. Dutch/German Workshop on Non-Monotonic Logics, W. Nejdl ed., Aachen, Dec. 1993.
16. P.-Y. Schobbens. Second-Order Proof Systems for Algebraic Specification Languages Proc. 9th Workshop on Specification of Abstract Data Types, F. Orejas ed., to appear.
17. P.-Y. Schobbens. Exceptions for software specification: on the meaning of "but". Technical report, Univ. Cath. de Louvain, 1989.
18. M. Wirsing. Structured algebraic specifications: A kernel language. *Theoretical Computer Science*, 42:123–249, 1986.

Springer-Verlag and the Environment

We at Springer-Verlag firmly believe that an international science publisher has a special obligation to the environment, and our corporate policies consistently reflect this conviction.

We also expect our business partners – paper mills, printers, packaging manufacturers, etc. – to commit themselves to using environmentally friendly materials and production processes.

The paper in this book is made from low- or no-chlorine pulp and is acid free, in conformance with international standards for paper permanency.

Printing: Weihert-Druck GmbH, Darmstadt
Binding: Theo Gansert Buchbinderei GmbH, Weinheim

Lecture Notes in Computer Science

For information about Vols. 1–745
please contact your bookseller or Springer-Verlag

Vol. 746: A. S. Tanguiane, Artificial Perception and Music Recognition. XV, 210 pages. 1993. (Subseries LNAI).

Vol. 747: M. Clarke, R. Kruse, S. Moral (Eds.), Symbolic and Quantitative Approaches to Reasoning and Uncertainty. Proceedings, 1993. X, 390 pages. 1993.

Vol. 748: R. H. Halstead Jr., T. Ito (Eds.), Parallel Symbolic Computing: Languages, Systems, and Applications. Proceedings, 1992. X, 419 pages. 1993.

Vol. 749: P. A. Fritzson (Ed.), Automated and Algorithmic Debugging. Proceedings, 1993. VIII, 369 pages. 1993.

Vol. 750: J. L. Díaz-Herrera (Ed.), Software Engineering Education. Proceedings, 1994. XII, 601 pages. 1994.

Vol. 751: B. Jähne, Spatio-Temporal Image Processing. XII, 208 pages. 1993.

Vol. 752: T. W. Finin, C. K. Nicholas, Y. Yesha (Eds.), Information and Knowledge Management. Proceedings, 1992. VII, 142 pages. 1993.

Vol. 753: L. J. Bass, J. Gornostaev, C. Unger (Eds.), Human-Computer Interaction. Proceedings, 1993. X, 388 pages. 1993.

Vol. 754: H. D. Pfeiffer, T. E. Nagle (Eds.), Conceptual Structures: Theory and Implementation. Proceedings, 1992. IX, 327 pages. 1993. (Subseries LNAI).

Vol. 755: B. Möller, H. Partsch, S. Schuman (Eds.), Formal Program Development. Proceedings. VII, 371 pages. 1993.

Vol. 756: J. Pieprzyk, B. Sadeghiyan, Design of Hashing Algorithms. XV, 194 pages. 1993.

Vol. 757: U. Banerjee, D. Gelernter, A. Nicolau, D. Padua (Eds.), Languages and Compilers for Parallel Computing. Proceedings, 1992. X, 576 pages. 1993.

Vol. 758: M. Teillaud, Towards Dynamic Randomized Algorithms in Computational Geometry. IX, 157 pages. 1993.

Vol. 759: N. R. Adam, B. K. Bhargava (Eds.), Advanced Database Systems. XV, 451 pages. 1993.

Vol. 760: S. Ceri, K. Tanaka, S. Tsur (Eds.), Deductive and Object-Oriented Databases. Proceedings, 1993. XII, 488 pages. 1993.

Vol. 761: R. K. Shyamasundar (Ed.), Foundations of Software Technology and Theoretical Computer Science. Proceedings, 1993. XIV, 456 pages. 1993.

Vol. 762: K. W. Ng, P. Raghavan, N. V. Balasubramanian, F. Y. L. Chin (Eds.), Algorithms and Computation. Proceedings, 1993. XIII, 542 pages. 1993.

Vol. 763: F. Pichler, R. Moreno Díaz (Eds.), Computer Aided Systems Theory – EUROCAST '93. Proceedings, 1993. IX, 451 pages. 1994.

Vol. 764: G. Wagner, Vivid Logic. XII, 148 pages. 1994. (Subseries LNAI).

Vol. 765: T. Helleseth (Ed.), Advances in Cryptology – EUROCRYPT '93. Proceedings, 1993. X, 467 pages. 1994.

Vol. 766: P. R. Van Loocke, The Dynamics of Concepts. XI, 340 pages. 1994. (Subseries LNAI).

Vol. 767: M. Gogolla, An Extended Entity-Relationship Model. X, 136 pages. 1994.

Vol. 768: U. Banerjee, D. Gelernter, A. Nicolau, D. Padua (Eds.), Languages and Compilers for Parallel Computing. Proceedings, 1993. XI, 655 pages. 1994.

Vol. 769: J. L. Nazareth, The Newton-Cauchy Framework. XII, 101 pages. 1994.

Vol. 770: P. Haddawy (Representing Plans Under Uncertainty. X, 129 pages. 1994. (Subseries LNAI).

Vol. 771: G. Tomas, C. W. Ueberhuber, Visualization of Scientific Parallel Programs. XI, 310 pages. 1994.

Vol. 772: B. C. Warboys (Ed.),Software Process Technology. Proceedings, 1994. IX, 275 pages. 1994.

Vol. 773: D. R. Stinson (Ed.), Advances in Cryptology – CRYPTO '93. Proceedings, 1993. X, 492 pages. 1994.

Vol. 774: M. Banâtre, P. A. Lee (Eds.), Hardware and Software Architectures for Fault Tolerance. XIII, 311 pages. 1994.

Vol. 775: P. Enjalbert, E. W. Mayr, K. W. Wagner (Eds.), STACS 94. Proceedings, 1994. XIV, 782 pages. 1994.

Vol. 776: H. J. Schneider, H. Ehrig (Eds.), Graph Transformations in Computer Science. Proceedings, 1993. VIII, 395 pages. 1994.

Vol. 777: K. von Luck, H. Marburger (Eds.), Management and Processing of Complex Data Structures. Proceedings, 1994. VII, 220 pages. 1994.

Vol. 778: M. Bonuccelli, P. Crescenzi, R. Petreschi (Eds.), Algorithms and Complexity. Proceedings, 1994. VIII, 222 pages. 1994.

Vol. 779: M. Jarke, J. Bubenko, K. Jeffery (Eds.), Advances in Database Technology — EDBT '94. Proceedings, 1994. XII, 406 pages. 1994.

Vol. 780: J. J. Joyce, C.-J. H. Seger (Eds.), Higher Order Logic Theorem Proving and Its Applications. Proceedings, 1993. X, 518 pages. 1994.

Vol. 781: G. Cohen, S. Litsyn, A. Lobstein, G. Zémor (Eds.), Algebraic Coding. Proceedings, 1993. XII, 326 pages. 1994.

Vol. 782: J. Gutknecht (Ed.), Programming Languages and System Architectures. Proceedings, 1994. X, 344 pages. 1994.

Vol. 783: C. G. Günther (Ed.), Mobile Communications. Proceedings, 1994. XVI, 564 pages. 1994.

Vol. 784: F. Bergadano, L. De Raedt (Eds.), Machine Learning: ECML-94. Proceedings, 1994. XI, 439 pages. 1994. (Subseries LNAI).

Vol. 785: H. Ehrig, F. Orejas (Eds.), Recent Trends in Data Type Specification. Proceedings, 1992. VIII, 350 pages. 1994.

Vol. 786: P. A. Fritzson (Ed.), Compiler Construction. Proceedings, 1994. XI, 451 pages. 1994.

Vol. 787: S. Tison (Ed.), Trees in Algebra and Programming – CAAP '94. Proceedings, 1994. X, 351 pages. 1994.

Vol. 788: D. Sannella (Ed.), Programming Languages and Systems – ESOP '94. Proceedings, 1994. VIII, 516 pages. 1994.

Vol. 789: M. Hagiya, J. C. Mitchell (Eds.), Theoretical Aspects of Computer Software. Proceedings, 1994. XI, 887 pages. 1994.

Vol. 790: J. van Leeuwen (Ed.), Graph-Theoretic Concepts in Computer Science. Proceedings, 1993. IX, 431 pages. 1994.

Vol. 791: R. Guerraoui, O. Nierstrasz, M. Riveill (Eds.), Object-Based Distributed Programming. Proceedings, 1993. VII, 262 pages. 1994.

Vol. 792: N. D. Jones, M. Hagiya, M. Sato (Eds.), Logic, Language and Computation. XII, 269 pages. 1994.

Vol. 793: T. A. Gulliver, N. P. Secord (Eds.), Information Theory and Applications. Proceedings, 1993. XI, 394 pages. 1994.

Vol. 794: G. Haring, G. Kotsis (Eds.), Computer Performance Evaluation. Proceedings, 1994. X, 464 pages. 1994.

Vol. 795: W. A. Hunt, Jr., FM8501: A Verified Microprocessor. XIII, 333 pages. 1994.

Vol. 796: W. Gentzsch, U. Harms (Eds.), High-Performance Computing and Networking. Proceedings, 1994, Vol. I. XXI, 453 pages. 1994.

Vol. 797: W. Gentzsch, U. Harms (Eds.), High-Performance Computing and Networking. Proceedings, 1994, Vol. II. XXII, 519 pages. 1994.

Vol. 798: R. Dyckhoff (Ed.), Extensions of Logic Programming. Proceedings, 1993. VIII, 362 pages. 1994.

Vol. 799: M. P. Singh, Multiagent Systems. XXIII, 168 pages. 1994. (Subseries LNAI).

Vol. 800: J.-O. Eklundh (Ed.), Computer Vision – ECCV '94. Proceedings 1994, Vol. I. XVIII, 603 pages. 1994.

Vol. 801: J.-O. Eklundh (Ed.), Computer Vision – ECCV '94. Proceedings 1994, Vol. II. XV, 485 pages. 1994.

Vol. 802: S. Brookes, M. Main, A. Melton, M. Mislove, D. Schmidt (Eds.), Mathematical Foundations of Programming Semantics. Proceedings, 1993. IX, 647 pages. 1994.

Vol. 803: J. W. de Bakker, W.-P. de Roever, G. Rozenberg (Eds.), A Decade of Concurrency. Proceedings, 1993. VII, 683 pages. 1994.

Vol. 804: D. Hernández, Qualitative Representation of Spatial Knowledge. IX, 202 pages. 1994. (Subseries LNAI).

Vol. 805: M. Cosnard, A. Ferreira, J. Peters (Eds.), Parallel and Distributed Computing. Proceedings, 1994. X, 280 pages. 1994.

Vol. 806: H. Barendregt, T. Nipkow (Eds.), Types for Proofs and Programs. VIII, 383 pages. 1994.

Vol. 807: M. Crochemore, D. Gusfield (Eds.), Combinatorial Pattern Matching. Proceedings, 1994. VIII, 326 pages. 1994.

Vol. 808: M. Masuch, L. Pólos (Eds.), Knowledge Representation and Reasoning Under Uncertainty. VII, 237 pages. 1994. (Subseries LNAI).

Vol. 809: R. Anderson (Ed.), Fast Software Encryption. Proceedings, 1993. IX, 223 pages. 1994.

Vol. 810: G. Lakemeyer, B. Nebel (Eds.), Foundations of Knowledge Representation and Reasoning. VIII, 355 pages. 1994. (Subseries LNAI).

Vol. 811: G. Wijers, S. Brinkkemper, T. Wasserman (Eds.), Advanced Information Systems Engineering. Proceedings, 1994. XI, 420 pages. 1994.

Vol. 812: J. Karhumäki, H. Maurer, G. Rozenberg (Eds.), Results and Trends in Theoretical Computer Science. Proceedings, 1994. X, 445 pages. 1994.

Vol. 813: A. Nerode, Yu. N. Matiyasevich (Eds.), Logical Foundations of Computer Science. Proceedings, 1994. IX, 392 pages. 1994.

Vol. 814: A. Bundy (Ed.), Automated Deduction—CADE-12. Proceedings, 1994. XVI, 848 pages. 1994. (Subseries LNAI).

Vol. 815: R. Valette (Ed.), Application and Theory of Petri Nets 1994. Proceedings. IX, 587 pages. 1994.

Vol. 816: J. Heering, K. Meinke, B. Möller, T. Nipkow (Eds.), Higher-Order Algebra, Logic, and Term Rewriting. Proceedings, 1993. VII, 344 pages. 1994.

Vol. 817: C. Halatsis, D. Maritsas, G. Philokyprou, S. Theodoridis (Eds.), PARLE '94. Parallel Architectures and Languages Europe. Proceedings, 1994. XV, 837 pages. 1994.

Vol. 818: D. L. Dill (Ed.), Computer Aided Verification. Proceedings, 1994. IX, 480 pages. 1994.

Vol. 819: W. Litwin, T. Risch (Eds.), Applications of Databases. Proceedings, 1994. XII, 471 pages. 1994.

Vol. 820: S. Abiteboul, E. Shamir (Eds.), Automata, Languages and Programming. Proceedings, 1994. XIII, 644 pages. 1994.

Vol. 821: M. Tokoro, R. Pareschi (Eds.), Object-Oriented Programming. Proceedings, 1994. XI, 535 pages. 1994.

Vol. 822: F. Pfenning (Ed.), Logic Programming and Automated Reasoning. Proceedings, 1994. X, 345 pages. 1994. (Subseries LNAI).

Vol. 823: R. A. Elmasri, V. Kouramajian, B. Thalheim (Eds.), Entity-Relationship Approach — ER '93. Proceedings, 1993. X, 531 pages. 1994.

Vol. 824: E. M. Schmidt, S. Skyum (Eds.), Algorithm Theory - SWAT '94. Proceedings. IX, 383 pages. 1994.

Vol. 826: D. S. Bowers (Ed.), Directions in Databases. Proceedings, 1994. X, 234 pages. 1994.

Vol. 827: D. M. Gabbay, H. J. Ohlbach (Eds.), Temporal Logic. Proceedings, 1994. XI, 546 pages. 1994.

Vol. 828: L. C. Paulson, Isabelle. XVII, 321 pages. 1994.